Antimicrobial Pharmacodynamics in Theory and Clinical Practice

Antimicrobial Pharmacodynamics in Theory and Clinical Practice

Second Edition

Edited by

Charles H. Nightingale
Hartford Hospital
Hartford, Connecticut, USA

Paul G. Ambrose
Institute for Clinical Pharmacodynamics
Albany, New York, USA

George L. Drusano
Ordway Research Institute
Albany, New York, USA

Takeo Murakawa
Kyoto University
Kyoto, Japan

CRC Press
Taylor & Francis Group
Boca Raton London New York

CRC Press is an imprint of the
Taylor & Francis Group, an **informa** business

CRC Press
Taylor & Francis Group
6000 Broken Sound Parkway NW, Suite 300
Boca Raton, FL 33487-2742

First issued in paperback 2019

© 2007 by Taylor & Francis Group, LLC
CRC Press is an imprint of Taylor & Francis Group, an Informa business

No claim to original U.S. Government works

ISBN-13: 978-0-367-38896-6 (pbk)

Library of Congress Cataloging-in-Publication Data

Antimicrobial pharmacodynamics in theory and clinical practice / [edited by]
    Charles H. Nightingale ... [et al.]. – 2nd ed.
      p. ; cm. – (Infectious disease and therapy ; v. 44)
    Includes bibliographical references and index

    1. Antibiotics–Dose-response relationship. I. Nightingale, C. H. II. Series.
    [DNLM: 1. Anti-Bacterial Agents–pharmacology. W1 IN406HMN
    v.44 2007 / QV 350 A6337 2007]

    RM267.A555 2007
    615'.329–dc22   200610346

Visit the Taylor & Francis Web site at
http://www.taylorandfrancis.com

and the CRC Press Web site at
http://www.crcpress.com

About the Editors

CHARLES H. NIGHTINGALE is currently Co-Director of the Center for Anti-infective Research and Development at Hartford Hospital, Connecticut, and is Research Professor at the University of Connecticut, Hartford, Connecticut. Dr. Nightingale is the author of over 560 peer reviewed articles with an additional 263 published abstracts, and continues to lecture internationally in his field of specialization. He is also an associate editor for various peer reviewed journals including *International Journal of Antimicrobial Agents, Journal of Pharmaceutical Care in Infectious Disease Management*, and *The Journal of Pharmacology*.

PAUL G. AMBROSE is Director of the Institute for Clinical Pharmacodynamics, Albany, New York. Dr. Ambrose is the author of over 70 peer-reviewed scientific publications, approximately 100 scientific abstracts, and has edited four textbooks. Dr. Ambrose is an associate editor of *Diagnostic Microbiology and Infectious Disease and Antimicrobial Agents* and *Chemotherapy*.

GEORGE L. DRUSANO is Director of Clinical Pharmacology Studies Unit, Albany Medical College, New York, and Co-Director and Senior Staff Scientist at Ordway Research Institute, Albany, New York. Dr. Drusano holds his B.S. in Physics from Boston College, Massachusetts, and received his M.D. from University of Maryland School of Medicine, College Park, Maryland.

TAKEO MURAKAWA is Director of Pharmaceutical Sciences, Kyoto University, Tokyo, Japan, and is Senior Lecturer at Tokushima University, Tokyo, Japan. Dr. Murakawa received his Ph.D. in Medical Science, Toho University, Tokyo, Japan, has worked for Fijisawa Pharmaceutical Co. Ltd. in drug development for over 35 years, and is now an independent consultant for drug development and regulatory affairs. He has published over 60 peer-reviewed articles, has written and edited three books, and is on the editorial boards of *International Journal of Antimicrobial Agents, International Journal of Antimicrobial Agents*, and *Asian J. Drug Metabolism and Pharmacokinetics*.

INFECTIOUS DISEASE AND THERAPY

Series Editor

Burke A. Cunha

Winthrop-University Hospital Mineola, and
State University of New York School of Medicine
Stony Brook, New York

Preface

The second edition of this book assumes that to use antibiotics properly, the clinician needs to understand fundamental concepts of pharmacodynamics. While many infectious disease physicians *do* understand these concepts, they are not the only physician group prescribing antibiotics. The objective of the second edition is to review the scientific and medical literature concerning antibiotics and pharmacodynamics, and then synthesize this information into an easy to understand discussion of the concepts and theory, along with application of these theories and concepts.

This book includes a discussion of the pharmacodynamics of all of the major classes of drugs. These include penicillins, cephalosporins, cephamycins, carbapenems, monobactams, aminoglycosides, quinolones, macrolides, azalides, ketolides, glycopeptides, metronidazole, clindamycin, tetracyclines, and antifungals. In addition to the topics on the antibiotics listed above, the second edition has added chapters on malaria, some antivirals, a section on non-clinical models of infection, a chapter on streptogramins and oxazolididones, and a section on pharmacodynamics in drug development. In addition to an updated introductary chapter, we include a chapter on the impact of pharmacodynamics on breakpoint selection for susceptibility testing, while retaining and updating the chapter on resistance and pharmacoeconomics.

This book is unique in that no other text of its kind currently exists and our chapter authors are among the leaders in the field. This book will find an audience in a large array of healthcare disciplines including college educators; medical, pharmacy, and microbiology students; infectious disease physicians; pharmacy specialists; medical house staff; clinical and staff pharmacists; clinical microbiologists; and other healthcare decision makers.

Charles H. Nightingale
Paul G. Ambrose
George L. Drusano
Takeo Murakawa

Contents

Contributors

Darren Abbanat Johnson & Johnson Pharmaceutical Research & Development, L.L.C., Raritan, New Jersey, U.S.A.

Paul G. Ambrose Institute for Clinical Pharmacodynamics, Ordway Research Institute, Albany, New York, and University of the Pacific School of Health Sciences, Stockton, California, U.S.A.

David Andes University of Wisconsin and William S. Middleton Memorial Veterans Hospital, Madison, Wisconsin, U.S.A.

Elizabeth A. Ashley Shoklo Malaria Research Unit, Mae Sot, Thailand; Faculty of Tropical Medicine, Mahidol University, Bangkok, Thailand; and Centre for Clinical Vaccinology and Tropical Medicine, Churchill Hospital, Headington, Oxford, U.K.

Sujata M. Bhavnani Institute for Clinical Pharmacodynamics, Ordway Research Institute, Albany, New York, U.S.A.

William Bishai Center for Tuberculosis Research, Johns Hopkins University School of Medicine, Baltimore, Maryland, U.S.A.

Karen Bush Johnson & Johnson Pharmaceutical Research & Development, L.L.C., Raritan, New Jersey, U.S.A.

Craig I. Coleman University of Connecticut School of Pharmacy, Storrs, Connecticut, and Pharmacoeconomics and Outcomes Studies Group, Hartford Hospital, Hartford, Connecticut, U.S.A.

William A. Craig University of Wisconsin and William S. Middleton Memorial Veterans Hospital, Madison, Wisconsin, U.S.A.

Jan G. den Hollander Department of Internal Medicine and Infectious Diseases, Medical Center Rotterdam Zuid, Rotterdam, The Netherlands

George L. Drusano Ordway Research Institute, Albany, New York, U.S.A.

Michael N. Dudley Mpex Pharmaceuticals, San Diego, California, U.S.A.

Anton F. Ehrhardt Department of Medical Affairs, Cubist Pharmaceuticals, Inc., Lexington, Massachusetts, U.S.A.

Alexander A. Firsov Gause Institute of New Antibiotics, Russian Academy of Medical Sciences, Moscow, Russia

Courtney V. Fletcher University of Colorado Health Sciences Center, Denver, Colorado, U.S.A.

Alan Forrest Ordway Research Institute, Albany, New York and University at Buffalo School of Pharmacy, Buffalo, New York, U.S.A

David Griffith Mpex Pharmaceuticals, San Diego, California, U.S.A.

Tawanda Gumbo University of Texas Southwestern Medical Center, Division of Infectious Diseases, Dallas, Texas, U.S.A.

Nancy D. Hanson Center for Research in Anti-infectives and Biotechnology, Department of Medical Microbiology and Immunology, Creighton University School of Medicine, Omaha, Nebraska, U.S.A.

Elizabeth D. Hermsen Department of Pharmaceutical and Nutrition Care, The Nebraska Medical Center, Omaha, Nebraska, U.S.A.

Fumiaki Ikeda Infectious Diseases Department, Pharmacology Research Laboratories, Drug Discovery Research, Astellas Pharma, Inc., Osaka, Japan

Sanjay Jain Center for Tuberculosis Research and Pediatric Infectious Diseases, Johns Hopkins University School of Medicine, Baltimore, Maryland, U.S.A.

Gunnar Kahlmeter Clinical Microbiology, Central Hospital, Växjö, Sweden

Myo-Kyoung Kim Pharmacy Practice, University of the Pacific Thomas J. Long School of Pharmacy, Stockton, California, U.S.A.

Jennifer J. Kiser University of Colorado Health Sciences Center, Denver, Colorado, U.S.A.

Effie L. Kuti University of Connecticut School of Pharmacy, Storrs, Connecticut, U.S.A.

Joseph L. Kuti Clinical and Economic Studies, Center for Anti-infective Research and Development, Hartford Hospital, Hartford, Connecticut, U.S.A.

Philip D. Lister Center for Research in Anti-infectives and Biotechnology, Department of Medical Microbiology and Immunology, Creighton University School of Medicine, Omaha, Nebraska, U.S.A.

Arnold Louie Emerging Infections and Pharmacodynamics Laboratory, Ordway Research Institute, Albany Medical College, Albany, New York, U.S.A.

Irene Y. Lubenko Gause Institute of New Antibiotics, Russian Academy of Medical Sciences, Moscow, Russia

Mark Macielag Johnson & Johnson Pharmaceutical Research & Development, L.L.C., Raritan, New Jersey, U.S.A.

Johan W. Mouton Department of Medical Microbiology and Infectious Diseases, Canisius Wilhelmina Hospital, Nijmegen, The Netherlands

Takeo Murakawa Graduate School of Pharmaceutical Sciences, Kyoto University, Kyoto, Japan

David P. Nicolau Center for Anti-infective Research and Development, Department of Medicine, Division of Infectious Diseases and Pharmacy, Hartford Hospital, Hartford, Connecticut, U.S.A.

Charles H. Nightingale Center for Anti-infective Research and Development, Hartford Hospital and University of Connecticut School of Pharmacy, Hartford, Connecticut, U.S.A.

Robert C. Owens, Jr. Maine Medical Center, Portland, Maine, and University of Vermont, College of Medicine, Burlington, Vermont, U.S.A.

Gigi H. Ross Ortho-McNeil Pharmaceutical, Inc., Raritan, New Jersey, U.S.A.

John C. Rotschafer Department of Experimental and Clinical Pharmacology, University of Minnesota College of Pharmacy, Minneapolis, Minnesota, U.S.A.

Zenzaburo Tozuka Exploratory ADME, JCL Bioassay Corporation, Hyogo, Japan

Nicholas J. White Faculty of Tropical Medicine, Mahidol University, Bangkok, Thailand, and Centre for Clinical Vaccinology and Tropical Medicine Churchill Hospital, Headington, Oxford, U.K.

Matt Wikler Pacific Beach Bioscience, San Diego, California, U.S.A.

Stephen H. Zinner Mount Auburn Hospital, Harvard Medical School, Cambridge, Massachusetts, U.S.A.

Pharmacodynamics of Antimicrobials: General Concepts and Applications

William A. Craig
University of Wisconsin and William S. Middleton Memorial Veterans Hospital, Madison, Wisconsin, U.S.A.

INTRODUCTION

"Pharmacodynamics" (PDs) is the term used to reflect the relationship between measurements of drug exposure in serum, tissues, and body fluids and the pharmacological and toxicological effects of drugs. With antimicrobials, PDs is focused on the relationship between concentrations and antimicrobial effect. Studies in the past have focused on pharmacokinetics (PKs) and descriptions of the time course of antimicrobials in serum, tissues, and body fluids. Much less emphasis has been placed on the time course of antimicrobial activity. Studies over the past 25 years have demonstrated marked differences in the time course of antimicrobial activity among antibacterials and antifungals (1–4). Furthermore, the pattern of antimicrobial activity over time is an important determinant of optimal dosage regimens (5). This chapter focuses on general concepts and the application of PDs to antimicrobial therapy.

MEASUREMENTS OF ANTIMICROBIAL ACTIVITY
Minimum Inhibitory and Minimum Bactericidal Concentrations

The minimum inhibitory concentration and minimum bactericidal concentration (MIC and MBC) have been the major parameters used to measure the in vitro activity of antimicrobials against various pathogens. Although MIC and MBC are excellent predictors of the potency of an antimicrobial against the infecting organism, they provide essentially no information on the time course of antimicrobial activity. For example, the MBC provides minimal information on the rate of bactericidal and fungicidal activity and on whether killing can be increased by higher drug concentrations. In addition, MIC provides no information on growth inhibitory effects that may persist after antimicrobial exposure. These persistent effects are due to three different phenomena: the postantibiotic effect (PAE), the postantibiotic sub-MIC effect (PAE-SME), and the postantibiotic leukocyte enhancement (PALE) (6–8). The killing effects of increasing concentrations on the bactericidal and fungicidal activity of antimicrobials combined with the magnitude of persistent effects give a much better description of the time course of antimicrobial activity than that provided by MIC and MBC.

Killing Activity

Antimicrobials exhibit two primary patterns of microbial killing. The first pattern is characterized by concentration-dependent killing over a wide range of concentrations. With this pattern, higher drug concentrations result in a greater rate and extent of microbial killing. This pattern is observed with the aminoglycosides, fluoroquinolones, daptomycin, ketolides, metronidazole, amphotericin B, and the echinocandins (2,9–13). The second pattern is characterized by minimal concentration-dependent killing. With this pattern, saturation of the killing rate occurs at low multiples of the MIC, usually around four to five times the MIC. Drug concentrations above these values do not kill microbes faster or more extensively. This pattern is also called time-dependent killing because the extent of microbial killing is primarily dependent on the duration of exposure. This pattern is observed with β-lactam antibiotics, macrolides, clindamycin, glycopeptides, tetracyclines, linezolid, and flucytosine (1,2,9,14,15).

The different patterns of bacterial killing are illustrated in Figure 1. by showing the effect of increasing drug concentrations on the in vitro antimicrobial activity of tobramycin, ciprofloxacin, and ticarcillin against a standard strain of *Pseudomonas aeruginosa* (2).

Increasing concentrations of tobramycin and ciprofloxacin produced more rapid and extensive bacterial killing, as exhibited by the steeper slopes of the killing curves. With ticarcillin, there was a change in slope as the concentration was increased from one to four times the MIC. However, higher concentrations did not alter the slope. The slight reduction in bacterial numbers at the higher doses is due to an earlier onset of bacterial killing. From two hour on, ticarcillin concentrations from 4 to 64 times the MIC produced the same rates of killing.

Persistent Effects

"Postantibiotic effect" is the term used to describe the persistent suppression of bacterial growth following antimicrobial exposure (1,8,16). It reflects the time it takes for an organism to recover from the effects of exposure to an antimicrobial and resume normal growth. This phenomenon was first observed in the 1940s in

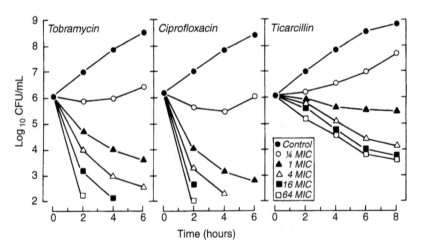

FIGURE 1 Time-kill curves of *Pseudomonas aeruginosa* ATCC 27853 with exposure to tobramycin, ciprofloxacin, and ticarcillin at concentrations from one-fourth to 64 times the MIC. *Abbreviations*: CFU; colony-forming units; MIC, minimum inhibitory concentration. *Source*: From Ref. 2.

early studies with penicillin against staphylococci and streptococci (17,18). Later studies starting in the 1970s extended this phenomenon to newer drugs and to gram-negative organisms. The PAE is demonstrated in vitro by following bacterial growth kinetics after drug removal.

Moderate-to-prolonged in vitro PAEs are observed for all antibacterials with susceptible gram-positive bacteria such as staphylococci and streptococci (16). Moderate-to-prolonged in vitro PAEs are also observed with gram-negative bacilli for drugs that are inhibitors of protein or nucleic acid synthesis. In contrast, short or no PAEs are observed for β-lactam antibiotics with gram-negative bacilli. The only exception has been for carbapenems, which exhibit moderate PAEs, primarily with strains of *P. aeruginosa* (19,20). In vitro postantifungal effects (PAFEs) have been observed with various yeasts following exposure to amphotericin B and flucytosine but not to triazoles such as fluconazole (3,21).

The PAE-SME demonstrates the additional effect sub-MICs scan have on the in vitro PAE. For example, exposure of streptococci in the PAE phase to macrolides at drug concentrations of one-tenth and three-tenths of the MIC increased the duration of the PAE by about 50% and 100%, respectively (16,22). The PAE phase can also make streptococci hypersensitive to the killing effects of sub-MICs of penicillin (6). The duration of the PAE-SMEs reported in the literature includes the duration of the PAE plus the enhanced duration due to sub-MICs. Morphological changes such as filaments can also be produced by sub-MICs (23).

PALE describes the effects of leukocytes on bacteria during the postantibiotic phase. Studies have demonstrated that such bacteria are more susceptible to intracellular killing or phagocytosis by leukocytes (7,16). This phenomenon can also prolong the duration of the in vitro PAE. Antimicrobials that produce the longest PAEs tend to exhibit the most prolonged effects when exposed to leukocytes.

The PAE has also been demonstrated in vivo in a variety of animal infection models (16,24). The in vivo phenomenon is actually a combination of the in vitro PAE and sub-MIC effects from gradually falling drug concentrations. The largest numbers of animal studies have used the neutropenic mouse thigh-infection model (4). When performed in non-neutropenic mice, the in vivo PAE would also include any PALE effects.

There are several important differences between in vivo and in vitro PAEs. In most cases, in vivo PAEs are longer than in vitro PAEs, most likely because of the additive effect of sub-MICs. Simulation of human PKs can further enhance the duration of the in vivo PAE by a similar mechanism. Prolongation of sub-MICs of amikacin by simulating the human drug half-life (two hours) extended the duration of in vivo PAEs by 40% to 100% over values observed with a dose producing the same area under the concentration-versus-time curve (AUC) but eliminated with a murine half-life of 20 minutes (25). In vivo PAEs with some drugs are further prolonged by the presence of leukocytes. In general, the presence of neutrophils tends to double the duration of the in vivo PAE for aminoglycosides and fluoroquinolones with gram-negative bacilli (16,25). However, leukocytes have no major effect on the minimal in vivo PAEs observed for β-lactams with gram-negative bacilli.

There are also some differences between in vitro and in vivo PAEs that question the value of measuring the in vitro PAE. First, the duration of the in vitro PAE is not predictive of the duration of the in vivo PAE (26). Second, prolonged PAEs for penicillin and cephalosporins with streptococci are observed in vitro but not in vivo (4,24,27). Third, in vitro studies that suggest that the PAE of aminoglycosides decreases and disappears over a prolonged dosing interval or with repeated doses have not been confirmed in vivo (28,29). Fourth, fluconazole exhibits a PAFE in vivo but not in vitro (3,14).

PATTERNS OF ANTIMICROBIAL ACTIVITY

The PD characteristics described above suggest that the time course of antimicrobial activity can vary markedly for different antibacterial and antifungal agents. As shown in Table 1, these drugs exhibit three major patterns of antimicrobial activity. The first pattern in characterized by concentration-dependent killing and moderate-to-prolonged persistent effects. Higher concentrations would kill organisms more rapidly and more extensively than lower levels. The prolonged persistent effects would allow for infrequent administration of large doses. This pattern is observed with aminoglycosides, fluoroquinolones, daptomycin, ketolides, metronidazole, and amphotericin B. The goal of a dosing regimen for these drugs would be to maximize concentrations. The peak level and the AUC should be the PK parameters that would determine in vivo efficacy.

The second pattern is characterized by time-dependent killing and minimal-to-no persistent effects. High drug levels would not kill organisms better than lower concentrations. Furthermore, organism regrowth would start very soon after serum levels fell below the MIC. This pattern is observed with β-lactams and flucytosine. The goal of a dosing regimen for these drags would be to optimize the duration of exposure. The duration of time that serum levels exceed some minimal value such as the MIC should be the major PK parameter determining the in vivo efficacy of these drugs.

The third pattern is also characterized by time-dependent killing, but the duration of the persistent effects is much prolonged. This can prevent any regrowth

TABLE 1 Three Patterns of Antimicrobial Activity

	Pattern 1	Pattern 2	Pattern 3
Pharmacodynamic characteristics	Concentration-dependent killing and moderate-to-prolonged persistent effects	Time-dependent killing and minimal-to-no persistent effects	Time-dependent killing and moderate-to-prolonged persistent effects
Antimicrobials included	Aminoglycosides, fluoroquinolones, daptomycin, ketolides, metronidazole, amphotericin B echinocandins	β-lactams, flucytosine	Azithromycin, macrolides, clindamycin, tetracyclines, glycylcyclines, streptogramins, oxazolidinones, glycopeptides, triazoles
Goal of dosing regimen	Maximize concentrations	Maximize duration of exposure	Optimize amount of drug
Pharmacokinetic parameter(s) determining efficacy	Peak level and AUC	Time above some threshold amount (e.g., minimum inhibitory concentration)	AUC

Abbreviation: AUC, area under the concentration-versus-time curve.

during the dosing interval. This pattern is observed with azithromycin, macrolides, clindamycin, tetracyclines and glycylcyclines, streptogramins, oxazolidinones, glycopeptides, and triazoles. The goal of a dosing regimen is to optimize the amount of drug administered to ensure that killing occurs for part of the time and there is no regrowth during remainder of the dosing interval. The AUC should be the primary PK parameter that would determine in vivo efficacy.

PK/PD INDICES

By using the MIC as a measure of the potency of drug–organism interactions, the PK parameters determining efficacy can be converted to PK/PD indices (5). Serum (or plasma) concentrations are used for determining the PK/PD indices. Because most infections occur in tissues and the common bacterial pathogens are extracellular organisms, interstitial fluid concentrations at the site of infection should be the primary determinants of efficacy. Serum levels are much better predictors of interstitial fluid levels than tissue homogenate concentrations. Because tissue homogenates mix the interstitial, intracellular, and vascular compartments together, they tend to underestimate or overestimate the interstitial fluid concentration depending on the ability of the drug to accumulate intracellularly (30).

Identification of the primary PK/PD indice that determines efficacy is complicated by the high degree of interdependence among the various indices. For example, a larger dose produces a higher peak/MIC ratio, a higher AUC/MIC ratio, and a longer duration of time above MIC. If the higher dose produces a better therapeutic effect than a lower dose, it is difficult to determine which PK/PD indice is of major importance, because all three increased. However, comparing the effects of dosage regimens that include different dosing intervals can reduce much of the interdependence among PK/PD indices. Such studies are often referred to as dose-fractionation studies (31–34). For example, dividing several total doses into 1, 2, 4, 8, and 24 doses administered at 24-, 12-, 6-, 3-, and 1-hour intervals, respectively, can allow one to identify which PK/PD indice is most important for in vivo efficacy.

Several investigators have used this study design in animal infection models to correlate specific PK/PD indices with efficacy for various antimicrobials against gram-positive cocci, gram-negative bacilli, and *Candida* species (9,12,14,25,31–37). This is demonstrated graphically in Figure 2 for ceftazidime against *Klebsiella pneumoniae* and in Figure 3 for temafloxacin against *Streptococcus pneumoniae*.

In these studies, pairs of mice were treated with multiple dosage regimens that varied both the dose and the dosing interval. The number of colony-forming units (CFUs) remaining in the thigh after 24 hours of therapy was plotted against the peak/MIC and 24-hour AUC/MIC ratios and the percentage of time that serum levels exceeded the MIC that was calculated for each dosage regimen from PK parameters. As shown in Figure 2, there was a very poor relationship between CFUs/thigh and peak/MIC and 24-hour AUC/MIC ratios. On the other hand, an excellent correlation was observed between the number of bacteria in the thighs and the percentage of time that serum levels exceeded the MIC. In contrast, the best correlation in Figure 3 was observed with the 24-hour AUC/MIC ratio followed by the peak/MIC ratio.

The specific PK/PD indices correlating with efficacy in animal infection models for different antibacterials and antifungals are listed in Table 2. As expected, time above MIC has consistently been the only PK/PD parameter correlating with

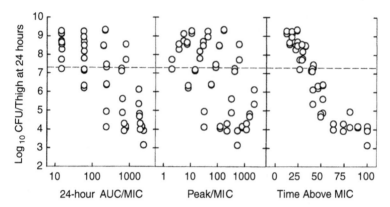

FIGURE 2 Relationship between three pharmacokinetic/pharmacodynamic indices (peak/MIC ratio, 24-hour AUC/MIC ratio, and percentage of time that serum levels exceed the MIC) and the number of *Klebsiella pneumoniae* ATCC 53816 in the thighs of neutropenic mice after 24-hour of therapy with ceftazidime. Each point represents data for one mouse. The dotted line reflects the number of bacteria at the beginning of therapy. *Abbreviations*: AUC, area under the concentration-versus-time curve; CFU, colony-forming unit; MIC, minimum inhibitory concentration.

the therapeutic efficacy of β-lactam antibiotics. Time above MIC is also the parameter correlating with efficacy of flucytosine.

The AUC/MIC and peak/MIC ratios have been the PK/PD indices that correlate with efficacy for aminoglycosides and fluoroquinolones. Most studies have shown slightly better correlation with the AUC/MIC ratio than with the peak/MIC ratio. Peak/MIC ratios appear to be more important in infections

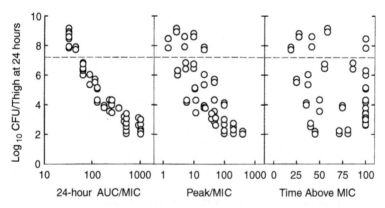

FIGURE 3 Relationship between three pharmacokinetic/pharmacodynamic indices (peak/MIC ratio, 24-hour AUC/MIC ratio, and percentage of time that serum levels exceed the MIC) and the number of *Streptococcus pneumoniae* ATCC 10813 in the thighs of neutropenic mice after 24 hours of therapy with temafloxacin. Each point represents data for one mouse. The dotted line reflects the number of bacteria at the beginning of therapy. *Abbreviations*: AUC, area under the concentration-versus-time curve CFU, colony-forming unit; MIC, minimum inhibitory concentration.

TABLE 2 Pharmacokinetic/Pharmacodynamic Parameters Determining Efficacy for Different Antimicrobials

PK/PD parameter	Antimicrobial
Time above MIC	Penicillins, cephalosporins, aztreonam, carbapenems, tribactams, and flucytosine
Peak/MIC ratio	Aminoglycosides, fluoroquinolones, daptomycin, glycopeptides, amphotericin B, and echinocandins
AUC/MIC ratio	Aminoglycosides, fluoroquinolones, daptomycin, glycopeptides, ketolides, macrolides, clindamycin, streptogramins, oxazolidinones, tetracyclines, glycylcyclines, triazoles, and echinocandins

Abbreviations: AUC, area under the concentration-versus-time curve; MIC, minimum inhibitory concentration.

where the emergence of resistant subpopulations is a significant risk and for drugs that act on the cell membrane, such as daptomycin and amphotericin B (13,31,36).

Although vancomycin, tetracyclines, azithromycin, macrolides, clindamycin, streptogramins, and oxazolidinones do not exhibit concentration-dependent killing, the AUC/MIC ratio has been the major PK/PD indice correlating with therapeutic efficacy of these drugs in neutropenic animals (5,9,15,38,39). A study in normal mice with vancomycin and teicoplanin against a strain of *S. pneumoniae* that used mortality as an endpoint demonstrated that the peak/MIC ratio was the most important indice (40).

MAGNITUDE OF PK/PD INDICES REQUIRED FOR EFFICACY

Because PK/PD indices can correct for differences in a drug's PKs and intrinsic antimicrobial activity, one would expect that the magnitude of the PK/PD indices required for efficacy would be similar in different animal species. Thus, results from studies in animal infection models could be predictive of the activity of drugs in humans. This would be especially helpful in designing dosage regimens for both old and new antibacterials in situations where it is difficult to obtain sufficient clinical data, such as with newly emerging resistant organisms.

Animal Infection Models

The largest number of studies addressing the magnitude of the PK/PD indices with various drugs, dosing regimens, pathogens, sites of infection, and animal species have been performed with β-lactams and fluoroquinolones. Time above MIC is the PK/PD indice that correlates with the therapeutic efficacy of the various β-lactam antibiotics. Studies in animal infection models demonstrate that antibiotic concentrations do not need to exceed the MIC for 100% of the dosing interval to obtain a significant antibacterial effect (24,32–34,38). In fact, an in vivo bacteriostatic effect is observed when serum levels exceed the MIC for about 30% to 40% of the dosing interval. If one uses survival after several days of therapy as the endpoint for efficacy of β-lactams in animal infection models, then slightly higher percentages of time above MIC are necessary (38,41,42). Figure 4 illustrates the relationship between time above MIC and mortality for animals infected with *S. pneumoniae* that were treated for several days with penicillins or cephalosporins.

Several studies included penicillin-intermediate and penicillin-resistant strains. The mortality was close to 100% if serum levels were above the MIC for

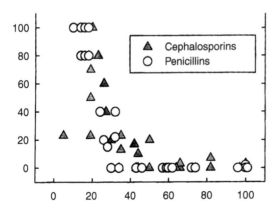

FIGURE 4 Relationship between percentage of time serum levels of β-lactams that exceed the MIC and survival in animal models infected with *Streptococcus pneumoniae*. *Abbreviation*: FPO. *Source*: From Refs. 41, 42.

20% or less of the dosing interval. As soon as the percentage of time above MIC reached 40% to 50% or higher, survival was in the order of 90% to 100%.

The PK/PD indice that best correlates with the efficacy of the fluoroquinolones is the 24-hour AUC/MIC ratio (12,38,43). The magnitude of this PK/PD indice required to produce a bacteriostatic effect in animal infection models varied for most organisms from 25 to 50 (12). These values are equivalent to averaging one to two times the MIC over a 24-hour period [i.e., $(1–2 \times MIC) \times 24$ hour = 24–48]. The relationship between 24-hour AUC/MIC values and outcome for fluoroquinolones as reported in the literature from studies that treated animals for at least two days, reported survival results at the end of therapy, and provided PK data is illustrated in Figure 5 (12,38).

The infections in these studies included pneumonia, peritonitis, and sepsis produced by gram-negative bacilli and a few gram-positive cocci in immunosuppressed mice, rats, and guinea pigs. In general, 24-hour AUC/MIC ratios less than 30 were associated with greater than 50% mortality, whereas AUC/MIC values of 100 or greater were associated with almost no mortality. A value of 100 is

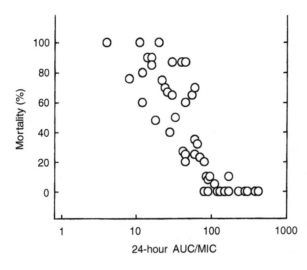

FIGURE 5 Relationship between 24-hour AUC/MIC for fluoroquinolones and survival in immunosuppressed animals infected with gram-negative bacilli and a few gram-positive cocci. *Abbreviations*: AUC, area under the concentration-versus-time curve; FPO; MIC, minimum inhibitory concentration. *Source*: From Refs. 12, 38.

equivalent to having serum concentrations average about four times the MIC over a 24-hour period [i.e., $(4 \times MIC) \times 24$ hour = 96].

Factors Affecting Magnitudes

Studies in animals with a large number of different antimicrobials would allow one to determine if the magnitude of the PK/PD indice required for efficacy was similar for (*i*) different dosing regimens, (*ii*) different drugs within the same antimicrobial class, (*iii*) different pathogens, and (*iv*) different sites of infection. As shown in Figures 2 and 3, the magnitude for the correct PK/PD parameter does not change with different dosing regimens. However, the rate of elimination for some drugs is so rapid that it becomes difficult to select the correct PK/PD indice. For example, time above MIC was the parameter for amikacin in neutropenic mice with normal renal function and a drug half-life of 20 minutes (25). However, when the half-life was lengthened to about two hours by transient renal impairment, the 24-hour AUC/MIC and the peak/MIC became the important PK/PD indices. In earlier reviews, time above MIC was thought to be the PK/PD indice for macrolides, clindamycin, and linezolid (5). However, these drugs all have rapid elimination in mice. If one eliminates 12- and 24-hourly dosing regimens from the analysis, then the 24-hour AUC/MIC is the important PK/PD indice.

Differences in the magnitude of the PK/PD indice are observed with different classes of β-lactams. For the same types of organisms, the percentages of time above MIC for a bacteriostatic effect were slightly lower for penicillins than for cephalosporins, and even lower for carbapenems (39). These differences are due to the rate of killing, which is fastest with the carbapenems and slowest with the cephalosporins.

Drugs within the same class of antimicrobial can also show differences in the magnitude of the PK/PD target required for efficacy. However, these differences are eliminated if one uses free drug concentrations for calculating the PK/PD indices. This has been observed with β-lactams, fluoroquinolones, macrolides, tetracyclines, and triazoles (38,39,44). Thus, protein binding is an import factor that can modify the magnitude of the PK/PD indice.

There are a few major differences in the magnitude of the PK/PD target for different organisms. The percentage of time above MIC required for efficacy with staphylococci is less than observed with gram-negative bacilli and streptococci (24). This difference is due to the prolonged in vivo PAEs observed for β-lactams with staphylococci but not with gram-negative bacilli and streptococci (4,16,39). In non-neutropenic mice, the magnitude of the 24-hour AUC/MIC for fluoroquinolones required for efficacy was about three- to fourfold lower for *S. pneumoniae* than for *K. pneumoniae* (12,38). In vitro models have also demonstrated a lower AUC/MIC value for strains of *S. pneumoniae* (45,46).

The magnitude of the important PK/PD indice appears to be rather constant when studied against strains with various resistance mechanisms. For β-lactams against penicillin-susceptible, penicillin-intermediate, and penicillin-resistant strains of *S. pneumoniae*, the percentage of time above MIC for free drug was very similar as the MIC increased (Fig. 6) for three cephalosporins, two penicillins, and three carbapenems.

This figure also illustrates the shortened time above MIC required for a bacteriostatic effect for penicillins and especially for carbapenems when compared to cephalosporins. The time above MIC resulting in a bacteriostatic effect has also been similar for various Enterobacteriaceae, including strains with extended

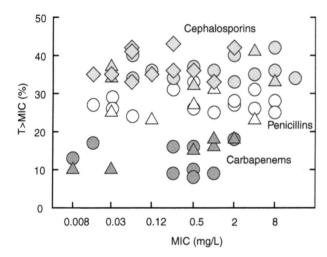

FIGURE 6 Relationship between time above MIC required for a bacteriostatic effect and in vitro MIC for cephalosporins, penicillins, and carbapenems against penicillin-susceptible, -intermediate, and -resistant strains of *Streptococcus pneumoniae*. *Abbreviation*: MIC, minimum inhibitory concentration. *Source*: From Refs. 24, 38, 39, 41.

spectrum β-lacatamases (39,47). Mutations primarily in *parC* and *gyrA* have elevated the in vitro MICs of different fluoroquinolones with *S. pneumoniae*, but the 24-hour AUC/MIC ratio for a bacteriostatic effect has remained very constant (48).

In general, the magnitude of the PK/PD indice required for efficacy is similar for different sites of infection. The one exception can be pneumonia where penetration into epithelial lining fluid (ELF) is the major determinant of efficacy. For example, vancomycin, which has decreased penetration into ELF, was about threefold less potent in a lung infection model compared to a thigh infection in the same neutropenic mice (49). Drugs like the macrolides, which have increased penetration into ELF have shown enhanced efficacy in pneumonia models in animals (50).

Human Infections

Bacteriological cure in patients with acute otitis media and acute maxillary sinusitis provides a sensitive model for determining the relationship between outcome and time above MIC for multiple β-lactam antibiotics. A variety of clinical trials have included pretherapy and repeat sinus puncture or tympanocentesis of middle ear fluid after two to seven days of therapy to determine whether the initial organism isolated had been eradicated (51–53). Figure 7demonstrates the relationship between time above MIC and bacteriological cure rate for many β-lactams against *S. pneumoniae* and *Haemophilus influenzae* in patients with these two infections.

Several of the recent studies have included penicillin-intermediate and penicillin-resistant strains. In general, percentages for time above MIC greater than 40% were required to achieve an 85% to 100% bacteriological cure rate for both organisms including resistant pneumococci.

Commonly used parenteral doses of ceftriaxone, cefotaxime, penicillin G, and ampicillin provide free drug concentrations above the MIC_{90} for penicillin-intermediate strains of *S. pneumoniae* for at least 40% to 50% of the dosing interval. A variety of clinical trials in severe pneumococcal pneumonia including bacteremic cases have demonstrated that these β-lactams are as effective against these

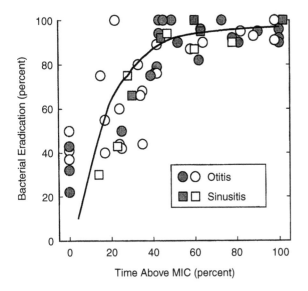

FIGURE 7 The relationship between time above MIC and bacteriological cure for various β-lactams against penicillin-susceptible, penicillin-intermediate, and penicillin-resistant *Streptococcus pneumoniae* and *Haemophilus influenzae* in patients with acute otitis media and acute maxillary sinusitis. *Abbreviations*: FPO; MIC, minimum inhibitory concentration. *Source*: From Refs. 51–53.

organisms as against fully susceptible strains (54–56). Thus, the magnitude of the PK/PD indice determining efficacy for β-lactams against pneumococci is very similar in animal infection models and in human infections such as pneumonia, sinusitis, and otitis media.

As illustrated in Figure 5, high survival in animal infection models treated with fluoroquinolones was observed with 24-hour AUC/MIC values of 100 or higher. Very similar values were observed in two clinical trials. Forrest et al. (57) found that a 24-hour AUC/MIC value of 125 or higher was associated with satisfactory outcome in seriously ill patients treated with intravenous ciprofloxacin. Lower values resulted in clinical and microbiological cure rates of less than 50%. Another study in patients with a variety of bacterial infections treated with levofloxacin found that a peak/MIC ratio of 12 or higher and a 24-hour AUC/MIC ratio of 100 or higher resulted in a statistically improved outcome (58). This and other studies have shown a lower value of around 34 for eradication of *S. pneumoniae* in respiratory infections (44,58,59). As stated earlier, a similar value has been observed with pneumococci in nonimmunocompromised animals. These studies further demonstrate that the magnitude of the PK/PD indice in animal infection models can be predictive of the magnitude of the parameter required for effective therapy in humans.

Drug Combinations

Very little is known about determining the magnitude of PK/PD parameters when drugs are used in combination. Some investigators have suggested that one can add the magnitude of the 24-hour AUC/MIC ratios for each of the drugs to estimate the PD activity of the combination (60). However, a study in neutropenic mice infected with *P. aeruginosa* demonstrated that the magnitudes of the PK/PD parameters required when a β-lactam, aminoglycoside, or fluoroquinolone is used alone are also important in predicting the efficacy of these drugs used in combination (61). Thus, adding the 24-hour AUC/MIC ratio of β-lactams to that of aminoglycosides or fluoroquinolones has a poor predictive value for their activity in combination. Instead, one must add the effect produced by the percentage of time above MIC for β-lactams to the effect resulting from the 24-hour AUC/MIC ratio of aminoglycosides and fluoroquinolones to accurately predict the activity of these combinations. Adding the 24-hour AUC/MIC ratios is appropriate for aminoglycoside-fluoroquinolone combinations, because that is the important PK/PD indice for both drugs.

PK/PD TARGETS FOR PREVENTING THE EMERGENCE OF RESISTANCE

The increasing incidence of antimicrobial resistance has challenged many researchers to identify the magnitude of the PK/PD indice that can prevent the emergence of resistance. Most studies have used in vitro models where large numbers of organisms can easily be exposed to different drug concentrations. In general, the magnitude of the PK/PD target is larger than required for efficacy but varies for different organisms. For example, the in vitro AUC/MIC exposure for garenoxacin required to prevent the emergence of resistant mutants in MSSA, MRSA, and *K. pneumoniae* ranged from 67 to 144 (44). Organisms that already demonstrated reduced susceptibility to the drug required much higher values to prevent further resistance. A peak/MIC ratio of 8 to 10 has also been shown in vitro to prevent the emergence of resistant mutants during therapy with aminoglycosides and fluoroquinolones (62). Some other studies with fluoroquinolones have suggested that it is the duration of time that serum concentrations persist in the mutant selection window that is most important for the emergence of resistant mutants (63).

Animal Models

Emergence of resistance in animal models has been more difficult to study because of the relative small number of bacteria or fungi at the beginning of therapy. The 24-hour AUC/MIC ratio of levofloxacin required to suppress the emergence of resistant subpopulations in a thigh-infection model was found initially to be 157 (64). This number was also prospectively validated in a second series of studies. Another study using the thigh-infection model observed that serum concentrations of ceftriaxone needed to exceed the MIC 64-fold for about 50% of the dosing interval to suppress the emergence of amp C derepressed mutants (65). Such concentrations would be impossible to obtain in humans.

Human Infections

A 24-hour AUC/MIC ratio greater than 100 has been associated with a significantly reduced risk for the emergence of resistance during therapy (60). The conclusion of this study was dependent almost entirely on the results with ciprofloxacin in patients with gram-negative bacillary infections, primarily those due to *P. aeruginosa* (66). On the other hand, a 24-hour AUC/MIC ratio as high as 1000 (mean value 40 times MIC) did not reduce the risk for the emergence of resistant organisms in patients treated with cephalosporins for infections due to gram-negative bacilli producing Type 1 β-lactamase (60). These data suggest that in vitro and animal models can be predictive of the magnitude of the PK/PD indice required to prevent the emergence of resistance.

APPLICATIONS OF PDs

Knowledge of the PDs of antimicrobials have proven useful for (*i*) establishing newer optimal dosing regimens for established drugs, (*ii*) developing new antimicrobials or new formulations, (*iii*) establishing susceptibility breakpoints, (*iv*) formulating guidelines for empirical therapy of infections, and (*v*) formulary development.

New Dosage Regimens

Administration of β-lactams by continuous infusion enhances their ability to maintain serum levels above the MIC. Despite many potential advantages of continuous infusion, only a few clinical trials have documented the success of this type of dosage regimen (67). For example, continuous infusion of ceftazidime has showed similar efficacy as with intermittent dosing (68). Initial results with continuous infusion of large doses of ampicillin have demonstrated success for moderately ampicillin-resistant strains (ampicillin MIC = 32–64 mg/L) of vancomycin-resistant *Enterococcus faecium* (5).

Prolonged infusions of β-lactams have also been proposed as a method for increasing the duration of time serum concentrations exceed the MIC for organisms with reduced susceptibility.

The peak/MIC ratio appears to be the major PK/PD parameter determining the clinical efficacy of aminoglycosides. High rates of clinical success in severe gram-negative bacillary infections and rapid resolution of fever and leukocytosis in gram-negative bacillary nosocomial pneumonia require a peak/MIC ratio of 8 to 10 (69,70). The once-daily dosage regimen for aminoglycosides was designed to enhance peak serum levels. In addition, once-daily dosing has the potential to decrease the nephro- and ototoxicity associated with these drugs. Uptake of aminoglycosides into renal tubular cells and middle ear endolymph is more efficient with low sustained concentrations than with high intermittent levels (71–73).

Most meta-analyses of clinical trials have demonstrated a small but significant increase in clinical outcome with once-daily dosing and a trend toward decreased nephrotoxicity (74–76). Studies have also demonstrated that the onset of nephrotoxicity occurs several days later when the drug is administered once daily than when multiple-daily dosage regimens are followed (77–79). Nevertheless, once-daily dosing may not be ideal for all indications. Studies in experimental enterococcal endocarditis have shown a greater reduction in bacterial vegetation titers when the aminoglycoside is administered by multiple-dosing regimens than by once-daily administration (80).

New Antimicrobials and Formulations

Identification of the PK/PD indice and its magnitude required for efficacy has proven useful for selecting the dosage regimen for Phase II to III clinical trials of new antimicrobials. For example, the 14:1 amoxicillin-clavulanate formulation was designed to enhance the time serum concentration exceeded MICs of 4 and 8 mg/L in both *S. pneumoniae* and *H. influenzae*. Double tap studies in young children have confirmed its efficacy against these strains (53).

Susceptibility Breakpoint Determinations

The Subcommittee on Antimicrobial Susceptibility Testing of the Clinical Laboratory Standards Institute (CLSI, formerly National Committee for Clinical Laboratory Standards) has incorporated PDs as one of the factors to consider when establishing susceptibility breakpoints (81). For example, PD breakpoints for

TABLE 3 Pharmacodynamic and New and Old Clinical Laboratory Standards Institute (or Food and Drug Administration) Susceptibility Breakpoints for Oral β-Lactams with *Streptococcus pneumoniae*

Drug	Old CLSI or FDA breakpoint (mg/L)	Pharmacodynamic ($T \leqslant MIC \geqslant 40\%$) breakpoint (mg/L)	New CLSI breakpoint (mg/L)
Amoxicillin	0.5	2	2
Cefaclor	8.0	0.5–1	1
Cefuroxime	0.5	1	1
Cefprozil	8.0	1–2	2
Cefpodoxime	0.5	0.5	0.5

Abbreviations: CLSI, Clinical Laboratory Standards Institute; FDA, Food and Drug Administration; MIC, minimum inhibitory concentration.

β-lactam antibiotics against *S. pneumoniae* would be defined as the highest MIC, following normal dosage that serum concentrations would exceed for at least 40% of the dosing interval. Table 3 compares the old and new susceptibility breakpoints of several oral β-lactams for *S. pneumoniae* with the PD breakpoints predicted from serum concentrations in children and adults. The two values for cefaclor and cefprozil reflect the difference in serum levels between different formulations or between children and adults. The CLSI breakpoints are identical to the PD breakpoints. A similar analysis was used to increase the breakpoint for ceftriaxone and cefotaxime for nonmeningitis infections produced by *S. pneumoniae*, such as pneumonia. More recent studies also suggest that the current third-generation cephalosporin breakpoints for Enterobactericeae are too high and need to be lowered two-to fourfold (39).

Guidelines for Empirical Therapy
Because the magnitude of PK/PD indices determined in animal infection models can be predictive of antimicrobial efficacy in human infections, it is easy to understand why PDs are being used more and more in establishing guidelines for empirical therapy. Recently published guidelines for otitis media, acute bacterial rhinosinusitis, and community-acquired and nosocomial pneumonia have used the ability of antimicrobials to reach the magnitude of PK/PD indices required for efficacy for both susceptible pathogens and those with decreased susceptibility to rank or select antimicrobials for empirical therapy of these respiratory infections (55,82,83).

Formulary Development
There are two major ways of using PD data for formulary development. In one method, PD susceptibility breakpoints are determined and then used to identify the percentage of potentially resistant strains. For example, the incidence of resistance with strains of *S. pneumoniae*, *H. influenzae*, and *Moraxella catarrhalis* for multiple oral antibiotics can be used to select certain drugs for the formulary to treat common respiratory infections (83). The use of local susceptibility data increases the accuracy of the drug decisions. In another method, Monte Carlo simulation, which integrates variability in PKs and MIC with the PK/PD index for efficacy, is used to estimate the probability of PK/PD target attainment (84). Drugs with dosing regimens that give high PK/PD target attainment are then selected for the formulary.

SUMMARY

Studies over the past 25 years have demonstrated that antibacterials and antifungals can vary markedly in their time course of antimicrobial activity. Three different patterns of antimicrobial activity are observed. Specific PK/PD indices, such as the peak/MIC and AUC/MIC ratios and the time above MIC, have also been shown to be major determinants of in vivo efficacy. The magnitude of the PK/PD indices required for efficacy are relatively similar in animal infection models and human infections and are largely independent of the dosing interval, the site of infection (except for pneumonia), the drug used within each antimicrobial class (providing free drug PK/PD indices are used), and the type of infecting pathogen. However, additional studies are needed to extend current observations to other antimicrobials and organisms and to correlate PK/PD indices with therapeutic efficacy in a variety of animal infection models and human infections. PDs has many applications including use for establishing optimal dosing regimens for old drugs, for developing new antimicrobials and formulations, for setting susceptibility breakpoints, and for providing guidelines for empirical therapy and formulary development.

REFERENCES

1. Bundtzen RW, Gerber AU, Cohn DL, Craig WA. Postantibiotic suppression of bacterial growth. Rev Infect Dis 1981; 3:28–37.
2. Craig WA, Ebert SC. Killing and regrowth of bacteria in vitro: a review. Scand J Infect Dis 1991; (suppl 74):63–70.
3. Turnidge JD, Gudmundsson S, Vogelman B, Craig WA. The postantibiotic effect of antifungal agents against common pathogenic yeast. J Antimicrob Chemother 1994; 34:83–92.
4. Vogelman B, Gudmundsson S, Turnidge J, Craig WA. The in vivo postantibiotic effect in a thigh infection in neutropenic mice. J Infect Dis 1988; 157:287–298.
5. Craig WA. Pharmacokinetic/pharmacodynamic parameters: rationale for antibacterial dosing of mice and men. Clin Infect Dis 1997; 26:1–12.
6. Cars O, Odenholt-Tornqvist I. The post-antibiotic sub-MIC effect in vitro and in vivo. J Antimicrob Chemother 1993; 31(suppl D):159–166.
7. McDonald PJ, Wetherall BL, Pruul H. Postantibiotic leukocyte enhancement: increased susceptibility of bacteria pretreated with antibiotics to activity of leukocytes. Rev Infect Dis 1981; 3:38–44.
8. McDonald PJ, Craig WA, Kunin CM. Persistent effect of antibiotics on *Staphylococcus aureus* after exposure for limited periods of time. J Infect Dis 1977; 135:217–223.
9. Andes D, Peng J, Craig WA. In vivo characterization of the pharmacodynamics of a new oxazolidinones (linezolid). Antimicrob Agents Chemother 2002; 46:3484–3489.
10. Andes D, Marchillo K, Lowther J, et al. In vivo characterization of HMR 3270, a glucan-synthase inhibitor in a murine candidiasis model. Antimicrob Agents Chemother 2003; 47:1187–1192.
11. Craig WA, Andes D. Differences in the in vivo pharmacodynamics of telithromycin and azithromycin against *Streptococcus pneumoniae*. Abstracts of 40th Interscience Conference on Antimicrobial Agents and Chemotherapy. Am Soc Microbiol, Washington, DC, 2000.
12. Craig WA, Dalhoff A. Pharmacodynamics of fluoroquinolones in experimental animals. In: Kuhknan J, Dalhoff A, Zeiler HJ, eds. Handbook of Experimental Pharmacology. Vol. 127. Quinolone Antibacterials. 207–232.
13. Safdar N, Andes D, Craig WA. In vivo pharmacodynamic activity of daptomycin. Antimicrob Agents Chemother 2004; 48:63–68.

14. Andes D, van Ogtrop M. Characterization and quantitation of the pharmacodynamics of fluconazole in a neutropenic murine disseminated candidiasis infection model. Antimicrob Agents Chemother 1999; 43:2116–2120.
15. Craig WA. Postantibiotic effects and the dosing of macrolides, azalides, and streptogramins. In: Zinner SH, Young LS, Acar JF, Neu HC, eds. Expanding Indications for the New Macrolides, Azalides, and Streptogramins. New York, NY: Marcel Dekker, 1997:27–38.
16. Craig WA, Gudmundsson S. Postantibiotic effect. In: Lorian V, ed. Antibiotics in Laboratory Medicine. 4th ed. Baltimore, MD: Williams and Wilkins, 1996:296–329.
17. Bigger JW. The bactericidal action of penicillin on *Staphylococcus pyogenes*. Irish J Med Sci 1994; 227:533–568.
18. Eagle H, Musselman AD. The slow recovery of bacteria from the toxic effects of penicillin. J Bacteriol 1949; 58:475–490.
19. Bustamante CL, Drusano GL, Tatem BA, Standiford HC. Post-antibiotic effect of imipenem on *Pseudomonas aeruginosa*. Antimicrob Agents Chemother 1984; 26: 678–682.
20. Gudmundsson S, Vogelman B, Craig WA. The in vivo postantibiotic effect of imipenem and other new antimicrobials. J Antimicrob Chemother 1986; 18(suppl E): 67–73.
21. Ernst EJ, Klepser ME, Pfaller MA. Postantifungal effects of echinocandin, azole, and polyene antifungal agents against *Candida albicans* and *Cryptococcus neoformans*. Antimicrob Agents Chemother 2000; 44:1108–1111.
22. Odenholt-Tornqvist I, Lowdin E, Cars O. Postantibiotic sub-MIC effects of vancomycin, roxithromycin, sparfloxacin, and amikacin. Antimicrob Agents Chemother 1992; 36:1852–1858.
23. Lorian V. Effect of low antibiotic concentrations on bacteria: effects on ultrastructure, virulence, and susceptibility to immunodefenses. In: Lorian V, ed. Antibiotics in Laboratory Medicine. 4th ed. Baltimore, MD: Williams and Wilkins, 1991:493–555.
24. Craig WA. Interrelationship between pharmacokinetics and pharmacodynamics in determining dosage regimens for broad-spectrum cephalosporins. Diagn Microbiol Infect Dis 1995; 21:1–8.
25. Craig WA, Redington J, Ebert SC. Pharmacodynamics of amikacin in-vitro and in mouse thigh and lung infections. J Antimicrob Chemother 1991; 27(suppl C):29–40.
26. Fantin B, Ebert S, Leggett J, et al. Factors affecting the duration of in vivo postantibiotic effect for aminoglycosides against gram-negative bacilli. J Antimicrob Chemother 1990; 27:829–836.
27. Tauber MG, Zak O, Scheld WM, Hengstler B, Sande MA. The postantibiotic effect in the treatment of experimental meningitis caused by *Streptococcus pneumoniae* in rabbits. J Infect Dis 1984; 149:575–583.
28. Den Hollander JG, Mouton JW, van Goor MP, et al. Alteration of postantibiotic effect during one dosing interval of tobramycin, simulated in an in vitro pharmacokinetic model. Antimicrob Agents Chemother 1996; 40:784–786.
29. McGrath BJ, Marchbanks CR, Gilbert D, Dudley MN. In vitro postantibiotic effect following repeated exposure to imipenem, temofloxacin, and tobramycin. Antimicrob Agents Chemother 1993; 37:1723–1725.
30. Redington J, Ebert SC, Craig WA. Role of antimicrobial pharmacokinetics and pharmacodynamics in surgical prophylaxis. Rev Infect Dis 1991; 13(suppl 10):S790–S799.
31. Drusano GL, Johnson DE, Rosen M. Pharmacodynamics of a fluoroquinolone antimicrobial agent in a neutropenic rat model of *Pseudomonas sepsis*. Antimicrob Agents Chemother 1993; 37:483–490.
32. Leggett JE, Fantin B, Ebert S, et al. Comparative antibiotic dose-effect relations at several dosing intervals in murine pneumonitis and thigh-infection models. J Infect Dis 1989; 159:281–292.
33. Leggett JE, Ebert S, Fantin B, Craig WA. Comparative dose-effect relations at several dosing intervals for beta-lactam, aminoglycoside and quinolone antibiotics against gram-negative bacilli in murine thigh-infection and pneumonitis models. Scand J Infect Dis 1991; (suppl 74):179–184.

34. Vogelman B, Gudmundsson S, Leggett J, Turnidge J, Ebert S, Craig WA. Correlation of antimicrobial pharmacokinetic parameters with therapeutic efficacy in an animal model. J Infect Dis 1988; 158:831–847.

35. Andes D, van Ogtrop ML. In vivo characterization of the pharmacodynamics of flucytosine in a neutropenic murine disseminated candidiasis model. Antimicrob Agents Chemother 2000; 44:938–942.

36. Andes D, Stamstad T, Conklin R. Pharmacodynamics of amphotericin B in a disseminated candidiasis model. Antimicrob Agents Chemother 2001; 45:922–926.

37. Louie A, Drusano GL, Banerjee P, et al. Pharmacodynamics of fluconazole in a murine model of systemic candidiasis. Antimicrob Agents Chemother 1998; 42: 1105–1109.

38. Andes D, Craig WA. Animal model pharmacokinetics and pharmacodynamics: a critical review. Int J Antimicrob Agents 2002; 19:261–268.

39. Craig WA. Basic pharmacodynamics of antibacterials with clinical applications to the use of β-lactams, glycopeptides, and linezolid. Infect Dis Clin N Am 2003; 17:479–501.

40. Knudsen JD, Fuursted K, Raber S, Espersen F, Fridmodt-Moller N. Pharmacodynamics of glycopeptides in the mouse peritonitis model of *Streptococcus pneumoniae* and *Staphylococcus aureus* infection. Antimicrob Agents Chemother 2000; 44:1247–1254.

41. Andes D, Craig WA. In vivo activities of amoxicillin and amoxicillin-clavulanate against *Streptococcus pneumonia*: application to breakpoint determinations. Antimicrob Agents Chemother 1998; 42:2375–2379.

42. Nicolau DP, Onyeji CO, Zhong M, Tessier PR, Banevicius MA, Nightingale CH. Pharmacodynamic assessment of cefprozil against *Streptococcus pneumoniae*: implications for breakpoint determinations. Antimicrob Agents Chemother 2000; 44: 1291–1295.

43. Andes DR, Craig WA. Pharmacodynamics of fluoroquinolones in experimental models of endocarditis. Clin Infect Dis 1998; 27:47–50.

44. Ambrose PG, Bhavnani SM, Owens Jr RC. Clinical pharmacodynamics of quinolones. Infect Dis Clin N Am 2003;17:529–543.

45. Lacy MK, Lu W, Xu X, et al. Pharmacodynamic comparisons of levofloxacin, ciprofloxacin, and ampicillin against *Streptococcus pneumoniae* in an in vitro model of infection. Antimicrob Agents Chemother 1999; 43:672–677.

46. Lister PD, Sanders CC. Pharmacodynamics of levofloxacin and ciprofloxacin against *Streptococcus pneumoniae*. J Antimicrob Chemother 1999; 43:79–86.

47. Andes D, Craig WA. Treatment of infections with ESBL-producing organisms: pharmacokinetics and pharmacodynamic considerations. Clin Microbiol Infect 2005; 11(suppl 6): 10–17.

48. Craig WA, Andes D. In vivo pharmacodynamic activity of gemifloxacin against multiple strains of *Streptococcus pneumoniae*. Abstracts of the 15th European Congress of Clinical Microbiology and Infectious Diseases, Copenhagen, Denmark, 2005.

49. Craig W, Andes D. Activity of oritavancin (O) versus vancomycin (V) in the neutropenic murine thigh-and lung-infection models. Abstracts of 44th Interscience Conference on Antimicrobial Agents and Chemotherapy. Am Soc Microbiol, Washington, DC, 2004.

50. Maglio D, Nocolau DP, Nightingale CH. Impact of pharmacodynamics on dosing of macrolides, azalides, and ketolides. Infect Dis Clin N Am 2003; 17:563–577.

51. Craig WA, Andes D. Pharmacokinetics and pharmacodynamics of antibiotics in otitis media. Pediatr Infect Dis J 1996; 15:255–259.

52. Dagan R, Klugman KP, Craig WA, Baquero F. Evidence to support the rationale that bacterial eradication in respiratory tract infection is an important aim of antimicrobial therapy. J Antimicrob Chemother 2001; 47:129–141.

53. Dagan R, Hobeman A, Johnson C, et al. Bacteriologic and clinical efficacy of high dose amoxicillin/clavulanate in children with acute otitis media. Pediatr Infect Dis J 2001; 20:829–837.

54. Feikin DR, Schuchat A, Kolczak M, et al. Mortality from invasive pneumococcal pneumonia in the era of antibiotic resistance, 1995–1997. Am J Public Health 2000; 90:223–229.

55. Heffelfinger JD, Dowell SF, Jorgensen JH, et al. Management of community-acquired pneumonia in the era of pneumococcal resistance. Arch Intern Med 2000; 160: 1399–1408.
56. Pallares R, Linares J, Vadillo M, et al. Resistance to penicillin and cephalosporin and mortality from severe pneumococcal pneumonia in Barcelona, Spain. N Engl J Med 1995; 333:474–480.
57. Forrest A, Nix DE, Ballow CH, Goss TF, Birmingham MC, Schentag JJ. Pharmacodynamics of intravenous ciprofloxacin in seriously ill patients. Antimicrob Agents Chemother 1993; 37:1073–1081.
58. Preston SL, Drusano GL, Berman AL, et al. Pharmacodynamics of levofloxacin: a new paradigm for early clinical trials. J Am Med Assoc 1998; 279:125–129.
59. Ambrose PG, Grasela DM, Grasela TH, Passarell J, Mayer HB, Pierce PF. Pharmacodynamics of fluoroquinolones against *Streptococcus pneumoniae* in patients with community-acquired respiratory tract infections. Antimicrob Agents Chemother 2001; 45: 2473–2477.
60. Thomas JK, Forrest A, Bhavnani SM, et al. Pharmacodynamic evaluation of factors associated with the development of bacterial resistance in acutely ill patients during therapy. Antimicrob Agents Chemother 1998; 42:521–527.
61. Mouton JW, van Ogtrop ML, Andes D, Craig WA. Use of pharmacodynamic indices to predict efficacy of combination therapy in vivo. Antimicrob Agents Chemother 1999; 43:2473–2478.
62. Blaser J, Stone BB, Groner MC, et al. Comparative study with enoxacin and netilmicin in a pharmacodynamic model to determine importance of ratio of antibiotic peak concentration to MIC for bactericidal activity and emergence of resistance. Antimicrob Agents Chemother 1987; 31:1054–1060.
63. Firsov AA, Vostrov SN, Lubenko IY, Drlica K, Portnoy YA, Zinner SH. In vitro pharmacodynamic evaluation of the mutant selection window hypothesis using four fluoroquinolones against *Staphylococcus aureus*. Antimicrob Agents Chemother 2003; 47:1604–1613.
64. Jumbe N, Louie A, Leary R, et al. Application of a mathematical model to prevent in vivo amplification of antibiotic-resistant bacterial populations during therapy. J Clin Invest 2003; 112:275–285.
65. Berkout J, Van Ogtrop ML, Van den Broek PJ, et al. Pharmacodynamics of ceftriaxone against cephalosporin-sensitive and –resistant *Enterobacter cloacae* in vivo. Program and Abstracts of the 39th Interscience Conference on Antimicrobial Agents and Chemotherapy, American Society for Microbiology, Washington, DC, 1999.
66. Craig WA. Does the dose matter? Clin Infect Dis 2001; 33(suppl 3):S233–S237.
67. Craig WA, Ebert SC. Continuous infusion of β-lactam antibiotics. Antimicrob Agents Chemother 1992; 36:2577–2583.
68. Nicolau D, MacNabb J, Lacy M. Continuous versus intermittent administration of ceftazidime in intensive care unit patients with nosocomial pneumonia. Int J Antimicrob Agents 2001; 17:497–504.
69. Kashuba AD, Nafziger AN, Drusano GL, Bertino JS. Optimizing aminoglycoside therapy for nosocomial pneumonia caused by gram-negative bacteria. Antimicrob Agents Chemother 1999; 43:623–629.
70. Moore RD, Lietman PS, Smith CR. Clinical response to aminoglycoside therapy: importance of the ratio of peak concentration to minimal inhibitory concentration. J Infect Dis 1987; 155:93–99.
71. De Broe ME, Verbist L, Verpooten GA. Influence of dosage schedule on renal cortical accumulation of amikacin and tobramycin in man. J Antimicrob Chemother 1991; 27(suppl C):41–47.
72. Tran BH, Deffrennes D. Aminoglycoside ototoxicity: influence of dosage regimen on drug uptake and correlation between membrane binding and some clinical features. Acta Otolaryngol (Stockholm) 1988; 105:511–515.
73. Verpooten GA, Giuliano RA, Verbist L, Estermans G, De Broe ME. Once-daily dosing decreases renal accumulation of gentamicin and netilmicin. Clin Pharmacol Ther 1989; 45:22–27.

74. Ali MZ, Goetz MB. A meta-analysis of the relative efficacy and toxicity of single daily dosing versus multiple daily dosing of aminoglycosides. Clin Infect Dis 1997; 24: 796–809.

75. Turnidge J. Pharmacodynamics and dosing of aminoglycosides. Infect Dis Clin N Amer 2003; 17:503–528.

76. Urban A, Craig WA. Daily dosing of aminoglycosides. Curr Clin Topics Infect Dis 1997; 17:236–255.

77. Int Antimicrob Therap Coop Group of EORTC. Efficacy and toxicity of single daily doses of amikacin and ceftriaxone versus multiple daily doses of amikacin and ceftazidime for infection in patients with cancer and granulocytopenia. Ann Intern Med 1993; 119:584–592.

78. Rybak MJ, Abate BJ, Kang SL, Ruffing MJ, Lerner SA, Drusano GL. Prospective evaluation of the effect of an aminoglycoside dosing regimen on rates of observed nephrotoxicity and ototoxicity. Antimicrob Agents Chemother 1999; 43:1549–1555.

79. Ter Braak EW, De Vries PJ, Bouter KP, et al. Once-daily dosing regimen for aminoglycoside plus β-lactam combination therapy of serious bacterial infections: comparative trial with netilmicin plus ceftriaxone. Am J Med 1990; 89:58–66.

80. Fantin B, Carbon C. Importance of the aminoglycoside dosing regimen in the penicillin-netilimicin combination for treatment of *Enterococcus faecalis*-induced experimental endocarditis. Antimicrob Agents Chemother 1990; 34:2387–2389.

81. National Committee for Clinical Laboratory Standards. Development of In-Vitro Susceptibility Testing Criteria and Quality Control Parameters; Approved Guideline. 2nd ed. Document M23–ANCCLS, (January) 2000.

82. Dowell SF, Butler JC, Giebink GS, et al. Acute otitis media: management and surveillance in an era of pneumococcal resistance—a report from the drug-resistant *Streptococcus pneumoniae* therapeutic working group. Pediatr Infect Dis J 1999; 18:1–9.

83. Sinus and Allergy Health Partnership. Antimicrobial treatment guidelines for acute bacterial rhinosinusitis 2004. Otolaryngol Head Neck Surg 2000; 130(suppl 1):1–45.

84. Andes D, Craig WA. Understanding pharmacokinetics and pharmacodynamics: application to the antimicrobial formulary decision process. In: Owens RC, Ambrose PG, Nightingale H, eds. Antibiotic Optimization: Concepts and Strategies in Clinical Practice. New York, NY: Marcel Dekker, 2005:65–88.

2 | Applying Pharmacodynamics for Susceptibility Breakpoint Selection and Susceptibility Testing

Johan W. Mouton
Department of Medical Microbiology and Infectious Diseases, Canisius Wilhelmina Hospital, Nijmegen, The Netherlands

Paul G. Ambrose
Institute for Clinical Pharmacodynamics, Ordway Research Institute, Albany, New York, and University of the Pacific School of Health Sciences, Stockton, California, U.S.A.

Gunnar Kahlmeter
Clinical Microbiology, Central Hospital, Växjö, Sweden

Matt Wikler
Pacific Beach Bioscience, San Diego, California, U.S.A.

William A. Craig
University of Wisconsin and William S. Middleton Memorial Veterans Hospital, Madison, Wisconsin, U.S.A.

INTRODUCTION

The main parameters used today to indicate whether the use of an antimicrobial will have a reasonable probability of success are the classifications "resistant" (R) and "susceptible" (S), based on the minimum inhibitory concentration (MIC), either directly by various dilution methods or by disk diffusion. For clinicians, this categorization is important, because the choice of therapy is often guided by these reports from the clinical microbiology laboratory. Supposedly, if a microorganism is categorized as S to an antimicrobial agent, there is a reasonably good probability of success when the patient is treated with that antimicrobial agent, while failure of therapy is more likely when an isolate is categorized as R. The "intermediate" (I) category is used for various purposes, but mainly to indicate a degree of uncertainty in response or dose dependency. However, the criteria used for categorization are less clear, and the meanings of S, I, and R have varied over time and place. Over the last decade, pharmacodynamics (PDs) has started to play a major role in distinguishing the S, I, and R categories. Because concentration–effect relationships became increasingly apparent and could be described in a meaningful manner, drug exposures indexed to MIC that result in a high probability of clinical success could be ascertained. In addition, it was increasingly appreciated that not all patients are created equal and that large differences in pharmacokinetic (PK) behavior between patients do exist. The use of a statistical technique called Monte Carlo simulation (MCS) (1,2) is now utilized to account for PK variation inherent in human populations. In this chapter, we will discuss the meaning of S, I, and R in a historical context as well as provide the current view, based on PK–PD relationships. Both the European Committee on Antimicrobial Susceptibility Testing (EUCAST) (3) and the Clinical and Laboratory Standards Institute [CLSI, formally known as the National Committee for Clinical Laboratory Standards (NCCLS)] consider PK–PD data in the selection of susceptibility breakpoints (4).

HISTORY

Initially, isolates were categorized as S or R largely based on MIC frequency distributions. Oftentimes, MIC distributions were bimodal, with the MIC values of wild type (WT) or S subpopulations surrounding one mode, and the MIC values of the subpopulation with a resistance determinant surrounding the other mode. The term "susceptibility breakpoint," which was first used in the report of an International Collaborative Study on Antimicrobial Susceptibility Testing by Ericsson and Sherris in 1971 (5), is the drug concentration that separates the WT subpopulation from that with a resistance determinant. Although this report, a hallmark in methodology and interpretation of susceptibility testing, ended with recommendations on categorization of sensitivities, the views from various persons and societies differed too much to reach a global consensus on the clinical meaning of these categories.

United States
In the early 1970s, the first susceptibility breakpoint document was published by the NCCLS. As noted above, these susceptibility breakpoints tended to be based solely upon frequency distributions that allowed one to set a susceptibility breakpoint that would separate apparent S strains from R strains. As these susceptibility breakpoints were meant to assist physicians in the selection of antimicrobial therapies for patients, the NCCLS in the mid-1980s published a guidance document (M23-A) for the development of interpretive criteria that included the requirement to also evaluate clinical and microbiological outcomes of antimicrobial therapy by MIC of the causative organism. In 2001, an updated guidance document (M23-A2) was published that outlined the manner in which susceptibility breakpoints are currently determined by CLSI. There are three types of data considered: microbiologic, PK–PD, and clinical outcome data. The process utilized by the CLSI today is described in detail later in this chapter.

Europe
Following the report of Ericsson and Sherris, it soon became apparent that consensus on antimicrobial susceptibility testing (AST) could not be reached. Several national committees evolved in Europe. In general, the process of establishing susceptibility breakpoints in Europe has been more or less the opposite to that in the United States. Most of the committees were, apart from creating reproducible AST methods, primarily interested in providing susceptibility breakpoints, which could be used to predict clinical and bacteriological efficacy based on serum concentrations achieved in patients. In most cases, the duration of time that the free fraction (non–protein-bound, see below) remained above a certain concentration was considered to be the susceptibility breakpoint (Table 1). With few exceptions, notably the Swedish Reference Group of Antimicrobials (SRGA) in Sweden and the British Society of Antimicrobial Chemotherapy (BSAC) in the United Kingdom, the European committees did not take into account the fact that

TABLE 1 Susceptibility Breakpoint Systems Used in Various Countries in Europe Until 2001

Country	Committee		References
France	CASFM	Formula based on PK: $(C_{max}/3 + C_{t1/2} + C_{4h})/3 \times (1 - k)^a$	6
Great Britain	BSAC	Formula based on PK: $C_{max} \times f \times s/(e \times t)^b$	7
Netherlands	CRG	70–80% $T_{>MIC}$ for non–protein-bound fraction	8
Sweden	SRGA	PK profile and frequency distribution, species-specific breakpoints	9
Norway	NWGA	67% $T_{>MIC}$	10
Germany	DIN	PK profile, frequency distributions, efficacy	11

[a]C_{max}, maximum concentration; $C_{t1/2}$, concentration in serum after one half-life; C_{4h}, minimum quantity obtained over 4 hr period that corresponds approximately to 10 bacterial generations; k, degree of protein binding.
[b]C_{max}, maximum serum concentration at steady state, usually 1 hr postdose; e, factor by which C_{max} should exceed MIC (usually 4); t, factor to allow for serum half-life; f, factor to allow for protein binding; s, shift factor to allow for reproducibility and frequency distributions (usually 1).
Abbreviations: BSAC, British Society of Antimicrobial Chemotherapy; CRG, Commissie Richtlijnen Gevoeligheidsbepalingen; CASFM, Committee for Antimicrobial Testing of the French Society of Microbiology; DIN, Deutsches Institut fur Normung; MIC, minimum inhibitory concentration; NWGA, Norwegian Working Group on Antimicrobials; PK, pharmacokinetic; SRGA, Swedish Reference Group of Antimicrobials; $T_{>MIC}$, duration of time the drug concentration remains above the MIC value.

the splitting by susceptibility breakpoints of natural population MIC distributions in combination with the limited reproducibility of the assays resulted in poor reproducibility of the S, I, and R categorization and thus in potential errors in clinical practice. Because of the different approaches in Europe, susceptibility breakpoints differed between countries. Table 1 compares criteria used until 2001 by six national committees in Europe.

Despite differences in approach, susceptibility breakpoints often do not differ more than by a factor of two and are in general more similar to each other than to those of the CLSI. Since 2002, the EUCAST is in the process of systematically harmonizing breakpoints in Europe (3), and since 2005, it determined susceptibility breakpoints for new drugs as part of the European regulatory process. This is part of an agreement with the European Medicines Agency (EMEA). Importantly, this precludes differences between susceptibility breakpoints set by regulatory authorities and professional bodies (12). For each antimicrobial or group of antimicrobials, EUCAST describes the background and rationale for the breakpoints. European breakpoint tables, decisions, and rationale documents are published on the EUCAST (13) and in *Clinical Microbiology and Infection* as "EUCAST Technical Notes" (12).

CONCENTRATION–EFFECT RELATIONSHIPS OF ANTIMICROBIALS

The relationship between antimicrobial concentrations and effect has been elucidated over the last three decades. In the 1970s and 1980s, it was recognized that there are two major groups of antimicrobial agents, each displaying different patterns of bactericidal activity in in vitro time–kill curve experiments (14,15). The first pattern was characterized by time-dependent killing, which is maximal at relatively low drug concentrations. Concentrations much higher than the MIC value did not result in increased bacterial killing over the time course of the experiment, which is usually 24 hours. These antimicrobials are sometimes referred to as "concentration-independent" agents. The second pattern of bactericidal activity consists of concentration-dependent killing over a wide range of drug concentrations ("concentration-dependent antimicrobials"). The major impasse before the 1990s had been that these concentration–effect relationships in vitro were determined at static drug concentrations, while during therapy, systemic drug concentrations change over time due to PK processes (i.e., absorption, distribution, metabolism, and elimination of drugs from the body). By integrating PK with PD properties of the drug–microorganism interaction, PK–PD relationships could be established and subsequently used for evaluating dosing regimens, development of new drugs, and the setting of susceptibility breakpoints. Thus, today, we recognize that the effect of changing drug concentration over time is fundamentally different for various classes of antibiotics, and in most cases, is directly related to the pattern of bactericidal activity (16).

As can be observed from Figure 1, the concentration–time curve of a drug possesses three major characteristics or PK–PD indices: the ratio of the peak or maximal concentration (C_{max}) to the MIC value of the drug to the pathogen (C_{max}:MIC), the ratio of the area under the concentration–time curve at 24 hours (AUC) to the MIC value of the drug to the pathogen (AUC:MIC), and the duration of time the drug concentration remains above the MIC value of the drug to the pathogen ($T_{>MIC}$).

By using different dosing regimens in animal models of infection and in dynamic in vitro PK–PD models, and by varying both the frequency and the dose of the drug, it has been shown that there is a clear relationship between a PK–PD index and efficacy. A more extensive discussion of this subject can be found in Chapter 1. In general, for concentration-dependent drugs, there is a clear relationship between the AUC:MIC ratio and/or C_{max}:MIC ratio and efficacy, while for time-dependent drugs, it is the $T_{>MIC}$ that is best correlated with effect. However, for drugs that exhibit prolonged delays in regrowth after drug exposure, the AUC:MIC ratio correlates best with efficacy. An example is shown in Figure 2.

For levofloxacin, there is a clear relationship between AUC:MIC ratio and effect, while hardly any relationship exists for the $T_{>MIC}$.

As can be observed from Figure 2, the relationship between PK–PD index and effect (AUC:MIC ratio value in this example) can be described by a sigmoid curve. One of the characteristics of a sigmoid curve is that at some exposure, there is maximum effect: Higher exposures do not result in significantly larger effects. Alternatively, little is gained by increasing values of 90% of the E_{max}, and this value therefore is also often used as an indicator value. Numerous studies in vitro (18–22) and in animals (23–32) and clinical studies (33–37) have shown that a maximum effect or 90% of E_{max} for quinolones is reached at AUC:MIC ratios of

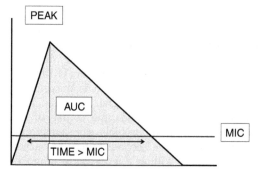

FIGURE 1 Diagram of a concentration–time curve showing the PK parameter's peak (or C_{max}) and AUC. The PK–PD indices are derived by relating the PK parameter to the MIC. AUC/MIC, C_{max}/MIC, and $T_{>MIC}$. *Abbreviations*: AUC, area under the concentration–time curve at 24 hours; MIC, minimum inhibitory concentration; PD, pharmacodynamic; PK, pharmacokinetic; $T_{>MIC}$, duration of time the drug concentration remains above the MIC value.

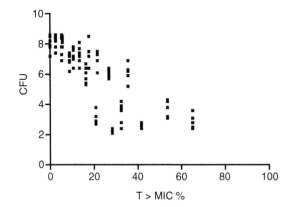

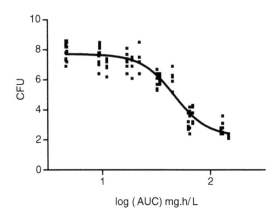

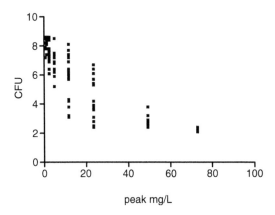

FIGURE 2 Relationship between $T_{>MIC}$ (*upper panel*), AUC (*middle*), and peak (*lower*) of levofloxacin in a mouse model of infection with *Streptococcus pneumoniae* as obtained by various dosing regimens and efficacy, expressed as cfu. The best relationship is obtained with the AUC; the curve drawn represents a model fit of the Hill equation with variable slope to the data. *Abbreviations*: AUC, area under the concentration–time curve at 24 hours; CFU, colony-forming; MIC, minimum inhibitory concentration; $T_{>MIC}$, duration of time the drug concentration remains above the MIC value. *Source*: From Ref. 17.

the unbound fraction of the drug of around 100 for infections caused by gram-negative rods and 30 to 40 for *Streptococcus pneumoniae*, and that these values are consistent within the class of quinolones. Since the objective of treating patients is to treat them optimally, this translates in, when administering antimicrobials to a patient to treat an infection, the AUC of the antimicrobial administered in relation to the MIC of the microorganism causing the infection, a value of at least 100 for gram-negative rods. Thus, when the dosing regimen of the antimicrobial is known, the average AUC in the average patient can be determined, and subsequently, the highest MIC a microorganism is allowed to have to result in an AUC/MIC ratio of at least 100 can be calculated. This MIC would be a reasonable estimate of the clinical susceptibility breakpoint. Higher MICs than this susceptibility breakpoint would result in lower AUC/MIC ratios and thus have a lower probability of successful treatment (R), while infections with microorganisms with lower MIC values would have a higher probability of successful treatment (S).

It has to be mentioned that the C_{max}:MIC ratio is at least as important in predicting efficacy of fluoroquinolones as the AUC:MIC ratio. Several studies in vitro, animal models, and human trials indicate that the minimum value for this ratio should be between 8 and 12 (28,35,38). Similar to the argument used for the AUC:MIC ratio, the susceptibility breakpoint would be the value where that ratio is reached. Depending on whether the dataset contains C_{max}:MIC ratios significantly exceeding 8 to 12, the AUC:MIC ratio or the C_{max}:MIC ratio prevails in best correlating with efficacy (2,17). Part of this relates to the emergence of R subpopulation. However, since most (the only exception being ciprofloxacin) quinolones are given once daily, the colinearity between the AUC:MIC ratio and C_{max}:MIC ratio is 1, and it is impossible to ascertain which of those two indices prevails in significance.

USE OF PK–PD TO SET SUSCEPTIBILITY BREAKPOINTS FOR QUINOLONES

An example of using PK–PD relationships in determining susceptibility breakpoints is shown for the fluoroquinolones against gram-negatives in Table 2 (39,40).

The table shows the AUC values of several fluoroquinolones based on dosing regimens commonly used. Importantly, the protein binding of the drug is indicated as well, since it is only the free fraction of the drug that is active (see below) (42,43). Using these figures, Table 2 shows the PD susceptibility breakpoints of these fluoroquinolones based on the assumption that the AUC:MIC ratio should be at least approximately 100 and the C_{max}:MIC ratio at least 8 to 12. For comparison, the CLSI susceptibility breakpoints are included as well (44). It is evident that for most fluoroquinolones, the current CLSI susceptibility breakpoints are relatively high when compared to PD susceptibility breakpoints. This is especially true for the older drugs, since PK–PD relationships are increasingly taken into account in setting susceptibility breakpoints for new drugs. As a consequence, there are inconsistencies when comparing the susceptibility breakpoints of various quinolones from the PK–PD point of view.

USE OF PK–PD TO SET SUSCEPTIBILITY BREAKPOINTS FOR β-LACTAMS

Another approach was taken by Andes and Craig (45). In a mouse model of infection with impaired renal function, they simulated human dosing regimens of

TABLE 2 PK Parameters and Susceptibility Breakpoints of Eight Fluoroquinolones Based on Dosing Regimens Generally Used[a]

Fluoroquinolone	Regimen	C_{max} (mg/L)	AUC (mg hr/L)	Protein binding (%)	S (susceptibility breakpoint) PD	CLSI
Ciprofloxacin	500 mg/12 hr	2.8	22.2	22	0.25	1
Sparfloxacin	200 mg/24 hr	0.6	16.4	45	0.125	0.5
Levofloxacin	500 mg/24 hr	5.2	61.1	30	0.5	2
Ofloxacin	200 mg/12 hr	2.2	29.2	30	0.25	2
Grepafloxacin	400 mg/24 hr	0.9	11.4	50	0.125	1
Trovafloxacin	200 mg/24 hr	2.2	30.4	70	0.125	1
Moxifloxacin	400 mg/24 hr	4.5	48.0	40	0.5	–
Gatifloxacin	400 mg/24 hr	3.5	32.8	20	0.5	–

[a]The susceptibility breakpoints based on an AUC/MIC ratio of 100 hr and a C_{max}/MIC ratio of 8 to 12 are indicated in the PD column.
Abbreviations: AUC, area under the concentration–time curve at 24 hours; CLSI, Clinical and Laboratory Standards Institute; C_{max}, maximum concentration; MIC, minimum inhibitory concentration; PD, pharmacodynamic.
Source: From Refs. 40, 41.

amoxicillin and amoxicillin/clavulanate, and looked at the effect of these dosing regimens on infections with *S. pneumoniae* with increasing MIC values. Infections with strains having an MIC of 2 mg/L or lower responded well to therapy, while infections with strains having MIC values of 4 mg/L or higher resulted in hardly any effect. In terms of time above MIC, the strains that responded well provided drug levels that exceeded the MIC for at least 40% of the dosing interval. Partly based on this study, the CLSI S breakpoint for amoxicillin was changed from 0.5 to 2 mg/L (44).

Free-Drug Level Is Used for Determination of the PK–PD Index: The Importance of Protein Binding and the Value of Tissue Levels

Although the importance of protein binding for clinical outcome of infections has been a matter of debate in the past, current evidence clearly demonstrates that it is only the unbound, free fraction of the drug that is active. In as early as 1947, it was shown that there was a relationship between the degree of protein binding and the ability of various penicillins to kill *Staphylococcus aureus* (46). Later, when the MIC became a standard measure of in vitro AST, Kunin in a classical experiment determined the MICs in broth with and without serum of various β-lactam antibiotics, with degrees of protein binding varying from 20% (ampicillin) to more than 90% (cloxacillin) (47). The MICs in broth with serum were higher than in broth alone in proportion to the degree of protein binding. In addition, calculating the free fraction of the drugs in the wells with serum added corresponded well to the MIC values in broth alone. Controversies remained until it became apparent that the efficacy of antimicrobial therapy could be correlated to PK/PD indices. The effect of protein binding can be studied in animal models of infection by determining the magnitude of the PK/PD index of an antimicrobial required to produce a certain effect over 24 hours based on both the total and the free concentrations of the drug. Examples are shown in Table 3 for several cephalosporins and Figure 3 for quinolones.

Both the table and the figure clearly show the concordance in the PK/PD index values needed to reach a bacteriostatic effect for the free fraction of the drug.

TABLE 3   Percentage over 24 hr of $T_{>MIC}$ [Mean (Range)] Required for a Static Effect After 24 hr of Therapy with Four Cephalosporins in a Mouse Model of Infection for *Enterobacteriaceae* and *Streptococcus pneumoniae*

Drug	*Enterobacteriaceae* (%$T_{>MIC}$)	*S. pneumoniae* (%$T_{>MIC}$)
Cefotaxime	38 (36–40)	38 (36–40)
Ceftazidime	36 (27–42)	39 (35–42)
Cefpirome	35 (29–40)	37 (33–39)
Ceftriaxone total	38 (34–42)	39 (37–41)
Ceftriaxone unbound fraction	72 (66–79)	64 (69–78)

Abbreviations: MIC, minimum inhibitory concentration; $T_{>MIC}$, duration of time the drug concentration remains above the MIC value.
Source: From Ref. 48.

In the case of the cephalosporins with very low protein binding, ceftazidime, cefpirome, and cefotaxime all display values of 38% to 40% $T_{>MIC}$, while if total concentrations of ceftriaxone, with a protein binding of 76% in mice, is considered, values needed are in the order of 70%. If the unbound fraction of ceftriaxone is considered, the effect is consistent with the other cephalosporins. Similarly, the AUC/MIC ratio needed for static effect of the highly protein-bound quinolones gemifloxacin and garenoxacin is comparable to that of the other quinolones only when the unbound fraction of the drug is taken into account.

The value of tissue levels depends on the methods used to measure them. The most important aspect is that tissue consists of different compartments, in general the intracellular and extracellular compartments, while the intracellular compartment can be further categorized by the various cell organelles. Thus, when tissue levels are determined by grinding up tissue and measuring the overall concentration in the tissue homogenate, the concentrations found are not informative with respect to the concentration of the antimicrobial at the site of infection (49,50). Since most bacterial infections are located in the extracellular compartment, it is those concentrations that are of primary interest. For drugs that are primarily distributed in the extracellular compartment, such as the β-lactams

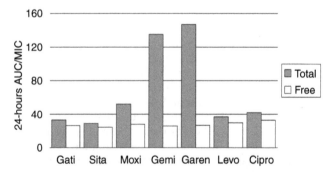

FIGURE 3   AUC/MIC ratios needed for static effect for eight quinolones. Gemifloxacin and garenoxacin are highly protein bound while the other six quinolones have a much lower protein binding. The effect of the free fractions and not total drug compare with each other. *Abbreviations*: AUC, area under the concentration–time curve at 24 hours; MIC, minimum inhibitory concentration.

and aminoglycosides, concentrations in the extracellular compartment will be underestimated by concentrations in tissue homogenates, while concentrations in extracellular fluid of drugs, which are taken up by cells to a relatively low (fluoroquinolones) or high (macrolides, azithromycin) extent, are overestimated by tissue homogenates. Most antibiotics have been shown to reach the extracellular fluid rapidly, and concentrations in extracellular fluid to be comparable to the non–protein-bound concentration in serum or plasma (51,52), although there seem to be some exceptions such as cerebrospinal fluid (CSF) and epithelial lining fluid (ELF) concentrations (53). Methods by which extracellular concentrations have been determined include suction blisters (54), inflammatory blisters (55), and threads (49). Recently, microdialysis techniques that measure only unbound-drug concentrations are increasingly being used to obtain concentration–time profiles in interstitial fluid (56–58). Thus, the strong relationship between unbound-drug concentrations in serum or plasma and those in extracellular fluid explains the good correlation found between unbound serum concentrations and in vivo effects and the lack of correlation of these effects with tissue concentrations obtained in homogenates. For intracellular infections, it is much less clear which concentrations correlate with effect. Drugs may be bound or trapped within the cell, while the activity of a drug intracellularly can also be distinctly different from that observed extracellularly (59).

In conclusion, except for some intracellular infections, CSF infections, and some lung infections where ELF concentrations are much higher than in serum or plasma, it is the free fraction of the drug in serum or plasma that correlates best with measures of in vivo efficacy. It should be emphasized that concentrations measured locally at sites of infection cannot be used to establish breakpoints since the quantitative relationship between local concentrations and effect is largely unknown.

ONE SIZE DOES NOT FIT ALL: VARIATIONS IN MIC AND PKs

In the discussion above, the AUC:MIC ratio of 100 was used as a reference value to determine tentative PK–PD susceptibility breakpoints for quinolones. However, the AUC values used to calculate the susceptibility breakpoint were mean values of the population, and within the population, there is variability in PK profiles. Thus, approximately half the population will have AUCs lower than this value (for instance, because of a higher than average clearance) and the other half will have a higher value, the extent being described by the variance. In a similar fashion, the MIC measurement has a certain margin of error, and in general is at least one twofold dilution. Both these variations result in PI values that will differ for each specific case and should be accounted for when determining susceptibility breakpoint values.

THE MIC

For over 50 years, no agreement could be reached over a standard methodology for MIC testing. In the earlier mentioned reference of Ericsson and Sherris (5), there appeared to be some consensus on using twofold dilutions and on including the concentration 1.0 mg/L. However, the medium, incubation times, volume, temperature, and other variables were still a matter of debate. In addition, because not all bacteria grow in standard media, numerous variations have been listed for various microorganisms. These methods are described in various countries by their organizations in conjunction with breakpoint tables from the official organization in the various countries. As an example of the differences, the CLSI method included an incubation temperature of 35°C in Mueller Hinton Medium and an inoculum of 5×10^5 cfu/mL, while the BSAC uses Iso Sensitest, 35°C to 37°C and an inoculum of 10^5 cfu/mL (60). For some microorganisms and antimicrobials, the conditions under which the tests are performed matter more than for others. For example, aminoglycoside and macrolide activity is influenced by pH and incubation atmosphere, and aminoglycoside activity also by the concentration of cations. It is therefore sometimes difficult or even impossible to compare MIC results.

In 2002, an initiative was taken by the European Committee on Standardization (CEN) to agree on a reference method for susceptibility testing. Through the Vienna Agreement, this was brought to the International Standards Organization (ISO) in 2004, and in 2006, all member states had agreed on a reference method for susceptibility testing of rapidly growing aerobic bacteria. This method, a microdilution method, is currently available from the ISO as ISO 25572-1. The document is based on earlier published methods of CLSI (4) and EUCAST (61). Thus, all future results of susceptibility testing should be calibrated to the reference method. Currently, a separate document that describes the calibration procedure is in the final stages of preparation (ISO 25572-2).

Irrespective of species (including *Candidae*), of where (country, point in time, biological niche, etc.) the strains are obtained, and of antimicrobial drug (including antifungal drugs), distributions of MIC values for strains without resistance mechanisms to the drug in question look very similar. MICs distribute over three to five dilution steps, typically as for *Escherichia coli* and ampicillin, 0.5 to 8 mg/L, as for *S. pneumoniae* versus benzylpenicillin, 0.008 to 0.064 mg/L, or as for *S. aureus* and vancomycin, 0.5 to 2 mg/L. The EUCAST has described this particular feature as the "WT MIC distribution." An example is shown in Figure 4 [and MIC distributions for most antimicrobials and microorganisms can be found on the website of EUCAST (13)].

Repeated MIC determination of any such strain produces an MIC distribution very similar to the distribution of consecutive different strains, indicating that the biological variation in the susceptibility to an antibiotic among strains without resistance mechanisms to the antibiotic is limited. This, taken together with the characteristic reproducibility of MIC determination, that is, plus–minus one dilution step around the mean, underlines the need to avoid setting breakpoints that divide WT distributions of important target microorganisms. Such breakpoints invite poor reproducibility of S, I, and/or R categorization.

Microorganisms that have higher MICs than those in the WT distribution can be expected to possess a resistance mechanism that may or may not be clinically significant. EUCAST distinguishes between WT and nonwild type (NWT) by defining an epidemiological cutoff (ECOFF) value (WT $\leq X$ mg/L; NWT $> X$ mg/L). The ECOFF value is defined for the drug and the species and relates to the graphic representation of WT MIC distribution. In the EUCAST graphs, the ECOFF is shown in the lower left hand corner (Fig. 4). By making a distinction between ECOFFs and clinical breakpoints, one of the old controversies of susceptibility breakpoint determination was solved (3).

To summarize this part of determining susceptibility breakpoints: (*i*) decide on target species for the drug, (*ii*) define the WT MIC distributions for target species, and (*iii*) for each species, determine which concentrations would constitute inappropriate breakpoints since they would split target species MIC distributions.

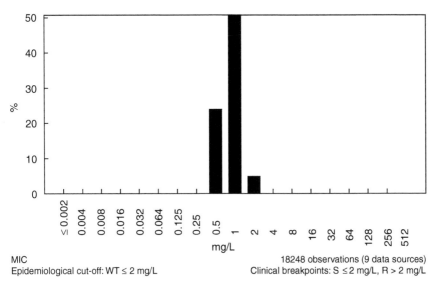

FIGURE 4 MIC distribution of levofloxacin. *Abbreviation*: MIC, minimum inhibitory concentration. *Source*: From Ref. 13.

The second issue is the indication of the drug. At present, this is not yet consistently implemented in the registration processes of the Food and Drug Administration (FDA) and the EMEA. It is clear, however, that "if" the PD relationship is apparent "and" the MIC distribution is known "and" the dosing regimen is known by default in the registration process, a conclusion can and should be drawn as to the possible indication of the antimicrobial. If the probability of target attainment is too low under the conditions mentioned, it should not be used unless there is compelling other evidence, for instance proven clinical efficacy.

THE APPLICATION FOR MCS IN SETTING SUSCEPTIBILITY BREAKPOINTS

When PK–PD indices are being used as values for the determination of susceptibility breakpoints to predict the probability of success of treatment, this should be true not only for the population mean, but also for each individual within the population. Since the PK behavior differs for each individual, the PK part of the PD index differs as well. An example is given in Figure 5.

The figure shows the proportion of the population reaching a certain concentration of ceftazidime after a 1 g dose. It is apparent from Figure 5 that there are individuals with a $T_{>MIC}$ of 50%, while others have, with the same dosing regimen, a $T_{>MIC}$ of more than 80%. One should be concerned about particularly individuals who have lower values than average reaching a PD target, since if these patients have an infection with a microorganism having an MIC at the susceptibility

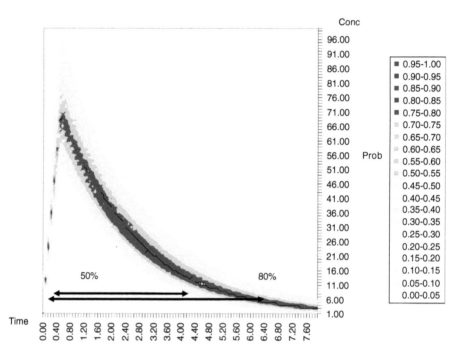

FIGURE 5 Simulation of ceftazidime after a 1 g dose using data from the authors' study. The gray scale indicates the probability of presence of a certain concentration. Due to interindividual variability, some individuals in the population will have a $T_{>MIC}$ of 50%, while others will have a value of 80%. The population mean is in the middle of the black area. *Abbreviations*: MIC, minimum inhibitory concentration; $T_{>MIC}$, duration of time the drug concentration remains above the MIC value. *Source*: From Ref. 41.

breakpoint level, it may be reported as susceptible, while the PK–PD index value for that particular individual may be suboptimal or even result in a different categorization. Thus, when using the relationship between PK–PD index and efficacy, the susceptibility breakpoint should be chosen after taking this interindividual variation into account.

Realizing this problem, Drusano et al. presented an integrated approach of population PKs and microbiological susceptibility information, at the FDA anti-infectives product advisory committee (1,2). The first step in that approach is to obtain estimates of the PK parameters of the population, using population PK analysis. These data can be obtained, for instance, from participants in Phase 1, 2, and/or 3 studies. Importantly, not only the estimates of the parameters are obtained, but also the estimates of variation. When these estimates are obtained, they are applied to simulate multiple concentration–time curves by performing MCS. This is a method that takes the variability in the input variables into consideration in the simulations (62). For each of these curves generated, which are all slightly different because the input parameters vary to a degree in relation to the variance of the parameters, the value of the PK–PD index is determined. The last step is to calculate the probability of target-attainment rate for various values of the PK–PD index. If the PK–PD index–effect relationship is known, this provides a tool for setting susceptibility breakpoints. This approach has been used now by several authors (2,63–71). An example is shown in Table 4 for two dosing regimens of ceftobiprole (BAL9141), a cephalosporin recently under clinical investigation.

The table shows the target-attainment rates of ceftobiprole for several values of $T_{>MIC}$. For instance, the simulation shows that for the 750 mg every 12 hour regimen, a probability of target attainment of 100% is achieved at 40% $fT_{>MIC}$ for microorganisms with an MIC of 4 mg/L. Since experimental studies have shown that 40% $fT_{>MIC}$ results in adequate efficacy, it is inferred that infections caused by microorganisms with an MIC of 4 mg/L or lower should be adequately treated with this particular dosing regimen, at least if they are devoid of resistance mechanisms.

The approach can be taken one step further by incorporating the frequency distribution of MIC values of the target pathogen. By multiplying the target-attainment rates and relative frequency of target pathogens, the fraction of response is obtained at each MIC, and by cumulating these, the cumulative fraction of target attainment is obtained. In this fashion, not only is the variability in PK parameters considered, but also the variance in susceptibility in the target pathogen population. The major drawback of this approach is that the MIC frequency distribution of the target microorganism population has to be unbiased and this almost never is the case. The cumulative frequency of target attainment can be very useful, however, in the development phase of a drug to determine whether the response is sufficiently adequate for further follow-up. For instance, Drusano et al. showed that the cumulative fraction target attainment for a 6 mg/ kg dose of everninomycin would be 34% given the priors in the simulations, and thereby concluded that further development of the drug was not justified (2).

After MCS has been performed, the results need to be interpreted. This is a somewhat gray area where opinions differ. Importantly, it should be realized that MCS is a tool that can be used in the decision-making process and does not give a final answer or outcome that cannot be reasonably disputed (see below). The first and foremost question that can be asked is: what is the probability of cure that one would hope to attain, given the susceptibility breakpoint? Ideally, the susceptibility breakpoint is set in such a way that MIC values at the susceptibility

TABLE 4 Probability of Target Attainment (%) for Two Dosing Regimens of Ceftobiprole Using Data from Human Volunteers[a]

Dosing regimen (MIC mg/L)	250 mg q12h				750 mg q12h			
	30%	40%	50%	60%	30%	40%	50%	60%
0.5	–	–	100	100	–	–	–	–
1	–	100	99	71	–	–	–	100
2	100	59	3	0	–	–	100	99
4	0	0	0	–	100	100	78	15
8	–	–	–	–	69	3	0	0
16	–	–	–	–	0	0	–	–
32	–	–	–	–	–	–	–	–
PTA 100%	2	1	0.5	0.5	4	4	2	1

[a]Probability of target attainments (PTA) are displayed for 30, 40, 50, and 60% $fT_{>MIC}$.
Abbreviations: MIC, minimum inhibitory concentration; $T_{>MIC}$, duration of time the drug concentration remains above the MIC value.
Source: Modified from Ref. 66.

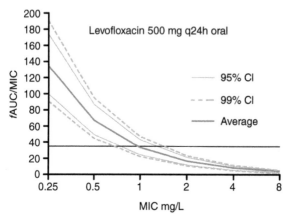

FIGURE 6 Probabilities of target attainment for levofloxacin orally for 500 mg q24h. The horizontal line indicates the PK/PD target for *Streptococcus pneumoniae*. The following PK parameters were used to obtain the probability of target attainment: Vd, 101 L; CV, 22%; Cl, 10.5 L/hr; CV, 14%; Fu, 70%; Ka, 1 hr^{-1}; F, 1. *Abbreviations*: AUC, area under the concentration–time curve at 24 hours; MIC, minimum inhibitory concentration; PD, pharmacodynamic; PK, pharmacokinetic. *Source*: From Ref. 13.

breakpoint or below (for CLSI susceptibility breakpoints, below the susceptibility breakpoint) indicate 100% probability of cure for the target population. Taking the WT population distribution into account, the minimum requirement thus would be a 100% probability of target attainment at the WT cut-off. This is almost never the case for the antimicrobials currently available. However, this approach can be used in the selection of doses to be used during the development of a drug (see above).

Another approach that is used by the EUCAST is to construct a plot of the probability of PK–PD target-attainment distribution. This comprises a graphical representation of the MIC as a function of the PK–PD index with the median as well as 95% and 99% confidence intervals (Fig. 6).

The susceptibility breakpoints follow from the intersection of the horizontal line representing the MIC and the lower limit of the confidence interval. One of the advantages of this method over using tables is that decisions can easier be made on the distribution as a whole and consequences of lower or higher susceptibility breakpoints can be observed immediately. This also applies to varying the target values that are used. Using a graphical representation of the table in conjunction, for one or two targets, with an MIC distribution, as is often done, precludes using different targets. Whatever one decides, the outcome of that decision is as good as the premise (see below).

The Weaknesses of MCS

Although MCSs provide insight into the effects of the variability in the population on the validity of susceptibility breakpoints and provide a tool to take this variation into account in setting susceptibility breakpoints, they have two distinct shortcomings. The first one is the assumption that the PK–PD value used to determine the target-attainment rates and thereby select the susceptibility breakpoint is the true value, which of course it is not, because the true value is unknown. For instance, the target value for the AUC is usually taken as 100 to 125 for gram-negatives, because that value has been found in various studies as discriminative between groups of patients responding to therapy and those who did not for infections. However, there are several reports that in some cases, higher values are clearly necessary, while lower values have also been described. While in the first study, published by Forrest et al. (34), 125 (notably, total drug) was the cutoff value below which the probability of cure was distinctly lower, values above 250 hours resulted in a faster cure rate. Thus, although the final effect was more or less equal for patients with AUC/MIC values of 125 and above, the rate at which the effect was achieved differed. Similarly, although the current assumption is that the PK–PD index value necessary for (bacteriological) cure is similar for most infections, this is not necessarily true. For instance, it has been shown that PK–PD index values needed to reach a maximum effect in sustained abscesses are higher (72). Thus, the target value may be different by microorganism as well as by clinical indication.

When results of animal infections are used for breakpoint determinations, the target resulting in bacteriostasis in neutropenic mice is usually used. This is especially true for drugs and microorganisms where the magnitude of the PK/PD index is significantly reduced by the presence of neutrophils. When this does not occur, as for β-lactams with pneumococci and for most drugs with gram-negative bacilli, a target that results in more killing, usually 1 to 2 $\log_{10}$ cfus over 24 hours, is used.

The second shortcoming is the input of the PK data used for simulation. The output of the simulations is directly dependent on the PK parameter values and their measures of dispersion used for input. Thus, if PK data are used from a small group of healthy young male volunteers obtained in Phase 1 or Phase 2 studies, the simulations will be biased toward relatively low target-attainment rates, because the elimination rate of most drugs is higher in volunteers than in the average patient. On the other hand, there are patient groups such as patients with cystic fibrosis who are known to have higher clearances for most drugs, and specific analyses have been made for such specific patient groups (65). Mouton et al. compared the results of MCSs of ceftazidime for three different populations: healthy volunteers, patients with cystic fibrosis, and intensive-care patients. Although there were differences at the extremes of the distribution, the general conclusion was similar for the patients and the volunteers.

THE USE OF CART® ANALYSIS AND OTHER TOOLS TO DETERMINE THE PI SUSCEPTIBILITY BREAKPOINT VALUE

The rate at which we are accumulating knowledge from analyses of PK–PD data derived from well-controlled clinical studies is accelerating. Typically, exposure–response analyses utilize a variety of statistical techniques such as univariate and multivariate logistic regression, as well as recursive partitioning (also known as tree-based modeling or classification and regression tree analysis). The latter tool is an exploratory nonparametric statistical algorithm that can accommodate continuous numeric data, such as AUC:MIC ratio values, or categorical data, such as clinical success or failure, as either independent or dependent variables.

For a categorical-dependent variable such as clinical response, recursive partitioning can be used to identify threshold values in an independent continuous variable such as AUC:MIC ratio such that impressive differences in response for patients with AUC:MIC values above and below a threshold can be seen. The results of recursive partitioning can then be used to inform decisions for constructing categorical-independent variables, the significance of which can then be tested in univariate or multivariate logistic regression analyses.

One of the first exposure–response analyses of clinical data that utilized both multivariate logistic regression and recursive partitioning was done by Forrest et al. (34). Intravenous ciprofloxacin was studied in critically ill patients with pneumonia involving predominantly *Enterobacteriaceae* and *Pseudomonas aeruginosa*. Multivariate logistic regression analyses identified the AUC_{0-24}:MIC ratio as being predictive of clinical and microbiological response ($P < 0.003$). Recursive partitioning identified a threshold AUC_{0-24}:MIC ratio value of 125. Patients in whom an AUC_{0-24}:MIC ratio of 125 or greater had a significantly higher probability of a positive therapeutic response than those patients in whom lesser exposures were attained. As ciprofloxacin is approximately 25% bound to serum proteins, this value corresponds to a free-drug AUC_{0-24}:MIC ratio of 90 to 100.

Similar relationships have been found between the AUC_{0-24}:MIC ratio and response in the neutropenic murine infection model. For instance, Drusano et al. demonstrated that for levofloxacin and *P. aeruginosa*, a total-drug AUC_{0-24}:MIC ratio of 88 in immunosuppressed mice was associated with a 99% reduction in bacterial burden (73); Craig et al. showed that for fluoroquinolones and primarily gram-negative bacilli in immunosuppressed animals, the AUC_{0-24}:MIC ratio was predictive of survival (Fig. 4) (74).

Although ultimately the best arbiter of the appropriateness of a dose regimen is therapeutic response data obtained from well-controlled clinical studies, as can be seen in Table 5, there has been rather good concordance between PK–PD animal studies and data from infected patients (75).

With the exception of telithromycin, the magnitudes of the PK–PD measure necessary for clinical effectiveness were similar to those identified from animal data, even across drug classes and across multiple clinical indications. As illustrated in Table 5, the magnitude of exposure identified for a two log-unit reduction in bacterial burden in immunocompromised animals was similar to the exposure threshold associated with good clinical outcomes for patients with hospital-acquired pneumoniae associated with gram-negative bacilli treated with ciprofloxacin or levofloxacin. Thus, it can be inferred that the exposure target in immunocompromised animals predictive of an adequate response in humans with such pneumonias is, at a minimum, two log-unit reduction in bacterial burden. This means that where human exposure–response data are unavailable, we understand the PK–PD profile needed in animals to attain clinical effectiveness in humans. Consequently, when such data are missing, it may be reasonable to use animal data to inform decisions such as susceptibility breakpoint recommendations.

USE OF PK–PD IN THE DEVELOPMENT OF ANTIMICROBIALS AND SETTING OF TENTATIVE SUSCEPTIBILITY BREAKPOINTS

Since the importance of concentration–effect relationships is increasingly appreciated, antimicrobials are now being developed using these relationships in two ways. The first is in the early phase to determine the PK–PD index that best correlates with effect in in vitro PK models and in animal models of infection, and whether the values obtained to ensure a maximum effect are possibly achieved in humans for those microorganisms that therapy is directed at, based on the frequency distributions for those microorganisms. In the second phase, PK studies in volunteers will generate estimates of PK parameters and measures of dispersion. These values can then be used to simulate dosing regimens using MCS and obtain target-attainment rates for various dosing regimens. Finally, the probability of target attainment for each dosing regimen is compared to the frequency distributions of the target pathogens. The dosing regimen to follow-up in subsequent

TABLE 5 PK–PD Targets Derived from Animal Infection Models and Clinical Data

Disease state	Drug	Clinically derived PK–PD target	Animal infection model; organism studied	Animal-derived PK–PD target
Hospital-acquired pneumonia	Quinolones	$fAUC_{0-24}$:MIC ratio: 62–75	Neutropenic mouse-thigh; gram-negative bacilli	$fAUC_{0-24}$:MIC ratio: 70–90 for 90% animal survival or two log-unit kill
Community-acquired respiratory tract infections	Quinolones	$fAUC_{0-24}$:MIC ratio: 34	Immunocompetent mouse-thigh; *Streptococcus pneumoniae*	$fAUC_{0-24}$:MIC ratio: 25–34 for 90% animal survival or 2 log-unit kill
	β-lactams	$T_{>MIC}$: 40% of the dosing interval	Immunocompetent mouse-thigh; *S. pneumoniae*	$T_{>MIC}$: 30–40% of the dosing interval for 90% animal survival
	Telithromycin	AUC_{0-24}:MIC ratio: 3.375	Neutropenic mouse-thigh; *S. pneumoniae*	AUC_{0-24}:MIC ratio: 1000 for stasis
Bacteremia	Oritavancin	$fT_{>MIC}$:22% of the dosing interval for *Staphylococcus aureus*	Neutropenic mouse-thigh; *S. aureus*	$fT_{>MIC}$: 20% of the dosing interval for a 0.5 log-unit kill
	Linezolid	AUC_{0-24}:MIC ratio: 85 for *S. aureus* or *Enterococcus faecium*	Neutropenic mouse-thigh; *S. aureus*	AUC_{0-24}:MIC ratio: 83 for stasis
Complicated skin and skin structure infections	Tigecycline	AUC_{0-24}:MIC ratio: 17.9	Neutropenic mouse-thigh; *S. aureus*	AUC_{0-24}:MIC ratio: 20 for stasis
	Linezolid	AUC_{0-24}:MIC ratio: 110	Neutropenic mouse-thigh; *S. aureus*	AUC_{0-24}:MIC ratio: 83 for stasis

Abbreviations: AUC, area under the concentration–time curve at 24 hours; MIC, minimum inhibitory concentration; PD, pharmacodynamic; PK, pharmacokinetic; $T_{>MIC}$, duration of time the drug concentration remains above the MIC value.

Phase 2/3 studies is the one that yields the greatest delta (difference) between the probabilities of target attainment and dose-limiting toxicity. An example here is the approach taken in the development of ceftobiprole, a new cephalosporin with anti-MRSA activity (76). Table 4 shows the probabilities of target-attainment for two simulated dosing regimens, 250 mg every 12 hours and 750 mg every 12 hours for several values of $T_{>MIC}$. Since the frequency distributions of the target pathogens indicate that the highest MIC is 2 mg/L for most species and only rare isolates of 4 mg/L, the dosing regimen of 250 mg every 12 hours is clearly insufficient to obtain target-attainment ratios nearing 100% for $\%T_{>MIC}$ as low as 30%. Of the two regimens, it is recommended that the 750 mg every 12 hours be used for follow-up in clinical trials and 4 mg/L be the susceptibility breakpoint value in that situation.

This approach not only speeds up the development phase of the drug, but also maximizes the probability that the dosing regimens to be used in Phase 2/3 clinical trials are effective and precludes in a certain respect the need for classical dose-finding studies. Importantly, the dosing regimen is chosen on the basis of an adequate probability of target attainment for target pathogens and thus radically changes the way breakpoints are set. Earlier, dosing regimens were applied and a susceptibility breakpoint was determined based on dosing regimens. Now, a susceptibility breakpoint is set and the dosing regimen is selected based on the susceptibility breakpoint.

THE CURRENT APPROACHES OF THE EUCAST AND CLSI
TO SUSCEPTIBILITY BREAKPOINT SETTING
The Current Approach of CLSI

The current process utilized by CLSI to set susceptibility breakpoints is outlined in CLSI document M23-A2 (4). As previously noted, the manner in which susceptibility breakpoints are currently determined has evolved over the past three decades and undoubtedly will continue to evolve in the future as we gain greater knowledge of the interactions of antimicrobial agents, microorganisms, and the patient. Unlike EUCAST, CLSI publishes only clinical susceptibility breakpoints and does not publish epidemiologic susceptibility breakpoints. As stated in the CLSI Subcommittee on Antimicrobial Susceptibility Testing mission statement, "The ultimate purpose of the subcommittee's mission is to provide useful information to enable laboratories to assist the clinician in the selection of appropriate antimicrobial therapy for patient care."

In an ideal world, one would be able to conduct well-designed clinical trials, evaluate the response of the patient to therapy, and sort these data by the MIC of the microorganism responsible for the infection. If one could do this, then the determination of the susceptibility breakpoint could be determined almost exclusively utilizing clinical data. Unfortunately, getting the answer solely from clinical studies is unlikely for many reasons. First, there are many reasons why a patient recovers from an infectious process and only one factor is the antimicrobial agent. Second, institutional review boards are unwilling to allow subjects into a study if the organism being treated is likely to be R to the antimicrobial being studied. As a result of this, patients with infections associated with organisms with somewhat higher MIC values are usually not included in clinical studies, making it impossible to determine the true "susceptible" susceptibility breakpoint. For many of the newer antimicrobial agents, there are few organisms that have higher MIC values; therefore, it is difficult to gain valuable clinical experience in the treatment of infections with those organisms that would provide the most useful data in determining the true susceptibility breakpoint.

As clinical data are unlikely to provide the answer, CLSI depends on microbiologic data, animal modeling data, and PK–PD modeling data to determine the likely susceptibility breakpoint. The clinical data are then utilized primarily as a means of confirming or fine tuning the susceptibility breakpoint. The microbiologic data consist of distributions of organisms and their MIC values, including a broad representation of organisms against which the antimicrobial agent is likely to be used, strains with known resistance mechanisms, and strains gathered from clinical studies. Animal modeling, which has been correlated with clinical outcomes, allows one to determine the PK–PD target that would be predictive of clinical outcomes. Once PK data are available and the targets identified, one can utilize PD modeling to predict a susceptibility breakpoint.

The Current Approach of EUCAST

EUCAST has defined a procedure for determining breakpoints for new antimicrobials, which is also used for harmonizing and/or revising breakpoints for existing antimicrobials.

1. Define dose or dosages. Consider an "intermediately susceptible" category only for drugs with more than one dose.
2. Define target microorganisms.
3. Define WT MIC distributions for target microorganisms and their ECOFF values.
4. Evaluate the PK properties of the drug and define a PK profile.
5. Evaluate the PD properties of the drug and define a PD profile.
6. Perform PK/PD modeling, including MCSs, to determine tentative breakpoints.
7. Evaluate the tentative breakpoints in relation to known clinical efficacy and WTs of target microorganisms. When necessary, to avoid dividing WT MIC distributions, adjust the breakpoint one MIC concentration up or down and explain in footnote.
8. Consult national breakpoint committees in Europe on EUCAST tentative breakpoints.
9. For new drugs, consult EMEA rapporteur and experts; for existing drugs, consult EUCAST General Committee (one representative per European country), expert groups, and pharmaceutical and AST industry.
10. Get final decision on EUCAST breakpoints from the EUCAST Steering Committee.
11. Finalize a rationale document for the antimicrobial drug or group of drugs: (*i*) for new drugs, send to EMEA for formal decision; and (*ii*) for existing drugs, get EUCAST steering committee decision.
12. Look up breakpoint table and rationale document on EUCAST website and EUCAST Technical Note in CMI.

CONCLUDING REMARKS

In this chapter, we discussed the methods and approaches currently used to establish clinical breakpoints of antimicrobials. From a historical perspective, the methods and interpretations used within various countries—very different a few decades ago—tend to approach each other. This is not surprising, since the scientific background is more and more elucidated and a consensus is therefore consistent with the developments in the field. While differences within Europe have largely disappeared, some dissimilarity between EUCAST and CLSI remains. However, as science moves forward, a worldwide consensus on breakpoints will be reached eventually.

REFERENCES

1. Drusano GL, D'Argenio DZ, Preston SL, et al. Use of drug effect interaction modeling with Monte Carlo simulation to examine the impact of dosing interval on the projected antiviral activity of the combination of abacavir and amprenavir. Antimicrob Agents Chemother 2000; 44(6):1655–1659.
2. Drusano GL, Preston SL, Hardalo C, et al. Use of preclinical data for selection of a phase II/III dose for evernimicin and identification of a preclinical MIC breakpoint. Antimicrob Agents Chemother 2001; 45(1):13–22.
3. Kahlmeter G, Brown DFJ, Goldstein FW, et al. European harmonisation of MIC breakpoints for antimicrobial susceptibility testing of bacteria. J Antimicrobial Chemotherapy 2003; 52(2):145–148.
4. NCCLS. Development of In Vitro Susceptibility Testing Criteria and Quality Control Parameters; Approved Guideline-Second Edition. NCCLS document M23-A2. Wayne: NCCLS, 2001.
5. Ericsson HM, Sherris JC. Antibiotic sensitivity testing. Report of an international collaborative study. Acta Pathol Microbiol Scand [B] Microbiol Immunol 1971; 217(suppl B):1–90.
6. Soussy CJ, Cluzel R, Courvalin P. Definition and determination of in vitro antibiotic susceptibility breakpoints for bacteria in France. The Comite de l'Antibiogramme de la Societe Francaise de Microbiologie. Eur J Clin Microbiol Infect Dis 1994; 13(3):238–246.
7. MacGowan AP, Wise R. Establishing MIC breakpoints and the interpretation of in vitro susceptibility tests. J Antimicrob Chemother 2001; 48(suppl 1):17–28.
8. Mouton JW, van Klingeren B, de Neeling AJ, Degener JE, Commissie Richtlijnen Gevoeligheidsbepalingen. Het vaststellen van gevoeligheidskriteria voor antibacteriele middelen in Nederland: verleden, heden en toekomst. Nederlands Tijdschrift voor Med Microbiol 2000; 8(3):73–78.
9. Swedish Reference Group of Antibiotics. Antimicrobial susceptibility testing in Sweden. Scandinav J Infect Dis 1997; 105S:5–31.
10. Bergan T, Bruun JN, Digranes A, Lingaas E, Melby KK, Sander J. Susceptibility testing of bacteria and fungi. Report from "the Norwegian Working Group on Antibiotics". Scand J Infect Dis Suppl 1997; 103:1–36.
11. Deutsches Institut fur Normung. DIN 58 940-10, Criteria for the Categorization of the In-Vitro Activity of Antimicrobial Agents. Beuth Verlag: Deutsches Institut fur Normung, 1990.
12. Kahlmeter G, Brown DF, Goldstein FW, et al. European Committee on Antimicrobial Susceptibility Testing (EUCAST) technical notes on antimicrobial susceptibility testing. Clin Microbiol Infect 2006; 12(6):501–503.
13. Website www.eucast.org
14. Vogelman B, Craig WA. Kinetics of antimicrobial activity. J Pediatr 1986; 108(5 Pt 2): 835–840.
15. Shah PM, Junghanns, Stille W. Bactericidal dosie-activity relationships with *E. coli*, *K. pneumoniae* and *Staph. aureus* (author's transl.). Dtsch Med Wochenschr 1976; 101(9):325–328.
16. Craig WA. Pharmacokinetic/pharmacodynamic parameters: rationale for antibacterial dosing of mice and men. Clin Infect Dis 1998; 26(1):1–10; quiz 11–2.
17. Scaglione F, Mouton JW, Mattina R. Pharmacodynamics of levofloxacin in a murine pneumonia model: importance of peak to MIC ratio versus AUC. Interscience

Conference Antimicrobial Agents Chemotherapy, 1999. San Francisco: American Society Microbiology, 1999:6.

18. Madaras-Kelly KJ, Ostergaard BE, Hovde LB, Rotschafer JC. Twenty-four-hour area under the concentration-time curve/MIC ratio as a generic predictor of fluoroquinolone antimicrobial effect by using three strains of *Pseudomonas aeruginosa* and an in vitro pharmacodynamic model. Antimicrob Agents Chemother 1996; 40(3):627–632.

19. Lister PD, Sanders CC. Pharmacodynamics of levofloxacin and ciprofloxacin against *Streptococcus pneumoniae*. J Antimicrob Chemother 1999; 43(1):79–86.

20. Lacy MK, Lu W, Xu X, et al. Pharmacodynamic comparisons of levofloxacin, ciprofloxacin, and ampicillin against *Streptococcus pneumoniae* in an in vitro model of infection. Antimicrob Agents Chemother 1999; 43(3):672–677.

21. Lister PD, Sanders CC. Pharmacodynamics of trovafloxacin, ofloxacin, and ciprofloxacin against *Streptococcus pneumoniae* in an in vitro pharmacokinetic model. Antimicrob Agents Chemother 1999; 43(5):1118–1123.

22. Lister PD. Pharmacodynamics of gatifloxacin against *Streptococcus pneumoniae* in an in vitro pharmacokinetic model: impact of area under the curve/MIC ratios on eradication. Antimicrob Agents Chemother 2002; 46(1):69–74.

23. Fernandez J, Barrett JF, Licata L, Amaratunga D, Frosco M. Comparison of efficacies of oral levofloxacin and oral ciprofloxacin in a rabbit model of a staphylococcal abscess. Antimicrob Agents Chemother 1999; 43(3):667–671.

24. Ng W, Lutsar I, Wubbel L, et al. Pharmacodynamics of trovafloxacin in a mouse model of cephalosporin- resistant *Streptococcus pneumoniae* pneumonia. J Antimicrob Chemother 1999; 43(6):811–816.

25. Onyeji CO, Bui KQ, Owens RC Jr, Nicolau DP, Quintiliani R, Nightingale CH. Comparative efficacies of levofloxacin and ciprofloxacin against *Streptococcus pneumoniae* in a mouse model of experimental septicaemia. Int J Antimicrob Agents 1999; 12(2):107–114.

26. Bedos JP, Rieux V, Bauchet J, Muffat-Joly M, Carbon C, Azoulay-Dupuis E. Efficacy of trovafloxacin against penicillin-susceptible and multiresistant strains of *Streptococcus pneumoniae* in a mouse pneumonia model. Antimicrob Agents Chemother 1998; 42(4):862–867.

27. Bedos JP, Azoulay-Dupuis E, Moine P, et al. Pharmacodynamic activities of ciprofloxacin and sparfloxacin in a murine pneumococcal pneumonia model: relevance for drug efficacy. J Pharmacol Exp Ther 1998; 286(1):29–35.

28. Drusano GL, Johnson DE, Rosen M, Standiford HC. Pharmacodynamics of a fluoroquinolone antimicrobial agent in a neutropenic rat model of *Pseudomonas sepsis*. Antimicrob Agents Chemother 1993; 37(3):483–490.

29. Andes DR, Craig WA. Pharmacodynamics of fluoroquinolones in experimental models of endocarditis. Clin Infect Dis 1998; 27(1):47–50.

30. Ernst EJ, Klepser ME, Petzold CR, Doern GV. Evaluation of survival and pharmacodynamic relationships for five fluoroquinolones in a neutropenic murine model of pneumococcal lung infection. Pharmacotherapy 2002; 22(4):463–470.

31. Croisier D, Chavanet P, Lequeu C, et al. Efficacy and pharmacodynamics of simulated human-like treatment with levofloxacin on experimental pneumonia induced with penicillin-resistant pneumococci with various susceptibilities to fluoroquinolones. J Antimicrob Chemother 2002; 50(3):349–360.

32. Mattoes HM, Banevicius M, Li D, et al. Pharmacodynamic assessment of gatifloxacin against *Streptococcus pneumoniae*. Antimicrob Agents Chemother 2001; 45(7):2092–2097.

33. Forrest A, Chodosh S, Amantea MA, Collins DA, Schentag JJ. Pharmacokinetics and pharmacodynamics of oral grepafloxacin in patients with acute bacterial exacerbations of chronic bronchitis. J Antimicrob Chemother 1997; 40(suppl A):45–57.

34. Forrest A, Nix DE, Ballow CH, Goss TF, Birmingham MC, Schentag JJ. Pharmacodynamics of intravenous ciprofloxacin in seriously ill patients. Antimicrob Agents Chemother 1993; 37(5):1073–1081.

35. Preston SL, Drusano GL, Berman AL, et al. Pharmacodynamics of levofloxacin: a new paradigm for early clinical trials (see comments). J Am Med Assoc 1998; 279(2):125–129.

36. Highet VS, Forrest A, Ballow CH, Schentag JJ. Antibiotic dosing issues in lower respiratory tract infection: population-derived area under inhibitory curve is predictive of efficacy. J Antimicrob Chemother 1999; 43(suppl A):55–63.

37. Ambrose PG, Grasela DM, Grasela TH, Passarell J, Mayer HB, Pierce PF. Pharmacodynamics of fluoroquinolones against *Streptococcus pneumoniae* in patients with community-acquired respiratory tract infections. Antimicrob Agents Chemother 2001; 45(10):2793–2797.

38. Blaser J, Stone BB, Groner MC, Zinner SH. Comparative study with enoxacin and netilmicin in a pharmacodynamic model to determine importance of ratio of antibiotic peak concentration to MIC for bactericidal activity and emergence of resistance. Antimicrob Agents Chemother 1987; 31(7):1054–1060.

39. Mouton JW. Breakpoints: current practice and future perspectives. Int J Antimicrob Agents 2002; 19(4):323–331.

40. Mouton JW. Het gebruik van farmcodynamische parameters voor het vaststellen van gevoeligheidscriteria van fluorchinolonen. Nederlands Tijdschrift voor Med Microbiol 2000; 8(3):82–86.

41. Mouton JW. Impact of pharmacodynamics on breakpoint selection for susceptibility testing. Infect Dis Clin North Am 2003; 17(3):579–598.

42. Cars O. Pharmacokinetics of antibiotics in tissues and tissue fluids: a review. Scand J Infect Dis Suppl 1990; 74:23–33.

43. Liu P, Muller M, Derendorf H. Rational dosing of antibiotics: the use of plasma concentrations vs. tissue concentrations. Int J Antimicrob Agents 2002; 19(4): 285–290.

44. NCCLS. Performance Standards for Antimicrobial Susceptibility Testing; 10th International Supplement. M100-S10 (M7). Wayne: NCCLS, 2000.

45. Andes D, Craig WA. In vivo activities of amoxicillin and amoxicillin-clavulanate against *Streptococcus pneumoniae*: application to breakpoint determinations. Antimicrob Agents Chemother 1998; 42(9):2375–2379.

46. Tompsett R, Shultz S, McDermott W. The relation of protein binding to the pharmacology and antibacterial activity of penicillins X, G, dhydro F, and K. J Bacteriol 1947; 53:581–595.

47. Kunin CM, Craig WA, Kornguth M, Monson R. Influence of binding on the pharmacologic activity of antibiotics. Ann N Y Acad Sci 1973; 226:214–224.

48. Craig WA. Interrelationship between pharmacokinetics and pharmacodynamics in determining dosage regimens for broad-spectrum cephalosporins. Diagn Microbiol Infect Dis 1995; 22(1–2):89–96.

49. Ryan DM, Cars O. Antibiotic assays in muscle: are conventional tissue levels misleading as indicator of the antibacterial activity? Scand J Infect Dis 1980; 12(4):307–309.

50. Ryan DM, Cars O. A problem in the interpretation of beta-lactam antibiotic levels in tissues. J Antimicrob Chemother 1983; 12(3):281–284.

51. Bergan T, Engeset A, Olszewski W. Does serum protein binding inhibit tissue penetration of antibiotics? Rev Infect Dis 1987; 9(4):713–718.

52. Craig WA, Suh B. Theory and practical impact of binding of antimicrobials to serum proteins and tissue. Scand J Infect Dis Suppl 1978; 14:92–99.

53. Drusano GL, Preston SL, Gotfried MH, Danziger LH, Rodvold KA. Levofloxacin penetration into epithelial lining fluid as determined by population pharmacokinetic modeling and monte carlo simulation. Antimicrob Agents Chemother 2002; 46(2):586–589.

54. Mouton JW, Horrevorts AM, Mulder PG, Prens EP, Michel MF. Pharmacokinetics of ceftazidime in serum and suction blister fluid during continuous and intermittent infusions in healthy volunteers. Antimicrob Agents Chemother 1990; 34(12):2307–2311.

55. Wise R, Baker S, Livingston R. Comparison of cefotaxime and moxalactam pharmacokinetics and tissue levels. Antimicrob Agents Chemother 1980; 18(3):369–371.

56. Muller M, Haag O, Burgdorff T, et al. Characterization of peripheral-compartment kinetics of antibiotics by in vivo microdialysis in humans. Antimicrob Agents Chemother 1996; 40(12):2703–2709.

57. Marchand S, Chenel M, Lamarche I, Couet W. Pharmacokinetic modeling of free amoxicillin concentrations in rat muscle extracellular fluids determined by microdialysis. Antimicrob Agents Chemother 2005; 49(9):3702–3706.

58. Buerger C, Plock N, Dehghanyar P, Joukhadar C, Kloft C. Pharmacokinetics of unbound linezolid in plasma and tissue interstitium of critically ill patients after multiple dosing using microdialysis. Antimicrob Agents Chemother 2006; 50(7):2455–2463.

59. Carryn S, Van Bambeke F, Mingeot-Leclercq MP, Tulkens PM. Comparative intracellular (THP-1 macrophage) and extracellular activities of beta-lactams, azithromycin, gentamicin, and fluoroquinolones against *Listeria monocytogenes* at clinically relevant concentrations. Antimicrob Agents Chemother 2002; 46(7):2095–2103.

60. Andrews JM. Determination of minimum inhibitory concentrations. J Antimicrob Chemother 2001; 48(suppl 1):5–16.

61. EUCAST. Determination of minimum inhibitory concentrations (MICs) of antibacterial agents by broth dilution. Discussion Document E. Dis 5.1. Munich, Germany: European Society Clinical Microbiology and Infectious Diseases, 2003.

62. Bonate PL. A brief introduction to Monte Carlo simulation. Clin pharmacokinetic 2001; 40:15–22.

63. Ambrose PG, Grasela DM. The use of Monte Carlo simulation to examine pharmacodynamic variance of drugs: fluoroquinolone pharmacodynamics against *Streptococcus pneumoniae*. Diagn Microbiol Infect Dis 2000; 38(3):151–157.

64. Nicolau DP, Ambrose PG. Pharmacodynamic profiling of levofloxacin and gatifloxacin using Monte Carlo simulation for community-acquired isolates of *Streptococcus pneumoniae*. Am J Med 2001; 111(suppl 9A):13S–18S; discussion 36S–38S.

65. Montgomery MJ, Beringer PM, Aminimanizani A, et al. Population pharmacokinetics and use of Monte Carlo simulation to evaluate currently recommended dosing regimens of ciprofloxacin in adult patients with cystic fibrosis. Antimicrob Agents Chemother 2001; 45(12):3468–3473.

66. Mouton JW, Schmitt-Hoffmann A, Shapiro S, Nashed N, Punt NC. Monte Carlo predictions of dosage regimens for BAL5788, a new broad-spectrum cephalosporin active against MRSA. Interscience Conference Antimicrobial Agents Chemotherapy, 2002. San Diego: American Society Microbiology, 2002.

67. Mouton JW, Punt N, Vinks AA. A retrospective analysis using Monte Carlo simulation to evaluate recommended ceftazidime dosing regimens in healthy volunteers, patients with cystic fibrosis, and patients in the intensive care unit. Clin Ther 2005; 27(6):762–772.

68. Ambrose PG. Antimicrobial susceptibility breakpoints: PK–PD and susceptibility breakpoints. Treat Respir Med 2005; 4(suppl 1):5–11.

69. Ambrose PG. Monte Carlo simulation in the evaluation of susceptibility breakpoints: predicting the future: insights from the society of infectious diseases pharmacists. Pharmacotherapy 2006; 26(1):129–134.

70. Ambrose PG, Bhavnani SM, Jones RN, Craig WA. Use of Pharmacokinetics-pharmacodynamics and Monte-Carlo simulation as decision support for the reevaluation of NCCLS cephem susceptibility breakpoints for *Enterobacteriaceae*. 44th Interscience Conference Antimicrobial Agents and Chemotherapy; Oct 12th–Nov 2nd, 2004, Washington, D.C.: ASM, 2004:12.

71. Kuti JL, Dandekar PK, Nightingale CH, Nicolau DP. Use of Monte Carlo simulation to design an optimized pharmacodynamic dosing strategy for meropenem. J Clin Pharmacol 2003; 43(10):1116–1123.

72. Stearne LE, Gyssens IC, Goessens WH, et al. In vivo efficacy of trovafloxacin against bacteroides fragilis in mixed infection with either *Escherichia coli* or a vancomycin-resistant strain of *Enterococcus faecium* in an established-abscess murine model. Antimicrob Agents Chemother 2001; 45(5):1394–1401.

73. Jumbe N, Louie A, Leary R, et al. Application of a mathematical model to prevent in vivo amplification of antibiotic-resistant bacterial populations during therapy. J Clin Invest 2003; 112(2):275–285.

74. Craig WA. Pharmacodynamics of antimicrobials: general concepts and applications. In: Nightingale CH, Murakawa T, Ambrose PG, eds. Antimicrobial Pharmacodynamics in Theory and Clinical Practice. New York: Marcel Dekker, Inc., 2002.

75. Ambrose PG, Bhavnani SM, Rubino CM, et al. Pharmacokinetics-pharmacodynamics of antimicrobial therapy: it's not just for mice anymore. Clin Infect Dis 2007; 44(1):79–86.
76. Jones RN, Deshpande LM, Mutnick AH, Biedenbach DJ. In vitro evaluation of BAL9141, a novel parenteral cephalosporin active against oxacillin-resistant staphylococci. J Antimicrob Chemother 2002; 50(6):915–932.

3 In Vitro Dynamic Models as Tools to Predict Antibiotic Pharmacodynamics

Alexander A. Firsov
Gause Institute of New Antibiotics, Russian Academy of Medical Sciences, Moscow, Russia

Stephen H. Zinner
Mount Auburn Hospital, Harvard Medical School, Cambridge, Massachusetts, U.S.A.

Irene Y. Lubenko
Gause Institute of New Antibiotics, Russian Academy of Medical Sciences, Moscow, Russia

INTRODUCTION

There is a gap between antibiotic effects in the clinical setting and routine suscept-ibility testing and/or traditional time-kill studies at fixed concentrations of anti-biotics: both reflect the intrinsic antibiotic activity but ignore pharmacokinetics. Being aware of these limitations, a way to simultaneously consider both factors was first suggested in 1966 by O'Grady and Pennington (1). They developed an in vitro model that simulates urinary excretion of antibiotics and exposes a bacterial culture to clinically achievable urinary drug concentrations. Soon after, in 1968, Sanfilippo and Morvillo (2) described another model that allows pharmacodynamic evaluation of sulfonamides by in vitro simulation of their pharmacokinetics in human serum. These innovative studies provided the impetus to further progress in this kind of modeling over the subsequent 15 years. The achievements were so imposing, especially in the area of model design, that a special meeting on the methodology and evaluation of in vitro dynamic models was organized by the Paul Ehrlich Society for Chemotherapy and the British Society for Antimicrobial Chemotherapy (Bad Honnef, 1984). Over the next 10 years, in vitro dynamic models were used in pharmacodynamic studies with many novel antibiotics, in particular those that exhibit different pharmacokinetics but similar intrinsic activities. Most of these studies demonstrated the very good potential of in vitro dynamic models rather than providing clinically useful findings. However, this experience led to the sub-sequent displacement of purely phenomenological studies by better designed, more predictive studies that have been reported over the past 10 years.

Studies that utilize in vitro dynamic models have been reviewed with special emphasis on the technical aspects of modeling (3–8). These aspects were discussed in detail in an excellent recently published review (9). Given the fact that most widely used models were designed in the 1970s and 1980s [early designs supplied with processor-controlled pumps (10–13) can hardly be considered original], a thorough review of technical issues is beyond the scope of this review. On the other hand, the *methodological* aspects of pharmacodynamic studies with dynamic models have been reviewed much less frequently (5,6,14–16). For this reason, methodology of pharmacodynamic studies using these models rather than their design is the primary subject of this review.

PHARMACOKINETIC SIMULATIONS IN VITRO
Principles

The general principle of in vitro simulations of the desired pharmacokinetic profile may be illustrated using the simplest example of monoexponential elimination of drug given as a bolus. As with in vivo experience, when continuously diluted drug in the blood is eliminated from the systemic circulation, a monoexponential concentration decay in vitro may be provided by continuous dilution of drug-containing medium keeping its volume (V) constant. Assuming first-order kinetics, the rate of time (t)-dependent changes in the amount (A) of drug is directly proportional to its concentration (C):

$$dA/dt = -FC \tag{1}$$

where F is the flow rate.

Dividing both parts of Eq. (1) by V gives the respective differential equation for the concentration rate:

$$dC/dt = -FC/V \tag{2}$$

Its integration at the initial condition of $C(0) = C_0$ yields a monoexponential equation:

$$C(t) = C_0 \exp\left(-(F/V)t\right) \tag{3}$$

Equation (3) is similar to the equation that fits drug pharmacokinetics in vivo:

$$C(t) = C_0 \exp\left(-(Cl/V_d)t\right) \tag{4}$$

where Cl is the total clearance and V_d is the volume of distribution.

Thus, the flow rate and the volume of antibiotic-containing medium, i.e., terms F and V in Eq. (3), represent the total clearance and the volume of distribution, respectively, terms Cl and V_d in Eq. (4). Therefore, to simulate a monoexponential concentration–time course with the desired half-life ($T_{1/2}$):

$$T_{1/2} = \ln 2 V_d/Cl = \ln 2 V/F \tag{5}$$

V, F or both V and F may be varied according to the following equations:

$$V = FT_{1/2}/\ln 2 \tag{6}$$

$$F = V \ln 2/T_{1/2} \tag{7}$$

Drug amount (A_0 or dose) that provides the desired initial concentration is defined by the equation:

$$A_0 = V C_0 \tag{8}$$

More complicated polyexponential pharmacokinetic profiles may be simulated in a similar way; moreover, there is a strong link between the number of compartments of the pharmacokinetic model and the number of units of the respective in vitro dynamic model (17). However, the target polyexponential concentration decay may be approximated by quasimonoexponential fragments.

This is easily achieved by stepwise changing the flow rates, without complicating the in vitro model. For example, biexponential concentration decay of cefuroxime was approximated by four monoexponential profiles (18). Later on, biexponential pharmacokinetics of a novel fluoroquinolone ABT-492, oritavancin, amoxicillin, and levofloxacin as well as were approximated by two monoexponents (19–22), and three-exponential pharmacokinetics of telithromycin were simulated by three monoexponents (23). So, a relatively simple in vitro model may be used to simulate quite complex pharmacokinetic patterns. In this light, attempts to classify in vitro dynamic models by the number of units (8) may be confusing.

The above approximations should not be confused with the stepwise approximation of monoexponential pharmacokinetics (24–32) as an alternative to simulations of continuously changing antibiotic concentrations. Although such stepwise approximations may provide quite smooth changes in concentrations, this kind of simulation is less elegant than that described in the beginning of this section and is rarely used in modern studies.

To simulate first-order absorption, drug is administered in an additional subunit (SU) of volume V_{SU} rather than in the central unit (CU) of volume V_{CU}. Drug dilution in the subunit at a constant flow rate F yields its monoexponential input into the central unit. The same flow rate provides concomitant dilution of drug entering the central unit. Integration of the respective system of differential equations:

$$dA_{SU}/dt = -F\, C_{SU} \tag{9}$$

$$dA_{CU}/dt = F\, CU_{SU} - F\, C_{CU} \tag{10}$$

at the initial conditions of $C_{SU}(0) = C_{SU,0}$ and $C_{CU}(0) = 0$ yields a biexponential equation (similar to the Bateman function used in pharmacokinetics) that describes the time course of drug concentration in the central unit:

$$C_{CU}(t) = B\{\exp(-(F/V_{CU})t) - \exp(-(F/V_{SU})t)\} \tag{11}$$

where

$$B = A_{SU,0}/(V_{CU} - V_{SU}) \tag{12}$$

To simulate the desired value of $T_{\frac{1}{2}}$, both V_{CU} and F or one of them may be varied according to the following equations:

$$V_{CU} = F\, T_{1/2}/\ln 2 \tag{13}$$

$$F = V_{CU} \ln 2/T_{1/2} \tag{14}$$

To simulate the desired value of the absorption half-life ($T_{\frac{1}{2}\text{abs}}$), the respective V_{SU} may be calculated for a chosen F:

$$V_{SU} = F\, T_{1/2'\text{abs}}/\ln 2 \tag{15}$$

Drug amount (dose) in the subunit that provides the desired concentration–time course in the central unit is defined by the product of B and the difference between V_{CU} and V_{SU} (17):

$$A_{SU,0} = B(V_{CU} - V_{SU}) \tag{16}$$

Various dynamic models have been designed based on these principles.

Embodiment

As mentioned in the previous section, continuous dilution of antibiotic-containing medium at a constant volume is the fundamental principle of in vitro simulations. Obviously, the elimination of antibiotic from a dynamic model will be accompanied by inevitable clearance of bacteria unless it is prevented in one way or another.

This clearance may interfere with the determination of antibiotic-induced bacterial killing. To minimize the possible overestimation of the antimicrobial effect, time-kill curves observed in a dilution model can be corrected by the dilution factor (φ): The respective equations were derived for mono- and biexponential elimination kinetics (33,34). However, thorough examination of one of the proposed algorithms, i.e.,

$$\varphi(t) = \exp((F/V)t) \tag{17}$$

as applied to growth kinetics of *Escherichia coli* at different flow rates (5) highlighted its drawbacks. The corrected bacterial counts were greater than those without dilution: the higher F, the more pronounced the bias. As a result, the corrected curves that were obtained at different flow rates were not superimposed. More sophisticated procedures (5,35) were subsequently shown to provide better correction of the time-kill curves. Unfortunately, both procedures are too cumbersome for routine use.

Fortunately, the problem of bacterial clearance is more apparent than real, at least for long-acting antibiotics whose half-lives considerably exceed the generation times of bacteria. At these conditions, i.e., at very high V/F ratios, the loss of organisms from the system may only have a negligible effect on accurate determination of the antimicrobial effect (5). For example, control growth curves of *Moraxella catarrhalis* at threefold different flow rates corresponding to the half-lives of four fluoroquinolones (from 4 hours for ciprofloxacin to 12 hours for moxifloxacin) were superimposed (36). Similar superimposed curves were reported with *Streptococcus pyogenes* and *Streptococcus pneumoniae* at flow rates mimicking different pharmacokinetics of azithromycin and roxithromycin (37), with *Staphylococcus aureus*, *E. coli*, and *Klebsiella pneumoniae* at flow rates simulating pharmacokinetics of moxifloxacin and levofloxacin (38), etc. No dilution effects were reported in a recent study that compared killing kinetics of moxifloxacin-exposed *E. coli* and *S. pneumoniae* in different type dynamic models (39).

For these reasons, simple dilution models, usually without correction by the dilution factor, are used more frequently than models that prevent clearance of bacteria. Selective filtration (40–45) or dialysis (46–50) of antibiotic through bacterial-impermeable membranes is used in these models. The dialysis models usually employ hollow fibers or artificial capillary units. Unfortunately, successful prevention of bacterial clearance may be fraught with less successful pharmacokinetic simulations than those provided by simple dilution models. Permeability of the membrane for antibiotic may be impaired because of partial occlusion of the pores by the growing bacterial culture during the experiments. That is why actual antibiotic concentrations have to be determined in each run.

Most models that prevent bacterial clearance consist of two units, one of which contains only antibiotic (the central unit) and the second that contains antibiotic with a bacterial culture [peripheral unit or units (47)]. As a rule, an equilibrium in the drug exchange between the central and peripheral units is reached quite rapidly, so that soon after the beginning of the experiment, both units are kinetically homogeneous. Therefore, common attempts to identify the peripheral unit of these in vitro models with the peripheral compartment of the pharmacokinetic model are inappropriate. In this light, direct in vitro simulations of peripheral tissue pharmacokinetics are more relevant, as was done for example, in a pharmacodynamic study with azithromycin and roxithromycin against *S. pyogenes* and *S. pneumoniae* that were exposed to tonsillar concentrations of both antibiotics (37).

END POINTS OF THE ANTIMICROBIAL EFFECT AND THEIR RELATIONSHIPS TO PHARMACOKINETICS

Although not specific to studies in in vitro models, the problem of adequate quantitative evaluation of the antimicrobial effect is often highlighted by such studies. Indeed, striking contrasts can be seen between the use of inappropriate end points and/or purely visual descriptions of time-kill curves and more sophisticated pharmacokinetic simulations. Accurate quantitative analysis of concentration– or dose–response relationships as well as accurate predictions of the efficient concentrations, doses, and dose regimens is impossible without the use of appropriate end points of the antimicrobial effect. Actually, the primary criteria for reliability of an end point are the ability to provide both reasonable relationships between the effect and its predictor(s) and clinically relevant predictions of antibiotic efficacy.

Classification of End Points of the Antimicrobial Effect

A recent paper examined the predictive value of 12 currently relevant end points (51). The end points were classified as those indices that reflected initial bacterial killing and those of that reflect the entire antimicrobial effect (Fig. 1).

Initial killing indices included $T_{90\%}$ (52), $T_{99\%}$ (53), $T_{99.9\%}$ (18), the time to achieve 10-, 100-, and 1000-fold reductions of the initial inoculum (N_0) and k_{elb}, the "bacterial elimination rate constant," reflected by the slope of time-dependent changes in the difference between logarithms of viable counts with (N_A) and without antibiotic (N_C — control growth) (54). Only k_{elb} accurately reflects the initial killing process because the observed reduction of viable counts is really the net result of two competing processes — continuing growth of a certain portion of the bacterial population and actual killing of another portion when bacterial killing predominates over growth (55).

Indices that reflect the entire antimicrobial effect (Fig. 1) (51) include T_E, the time shift between control growth and the regrowth curves after antibiotic exposure, i.e., a measure of the effect duration (56); $\Delta \log N_t$, the viable bacteria count at an arbitrarily chosen time near the end of the observation period ($N_{A,t}$) that may or may not correspond to the dosing interval (τ)—($N_{A,\tau}$), usually presented as a difference between $\log N_0$ and $\log N_{A,t}$ (57); AUBC,[a] the area under $\log N_A$-time curve (59,60); AAC, the area above this curve and under the baseline drawn at the level of $N_A = N_0$ (61) that is the algebraic sum of the areas around the N_0 level; ABBC, the area between control growth and bacterial killing/regrowth curves (62) over the dosing interval τ, and ABBC determined up to the

---

[a]AUBKC abbreviation (58) was used instead of AUBC in some publications.

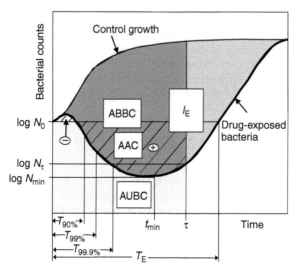

FIGURE 1 Different end points of the antimicrobial effect. *Source:* Adapted from Ref. 51.

end of the regrowth phase, which is also referred to as the intensity of the antimicrobial effect [I_E (42)].

Unlike T_E, $\Delta \log N_t$, AUBC, AAC, ABBC, and I_E, the minimal number of bacteria resulting from exposure to antibiotic [N_{min} (63)—often represented by the difference between logarithms of N_0 and N_{min} ($\Delta \log N_{min}$)] and the time to reach this nadir [t_{min} (63)] should not be considered indices of the entire effect because they actually reflect the state of equilibrium between bacterial growth and killing, i.e., "intermediate" killing (51).

On the other hand, some indices may be classified as point estimates including $T_{90\%}$, $T_{99\%}$, $T_{99.9\%}$, $\Delta \log N_{min}$, t_{min}, $\Delta \log N_t$, and T_E. This is in contrast to k_{elb} and the integral indices AUBC, AAC, ABBC, and I_E that are based on analysis of all points on the time-kill curve. Obviously, from the viewpoint of accuracy, the latter indices are more robust solid than the former.

Ability of the End Points to Provide Reasonable Concentration–Response Relationships

Each of the three point indices of initial killing, i.e., $T_{90\%}$, $T_{99\%}$, and $T_{99.9\%}$ usually exhibit erratic AUC (area under the curve)/MIC (minimum inhibitory response) relationships, without systematic shortening in response to increased AUC/MICs (Fig. 2, upper panel). A more reasonable relationship may be seen with k_{elb} (Fig. 2, bottom panel), although the AUC/MIC curve may contain "hills" and "ravines" resulting from inaccurate approximation of the time courses of the difference between $\log N_A$ and $\log N_C$. As seen in Figure 3, these time courses may show complicated shapes with only short time periods where quasilinear decay is observed. Estimates of k_{elb} estimates derived from such quasilinear plots may be unreliable.

At least one of the indices of "intermediate" killing, $\Delta \log N_{min}$, may exhibit a reasonable AUC/MIC relationship (Fig. 4, upper panel), but such a relationship may not be seen at relatively low and high AUC/MIC ratios. Unlike N_{min}, no reasonable relationships usually exist between t_{min} and AUC/MIC (Fig. 4, bottom panel).

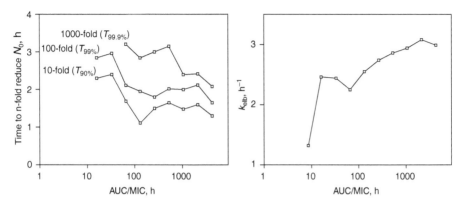

FIGURE 2 AUC/MIC-dependent changes in different indices of initial killing of *Escherichia coli* exposed to ciprofloxacin. *Source*: Adapted from Ref. 51.

Using both point indices of the entire effect, i.e., $\Delta \log N_t$ and T_E, reasonable AUC/MIC relationships are possible (Fig. 5), although the curve representing $\Delta \log N_t$ versus AUC/MIC may contain "hills" and "ravines" similar to those seen with k_{elb}. It should be noted that the AUC/MIC relationship with $\Delta \log N_t$ may not be seen at the relatively low and high AUC/MIC ratios. For this reason, different $N_{A,t}$s observed at different ts are listed in some studies reserving a preference for one time section over others. The use of T_E is free of such uncertainties. In addition, more systematic increases in T_E than in $\Delta \log N_t$ may be seen in response to increasing AUC/MIC ratios.

All four integral indices of the entire effect, AUBC, AAC, ABBC, and I_E, usually provide quite accurate but different AUC/MIC–response relationships (Fig. 6). A systematic decrease in AUBC and concomitant increases in AAC and ABBC at relatively small AUC/MIC ratios are followed by a plateau at relatively large AUC/MICs. Unlike AUBC, AAC, and ABBC, a less pronounced if any saturation is seen on the curve representing I_E versus AUC/MIC, derived from both single- and multiple-dose studies. For example, with *S. aureus* exposed to three consecutive

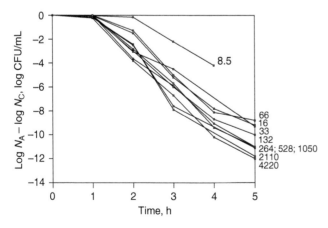

FIGURE 3 Time courses of the difference between $\log N_A$ and $\log N_C$: killing kinetics of *Escherichia coli* exposed to ciprofloxacin at different AUC/MIC ratios. AUC/MICs (in hours) simulated in an in vitro dynamic model are indicated at each curve. *Source*: Adapted from Ref. 51.

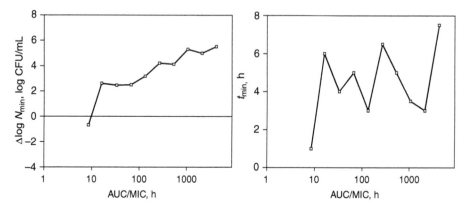

FIGURE 4 AUC/MIC-dependent changes in the indices of "intermediate" killing of *Escherichia coli* exposed to ciprofloxacin. *Source*: Adapted from Ref. 51.

daily doses of moxifloxacin and levofloxacin (64), AUC_t/MIC-associated changes in AUBC, AAC, and ABBC were seen at AUC_t/MICs < 100 hours ($t = 24$ hours), i.e., at suboptimal conditions, but not at AUC_t/MICs > 100 hours (Fig. 7)—superoptimal conditions. The plateau observed at clinically achievable AUC_t/MIC ratios actually precludes accurate comparison of the antibiotics. Use of I_E is free of this problem: Curves that represent the relationship of AUC_t/MIC to I_E remain different for moxifloxacin and levofloxacin even at high AUC_t/MIC ratios.

The reasons for different patterns of these relationships were discussed in a recent paper (65). AUBC and AAC but not ABBC underestimated the true effect at small AUC/MIC ratios, and all three end points also underestimate the effect at relatively large AUC/MICs. As seen in Figure 8, these underestimations result from the fact that AUBC and AAC measurements include the zones of "vanished" effect and AUBC, AAC, and ABBC also exclude the zones of persisting effect. So, the saturable patterns of the ABBC-, AAC-, and AUBC-log AUC/MIC curves result from inherent limitations of these end points (65) and lead to inappropriate claims about "AUC/MIC-independent" antimicrobial effect. Unlike these three end points, I_E considers the actual duration of the effect—from time zero to the time when

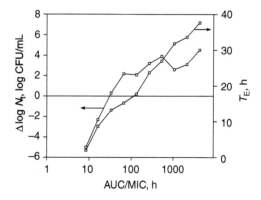

FIGURE 5 AUC/MIC-dependent changes in the point indices of the entire antimicrobial effect of ciprofloxacin on *Escherichia coli*. *Source*: Adapted from Ref. 51.

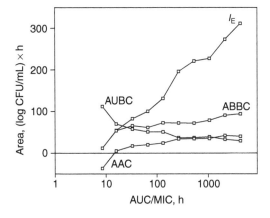

FIGURE 6 AUC/MIC-induced changes in the integral parameters of antimicrobial effect of ciprofloxacin on *Escherichia coli*. *Source*: Adapted from Ref. 51.

bacterial counts on the regrowth curve achieve the same maximal numbers as in the absence of antimicrobial, even though this occurs later than $t = \tau$.

The above limitations of AUBC, AAC, and ABBC do not exclude their use in situations when I_E cannot be accurately determined. Moreover, these indices are useful in studies that examine amplification or degradation of the antimicrobial effects in multiple-dose simulations. For example, changing effects of moxifloxacin on *S. pneumoniae* were demonstrated using the ABBC measured within each of three consecutive 24-hour dosing intervals (66).

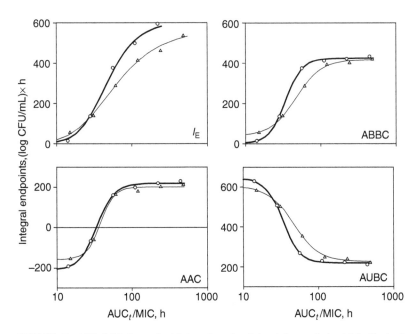

FIGURE 7 AUC_t/MIC-dependent integral end points of the antimicrobial effect of moxifloxacin (*bold lines, circles*) and levofloxacin (*thin lines, triangles*) on *Staphylococcus aureus* fitted by an E_{max} model. *Source*: Adapted from Ref. 64.

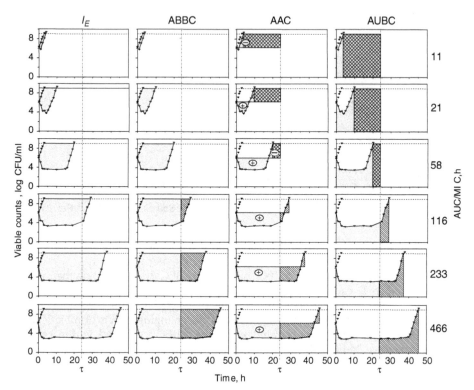

FIGURE 8 Time-kill curves of *Staphylococcus aureus* exposed to gemifloxacin and areas exploited by I_E, ABBC, AAC, and AUBC) (▨) including the zones of vanished effect (▨) and excluding the zones of persisting effect (▨). *Source*: From Ref. 65.

Sensitivity of the End Points to Changes in the AUC/MIC Ratio

All four indices of initial killing are narrowly sensitive to changes in the AUC/MIC ratio. For example, in the above study with ciprofloxacin-exposed *E. coli* (51), 1000-fold differences in the AUC/MIC ratio induced only twofold differences in $T_{90\%}$, $T_{99\%}$, $T_{99.9\%}$, and k_{elb} in contrast to a sixfold difference in $\Delta \log N_{min}$ as an index of "intermediate" killing. In the same study, a point index of the entire antimicrobial effect, $\Delta \log N_t$ ($t = 12$ hours), was even more sensitive to AUC/MIC increases: a 10-fold range of $\Delta \log N_t$ corresponded to a 450-fold range of AUC/MIC ratios.

As a rule, integral indices of the entire effect are more sensitive to changes in the AUC/MIC ratio than indices of initial killing. Similar eightfold changes in three t-dependent integral indices, AUBC, AAC, and ABBC, were reported with a 450-fold range of ciprofloxacin AUC/MICs against *E. coli* (51). Much greater sensitivity to the simulated AUC/MIC ratio was observed in the same study for T_E and I_E (20- and 30-fold changes, respectively).

Interrelations of Different End Points and Their Abilities to Predict the Entire Antimicrobial Effect

Being less precise, most point indices ($T_{90\%}$, $T_{99\%}$, $T_{99.9\%}$, t_{min}, $\Delta \log N_{min}$, and $\Delta \log N_t$) except for T_E often conflict with each other making evaluation of the antimicrobial effect quite difficult. For example, in the cited study with *E. coli* exposed to cipro-floxacin (51), $T_{99.9\%}$ did not correlate with $T_{90\%}$ (r^2 0.16) or $T_{99\%}$ (r^2 0.19), whereas $T_{90\%}$ did correlate with $T_{99\%}$ (r^2 0.87). Both $T_{90\%}$ and $T_{99\%}$ correlated poorly with Δ log N_{min} (r^2 0.35 and 0.48), although they did correlate fairly well with Δ log N_t (r^2 0.58 and 0.77). $T_{99.9\%}$ correlated relatively well with Δ log N_{min} (r^2 0.69) but not with Δ log N_t (r^2 0.16). These examples indirectly demonstrate insufficient robustness of point indices as end points of the antimicrobial effect.

Unlike the other point indices, T_E and indices that consider the totality of points on the time-kill curve (k_{elb}, AUBC, AAC, ABBC, and I_E) correlate better with each other (r^2 0.74–0.99).

The abilities of the above-mentioned indices to predict the entire antimicro-bial effect are different. As seen in Figure 9, only one of the indices of initial bacterial killing, k_{elb}, is able to predict the entire effect as expressed by I_E, whereas $T_{90\%}$, $T_{99\%}$, and $T_{99.9\%}$ correlate poorly with I_E. Although t_{min} does not correlate with I_E, there is a fairly good correlation between Δ log N_{min} and I_E (r^2 0.85). Less pronounced correlations were found between I_E and each of the four *t*-dependent integral indices measured at $t = 12$ hours (r^2 0.74–0.76).

Is I_E an Ideal End Point of the Antimicrobial Effect?

Based on the analysis described in sections "Ability of the End points to Provide Reasonable Concentration-Response Relationships," "Sensitivity of the End points to Changes in the AUC/MIC Ratio," and "Interrelations of Different End points and Their Abilities to Predict the Entire Antimicrobial Effect," I_E seems to be the most reliable end point of the entire antimicrobial effect. Being most sensitive to changes in the AUC/MIC ratio, I_E allows establishment of reasonable AUC/MIC–response relationships that provide accurate comparisons of antibiotic effects over clinically achievable AUC/MIC ranges when alternative end points that operate with truncated areas are not descriptive. Moreover, I_E is the only integral index that has a physical meaning: It reflects the total number of killed organisms.

However, I_E cannot be considered to be the ideal end point of the antimicro-bial effect. As mentioned above, I_E reflects both magnitude (extent of killing) and duration of the antimicrobial effect (from time zero to the time when the number of antibiotic-exposed organisms reaches the maximal number of antibio-tic-unexposed organisms). This intrinsic feature ensures an important advantage of I_E over other integral indices but it also was considered as a limitation (14). In particular, at large AUC/MICs, when N_{min} approaches the limit of detection, the contribution of effect duration to I_E may predominate over the magnitude of the effect. It is not by chance that I_E highly covaries with T_E for fluoroquinolones whose effect duration is defined by the time above MIC and, indirectly, by their half-lives (51,67). This may result in supposing that I_E might overestimate the true effect of long-circulating antibiotics especially in single-dose simulations, because it considers events that occur beyond the dosing interval (58).

To check if these concerns are real or apparent, the pharmacodynamics and pharmacokinetics of two hypothetical fluoroquinolones were analyzed with half-lives four times (Drug I) and two times (Drug II) shorter than the dosing interval (36). With a 24-hour dosing interval, Drug I should have a half-life of six hours and Drug II of 12 hours (gatifloxacin, gemifloxacin, and levofloxacin could be prototypes of Drug I and grepafloxacin, moxifloxacin, and trovafloxacin proto-types of Drug II). At the same AUC/MIC ratio (138.5 hours), trough concentration (C_{min}) of Drug I is equal to the MIC (C_{min}/MIC $= 1$) and the peak concentration

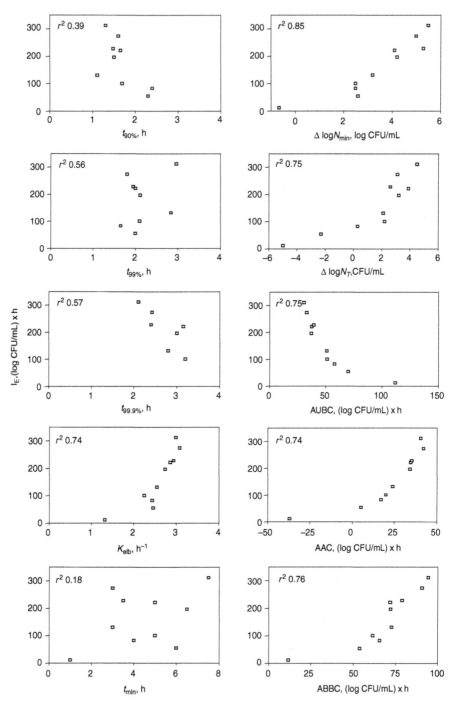

FIGURE 9 Prediction of the entire antimicrobial effect of ciprofloxacin on *Escherichia coli* by different parameters ($\Delta \log N_t$, AUBC, AAC, and ABBC were estimated at $t = 12$ hours). *Source:* Adapted from Ref. 51.

(C_{max})-to-MIC ratio is 16, whereas C_{min}/MIC for Drug II is 2 and C_{max}/MIC 8 (Fig. 10, upper panel). Assuming bacterial regrowth begins when fluoroquinolone concentration falls below the MIC, later regrowth should be expected after a single dose of Drug II compared to Drug I if no further dosing is to occur (Fig. 10, bottom panel). This difference is reflected by the greater I_E that reflects the effect of Drug II relative to Drug I despite similar time-kill curves observed with both drugs within the dosing interval.

At first glance, the different I_Es determined after single administration of Drug I and Drug II make no sense: the second and subsequent doses may whittle away the distinction between the drugs. For example, antibiotic-specific AUC/MIC relationships of I_E that reflect killing of daptomycin- and vancomycin-exposed *S. aureus* were seen better in single-dose simulations (68) than in multiple-dose simulations (69). However, with pharmacokinetically different fluoroquinolones, antibiotic-specific relationships between I_E and AUC/MIC were reported both in single- and multiple-dose studies. Moreover, the ratio of equiefficient AUC/MICs of moxifloxacin and levofloxacin against *S. aureus* reported in single-dose simulations (38) was similar to the respective ratio of AUC_t/MICs ($t = 24$ hours) in multiple-dose simulations with the same antibiotics (64).

Why then do differences in pharmacodynamics observed after single administration of the fluoroquinolones not disappear after subsequent dosing, and why are they predictive of multiple-dose fluoroquinolone pharmacodynamics? To answer these questions, let us return to the above example with two hypothetical antibiotics. As seen in Figure 10, the antimicrobial potential of Drug I at the end of the first dosing interval is completely exhausted (C_{min}/MIC = 1) whereas the respective potential of Drug II is not (C_{min}/MIC = 2). Therefore, multiple dosing of Drug II but not Drug I will be accompanied by accumulation of the

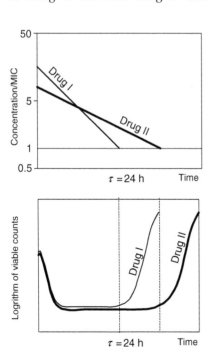

FIGURE 10 Pharmacokinetics and pharmacodynamics of hypothetical antibiotics. *Source:* Adapted from Ref. 36.

antimicrobial potential. For example, after the fifth dose of Drug II (steady-state conditions), its C_{max}/MIC ratio increases more than 30% in contrast to only 6% with Drug I. This analysis emphasizes that continued tracking of the time courses of antibiotic-exposed organisms beyond the dosing interval may be critical for accurate comparisons between different antibiotics.

As shown in this section, adequate experimental design and the choice of an appropriate end point of the antimicrobial effect may be decisive for delineation of concentration–response relationships. All too often, the lack of a reasonable relationship between the effect and its predictor(s) reported in some studies actually results from shortcomings of the design and/or data analysis. There are many such examples in the literature but we refer here to only one. Being "unable to find a...parameter that appeared to correlate with" $T_{99.9\%}$, Wright et al. (70) concluded that "fluoroquinolones do not appear to be concentration-dependent killers of *S. pneumoniae*"...[and]..."AUC/MIC cannot be extrapolated as a clinical outcome predictor for *S. pneumoniae* isolates." In this case, the use of a narrowly descriptive end point resulted in far-reaching conclusions, which conflict with the commonly accepted notion of AUC/MIC as a most reliable predictor of fluoroquinolone effects.

PREDICTION OF THE ANTIMICROBIAL EFFECT AND
ANTIBIOTIC OPTIMAL DOSING
Predictive Pharmacodynamics Using In Vitro Dynamic Models

Compared with routine susceptibility testing and static time-kill studies, dynamic studies that utilize in vitro models are much more complicated, time consuming, and labor intensive and therefore, only a relatively small numbers of organisms can be realistically studied in these models. Especially for this reason, it is vitally important that in vitro model experiments can comprehensively predict antibiotic pharmacodynamics. Moreover, the prediction of optimal antibiotic dosing is implied as a primary goal of these models. Despite this common understanding, surprisingly few studies really provide accurate predictions of equiefficient doses and AUC/MIC and C_{max}/MIC breakpoints. In contrast to these predictive studies, some attempts to directly extrapolate in vitro model data to clinical conditions continue, ignoring the fact that any model is only a model. On the other hand, there are many studies that do not provide any predictions at all or that only predict antibiotic effects, which are quite expected from ordinary susceptibility test results. The reasons for this often arise from inadequate experimental design or in particular from study designs that do not provide delineation of concentration–response relationships.

Concentration–Response Relationship as a Basis to Predict the
Antimicrobial Effect: Generalization from Organism-Specific Data

As with any pharmacological agent, the only basis to predict the antimicrobial effect is antibiotic concentration–response relationship. However, because of different susceptibilities of different pathogens ("diversified targets"), a specific concentration–response relationship is inherent in every antibiotic–bacteria pair. Therefore, unlike most pharmacological agents, with antibiotics a set of the relationships should be established with each drug. To reduce the magnitude of this task, the antimicrobial effect is usually related not to antibiotic concentrations (more specifically, to AUC, C_{max}, etc.) as such but their complexes that incorporate indices of bacterial susceptibility (MIC, rarely, MBC). This allows generalization of data obtained with specific organisms and extrapolation of these specific data to other members of the same species. If one is particularly fortunate, this extrapolation can extend to other bacterial species, assuming that AUC/MIC or C_{max}/MIC relationships with the antimicrobial effect are strain- and species-independent.

MIC-related pharmacokinetic variables (AUC/MIC, C_{max}/MIC, etc.) are considered to be potential predictors of antibiotic pharmacodynamics. In this light, prevalent referring AUC/MIC, C_{max}/MIC, etc., to the "pharmacodynamic parameters" or "pharmacokinetic/pharmacodynamic (PK/PD) indices" (71) may be confusing.

Design of Studies Targeted to Delineate Concentration–Response Relationships

Constant Antibiotic Dose Against Differentially Susceptible Organisms Vs. Different Doses Against a Specific Organism

AUC/MIC and C_{max}/MIC relationships with the antimicrobial effect can be established by using either differentially susceptible organisms exposed to a given antibiotic dose (usually, the clinical dose) or a specific organism exposed to a range of doses.

In vitro simulation of human antibiotic pharmacokinetics at a given dose against one or more similarly susceptible organisms is a specific feature of the design used in most early and some more recent studies. At first glance, such a design seems quite logical because by default, it does provide prediction of the antimicrobial effect at clinically attainable antibiotic concentrations. However, AUC/MIC ratios simulated in this type of studies vary over a very narrow range or do not vary at all. Therefore, an AUC/MIC relationship with the antimicrobial effect might not be observed. As a result, the effects predicted for the studied organisms cannot be extrapolated to other organisms or to other AUC/MIC ratios. Because of these shortcomings, the predictive value is usually low in studies designed in this manner.

However, mimicking clinical antibiotic doses against differentially susceptible bacteria can be useful if the resulting AUC/MIC ratios vary over a wide range. For example, in a study with four strains of *S. pneumoniae* exposed to a 400-mg dose of moxifloxacin (72), where MICs varied from 0.08 to 3.6 mg/L and the respective AUC_{24}/MIC ratios from 7 to 305 hours, a good correlation (r^2 0.8) was established between $AUBC_{24}$ ($AUBKC_{24}$) and log AUC_{24}/MIC for the combined data (Fig. 11). Unfortunately, it is sometimes difficult to find a wide range of differentially susceptible organisms against a new antibiotic. In this case, the AUC/MIC range would be too narrow to show reasonable AUC/MIC–response relationship. To expand the AUC/MIC range, data obtained with different dosing regimens of a given antibiotic, pharmacokinetically different antibiotics, or a given antibiotic simulated at different half-lives (pharmacokinetic hybrids) have been combined in

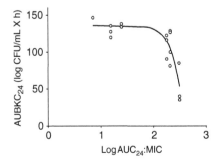

FIGURE 11 AUC_{24}/MIC-dependent $AUBC_{24}$ ($AUBKC_{24}$): moxifloxacin against seven strains of *Streptococcus pneumoniae* with various MICs. *Source*: Adapted from Ref. 72.

some studies (73), even though a specific AUC/MIC relationship with the antimicrobial effect may be inherent in each antibiotic (38,74–77) or each dosing regimen (78–80). Exactly for this reason, the antipneumococcal effect correlated better with the AUC_{24}/MIC of an individual fluoroquinolone (levofloxacin—r^2 0.57) than with the AUC_{24}/MIC of three quinolones taken together (r^2 0.15) (81).

More thoroughly designed studies that expose each organism to widely ranging doses (and by that, different AUC/MIC ratios) are free of the above problems. In particular, there is no necessity to assume *a priori* strain- or species-independent pharmacodynamics. This hypothesis can be tested using AUC/MIC relationships with the antibiotic effect for each individual organism. For example, the lack of systematic differences between I_E-log AUC/MIC plots for *E. coli* and *Pseudomonas aeruginosa* exposed to ciprofloxacin allowed combination of these data so they could be represented by one AUC/MIC–response relationship with r^2 of 0.95 (77). Essentially, such a design provides an additional opportunity to examine reliability of a chosen end point of the antimicrobial effect.

While both strategies can provide a wide range of AUC/MIC (or C_{max}/MIC) ratios, dose-escalation studies with individual organisms are preferable.

Comparative Pharmacodynamics As a Tool to Avoid Unjustified In Vitro–In Vivo Extrapolations

Time-kill studies that utilize in vitro dynamic models approximate actual antibiotic–pathogen interactions as they occur in vivo more or less roughly. Therefore, the effects observed in vitro cannot be directly extrapolated to in vivo conditions. For example, it is impossible to be sure that an n-fold reduction of the starting inoculum or a certain value of I_E or ABBC achieved in vitro would guarantee acceptable antibiotic efficacy in patients. Moreover, data obtained using different models may differ substantially. For example, at comparable AUC/MIC ratios (around 30 hours), one study (82) reported eradication of *S. pneumoniae* exposed to levofloxacin whereas another study (83) reported pronounced bacterial regrowth. Pronounced differences in the time-kill curves were also seen in seven different dynamic models that exposed the same strains each of *S. pneumoniae* and *E. coli* to moxifloxacin (39), regardless of whether bacterial cells were or were not eliminated from the model. To avoid unjustified in vitro–in vivo extrapolations, strictly comparative in vitro studies are needed with each organism.

Choosing a Reference Drug

Ideally, the reference agent should be an antibiotic with a clinically established AUC/MIC or C_{max}/MIC breakpoint. Because of the limited number of antibiotics that meet this requirement, any antibiotic whose dose was proven clinically may be used as a reference drug, assuming its clinical dose provides acceptable efficacy in patients.

Selecting Appropriate Organisms for Study

Representative members of each bacterial species that are potential targets for the antibiotic in question should be studied. Organisms of a given species may be considered representative if their MICs are comparable to the respective MIC_{50}s [not MIC_{90} (84)].

When considering a new antibiotic with broad spectrum of activity, both susceptible and less susceptible target organisms should be included in the study. In fact, less susceptible bacteria may be even more important because these studies are decisive in the prediction of optimal dosing. A newly developed antibiotic must be at least as efficient as its comparator(s) in infections caused by the less susceptible pathogens. It goes without saying that an efficient antibiotic dose against a less susceptible organism will be far in excess of what is needed by patients infected with a more susceptible pathogen. In contrast, many in vitro model studies of new antibiotics are directed at the most susceptible organisms. For example, almost 50% of studies with new fluoroquinolones involved highly susceptible pneumococci. Because of very low MICs, the clinically attainable AUC/MIC ratios simulated in many of these studies were large enough to demonstrate the expected sterilization of the model but they were too large (superoptimal dosing) to provide accurate effect measurements.

Range of Simulated MIC-Related Pharmacokinetic Variables

To thoroughly compare the antimicrobial effects, simulated ranges of AUC/MIC for antibiotic in question and those for the comparator should overlap, even though their clinically achievable values differ. Without overlapping ranges, it is impossible to compare antibiotic effects at the same AUC/MIC ratio as it occurred, for example, in a study with *S. pneumoniae* exposed to ciprofloxacin (AUC_{24}/MIC from 6–13 hours) and levofloxacin (AUC_{24}/MIC from 44–126 hours) (85).

Simulated Pharmacokinetics

Human antibiotic pharmacokinetics has been simulated in the vast majority of studies.[b] Being aware of the decisive role of antibiotic elimination, most investigators attempt to accurately simulate the terminal half-life reported in humans. As a rule, the half-lives in healthy volunteers are the targeted values, but some studies mimic half-lives in patients with impaired elimination of antibiotics. For example, aminoglycoside and fluoroquinolone half-lives mimicking the pharmacokinetics reported in patients with impaired renal function were simulated in in vitro studies with *P. aeruginosa* and *S. aureus* (78,89) to predict specific AUC or AUC/MIC relationships of I_E in "renal insufficiency." These simulations are strongly linked to clinically relevant situations and should not be confused with studies that artificially extend the AUC/MIC range by simulating clinically irrelevant half-lives of an antibiotic (90,91). Although these particular studies allow demonstration of the impact of half-life on antibiotic pharmacodynamics, their clinical interpretation is hardly possible. Indeed, gatifloxacin simulated at half-lives of three, four, and five hours (90) is no longer comparable to "real" gatifloxacin (half-life in humans seven hours) and garenoxacin at simulated half-lives of three, five, six, and eight hours (91) is not "real" garenoxacin (half-life in humans 13–18 hours).

Multiple Dosing Vs. Single-Dose Simulations

Generally, multiple-dose simulations that mimic clinical antibiotic therapy are preferable to single-dose studies. The latter might or might not accurately predict events that occur with multiple dosing (see section "Is I_E an Ideal End point of the Antimicrobial Effect?"). However, few studies properly simulate multiple-dose pharmacokinetics of new long-acting antibiotics that are given once-daily (64,92–95). The widely used two-dose design does not adequately represent multiple-dose regimens of these antibiotics in humans, especially when the observations are discontinued soon after the second dose. There are many examples of this design;

---

[b]Some studies have simulated animal pharmacokinetics to correlate the in vitro model pharmacodynamics with that in infected animals (12,86–88).

for example, a study with *S. pneumoniae* exposed to levofloxacin followed time-kill curves for only 30 hours (96).

Prediction of Antibiotic Effects on Susceptible Subpopulations

Most time-kill studies using in vitro dynamic models are performed at a starting inoculum of 10^5 to 10^6 CFU/mL (10^6–10^8 CFU per volume of the central unit of a dynamic model) where the presence of spontaneous resistant mutants is unlikely. In this case, the primary antibiotic target is the susceptible subpopulation, and, in some cases the observed antimicrobial effects as well as breakpoint values of simulated AUC/MIC or C_{max}/MIC ratios relate exclusively to this subpopulation.

Equiefficient MIC-Related Variables and Breakpoint Values

A method to predict equiefficient AUC/MIC ratios for two or more antibiotics was first suggested in an in vitro study with *E. coli*, *P. aeruginosa*, and *K. pneumoniae* exposed to an eightfold AUC/MIC range of trovafloxacin and ciprofloxacin (97). The same approach was used subsequently in in vitro pharmacodynamic studies with gatifloxacin (76), gemifloxacin (75), grepafloxacin (98), and moxifloxacin (38). As seen in Figure 12, a specific relationship between I_E and log AUC/MIC is inherent in each fluoroquinolone–pathogen pair. Being bacterial strain-independent, the I_E-log AUC/MIC plots are of different slope. For example, at AUC/MIC of 250 hours, trovafloxacin was 1.5 times more efficient and at AUC/MIC of 125 hours, it was 1.4 times more efficient than ciprofloxacin. To avoid the uncertainty resulting from the different slopes of the I_E-log AUC/MIC plots, an AUC/MIC ratio of 125 hours that has been reported as a breakpoint value in a clinical study with ciprofloxacin (99) was used as a reference point to compare the observed in vitro quinolone effects. Based on the quinolone-specific I_E-log AUC/MIC relationships, breakpoints that are equivalent to a ciprofloxacin AUC/MIC ratio of 125 hours were predicted for each of these drugs as well as levofloxacin that was used as an additional comparator (67). As seen in Figure 12, AUC/MIC breakpoints for grepafloxacin, moxifloxacin, and trovafloxacin (75–78 hours) are significantly lower than that of ciprofloxacin, whereas the breakpoints for gatifloxacin, gemifloxacin, and levofloxacin (95–115 hours) are comparable to the breakpoint for ciprofloxacin.

To verify the clinical relevance of these predictions, the in vitro predicted breakpoints were compared with proven breakpoints that have been reported in two clinical studies with grepafloxacin [AUC/MIC 75 hours (100)] and with levofloxacin [C_{max}/MIC of 12.2 (101), which correspond to an AUC/MIC of 110 hours (102)]. Based on I_E-log AUC/MIC relationships established in an in vitro study (67), the AUC/MIC breakpoint for grepafloxacin (78 hours) is very close to the 75-hour value determined clinically. The AUC/MIC breakpoint predicted for levofloxacin (115 hours) was close to the 110-hour value established in a clinical setting. Unfortunately, these are the only two encouraging published examples of in vitro–in vivo qualitative correlations. Breakpoints predicted in some other in vitro studies, for example with ofloxacin [AUC/MIC 100 hours (73), moxifloxacin 75 hours (38), 150 to 200 hours (103), 135 hours (104)] and gemifloxacin [103 hours (75), 150–200 hours (105)], gatifloxacin [95 hours (76), 105 hours (106)] and trovafloxacin [78 hours (97)], could not be compared with the clinical values because they were not reported.[c]

---

[c]Breakpoint value reported for gatifloxacin in pneumococcal infection (107) cannot be considered because organisms other than *S. pneumoniae* were exposed to gatifloxacin in vitro (106).

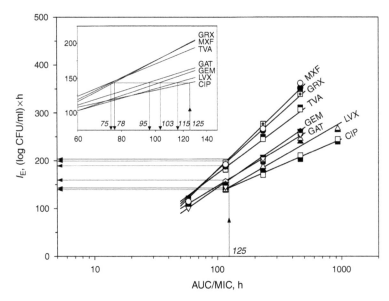

FIGURE 12 AUC$_{24}$/MIC-dependent effects of seven fluoroquinolones on two strains of *Staphylococcus aureus* (indicated by filled and open symbols). The equivalent AUC$_{24}$/MIC breakpoints are indicated by the italicized numbers in the inserted figure. *Source*: From Ref. 67.

Therefore, further evidence is needed to confirm the clinical relevance of AUC/MIC breakpoints predicted in in vitro studies using dynamic models.

Equiefficient Doses

Based on AUC-dose relationships, AUC/MIC relationships with I_E may be easily converted into the respective dose–response curves. The first example of this kind is presented in a study with trovafloxacin and ciprofloxacin (74). Using dose–response curves plotted for each organism, doses of the newer quinolone that provide the same effect as two 12-hour doses of the older quinolone (AUC/MIC 125 hours) were predicted. These equiefficient doses (199–226 mg) are very close to a 200-mg dose of trovafloxacin that was recommended in clinical trials. A gatifloxacin versus ciprofloxacin study with *E. coli, K. pneumoniae,* and *S. marcescens* (76) provides another example of successful dose predictions. Here, the equiefficient doses (330–380 mg) were close to the clinically proven 400-mg dose of gatifloxacin.

To make these predictions more thorough, dose–response curves may be plotted for a hypothetical organism with MIC equal to MIC$_{50}$, assuming bacterial strain- and/or species-independent patterns of the AUC/MIC–response relationships. For example, based on data obtained with three differentially susceptible staphylococci exposed to trovafloxacin and ciprofloxacin, their equiefficient doses were predicted using dose relationships with I_E (74). Similar analysis was later reported in studies with gatifloxacin (76), gemifloxacin (75), and moxifloxacin (38).

Prediction of Antibiotic Effects on Resistant Subpopulations

As mentioned above, most time-kill studies that use in vitro dynamic models are focused on the susceptible subpopulation, without special reference to resistant subpopulation(s). For this reason, the reported antimicrobial effects and the respective breakpoint values of the simulated AUC/MIC or C_{max}/MIC ratios might not be directly extrapolated to less susceptible subpopulations. Until recently, bacterial resistance has been studied only infrequently with in vitro dynamic models. Limited observations reported in earlier time-kill studies (73,76,108–111) precluded delineation of AUC/MIC relationships with resistance because the ranges of the simulated AUC/MIC ratios were too narrow. In fact, the first attempts to relate resistance to AUC/MIC or C_{max}/MIC ratios were reported quite recently in studies that declared resistance analysis as a primary goal (66,81,83,89,95,112–122).

Failures to Relate Bacterial Resistance to Antibiotic Concentrations
Despite wide ranges of AUC/MIC ratios simulated in some recent in vitro model studies (81,95,115,116,119,121), reasonable relationships with resistance were not established. Studies of these relationships can be classified as those that clearly state failure to relate resistance to simulated pharmacokinetics (81,115), and those that imply the existence of AUC_{24}/MIC- or C_{max}/MIC-resistance relationships but do not actually report them (118–120). One study did report a complex effect of AUC_{24}/MIC and duration of moxifloxacin treatment on bacterial resistance, but only duration of treatment was significant: the longer the treatment, the greater the resistance (113).

Without AUC/MIC and C_{max}/MIC relationships with resistance, AUC/MICs and C_{max}/MICs reported to protect against the selection of resistant mutants appear to be contradictory. For example, with *S. pneumoniae*, "protective" AUC/MIC ratios of grepafloxacin varied from 32 hours (115) to 80 hours (83) and those of levofloxacin from nine hours (115) to 26 hours (81) and 35 hours (122). Furthermore, although moxifloxacin-resistant *S. pneumoniae* was not found at AUC/MIC ratios of 60 hours (122) and 107 hours (83), significant losses in susceptibility were seen at AUC/MICs as high as 43,500 hours (115).

Possible reasons for these contradictions have been analysed recently (113). As in pharmacodynamic studies discussed in section "Constant Antibiotic Dose Against Differentially Susceptible Organisms Vs. Different Doses Against a Specific Organism," unsuccessful attempts to delineate concentration-resistance relationships resulted from insufficient study design. Most resistance studies exposed one strain (118,120) or a few similarly susceptible strains (83,81,112,122) to clinical antibiotic doses, so that only one or two values of the AUC_{24}/MIC for each antibiotic could be related to the observed resistance. Moreover, the majority of the simulated AUC_{24}/MICs were high enough to completely sterilize the model and neither population analysis of antibiotic-exposed organisms nor repeated susceptibility testing was possible. For example, in experiments with *S. pneumoniae*, repeated MIC determinations could be made for only one or two of six fluoroquinolones (83,122). Overall, only 30% to 50% of the observations in these studies provided useful information.

It is fair to say that similar problems also are inherent in more rigorously designed dose (AUC/MIC)-ranging studies (115–117,119,121). For example, in studies where *S. pneumoniae* (115), *Bacteroides fragilis* (117), and *Bacteroides thetaiotaomicron* (119) were exposed to wide ranges of quinolone AUC_{24}/MICs, quantitative data could be obtained in only 10% to 66% of experiments. As a result, a "correspondence" between AUC/MIC ≤44 hours (25) and AUC/MIC of <100 hours (29), which are associated with the selection of resistant mutants was stated, adding further confusion to the picture. Given these limitations, reported "protective" AUC/MICs or C_{max}/MICs (83,117–120,122) should be considered cautiously.

Together with limited quantitative data, short-term observations [typically, one-day (81,117–119,121) or two-day courses (83,115,116,122)] may contribute to the controversial results. As shown in studies with quinolone-exposed *S. aureus* (89,113,114), enrichment of the resistant mutants was first observed on the third to fourth day of treatment, with *S. pneumoniae*—on the second-third day (66,95) and with *P. aeruginosa*—on the third day (95).

The use of a relatively low starting inoculum—10^7 to 10^8 CFU (83,116), with few if any resistant mutants also might result in uncertain findings, because these inocula may contain only one resistant cell (123). It is not by chance that resistance data obtained in our study with moxifloxacin- and levofloxacin-exposed *S. aureus* at a starting inoculum of 10^6 CFU/mL $\times$ 60-mL volume = 6×10^7 CFU (64) were less reproducible than those in a later study (113) where the starting inoculum was 6×10^9 CFU.

Concept of the Mutant Selection Window; Bell-Shaped Relationship of Resistance to Antibiotic Concentration

There might be an additional and more specific reason why the expected AUC/MIC relation of resistance—the greater the AUC/MIC ratio, the less pronounced enrichment of resistant mutants—is not seen. Indeed, the simulated concentrations might or might not fall into the "mutant selection window" (MSW), i.e., the concentration range from the MIC to the mutant prevention concentration (MPC), within which it is proposed that resistant mutants are selected (123). The MSW hypothesis was first tested in an in vitro model study with *S. aureus* exposed to three-day dosing of four fluoroquinolones at peak concentrations equal to the MIC, between the MIC and MPC, and above the MPC (113).

With each quinolone, loss in susceptibility of *S. aureus* occurred at concentrations that fell into the MSW but not at concentrations below the MIC or above the MPC, supporting the hypothesis. A quinolone-independent AUC_{24}/MIC relationship with resistance [expressed by the ratio of final MIC (MIC_{final}) to its initial value ($MIC_{initial}$)] was reflected by a bell-shaped curve fitted by the Gaussian function (Fig. 13). As seen in the figure, pronounced losses in susceptibility occurred at AUC_{24}/MIC ratios of 25 to 100 hours, whereas no differences between the final and initial MICs were seen at $AUC_{24}/MICs$ <15 hours, when minimal killing of the susceptible subpopulation was observed, or at $AUC_{24}/MICs$ >200 hours, when maximal killing was observed. Given the bell-shaped pattern of the AUC_{24}/MIC relationships with resistance, reported failures to correlate resistance with AUC/MIC and C_{max}/MIC using linear or log-linear regression are understandable.

Similar bell-shaped curves (using both susceptibility and population analysis data) have been reported with moxifloxacin-exposed *S. pneumoniae* (66) and daptomycin- and vancomycin-exposed *S. aureus* (124,125). Moreover, according to our analysis (113), the Gaussian function also fits resistance data on levofloxacin- and trovafloxacin-exposed *B. fragilis* reported by others (117) showing that the described pattern of the AUC_{24}/MIC-resistance curve may be general. Indirectly, this conclusion is supported by a study on norfloxacin- and ciprofloxacin-exposed *S. aureus* (112). As seen in Figure 14, these data are consistent with a bell-shaped curve despite the use of different end points of resistance (resistance frequency vs. susceptibility testing). So, the more pronounced resistance to norfloxacin at a relatively large AUC_{24}/MIC ratio (55 hours) compared to less pronounced resistance at a small AUC_{24}/MIC (three hours) no longer seems "paradoxical." Moreover,

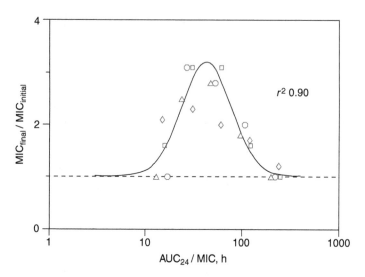

FIGURE 13 Resistance of *Staphylococcus aureus* related to the simulated AUC_{24}/MIC ratio for four fluoroquinolones. Symbols are the same as in Figure 12. *Source*: Adapted from Ref. 113.

similar resistance frequencies at the 16-fold different $AUC_{24}/MICs$ of ciprofloxacin are also quite explainable.

As seen in Figure 13, an AUC_{24}/MIC of ≥ 200 hours can be considered to protect against enrichment of resistant staphylococci in treatments with four quinolones. The very same protective value was reported later for garenoxacin-exposed *P. aeruginosa* in a study that did not consider the MSW hypothesis (126). A much greater protective value (≥ 500 hours) with most pronounced selection of resistant mutants at AUC_{24}/MIC of 85 to 160 hours reported for ciprofloxacin-exposed *S. aureus* (127) might seem contradictory only at first glance. In fact, unlike the studies with four quinolones (113) and garenoxacin (128), there was a large gap in the set of simulated $AUC_{24}/MICs$ (no simulations at $AUC_{24}/MICs$ between 160 and 580 hours), so that the true protective value might be considerably less than reported. It should be noted that being designed as suggested earlier (113)—oscillating concentrations within and out of the MSW—in a ciprofloxacin study (127) also simulates ciprofloxacin continuous infusions with concentrations close to the MICs (concentration-to-MIC ratio of 1.1 during the first dosing interval and 1.3 at the steady-state conditions). These boundary conditions were put forth as presenting quinolone concentrations that fall into the MSW throughout the dosing interval, but actually they only reached the lower boundary of the MIC-MPC range. Exactly for this reason, changes in susceptibility observed in these simulations were much less than those in simulations of ciprofloxacin concentrations that really fell into the MSW.

The described AUC/MIC analysis of resistance makes sense if one assumes comparable ratios between the MIC and the MPC for different antibiotics [as in studies with quinolone- and lipopeptide/glycopeptide-exposed staphylococci (113,125)] or for a given antibiotic against different organisms [as in studies with ciprofloxacin- and lipopeptide/glycopeptide-exposed staphylococci (125,127)]. Generally, AUC/MPC analysis might be more appropriate. For example, in the

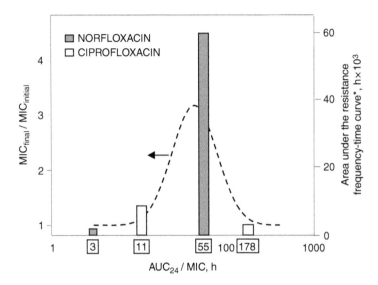

FIGURE 14 AUC_{24}/MIC-dependent resistance frequency of *Staphylococcus aureus* exposed to norfloxacin and ciprofloxacin (112) as compared with the $MIC_{final}/MIC_{initial}$ ratio observed with four fluoroquinolones (23). *Note*: * Reconstructed values. *Source*: Adapted from Ref. 113.

four quinolone study, protective AUC/MPC ratios were estimated at 60 to 70 hours (113) and in the ciprofloxacin study—at 90 to 100 hours (127). Again, given the different designs, this difference is not considered significant.

Together with these AUC/MIC or AUC/MPC analyses, resistance might be related to the time during which antibiotic concentrations are within the MSW (T_{MSW}) (113). Using T_{MSW}, the AUC_{24}/MIC-resistance curve transforms into a sigmoid curve (Fig. 15). As seen in the figure, the $MIC_{final}/MIC_{initial}$ ratio correlates with T_{MSW} regardless of whether quinolone concentrations were above or below the MPCs. The $MIC_{final}/MIC_{initial}$ relationship of T_{MSW} predicts the selection of resistant staphylococci when quinolone concentrations are within the MSW longer than 20% of the dosing interval. There is a gradual increase in the $MIC_{final}/MIC_{initial}$ ratio at T_{MSW} from 40% to 60% and no changes in this ratio at $T_{MSW} > 60\%$. Given this observation, the reported lack of a "clear relationship between … (T_{MSW} of >50–100%—A. F., S. Z., and I. L.) … and the degree of resistance" (127) supports rather than contradicts the earlier findings (113). As for the minimal loss in susceptibility observed at the constant ciprofloxacin concentrations with a "T_{MSW} of 100%" (127), the actual T_{MSW} in these simulations might better be described equal to zero.

Based on the predicted protective AUC/MIC, protective doses can be calculated. In the above example with *S. aureus* exposed to four quinolones (113), similar protective values of the AUC_{24}/MIC ratio (201 hours for levofloxacin, 222 hours for moxifloxacin, 241 hours for gatifloxacin, and 244 hours for ciprofloxacin) correspond to quite different protective *doses*. As a result, the clinically proven dose exceeds the protective dose only with moxifloxacin, at least against the studied organism.

In conclusion, the protective AUC/MIC and doses predicted using in vitro models should be taken with some reservations. Unlike AUC/MIC and C_{max}/MIC breakpoints predicting killing of susceptible subpopulations (see section

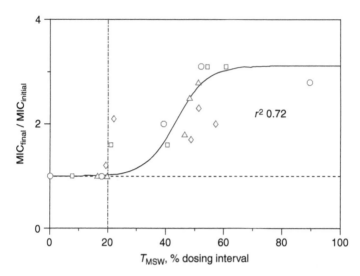

FIGURE 15 Resistance of *Staphylococcus aureus* related to the simulated T_{MSW} for four fluoroquinolones. Symbols are the same as in Figure 13. *Source*: Adapted from Ref. 113.

"Equiefficient MIC-Related Variables and Breakpoint Values"), the protective AUC/MIC or C_{max}/MIC has not been established in any clinical setting. To understand the clinical relevance of these predictions, further studies are needed in patients with special emphasis on the selection of resistant bacteria.

LIMITATIONS OF IN VITRO DYNAMIC MODELS AND FUTURE CHALLENGES

Current notions of the limitations of dynamic models are somewhat polarized and range from the naive belief that the killing effects observed in vitro can be directly extrapolated to patients to the accentuation of differences between the in vitro and in vivo conditions that question the very idea of relevance of these models. For example, up to 26 such differences were listed in a recent review (8) that associates them with drawbacks of in vitro dynamic models. Fortunately, many of the listed differences/drawbacks are more apparent than real. That most in vitro studies (*i*) use relatively small starting inocula that are lower than in many clinical infections, (*ii*) control only Ca^{2+} and Mg^{2+} concentrations but not other cations, (*iii*) simulate drug concentrations in the systemic circulation but not in peripheral tissues, (*iv*) simulate concentrations of parent compound but not its biologically active metabolite(s), and (*v*) provide treatment courses shorter than those in the clinical setting, etc., should not be considered as inherent limitations of dynamic models. All these factors can be incorporated in future studies and some of them have already been considered [higher inocula (66,95,113,129,130), tissue pharmacokinetics (21,37), and longer treatments (92,113,131,132)]. This completely applies to many other "drawbacks" of in vitro models except for the absence of host defense and protein-binding factors.

The infrequent attempts to consider host defense factors date back two decades when whole blood was used as a culture medium (49). This study did not provide a clear demonstration of the net impact of host defense factors on bacterial killing. When cultivating *E. coli* and *K. pneumoniae* in whole blood without antibiotic, time courses of viable counts observed in replicated experiments were distinctly different and ranged from continuous growth to transitory reduction of the starting inoculum. Moreover, killing of these similarly susceptible organisms in blood-containing imipenem at the same concentrations also was variable, although subsequent findings with synthetic nutrient media did demonstrate concentration/MIC-dependent imipenem antimicrobial effects (133). This suggests that using blood as a culture medium is an additional source of variability compared to more standardized experiments with synthetic media.

A recent study in a fibrin clot model that examined the host defense role of platelets (134) also appeared contradictory. Activated platelets limited colonization and proliferation of one of three strains of *S. aureus* in the simulated vegetation, but they either stimulated or did not influence growth of the same organisms in the vegetation environment (134). Platelets did enhance the killing effects of nafcillin in the vegetation environment on two of the strains of *S. aureus* and a GISA strain, whereas enhanced effects of vancomycin were seen only with GISA but not with the other organisms, whose killing was inhibited by platelets (135). Further studies are needed to evaluate the relevance of in vitro simulations that focus on host defense factors.

Protein binding of antibiotics is another factor that might influence the antimicrobial effect. To account for protein binding, it appears sufficient either to simulate antibiotic concentrations that correspond to the concentrations of unbound antibiotic or to simply recalculate the antibiotic dose from simulated concentrations using a reported ratio of free-to-total concentration. In other words, it is sufficient to consider simulated concentrations as free (9,136); moreover, protein-supplemented media are rarely used in in vitro models.

Using this approach with strongly bound antibiotics, there might be a dramatic difference between "free" and total concentration-based analyses. For example, based on total concentrations that correspond to clinical doses, ABT-492 was more pharmacodynamically efficient (Fig. 16, left upper panel) and more protective against loss of susceptibility of *S. aureus* (Fig. 16, left bottom panel) than levofloxacin (20). When these data were corrected for protein binding (84% for ABT-492 and 30% for levofloxacin), the differences were mitigated. As seen in Figure 16, the smaller effect on the susceptible subpopulation (right upper panel) and the protective potential similar to levofloxacin (right bottom panel) were predicted by accounting for differences in protein binding as described above.

Is the "free" concentration analysis perfect? This question was discussed in infrequent in vitro studies that examine the impact of protein binding on bacterial killing (26,137–139). Unfortunately, these findings appeared very conflicting. On one hand, albumin added to nutrient medium at physiological concentrations significantly diminished killing of *S. aureus* exposed to the 94% bound daptomycin (especially its lower 2 mg/kg dose) (138). Also, daptomycin-induced killing in the presence of albumin was less pronounced than in simulations of daptomycin-"free" concentrations (137). On the other hand, no albumin-associated changes on the antimicrobial effects of daptomycin on *Enterococcus faecium* (137) and 50% protein-bound vancomycin on *S. aureus* (138) were reported. Earlier, similar killing of *Enterococcus faecalis* exposed to 97% protein-bound teicoplanin was observed with and without albumin (26).

Recently, a significant impact of albumin on the killing of *S. aureus*, *S. pneumoniae*, and *Haemophilus influenzae* exposed to 94% protein-bound faropenem

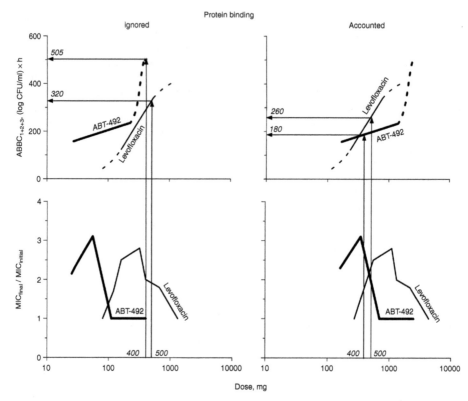

FIGURE 16 Dose-dependent ABBC$_{1+2+3}$ and the MIC$_{final}$/MIC$_{initial}$ ratio for quinolone-exposed *Staphylococcus aureus*: impact of protein binding. Note: ABBC$_{1+2+3}$—cumulative antimicrobial effect expressed as the sum of ABBCs determined within the first, second, and third dosing interval. *Source*: From Ref. 20.

was reported, but only minor changes were seen in killing kinetics of *S. aureus* exposed to 76% protein-bound trovafloxacin and in *S. aureus* and *H. influenzae* exposed to 70% bound cefoxitin. No changes in the antimicrobial effects of four ≤37% bound antibiotics were seen (139). This comprehensive study also demonstrates an impressive difference in time-kill curves with simulations of total faropenem concentrations in the presence of albumin and with "free" concentrations without albumin. With all three organisms, "free" concentrations exhibit lower antimicrobial effects than total concentration in the presence of albumin. This study suggests that simulations of "free" concentrations might overestimate the true impact of protein binding on bacterial killing and underestimate the antimicrobial potential of protein-bound antibiotics. Similar concerns were mentioned elsewhere (20), with special emphasis on the fact that "free" concentration analyses ignore the dynamic nature of the equilibrium between protein-bound and unbound fractions.

Despite these limitations, in vitro dynamic models present an essential and promising tool in drug development and experimental antimicrobial therapy. Any of the widely used in vitro models, regardless of their design, provides a simplified and rough approximation of antibiotic–pathogen interaction (not a bonafide infection!). Therefore, recurring disputes on the advantages of one type of model over others or concern over "artifacts" inherent in some models but not in others seem somewhat gratuitous. The only criterion of reliability of an in vitro model is its ability to accurately predict antimicrobial effects in humans.

With proper design and interpretation, pharmacodynamic studies using these models provide reasonable relationships of the antimicrobial effect to antibiotic concentrations that might be predictive of antibiotic pharmacodynamics in vivo. Unfortunately, there are only a few notable examples of in vitro–in vivo correlations that link in vitro dynamic models with clinical infections. Future predictive studies and clinical correlations using these models are warranted.

REFERENCES

1. O'Grady F, Pennington JH. Bacterial growth in an in vitro system simulating conditions in the urinary bladder. Br J Exp Pathol 1966; 47:152–157.
2. Sanfilippo A, Morvillo E. An experimental model for the study of the antibacterial activity of the sulfonamides. Chemotherapy 1968; 13:54–60.
3. Bergan T. In vitro models simulating in vivo pharmacokinetics and bacterial response to antibiotics. In: O'Grady F, Percival A, eds. Prediction and Assessment of Antibiotic Clinical Efficacy. London: Academic Press, 1986:27–53.
4. Blaser J, Zinner SH. In vitro models for the study of antibiotic activities. Progr Drug Res 1987; 31:349–381.
5. Firsov AA, Nazarov AD, Chernykh VM. Pharmacokinetic approaches to rational antibiotic therapy. Advances in Science and Technology (Itogi Nauki i Tekniki). Vol. 17. Moscow: VINITY, 1989:1–228.
6. Firsov AA, Zinner SH. Use of modeling techniques to aid in antibiotic selection. Curr Infect Dis Rep 2001; 3:35–43.
7. Murakawa T. In vitro/in vivo kinetic models for evaluating efficacy of antimicrobial agents. In: Kuemmerle H-P, Murakawa T, Nightingale CH, eds. Pharmacokinetics of Antimicrobial Agents: Principles, Methods, Application, Part IV, Pharmacokinetics and Therapeutic Efficacy, Section IV-1. Landsberg/Lech: Ecomed Verlags, 1993:165–177.
8. White RL. What in vitro models of infection can and cannot do. Pharmacotherapy 2001; 21:292–301.
9. Rybak MJ, Allen GP, Hershberger E. In vitro antibiotic pharmacodynamic models. In: Nightingale CH, Murakawa T, Ambrose PG, eds. Antimicrobial Pharmacodynamics in Theory and Clinical Practice. Marcell Dekker, Inc.: New York, 2001:41–65.
10. Aviles P, Falcoz C, Guillen MJ, San Roman R, Gomez De Las Heras F, Gargallo-Viola D. Correlation between in vitro and in vivo activities of GM 237354, a new sordarin derivative, against *Candida albicans* in an in vitro pharmacokinetic-pharmacodynamic model and influence of protein binding. Antimicrob Agents Chemother 2001; 45:2746–2754.
11. Ba BB, Bernard A, Iliadis A, et al. New approach for accurate simulation of human pharmacokinetics in an in vitro pharmacodynamic model: application to ciprofloxacin. J Antimicrob Chemother 2001; 47:223–227.
12. Bonapace CR, Friedrich LV, Bosso JA, White RL. Determination of antibiotic effect in an in vitro pharmacodynamic model: comparison with an established animal model of infection. Antimicrob Agents Chemother 2002; 46:3574–3579.
13. Ohya S, Shimada K, Kumamoto Y. Principles and application of in-vitro drug pharmacokinetic simulation systems controlled by stepwise method. J Infect Chemother 2001; 7:133–141.
14. Macgowan A, Bowker K. Developments in PK/PD: optimising efficacy and prevention of resistance. A critical review of PK/PD in vitro models. Int J Antimicrob Agents 2002; 19:291–298.
15. Macgowan A, Rogers C, Bowker K. The use of in vitro pharmacodynamic models of infection to optimize fluoroquinolone dosing regimens. J Antimicrob Chemother 2000; 46:163–170.
16. Macgowan A, Rogers C, Bowker K. In vitro models, in vivo models, and pharmacokinetics: what can we learn from in vitro models? Clin Infect Dis 2001; 33(suppl 3):S214–S220.

17. Navashin SM, Fomina IP, Firsov AA, Chernykh VM, Kuznetsova SM. A dynamic model for in-vitro evaluation of antimicrobial action by simulation of the pharmacokinetic profiles of antibiotics. J Antimicrob Chemother 1989; 23:389–399.

18. Xerri L, Orsolini P, Broggio R. Antibacterial activity of cefuroxime in an in vitro pharmacokinetic model. Drugs Exptl Clin Res 1981;VII:459–464.

19. Aeschlimann JR, Allen GP, Hershberger E, Rybak MJ. Activities of LY333328 and vancomycin administered alone or in combination with gentamicin against three strains of vancomycin- intermediate *Staphylococcus aureus* in an in vitro pharmacodynamic infection model. Antimicrob Agents Chemother 2000; 44:2991–2998.

20. Firsov AA, Vostrov SN, Lubenko IY, Arzamastsev AP, Portnoy YA, Zinner SH. ABT492 and levofloxacin: comparison of their pharmacodynamics and their abilities to prevent the selection of resistant *Staphylococcus aureus* in an in vitro dynamic model. J Antimicrob Chemother 2004; 54:178–186.

21. Florea NR, Tessier PR, Zhang C, Nightingale CH, Nicolau DP. Pharmacodynamics of moxifloxacin and levofloxacin at simulated epithelial lining fluid drug concentrations against *Streptococcus pneumoniae*. Antimicrob Agents Chemother 2004; 48:1215–1221.

22. Sevillano D, Calvo A, Giménez MJ, et al. Bactericidal activity of amoxicillin against non-susceptible *Streptococcus pneumoniae* in an in vitro pharmacodynamic model simulating the concentrations obtained with the 2000/125 mg sustained-release co-amoxiclav formulation. J Antimicrob Chemother 2004; 54:1148–1151.

23. Firsov A, Portnoy Y, Lubenko I. MIC50-related pharmacokinetic prediction of the antimicrobial effects of azithromycin and telithromycin on four respiratory pathogens. In Program and Abstracts of the 6th International Conference on the Macrolides, Azalides, Streptogramins, Ketolides and Oxazolidinones, 2001, abstr. 4.07.

24. Bauernfeind A. Pharmacodynamics of levofloxacin and ofloxacin against *Streptococcus pneumoniae*. J Antimicrob Chemother 1999; 43(suppl C):77–82.

25. Bauernfeind A, Eberlein E, Horl G. Bactericidal kinetics of various dosages of fleroxacin simulated in bacterial cultures. J Antimicrob Chemother 1988; 22(suppl D):81–89.

26. Bauernfeind A, Eberlein E, Schlesinger R, Preac-Mursic V. Bakterizidie von Teicoplanin, Mezlocillin, Netilmicin und Ciprofloxacin sowie ihrer Zweierkombinationen gegen *Enterococcus faecalis* in einem pharmakodynamischen Modell. Fortscr Antimikrob Antineoplast Chemother (FAC) 1992; 11:589–596.

27. Bauernfeind A, Jungwirth R, Petermüller C. Simultaneous simulation of the serum profiles of two antibiotics and analysis of the combined effect against a culture of *Pseudomonas aeruginosa*. Chemotherapy 1982; 28:334–340.

28. Haller I. The combined action of azlocillin and sisomicin in a model simulating the in vivo serum kinetics. Infection 1983; 11(suppl 2):S81–S82.

29. Kobayashi S, Arai S, Hayashi S, Sakaguchi T, Kawana R. In vitro synergy between cefotaxime and its main metabolite, desacetylcefotaxime. Jpn J Antibiot 1988; 41:594–601.

30. Leitner F, Goodhines RA, Buck RE, Price KE. Bactericidal activity of cefadroxil, cephalexin, and cephradine in an in vitro pharmacokinetic model. J Antibiot (Tokyo) 1979; 32:718–726.

31. Nishida M, Murakawa T, Kamimura T, Okada N. Bactericidal activity of cephalosporins in an in vitro model simulating serum levels. Antimicrob Agents Chemother 1978; 14:6–12.

32. Scaglione F, Demartini G, Dugnani S, Fraschini F. In vitro comparative dynamics of modified-release clarithromycin and of azithromycin. Chemotherapy 2000; 46:342–352.

33. Murakawa T, Hirose T, Nishida M. Bactericidal activity of antibiotics in in vitro model simulating antibiotic levels in body (2). A new system simulating serum levels by a two compartment open model. Chemotherapy 1980; 28:26–30.

34. Murakawa T, Sakamoto H, Hirose T, Nishida M. New in vitro kinetic model for evaluating bactericidal efficacy of antibiotics. Antimicrob Agents Chemother 1980; 18:377–381.

35. Keil S, Wiedemann B. Mathematical corrections for bacterial loss in pharmacodynamic in vitro dilution models. Antimicrob Agents Chemother 1995; 39:1054–1058.

36. Firsov AA, Zinner SH, Lubenko IY, Portnoy YA, Vostrov SN. Simulated in vitro quinolone pharmacodynamics at clinically achievable AUC/MIC ratios: advantage of I_E over other integral parameters. Chemotherapy 2002; 48:275–279.

37. Firsov AA, Zinner SH, Vostrov SN, et al. Comparative pharmacodynamics of azithromycin and roxithromycin with *S. pyogenes* and *S. pneumoniae* in a model that simulates in vitro pharmacokinetics in human tonsils. J Antimicrob Chemother 2002; 49:113–119.

38. Firsov AA, Lubenko IY, Vostrov SN, Kononenko OV, Zinner SH, Portnoy YA. Comparative pharmacodynamics of moxifloxacin and levofloxacin in an in vitro dynamic model: prediction of the equivalent AUC/MIC breakpoints and equiefficient doses. J Antimicrob Chemother 2000; 46:725–732.
39. Dalhoff A, Macgowan A, Cars O, et al. Comparative evaluation of seven different in vitro pharmacodynamic models of infection. In Program and Abstracts of the 43d Interscience Conference on Antimicrobial Agents and Chemotherapy, 2003, abstr. A-1147.
40. Al-Asadi MJ, Greenwood D, O'Grady F. In vitro model simulating the form of exposure of bacteria to antimicrobial drugs encountered in infection. Antimicrob Agents Chemother 1979; 16:77–80.
41. Firsov AA, Chernykh VM, Kuznetsova SM, Navashin SM. A dynamic system for the in vitro study of the kinetics of the antimicrobial effect of antibiotics in pharmacokinetic changes in their concentration. Antibiot Med Biotekhnol 1985; 30:36–43.
42. Firsov AA, Chernykh VM, Navashin SM. Quantitative analysis of antimicrobial effect kinetics in an in vitro dynamic model. Antimicrob Agents Chemother 1990; 34:1312–1317.
43. Schneider P, Tosch W, Maurer M, Zak O. Antibacterial effects of cefroxadine, cephalexin and cephradine in a new in vitro pharmacokinetic model. J Antibiot (Tokyo) 1982; 35:843–849.
44. Shah PM. An improved method to study antibacterial activity of antibiotics in an in vitro model simulating serum levels. Methods Find Exp Clin Pharmacol 1980; 2:171–176.
45. Shah PM. Simultaneous simulation of two different concentration time curves in vitro. J Antimicrob Chemother 1985; 15(suppl A):261–264.
46. Blaser J, Stone BB, Zinner SH. Efficacy of intermittent versus continuous administration of netilmicin in a two-compartment in vitro model. Antimicrob Agents Chemother 1985; 27:343–349.
47. Blaser J, Stone BB, Zinner SH. Two compartment kinetic model with multiple artificial capillary units. J Antimicrob Chemother 1985; 15(suppl A):131–137.
48. Sakamoto H, Hirose T, Murakawa T, Nishida M. Bactericidal activity of antibiotics in in vitro model simulating antibiotic levels in body (3). Incorporation of cefazolin into the dialysis cell. Chemotherapy 1980; 28:842–849.
49. Shah PM. Activity of imipenem in an in-vitro model simulating pharmacokinetic parameters in human blood. J Antimicrob Chemother 1985; 15(suppl A):153–157.
50. Toothaker RD, Welling PG, Craig WA. An in vitro model for the study of antibacterial dosage regimen design. J Pharm Sci 1982; 71:861–864.
51. Firsov AA, Vostrov SN, Shevchenko AA, Cornaglia G. Parameters of bacterial killing and regrowth kinetics and antimicrobial effect examined in terms of area under the concentration-time curve relationships: action of ciprofloxacin against *Escherichia coli* in an in vitro dynamic model. Antimicrob Agents Chemother 1997; 41:1281–1287.
52. Satta G, Cornaglia G, Foddis G, Pompei R. Evaluation of ceftriaxone and other antibiotics against *Escherichia coli*, *Pseudomonas aeruginosa*, and *Streptococcus pneumoniae* under in vitro conditions simulating those of serious infections. Antimicrob Agents Chemother 1988; 32:552–560.
53. Xerri L, Broggio R. Study of the antibacterial activity of ceftazidime in an in vitro pharmacokinetic model. Drugs Exptl Clin Res 1985;VII:49–54.
54. Duffull SB, Begg EJ, Chambers ST, Barclay ML. Efficacies of different vancomycin dosing regimens against *Staphylococcus aureus* determined with a dynamic in vitro model. Antimicrob Agents Chemother 1994; 38:2480–2482.
55. Jusko WJ. Pharmacodynamics of chemotherapeutic effects: dose-time-response relationships for phase-nonspecific agents. J Pharm Sci 1971; 60:892–895.
56. Firsov AA, Nazarov AD, Chernykh VM, Navashin SM. Validation of optimal ampicillin/sulbactam ratio in dosage forms using in-vitro dynamic model. Drug Dev Ind Pharm 1988; 14:2425–2442.
57. Wiedemann B, Seeberg AH. The activity of cefotiam on b-lactamase-producing bacteria in an in-vitro model. J Antimicrob Chemother 1984; 13:111–119.
58. Macgowan A, Rogers C, Holt HA, Wootton M, Bowker K. Assessment of different antibacterial effect measures used in in vitro models of infection and subsequent use in pharmacodynamic correlations for moxifloxacin. J Antimicrob Chemother 2000; 46:73–78.
59. Navashin SM, Fomina IP, Chernykh VM, Nazarov AD, Firsov AA. Microcalorimetric method in kinetic study of aminoglycoside antimicrobial action on gram-negative bacteria.

In Program and Abstracts of the IXth International Congress of Infectious and Parasitic Diseases, 1986, abstr. No.1394.

60. White CA, Toothaker RD. Influence of ampicillin elimination half-life on in-vitro bactericidal effect. J Antimicrob Chemother 1985; 15(suppl A):257–260.

61. Wiedemann B, Jansen A. Antibacterial activity of cefpodoxime proxetil in a pharmacokinetic in-vitro model. J Antimicrob Chemother 1990; 26:71–79.

62. Firsov AA, Savarino D, Ruble M, Gilbert D, Manzano B, Medeiros AA, Zinner SH. Predictors of effect of ampicillin-sulbactam against TEM-1 beta- lactamase-producing *Escherichia coli* in an in vitro dynamic model: enzyme activity versus MIC. Antimicrob Agents Chemother 1996; 40:734–738.

63. Shah PM. Bactericidal activity of ceftazidime against *Pseudomonas aeruginosa* under conditions simulating serum pharmacokinetics. J Antimicrob Chemother 1981; 8:135–140.

64. Firsov AA, Zinner SH, Vostrov SN, Portnoy YA, Lubenko IY. AUC/MIC relationships to different endpoints of the antimicrobial effect: multiple-dose in vitro simulations with moxifloxacin and levofloxacin. J Antimicrob Chemother 2002; 50:533–539.

65. Firsov AA, Lubenko IY, Portnoy YA, Zinner SH, Vostrov SN. Relationships of the area under the curve/MIC ratio to different integral endpoints of the antimicrobial effect: gemifloxacin pharmacodynamics in an in vitro dynamic model. Antimicrob Agents Chemother 2001; 45:927–931.

66. Zinner SH, Lubenko IY, Gilbert D, et al. Emergence of resistant *Streptococcus pneumoniae* in an in vitro dynamic model that simulates moxifloxacin concentrations inside and outside the mutant selection window: related changes in susceptibility, resistance frequency and bacterial killing. J Antimicrob Chemother 2003; 52:616–622.

67. Firsov AA, Lubenko IY, Vostrov SN, Portnoy YA, Zinner SH. Anti-staphylococcal effect related to the area under the curve/MIC ratio in an in vitro dynamic model: predicted breakpoints versus clinically achievable values for seven fluoroquinolones. Antimicrob Agents Chemother 2005; 49:2642–2647.

68. Firsov A, Lubenko I, Vostrov S, Alferova I, Smirnova M, Zinner S. Comparative pharmacodynamics of daptomycin and vancomycin with *Staphylococcus aureus* in an in vitro dynamic model: focus on the clinically achievable ratios of area under the curve (AUC) to MIC. Int J Antimicrob Agents 2004; 24:111.

69. Zinner S, Smirnova M, Alferova I, et al. Comparative pharmacodynamics of daptomycin and vancomycin with *Staphylococcus aureus*: multiple-dose simulations using an in vitro dynamic model. Clin Microbiol Infection 2005; 11:518.

70. Wright DH, Hovde LB, Peterson ML, Hoang AD, Rotschafer JC. In vitro dose-ranging evaluation of fluoroquinolone pharmacodynamic parameters against *Streptococcus pneumoniae*. J Antimicrob Chemother 2001; 47:42.

71. Mouton JW, Dudley MN, Cars O, Derendorf H, Drusano GL. Standardization of pharmacokinetic/pharmacodynamic (PK/PD) terminology for anti-infective drugs: an update. J Antimicrob Chemother 2005; 55:601–607.

72. Macgowan AP, Bowker KE, Wootton M, Holt HA. Activity of moxifloxacin, administered once a day, against *Streptococcus pneumoniae* in an in vitro pharmacodynamic model of infection. Antimicrob Agents Chemother 1999; 43:1560–1564.

73. Madaras-Kelly KJ, Ostergaard BE, Hovde LB, Rotschafer JC. Twenty-four-hour area under the concentration-time curve/MIC ratio as a generic predictor of fluoroquinolone antimicrobial effect by using three strains of *Pseudomonas aeruginosa* and an in vitro pharmacodynamic model. Antimicrob Agents Chemother 1996; 40:627–632.

74. Firsov AA, Vasilov RG, Vostrov SN, Kononenko OV, Lubenko IY, Zinner SH. Prediction of the antimicrobial effects of trovafloxacin and ciprofloxacin on staphylococci using an in-vitro dynamic model. J Antimicrob Chemother 1999; 43:483–490.

75. Firsov AA, Zinner SH, Lubenko IY, Vostrov SN. Gemifloxacin and ciprofloxacin pharmacodynamics in an in-vitro dynamic model: prediction of the equivalent AUC/MIC breakpoints and doses. Int J Antimicrob Agents 2000; 16:407–414.

76. Vostrov SN, Kononenko OV, Lubenko IY, Zinner SH, Firsov AA. Comparative pharmacodynamics of gatifloxacin and ciprofloxacin in an in vitro dynamic model: prediction of equiefficient doses and the breakpoints of the area under the curve/MIC ratio. Antimicrob Agents Chemother 2000; 44:879–884.

77. Zinner SH, Vostrov SN, Alferova IV, Lubenko IY, Portnoy YA, Firsov AA. Comparative pharmacodynamics of the new fluoroquinolone ABT492 and ciprofloxacin with *Escherichia coli* and *Pseudomonas aeruginosa* in an in vitro dynamic model. Int J Antimicrob Agents 2004; 24:173–177.

78. Firsov AA. In vitro simulated pharmacokinetics profiles: forecasting antibiotic optimal dosage. Eur J Drug Metab Pharmacokinet 1991; Spec:406–409.

79. Firsov AA, Mattie H. Relationships between antimicrobial effect and area under the concentration-time curve as a basis for comparison of modes of antibiotic administration: meropenem bolus injections versus continuous infusions. Antimicrob Agents Chemother 1997; 41:352–356.

80. Lubenko I, Vostrov S, Zinner S, Firsov A. Examination of fluoroquinolone dosing regimens using in vitro dynamic models: dose-ranging versus one-dose level studies. In Program and Abstracts of the 6th International Symposium on New Quinolones, 1998, abstr. S15.12.

81. Madaras-Kelly KJ, Demasters TA. In vitro characterization of fluoroquinolone concentration/MIC antimicrobial activity and resistance while simulating clinical pharmacokinetics of levofloxacin, ofloxacin, or ciprofloxacin against *Streptococcus pneumoniae*. Diagn Microbiol Infect Dis 2000; 37:253–260.

82. Lister PD, Sanders CC. Pharmacodynamics of moxifloxacin, levofloxacin and sparfloxacin against *Streptococcus pneumoniae*. J Antimicrob Chemother 2001; 47:811–818.

83. Coyle EA, Kaatz GW, Rybak MJ. Activities of newer fluoroquinolones against ciprofloxacin-resistant *Streptococcus pneumoniae*. Antimicrob Agents Chemother 2001; 45:1654–1659.

84. Wiedemann B. Evaluation of data from susceptibility testing. Int J Antimicrob Agents 1998; 10:89–90.

85. Garrison MW. Comparative antimicrobial activity of levofloxacin and ciprofloxacin against *Streptococcus pneumoniae*. J Antimicrob Chemother 2003; 52:503–506.

86. Den Hollander JG, Knudsen JD, Mouton JW, et al. Comparison of pharmacodynamics of azithromycin and erythromycin in vitro and in vivo. Antimicrob Agents Chemother 1998; 42:377–382.

87. Hershberger E, Coyle EA, Kaatz GW, Zervos MJ, Rybak MJ. Comparison of a rabbit model of bacterial endocarditis and an in vitro infection model with simulated endocardial vegetations. Antimicrob Agents Chemother 2000; 44:1921–1924.

88. Schwank S, Rajacic Z, Zimmerli W, Blaser J. Impact of bacterial biofilm formation on in vitro and in vivo activities of antibiotics. Antimicrob Agents Chemother 1998; 42:895–898.

89. Firsov AA, Vostrov SN, Lubenko IY, Zinner SH, Portnoy YA. Concentration-dependent changes in the susceptibility and killing of *Staphylococcus aureus* in an in vitro dynamic model that simulates normal and impaired gatifloxacin elimination. Int J Antimicrob Agents 2004; 23:60–66.

90. Lister PD. Pharmacodynamics of gatifloxacin against *Streptococcus pneumoniae* in an in vitro pharmacokinetic model: impact of area under the curve/MIC ratios on eradication. Antimicrob Agents Chemother 2002; 46:69–74.

91. Lister PD. Impact of AUC/MIC ratios on the pharmacodynamics of the des-F(6) quinolone garenoxacin (BMS-284756) is similar to other fluoroquinolones. J Antimicrob Chemother 2003; 51:199–202.

92. Gumbo T, Louie A, Deziel MR, Parsons LM, Salfinger M, Drusano GL. Selection of a moxifloxacin dose that suppresses drug resistance in *Mycobacterium tuberculosis*, by use of an in vitro pharmacodynamic infection model and mathematical modeling. J Infect Dis 2004; 190:1642–1651.

93. Laplante KL, Rybak MJ. Impact of high-inoculum *Staphylococcus aureus* on the activities of nafcillin, vancomycin, linezolid, and daptomycin, alone and in combination with gentamicin, in an in vitro pharmacodynamic model. Antimicrob Agents Chemother 2004; 48:4665–4672.

94. Macgowan AP, Bowker KE. Mechanism of fluoroquinolone resistance is an important factor in determining the antimicrobial effect of gemifloxacin against *Streptococcus pneumoniae* in an in vitro pharmacokinetic model. Antimicrob Agents Chemother 2003; 47:1096–1100.

95. Macgowan AP, Rogers CA, Holt HA, Bowker KE. Activities of moxifloxacin against, and emergence of resistance in, *Streptococcus pneumoniae* and *Pseudomonas aeruginosa* in an in vitro pharmacokinetic model. Antimicrob Agents Chemother 2003; 47:1088–1095.

96. Lister PD. Pharmacodynamics of 750 mg and 500 mg doses of levofloxacin against ciprofloxacin-resistant strains of *Streptococcus pneumoniae*. Diagn Microbiol Infect Dis 2002; 44:43–49.

97. Firsov AA, Vostrov SN, Shevchenko AA, Portnoy YA, Zinner SH. A new approach to in vitro comparisons of antibiotics in dynamic models: equivalent area under the curve/MIC breakpoints and equiefficient doses of trovafloxacin and ciprofloxacin against bacteria of similar susceptibilities. Antimicrob Agents Chemother 1998; 42:2841–2847.

98. Firsov A, Vostrov S, Lubenko I, Zinner S. Comparative pharmacodynamics of moxifloxacin and grepafloxacin with *Staphylococcus aureus* in an in vitro dynamic model. Antiinfect Drug Chemother 2000; 17:77.

99. Forrest A, Nix DE, Ballow CH, Goss TF, Birmingham MC, Schentag JJ. Pharmacodynamics of intravenous ciprofloxacin in seriously ill patients. Antimicrob Agents Chemother 1993; 37:1073–1081.

100. Forrest A, Chodosh S, Amantea MA, Collins DA, Schentag JJ. Pharmacokinetics and pharmacodynamics of oral grepafloxacin in patients with acute bacterial exacerbations of chronic bronchitis. J Antimicrob Chemother 1997; 40(suppl A):45–57.

101. Preston SL, Drusano GL, Berman AL, et al. Pharmacodynamics of levofloxacin: a new paradigm for early clinical trials. JAMA 1998; 279:125–129.

102. Schentag JJ, Meagher AK, Forrest A. Fluoroquinolone AUIC break points and the link to bacterial killing rates. Part 1: In vitro and animal models. Ann Pharmacother 2003; 37:1287–1298.

103. Macgowan AP, Bowker KE, Wootton M, Holt HA. Exploration of the in-vitro pharmacodynamic activity of moxifloxacin for *Staphylococcus aureus* and streptococci of lancefield groups A and G. J Antimicrob Chemother 1999; 44:761–766.

104. Zelenitsky SA, Ariano RE, Iacovides H, Sun S, Harding GK. AUC_{0-t}/MIC is a continuous index of fluoroquinolone exposure and predictive of antibacterial response for *Streptococcus pneumoniae* in an in vitro infection model. J Antimicrob Chemother 2003; 51:905–911.

105. Macgowan AP, Rogers CA, Holt HA, Wootton M, Bowker KE. Pharmacodynamics of gemifloxacin against *Streptococcus pneumoniae* in an in vitro pharmacokinetic model of infection. Antimicrob Agents Chemother 2001; 45:2916–2921.

106. Booker BM, Smith PF, Forrest A, et al. Application of an in vitro infection model and simulation for reevaluation of fluoroquinolone breakpoints for *Salmonella enterica* serotype typhi. Antimicrob Agents Chemother 2005; 49:1775–1781.

107. Ambrose PG, Grasela DM, Grasela TH, Passarell J, Mayer HB, Pierce PF. Pharmacodynamics of fluoroquinolones against *Streptococcus pneumoniae* in patients with community-acquired respiratory tract infections. Antimicrob Agents Chemother 2001; 45:2793–2797.

108. Blaser J, Stone BB, Groner MC, Zinner SH. Comparative study with enoxacin and netilmicin in a pharmacodynamic model to determine importance of ratio of antibiotic peak concentration to MIC for bactericidal activity and emergence of resistance. Antimicrob Agents Chemother 1987; 31:1054–1060.

109. Dudley MN, Mandler HD, Gilbert D, Ericson J, Mayer KH, Zinner SH. Pharmacokinetics and pharmacodynamics of intravenous ciprofloxacin. Studies in vivo and in vitro model. Am J Med 1987; 82:363–368.

110. Madaras-Kelly KJ, Larsson AJ, Rotschafer JC. A pharmacodynamic evaluation of ciprofloxacin and ofloxacin against two strains of *Pseudomonas aeruginosa*. J Antimicrob Chemother 1996; 37:703–710.

111. Marchbanks CR, Mckiel JR, Gilbert DH, et al. Dose ranging and fractionation of intravenous ciprofloxacin against *Pseudomonas aeruginosa* and *Staphylococcus aureus* in an in vitro model of infection. Antimicrob Agents Chemother 1993; 37:1756–1763.

112. Aeschlimann JR, Kaatz GW, Rybak MJ. The effects of NorA inhibition on the activities of levofloxacin, ciprofloxacin and norfloxacin against two genetically related strains of *Staphylococcus aureus* in an in-vitro infection model. J Antimicrob Chemother 1999; 44:343–349.

113. Firsov AA, Vostrov SN, Lubenko IY, Drlica K, Portnoy YA, Zinner SH. In vitro pharmacodynamic evaluation of the mutant selection window hypothesis using four fluoroquinolones against *Staphylococcus aureus*. Antimicrob Agents Chemother 2003; 47:1604–1613.

114. Firsov AA, Vostrov SN, Lubenko IY, Portnoy YA, Zinner SH. Prevention of the selection of resistant *Staphylococcus aureus* by moxifloxacin plus doxycycline in an in vitro dynamic model: an additive effect of the combination. Int J Antimicrob Agents 2004; 23:451–456.

115. Klepser ME, Ernst EJ, Petzold CR, Rhomberg P, Doern GV. Comparative bactericidal activities of ciprofloxacin, clinafloxacin, grepafloxacin, levofloxacin, moxifloxacin, and trovafloxacin against *Streptococcus pneumoniae* in a dynamic in vitro model. Antimicrob Agents Chemother 2001; 45:673–678.

116. Lacy MK, Lu W, Xu X, et al. Pharmacodynamic comparisons of levofloxacin, ciprofloxacin, and ampicillin against *Streptococcus pneumoniae* in an in vitro model of infection. Antimicrob Agents Chemother 1999; 43:672–677.

117. Peterson ML, Hovde LB, Wright DH, Brown GH, Hoang AD, Rotschafer JC. Pharmacodynamics of trovafloxacin and levofloxacin against *Bacteroides fragilis* in an in vitro pharmacodynamic model. Antimicrob Agents Chemother 2002; 46:203–210.

118. Peterson ML, Hovde LB, Wright DH, et al. Fluoroquinolone resistance in *Bacteroides fragilis* following sparfloxacin exposure. Antimicrob Agents Chemother 1999; 43:2251–2255.

119. Ross GH, Wright DH, Hovde LB, Peterson ML, Rotschafer JC. Fluoroquinolone resistance in anaerobic bacteria following exposure to levofloxacin, trovafloxacin, and sparfloxacin in an in vitro pharmacodynamic model. Antimicrob Agents Chemother 2001; 45:2136–2140.

120. Thorburn CE, Edwards DI. The effect of pharmacokinetics on the bactericidal activity of ciprofloxacin and sparfloxacin against *Streptococcus pneumoniae* and the emergence of resistance. J Antimicrob Chemother 2001; 48:15–22.

121. Wright DH, Gunderson SM, Hovde LB, Ross GH, Ibrahim AS, Rotschafer JC. Comparative pharmacodynamics of three newer fluoroquinolones versus six strains of staphylococci in an in vitro model under aerobic and anaerobic conditions. Antimicrob Agents Chemother 2002; 46:1561–1563.

122. Zhanel GG, Walters M, Laing N, Hoban DJ. In vitro pharmacodynamic modelling simulating free serum concentrations of fluoroquinolones against multidrug-resistant *Streptococcus pneumoniae*. J Antimicrob Chemother 2001; 47:435–440.

123. Zhao X, Drlica K. Restricting the selection of antibiotic-resistant mutants: a general strategy derived from fluoroquinolone studies. Clin Infect Dis 2001; 33:147–156.

124. Firsov AA, Smirnova MV, Lubenko JY, et al. Testing the mutant selection window hypothesis with *Staphylococcus aureus* exposed to daptomycin and vancomycin in an in vitro dynamic model. J Antimicrob Chemother 2006; 58:1185–1192.

125. Firsov A, Alferova I, Smirnova M, et al. Relative abilities of daptomycin and vancomycin to prevent the enrichment of resistant *Staphylococcus aureus* mutants in an in vitro dynamic model. Clin Microbiol Infect 2005; 11:518.

126. Tam VH, Louie A, Deziel MR, et al. Pharmacodynamics of BMS-284756 and ciprofloxacin against *Pseudomonas aeruginosa* and *Klebsiella pneumoniae* in hollow-fiber system. In Program and Abstracts of the 41th Interscience Conference on Antimicrobial Agents and Chemotherapy, 2001, abstr. A-443.

127. Campion JJ, Mcnamara PJ, Evans ME. Evolution of ciprofloxacin-resistant *Staphylococcus aureus* in vitro pharmacokinetic environments. Antimicrob Agents Chemother 2004; 48:4733–4744.

128. Tam VH, Louie A, Deziel MR, Liu W, Leary R, Drusano GL. Bacterial-population responses to drug-selective pressure: examination of garenoxacin's effect on *Pseudomonas aeruginosa*. J Infect Dis 2005; 192:420–428.

129. Allen GP, Kaatz GW, Rybak MJ. Activities of mutant prevention concentration-targeted moxifloxacin and levofloxacin against *Streptococcus pneumoniae* in an in vitro pharmacodynamic model. Antimicrob Agents Chemother 2003; 47:2606–2614.

130. Madaras-Kelly KJ, Daniels C, Hegbloom M, Thompson M. Pharmacodynamic characterization of efflux and topoisomerase IV-mediated fluoroquinolone resistance in *Streptococcus pneumoniae*. J Antimicrob Chemother 2002; 50:211–218.

131. Aeschlimann JR, Zervos MJ, Rybak MJ. Treatment of vancomycin-resistant *Enterococcus faecium* with RP 59500 (quinupristin-dalfopristin) administered by intermittent or continuous infusion, alone or in combination with doxycycline, in an in vitro pharmacodynamic infection model with simulated endocardial vegetations. Antimicrob Agents Chemother 1998; 42:2710–2717.

132. Cha R, Rybak MJ. Pulsatile delivery of amoxicillin against *Streptococcus pneumoniae*. J Antimicrob Chemother 2004; 54:1067–1071.

133. Mcgrath BJ, Lamp KC, Rybak MJ. Pharmacodynamic effects of extended dosing intervals of imipenem alone and in combination with amikacin against *Pseudomonas aeruginosa* in an in vitro model. Antimicrob Agents Chemother 1993; 37:1931–1937.
134. Mercier RC, Rybak MJ, Bayer AS, Yeaman MR. Influence of platelets and platelet microbicidal protein susceptibility on the fate of *Staphylococcus aureus* in an in vitro model of infective endocarditis. Infect Immun 2000; 68:4699–4705.
135. Mercier RC, Dietz RM, Mazzola JL, Bayer AS, Yeaman MR. Beneficial influence of platelets on antibiotic efficacy in an in vitro model of *Staphylococcus aureus*-induced endocarditis. Antimicrob Agents Chemother 2004; 48:2551–2557.
136. Drusano G. Antimicrobial pharmacodynamics: critical interactions of "bug and drug." Nat Rev Microbiol 2004; 2:289–300.
137. Cha R, Rybak MJ. Influence of protein binding under controlled conditions on the bactericidal activity of daptomycin in an in vitro pharmacodynamic model. J Antimicrob Chemother 2004; 54:259–262.
138. Garrison MW, Vance-Bryan K, Larson TA, Toscano JP, Rotschafer JC. Assessment of effects of protein binding on daptomycin and vancomycin killing of *Staphylococcus aureus* by using an in vitro pharmacodynamic model. Antimicrob Agents Chemother 1990; 34:1925–1931.
139. Wiedemann B, Fuhst C. Revising the effect of protein binding on the pharmacodynamics of antibiotics. In Program and Abstracts of the 44th Interscience Conference on Antimicrobial Agents and Chemotherapy, 2004, abstr. A-1466.

Animal Models of Infection for the Study of Antibiotic Pharmacodynamics

David Griffith and Michael N. Dudley
Mpex Pharmaceuticals, San Diego, California, U.S.A.

INTRODUCTION

Animal models of infection have had a prominent place in the evaluation of infection and its treatment. From early experiments demonstrating transmission of infectious agents to satisfy Koch's postulates to evaluation of chemotherapy in the antibiotic era, animal models have proven useful for understanding human diseases.

More recently, animal models for the study of infectious disease have been further refined to consider the importance of pharmacokinetic (PK) factors in the outcome of infection. This allows for the study of the relationship between drug exposure (PKs) and anti-infective activity [pharmacodynamics (PDs)]. Consideration of these aspects has enhanced the information provided by animal models of infection and made the results more predictive of the performance of drug regimens in humans. This has largely been accomplished through a more thorough understanding of the importance of PKs in the outcome of an infection, advances in quantitation of drug concentrations in biological matrices, and the development of metrics to quantify antibacterial effects in vivo.

This chapter will review approaches for design, analysis, and application of these approaches in animal models for the study of optimization of anti-infective therapy and the discovery and development of new agents.

The use of animal models of infection to study the relationship between drug exposure in vivo and antibacterial effects dates back to the very early studies in penicillin. These early investigators noted that the duration of efficacy of penicillin in the treatment of streptococcal infections was dependent on the length of time for which serum concentrations exceeded the minimum inhibitory concentration (MIC) (1–3). How these early observations ultimately impacted on penicillin therapy in humans is difficult to ascertain; however, it is clear that PD studies in animal models have a crucial role in the preclinical evaluation of new anti-infectives, optimization of marketed agents, and assessment of drug resistance.

RELEVANCE OF ANIMAL MODELS TO INFECTIONS IN HUMANS

It is largely assumed that the efficacy of a drug in an animal model will correspond to that in humans. However, in many cases, the induction and progression of infection in small animals does not correspond to that seen in the human setting. For example, humans rarely suffer from large, bolus challenges of bacteria by the intravenous route. A much more clinically relevant model for pathogenesis in humans involving translocation of bacteria from gastrointestinal membranes damaged by cytostatic agents has been developed (4). Despite this major difference, the model of sepsis is a mainstay for early preclinical evaluation of anti-infectives in rodent models.

A few animal models have been widely accepted for prediction of effects in humans. The rabbit model of endocarditis (see below) has proven faithful in mimicking damage to valvular surfaces similar to that reported in human patients, and colonization of these vegetations by bacteria and growth corresponds closely to that observed in humans. For several years, prophylaxis regimens for endocarditis were based largely on experiments in this model.

ENDPOINTS IN ANIMAL MODELS OF INFECTION

Examples of major endpoints used in PD studies in animal models of infection are shown in Table 1. For lethal infections, the proportion of animals surviving at each dose/exposure level is determined to generate typical dose–response curves. With greater awareness of the potential for suffering in test animals during the later stages of overwhelming infection from a failing drug regimen, most investigators have substituted rapid progression to a moribund condition as a more humane alternative endpoint. Both endpoints require careful monitoring by the investigator and consistent application of these definitions to ensure reproducible results. Further, only the potency of the antimicrobial is estimated, because the maximum response is generally bounded (100% survival). However, given that death from infection often arises due to a complex interplay of host factors, organ damage, etc., or even other factors that may be unrelated to antibacterial effects (e.g., a dropped cage), the investigator must be cautious in interpreting results.

In contrast, quantitation of bacteria in tissues at various time points allows for the measurement of changes in bacterial numbers over time. The antibacterial effects often correlate with meaningful clinical outcomes in patients [e.g., relation between time to sterilization of cerebrospinal fluid (CSF) in meningitis and residual neurological effects in children]. Quantification of bacterial counts in tissues or fluids detects more subtle changes in bacterial killing with dosage regimens. In addition, changes in susceptibility of the pathogens with treatment can be measured. The best approach is validation of the number of pathogens in tissues or fluids that correlates with survival in treated or untreated animals. Identification of an earlier endpoint that predicts survival in both treated and untreated animals meets the most rigorous definition for a true surrogate marker (i.e., a marker that predicts survival in treated and untreated groups).

Metrics for Quantifying Anti-infective Effects In Vivo

Quantitation of antimicrobial effects in vivo usually measures the tissue or host burden of organism at specified intervals after the initiation of treatment. In view of the differences among drugs in in vitro PD properties (e.g., rate of bacterial killing and postantibiotic and subinhibitory effects) as well as PK properties, several approaches have been developed to measure the time course of antibacterial effects in vivo. These metrics tend to focus either on the overall extent of bacterial killing over time or on the rate of bacterial killing, either by measuring the time to reach a particular level of bacteria (e.g., time to 3 log decrease) or by calculating a killing rate [e.g., reduction in log (colony-forming units (CFU)/hr]. Table 2 summarizes several representative metrics used for describing the PD effects on bacteria in vivo.

TABLE 1 Summary of Advantages and Disadvantages of Various Endpoints in Animal Models of Infection

Endpoint	Advantages	Disadvantages
Death/moribund condition	Clear endpoint (death) Comparable endpoint in humans Represents a difficult test of antibiotic/dosage regimen	Animal stress and suffering Short-term models require overwhelming inoculum that may trigger cytokine responses irrelevant to that seen in humans Nonspecific effects of drugs (bacterial killing vs. other pharmacological effects) Cannot differentiate between cidal and static effects in vivo Difficult to assess emergence of drug resistance during therapy Outcomes can be very dependent on inoculum and other adjuvants and can obscure PK–PD analysis Unanticipated changes in drug pharmacokinetics due to altered physiology during severe infection
Quantitation of pathogens in body tissues and fluids	Can determine static/cidal activity of agent Can assess emergence of drug resistance during treatment to test agents Measure postantibiotic, subinhibitory effects and correlate with in vitro properties May test strains that are avirulent in other models (e.g., sepsis) Assess relationship between drug concentrations in infected compartment with bacterial eradication (e.g., drug concentrations in cerebrospinal fluid)	Relevance of tissue burden to clinical setting in humans often not established ("fuzzy test-tube") Unrealistic introduction of pathogens into sterile body sites Need to control for possible antibiotic carryover in processing specimens for quantitative culture
Tissue damage and inflammation	Considers impact of both bacterial growth and resultant effects of antibiotics	Relevance to human infection uncertain

PK CONSIDERATIONS IN ANIMAL MODELS OF INFECTION

As described in earlier chapters, one objective of PD modeling is to relate the in vivo exposure of an anti-infective to the observed effects listed above. This requires careful study of antimicrobial PKs in the test species.

Our approach is to measure the PK properties of readily available drugs in the infected animals. Although literature data are often available, the results in a preclinical PK study (where the goal is to carefully measure PK properties) may differ from the values obtained in sick animals or with multiple doses. For novel compounds, the PKs are not known, and studies in both infected and uninfected animals are needed to characterize the PK properties of the agent as well as to determine the actual exposure to drug in the experimental model. Nonlinear

TABLE 2 Summary of Metrics Used to Describe Rate and Extent of Bacterial Killing Under In Vivo Conditions

Metric(s)	Typical units	Description	Comments	Ref.
Extent of bacterial killing				
Change in CFU	Log CFU per tissue weight, organ, or volume	Shows change in total bacterial numbers at sentinel time compared to starting inoculum or untreated control animals	Most easily used to determine effects when destructive sampling is required	Many
Area under the CFU vs. time curve	CFU hr/vol or weight of tissue	Quantifies the total burden of organisms over time by fitting function or by area estimation	Comparisons assume similar starting inoculum (can correct by ratio of observation with starting inoculum to calculate survival fraction). Used to determine the extent of anti-infective effect following single or multiple doses	5–7
Static dose (exposure); EC-50, EC-90~	Log CFU per tissue weight, organ, or volume	Static: No change in CFU over the duration of the experiment compared to challenge inoculum. EC-50, EC-90: Corresponds to 50%, 90% of maximum effects	Used for comparison of magnitude of pharmacokinetic–pharmacodynamic parameter for a given level of effect. Static dose observed to correlate with mortality in mice with some infections (e.g., pneumonia)	8
Rate and duration of bacterial killing				
Bactericidal rate	Log CFU/(mLh)	Depicts the rate of bacterial killing over time by fitting slope to log CFU/mL vs. time curve	Need complete depiction of curve to estimate accurate slope. Difficulty with simple models with data with biphasic killing pattern or bacterial regrowth	9, 10
Time to 99.9% decrease from starting inoculum	Hours	Describes the time necessary for a regimen to produce a specified level of reduction in CFU	Choice of endpoint (99.9% reduction) often arbitrary	11
Effective regrowth time	Hours	Determines the time needed for a treatment regimen to resume growth and return to CFU counts at the start of treatment	Adapted from in vitro PAE studies but may be applied to animal model results. Studies need to be prolonged to achieve a result. Evaluation of multiple dose regimens may not be possible	12
Fully parameterized methods				
Estimates of growth, bacterial killing rates	Various parameters (e.g., kill rate, growth rate)	Models using polynomials or (more appropriately) parameterized for bacterial killing and regrowth phases. EC-50, EC-90: 50% or 90% effective concentration, respectively	Most adapted from in vitro curves but can be applied to results in animals. Many models unable to consider multiple doses. Limited application outside models of infection	13–15

Abbreviation: CFU, colony-forming units; EC-50, 50% of maximum effective concentration; EC-90, 90% of maximum effective concentration.

changes in exposure versus dose (due to changes in drug clearance or bioavailability for extravascular administration) should be probed. When studying combinations of drugs, one must ensure that PK drug interactions do not occur; PK interactions could cause false interpretations of combination experiments (e.g., increased effects from antimicrobial synergism), where the effect was really due to elevated concentrations of an active component. We have also found that the formulation used for a second agent may also influence the PK properties of antibiotics [e.g., polyethylene glycol (PEG) plus levofloxacin in mice; unpublished observations].

Design Issues in PK Studies for Animal Models
Sampling
Appropriate blood or tissue sampling strategies are key to getting robust estimates of PK parameters in animals. In mice, it is usually necessary to euthanize the animal using CO_2 or another American Veterinary Medical Association (AVMA)-approved method to obtain adequate volumes of blood for drug assay. Cardiac puncture in a dead animal usually yields the greatest volume of blood. Harvest of tissues or other samples and immersion in a suitable matrix for sample processing (e.g., homogenization) can also be done to quantify tissue levels. In larger animal species (e.g., rats, rabbits), an indwelling intravenous or intraarterial catheter allows collection of multiple samples over time. In these cases, placement of two cannulas in separate vessels should be undertaken to avoid possible carryover of drug-containing infusate into samples.

One can use d-optimality criteria to select sampling points (16). However, we rarely use this, because the PKs of drugs are rapid in small animals and the slight advantage of narrowing sampling times is easily lost. Further, for novel compounds, there is no information upon which to base the selection of sampling times. The maximal volume of blood to be collected over a PK study session should be estimated and approved upon consultation with a veterinarian and the Institutional Animal Care and Use Committee.

After collection, samples can be centrifuged to separate cells from plasma or serum and aliquots of the sample transferred to screw-capped storage tubes. Pilot studies should consider possible loss of drug in polypropylene or other materials that may result in falsely low levels of drug. Administration of plasma expanders or volume replacement should be considered for prolonged studies in small animals. Patency of catheters can be maintained by gentle flushing of lines using heparin (10–100 U/mL) in 0.9% saline, but overzealous use can result in a systemic anticoagulant effect.

Drug Assay
Assay of serum and tissue concentrations of drug in the test animal species can be done using high performance liquid chromatographic (HPLC) or microbiological assays. The more widespread availability, precision, sensitivity, and rapidity of HPLC methods (particularly those using detection by tandem mass spectrometry) has resulted in a shift toward the use of these methods, particularly in the early discovery or preclinical setting. Metabolites and their antimicrobial activity should be considered in the final PK–PD analyses. Preparation of standards in the appropriate biological matrix (e.g., serum; tissue homogenate) should be undertaken and the sensitivity, specificity, reproducibility, and ruggedness be validated to ensure reproducibility of results.

Modeling PK Data

Serum or tissue concentration versus time data can be analyzed using a variety of methods. Although noncompartmental (SHAM—slope, height, area, moment) methods are used by many investigators, we have found that a more informative analysis results when compartmental models are used. In addition, full parameterization allows for better simulations. Critical analysis of the selection of the appropriate PK model and generation of PK parameters and their confidence intervals is also helpful for performing simulations to calculate indices of PK parameters related to the MIC. Compartmental modeling is also particularly important for estimating PK parameters from studies that generate a single composite "population" curve of single data points gathered from destructive sampling (e.g., serum from mice). Many investigators have adopted WinNONLIN as the fitting program, but several other suitable programs include ADAPT II and MKMODEL. Even population-modeling programs (e.g., NOMEM, NPEM) may be useful with some datasets.

After estimation of PK parameters, simulation of serum levels for various dosage regimens can be undertaken. Unbound drug concentrations should be simulated using protein-binding values generated in the same species. It is particularly important to correlate effects with unbound drug exposure for studies where development of PK–PD relationships will be extended to other species, particularly humans. As will be discussed below, several studies have shown that the free concentration of drug in serum is linked to the efficacy of several anti-infectives. Differences in serum protein binding between preclinical species and humans could lead to errors in conclusions concerning target levels of total drug to obtain acceptable levels of efficacy. Examples include experience with cefonicid in the treatment of endocarditis due to *Staphylococcus aureus* (17).

Animal Vs. Human PKs

The PKs of drugs are known to differ markedly between preclinical animal species (particularly small animals used in PK studies) and humans. These differences largely arise due to well-described allometric relationships that relate physiological variables measured within species to differences in body weight (18). However, the differences in PK properties between animals and humans may arise due to differences in metabolism, biliary, or renal transport, or a combination of these factors. For example, although meropenem is resistant to hydrolysis by human renal dehyropeptidases, it is highly susceptible to inactivation to a deydropeptidase produced in mouse tissues (19). Table 3 depicts examples of differences in PK properties between humans and small animals for several classes of agents. These differences are often most marked for β-lactam agents.

In conducting PK–PD investigations in animal models with drugs used in humans, it is crucial that the differences between humans and animals be considered in the design and interpretation of experiments. This is particularly important for compounds whose activity in vivo is dependent upon the length of time concentrations exceed the MIC (e.g., β-lactams and *Pseudomonas aeruginosa*). When PK differences between humans and small animals are ignored, the results from animal models of infection can underestimate the efficacy of compounds in humans.

A vivid example of the importance of consideration of human versus mouse PKs was shown in studies with ceftazidime treatment of *P. aeruginosa* infection in the neutropenic mouse thigh infection model by Gerber et al. in a neutropenic

TABLE 3 Comparison of Pharmacokinetic Parameters of Selected Antimicrobials in Small Animals and Humans

Drug	Clearance [L/(hr kg)]			Elimination $t_{1/2}$ (hr)			References
	Human	Rat	Mouse	Human	Rat	Mouse	
Ceftizoxime	0.12	1.16	3.67	1.27	0.26	0.15	18, 20
Vancomycin	1.89	0.36	0.75	1.5	2.8	0.63	21, 22
Ofloxacin	0.15	0.45	NA	7.0	1.84	NA	23, 24
Azithromycin	1.29	3.85	3.33	65.9	6.8	0.72	25–27

Abbreviation: NA, not available.

mouse model of infections (Fig. 1) (28). When human PKs were simulated in the mice, the bacterial killing was much more sustained than that observed under normal mouse PK conditions.

Generally, two methods exist for simulation of human PK parameters in animals. In the simplest terms, they involve changing drug input or drug elimination.

Changing Drug Input

Several investigators have used multiple, frequent administrations of drugs with rapid drug clearances and short serum elimination half-lives in mice or rats. This approach uses the principle of superposition of successive "bolus" doses given when concentrations are expected to have declined below target levels. The frequency of administration may be as short as every 30 min around the clock (Fig. 1).

The availability of computer-controlled pumps for delivery of intravenous fluids has enabled the use of these devices to use decreasing rates of drug deliver to simulate concentrations in humans (29). Two approaches may be used: continuous infusion of diluent into the infusate, which is delivered as an infusion into the animal (much like what is done in in vitro PD models), or a continuously changing intravenous infusion rate of a fixed concentration of drug in infusate. The former approach has particular usefulness if one wishes to simulate human concentrations of two drugs in the animal. Both of these approaches have been used in larger animal species (rats, rabbits) in the evaluation of treatment of endocarditis, peritonitis, pneumonia, and meningitis.

The theoretical advantage associated with continuous infusion of antibiotics with certain in vitro PD properties has been studied in animal models of infection. Continuous infusion studies are considerably simpler than intermittent administration. One simply needs to divide the desired steady-state drug concentration by the drug clearance in the animal species to determine the infusion rate. Studies have evaluated delivery by this route for up to 5 days. Continuous infusion of drugs may also be obtained in mice using Alzet peritoneal or subcutaneous micro-osmotic pumps.

Changing Drug Elimination

An alternative (and perhaps more direct) way to mimic human drug PKs in small animals is to slow drug elimination. For drugs largely excreted unchanged in urine, inducing renal impairments by administration of a nephrotoxin can result in slow excretion and allow for close simulation of human dosage regimens in mice. Several investigators have administered a single dose of uranyl nitrate

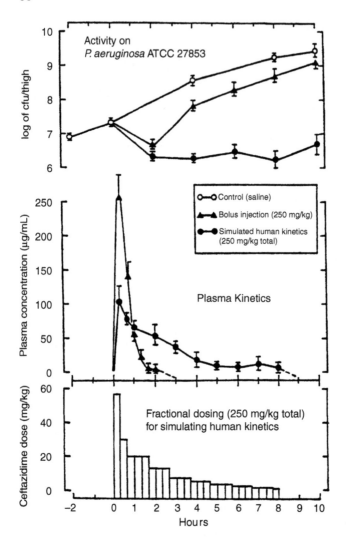

FIGURE 1 Comparison of the effects of a single dose of ceftazidime against a strain of *Pseudomonas aeruginosa* in the netropenic mouse thigh infections model. When a single dose was given to a mouse, bacterial killing occurred only transiently, followed by regrowth. In contrast, when multiple frequent doses of drug were given to simulate the PK profile in humans, much more sustained bacterial killing was observed. *Source*: From Ref. 28.

For example, 10 mg/kg given 2–3 days prior to antibiotic treatment) to induce temporary renal dysfunction for studies of short-term duration (e.g., 24–30 h) (8).

Although convenient, this substance is considered to have low-level radiation, and some states require special permits for its handling and disposal. Other substances have been reported to cause renal impairment in rodents (e.g., glycerol) (30); however, we did not find that it significantly altered the PKs of aztreonam in mice (Tembe, Chen, Griffith, and Dudley, unpublished observations).

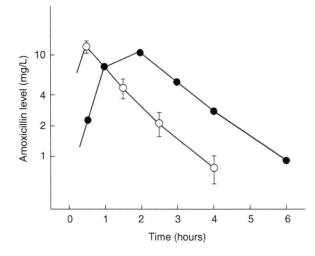

FIGURE 2 Comparison of serum amoxicillin concentrations found in renally impaired mice with those in human volunteers. *Source*: From Ref. 8.

Figure 2 compares the serum PK profile for amoxicillin in mice pretreated with a single dose of uranyl nitrate with concentrations in humans.

When proper doses are administered, the serum concentration versus time profile is comparable to that observed in humans. Simulation of human serum drug levels allowed for testing of the relationship between amoxicillin MIC and effects in a mouse model using drug exposures comparable to those observed in humans (8).

Other approaches for reducing drug clearance include administration of drugs that block renal tubular secretion. This requires that the agent of interest be excreted to a high degree by net renal tubular secretion in the animal species to be tested. Probenecid is one example; however, the magnitude of the effect may be variable and highly dose dependent. In addition, probenecid may have effects on other nonrenal pathways for elimination, including Phase II metabolism.

PD MODELING USING MATHEMATICAL INDICES FOR RELATING PK DATA WITH THE MIC
PK–PD Indices and Relationship to Dose, MIC, and Drug Half-Life
As described in earlier chapters, several indices for integrating PK data with the MIC have been applied for the study of anti-infective PDs. Figure 3 depicts several indices that have been used to express these relationships.

As shown in the figure, all the parameters are highly correlated; e.g., an increase in dose results in an increase in all parameters. However, the magnitude of changes in each parameter with different doses, dosing intervals, drug clearance, and MIC will not increase proportionately with all the PK–PD indices. This is especially important in the design and interpretation of studies of drugs with short half-lives in small animals where the percent of the dosing interval drug concentrations that exceed the MIC is the most important parameter for describing in vivo antimicrobial effects.

A brief consideration of the effects of dose on PK–PD parameters is helpful in recognizing the relative importance of each of these variables. For the ratios C_{max}/MIC or area under the concentration–time curve (AUC)/MIC, a change in dose results in proportional and linear (assuming linear PKs) changes in AUC/MIC.

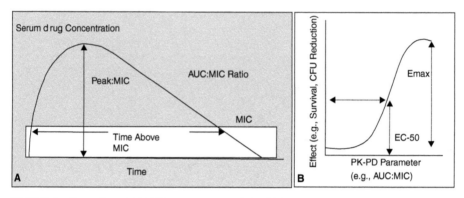

FIGURE 3 Examples of (**A**) PK parameters and the MIC and (**B**) PD relationship between a PK–PD parameter and effects in vivo. *Abbreviations*: MIC, minimum inhibitory concentration; PD, pharmacodynamics; PK, pharmacokinetics.

Similarly, the MIC affects these parameters (inversely) but the proportion of change is linear.

The picture is considerably more complicated in the calculation of the number of hours drug concentrations exceed the MIC. For a one-compartment PK model and bolus drug input,

$$\text{Hours concn. exceeds MIC}\,(T > \text{MIC}) = \frac{\ln \text{dose}/V - \ln \text{MIC}}{Cl/V}$$

In contrast to the ratio metrics C_{max}/MIC and AUC/MIC, the magnitude of change in $T > \text{MIC}$ with dose or MIC is not directly proportional but changes as the natural logarithm of these values. Changes in drug clearance (half-life) produce the most marked change in this parameter.

The practical implications of this relationship for the design of studies and the evaluation of PK–PD data in animal models of infection for drugs where activity in vivo is dependent upon $T > \text{MIC}$ are profound. Figure 4 depicts changes in $T > \text{MIC}$ according to MIC or drug dose. When drug half-life is very fast (Cl/V is large), even high doses of drug will produce only minor changes in $T > \text{MIC}$. Similarly, differences in in vitro potency (MIC) will have little effect on $T > \text{MIC}$.

Failure to incorporate these considerations into experiments results in erroneous conclusions concerning the effects of changes in MIC or drug dose and in vivo response for drugs where in vivo activity is linked to $T > \text{MIC}$.

Designing Experiments to Identify Relationships Between PKs, MIC, and In Vivo Effects

In planning experiments in which many regimens can be evaluated (e.g., short-term studies in mouse models), simulations of serum drug concentrations versus time using PK parameters derived from infected animals should be performed. The parameters, MICs, and dosage regimens can easily be programmed into a spreadsheet to generate PK–PD data. One can assess a number of "what if" scenarios for several doses, dosage regimens, and MICs and generate two- or three-dimensional

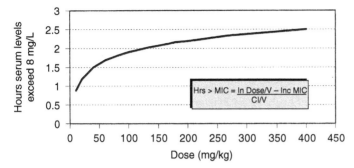

FIGURE 4 Relationship between MIC and *T*>MIC and dose for vancomycin administered subcutaneously in mice. *Abbreviation*: MIC, minimum inhibitory concentration.

plots showing the correlation for each regimen. An example of uncorrelated PK–PD parameters for several vancomycin regimens in a mouse is shown in Figure 5.

One can finalize the selection of dosage regimens that will minimize the covariance among different PK–PD parameters. In selecting regimens, it is also important to be mindful of the relationships of dose and MIC with the number of hours serum concentrations will exceed the MIC. In some cases, it may be virtually impossible to design dosage regimens that produce a range of PK–PD parameters that correspond to those possible in humans. In these cases, the use of techniques to alter drug input or retard drug clearance may be required.

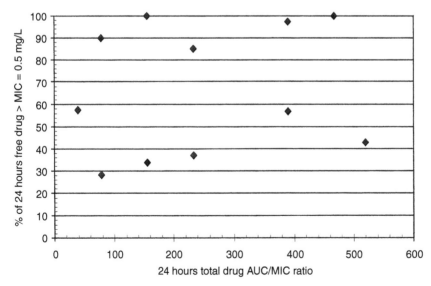

FIGURE 5 Lack of correlation between *T*>MIC and AUC/MIC ratio for several regimens of vancomycin administered every 2 to 12 hours in a mouse model of infection. *Abbreviations*: AUC, area under the concentration–time curve; MIC, minimum inhibitory concentration.

ANIMAL MODELS USED IN PK–PD STUDIES
Selection of Appropriate Models for Study of PK–PD Issues

For the study of PKs–PDs in an animal model, one must take into considera-tion that the time course of antimicrobial activity will vary between different antimicrobial agents. For example, β-lactam antibacterials exhibit very little con-centration-dependent killing and, in the case of staphylococci, long in vivo postantibiotic effects (PABs). For these antibacterials, high drug levels will not kill bacteria more effectively or more rapidly than lower drug levels, because these agents should be bactericidal as long as the drug concentrations exceed the MIC. In contrast, fluoroquinolones and aminoglycosides show concentration-dependent killing. For these agents, the peak/MIC or AUC/MIC ratios should be the PK–PD parameters that most effectively describe their efficacy (1). With these caveats in mind, the choice of an animal model to study PK–PD parameters will depend on the organism one wishes to study, the PKs of the antimicrobial agent, and the type of infection one ultimately intends to treat in humans.

Of the animal models used to study PK–PD relationships, the most commonly used is the neutropenic mouse thigh model, fully described by Gerber and Craig in 1982 (20). Other animal models used to study PK–PD relationships include animal models of pneumonia (21–23), kidney infections/pyelonephritis (24), peritonitis/septicemia (25,26), meningitis (27–31), osteomyelitis (32), and endocarditis (33–35). In fact, almost any model can be used as long as drug exposure can be correlated with organism recovery or response.

Factors Influencing Endpoints

Factors that affect the results of all endpoints include the test strain and initial challenge inoculum, treatment-free interval (i.e., interval between bacterial inocula-tion and initiation of treatment), immunocompetence of test animal species, and administration of adjuvants. All these can contribute to the outcome and the conclusion concerning the magnitude of the PK–PD parameter required for efficacy. Gerber et al. (36) showed that the efficacy of aminoglycosides and β-lactams was markedly affected by delaying treatment, requiring a higher dose. In fluoroquinolone treatment of pneumococcal infection in mice, the AUC/MIC ratio associated with the static effect is similar in immunocompetent or neutropenic mice. In contrast, AUC/MIC is considerably different between neutropenic and nonneutropenic mice in infections due to *Klebsiella pneumoniae* (37).

Animal Models

To study the PDs of an antimicrobial agent, the PKs of the agent in question should be determined in the animal species and under the same conditions that will be used in the infection model (e.g., neutropenic mice in the case of the neutropenic thigh model). The MICs of the organisms to be tested should be determined by a National Committee for Clinical Laboratory Standards (NCCLS) reference method.

Thigh Infection Model

The thigh infection model has been used extensively to describe the PDs of several classes of drugs, particularly fluoroquinolones, macrolides, β-lactams, and amino-glycosides. Most investigators have used Swiss albino mice for this model, although any mouse strain can be used. Although occasional studies are done in normal (nonneutropenic) animals, most of the experience with this model has been obtained in neutropenic mice (38). Most investigators induce neutropenia by the administration of 150 and 100 mg/kg of cyclophosphamide on days 0 and 3, respectively. This results in severe neutropenia by day 4, which is sustained over at least a 24- to 48-hour period (38). Inocula should be prepared in a suitable medium and allowed to grow into log phase. After dilution, 0.1 mL of bacterial suspension ($10^5–10^6$ CFU) is injected into each thigh while the animals are under light anesthesia (e.g., isoflurane, sodium pentobarbital, ether). For some experimental designs, a different organism may be injected in the contralateral thigh (e.g., when studying resistance mechanisms in isogenic strains), or mixtures of organisms can be used. The organisms are usually allowed to grow for two hours prior to the start of treatment. Treatment is given in multiple dosing regimens over 24 hours on the basis of the PKs of the drug and PK–PD parameters being tested. After 24 hours of treatment, the animals are euthanized and their thighs are removed, homogenized in sterile saline, and serially diluted and cultured on a suitable medium to perform colony counts. In the event of agents with long half-lives (e.g., azithromycin), thigh homogenates can be assayed for the drug to exclude the possibility of drug carryover interfering with the results, or a substance to inactivate drug (e.g., β-lactamase) may be added. The CFU/thigh determined for each dosing regimen can be quantified, and relationships to drug exposure fit to a suitable PD model.

Pneumonia Models

The pneumonia model described below is a mouse model; however, there are a number of different models in rats (21,39), hamsters (40), or rabbits (41). Organisms used in pneumonia models include, but are not limited to, *Streptococcus pneumoniae, K. pneumoniae, P. aeruginosa, S. aureus, Pseudomonas mirabilis, Fecal coli, Legionella pneumophila*, and *Aspergillus fumigatus*. Adjuvants are often used to increase virulence. In the mouse model, most Swiss albino mice of either sex can be used.

Depending upon the organism, the mice may or may not need to be rendered neutropenic. Inocula should be prepared in a suitable medium and allowed to grow into log phase. After dilution, the animals are infected by intranasal instillation of 0.05 mL of bacterial suspension ($\sim10^6–10^7$ CFU) while the animals are under anesthesia (isoflourane, sodium pentobarbital, ether, etc.). Treatment is given in multiple dosing regimens for up to two days starting 24 hours after infection. After 24 to 48 hours of treatment, the animals are killed and their lungs are removed, homogenized, serially 10-fold diluted, and then cultured on a suitable medium to perform colony counts. The CFU/lungs determined for each dosing regimen can then be fit to a suitable PD model. The advantages of this model are that the prolonged endpoint (24–48 hours) allows for several cycles of bacterial growth.

Using this model, Woodnutt and Berry (21) found that the efficacy of amoxicillin–clavulanate (a combination of the two drugs) was best described by percentage time above the MIC, with the maximum bactericidal effect being achieved at a $T > $ MIC of 35% to 40%.

Pyelonephritis/Kidney Infection Models

Kidney infection models for study of antifungal activity have been described in both mice (24) and rats (42). Organisms used in kidney infection models include, but are not limited to, *Bacterium faecalis, F. coli, S. aureus, Candida albicans*, and *A. fumigatus*.

For the mouse model, most Swiss albino, Balb/c, and DBA/2 mice can be used. Swiss albino mice can be rendered neutropenic to support growth of the organism. Inocula should be prepared in a suitable medium and allowed to grow into log phase. After dilution, the animals are infected by intravenous injection. Treatment is started two hours after infection in the case of *C. albicans* but can be started as late as 24 hours after infection in the case of *F. faecalis*. Treatment is given in multiple dosing regimens for 24 hours or longer. After 24 to 120 hours of treatment, the animals are killed and their kidneys are removed, homogenized, serially 10-fold diluted, and then cultured on a suitable medium to perform colony counts. The CFU/kidneys determined for each dosing regimen can then be fit to a suitable PD model. The advantages of this model are endpoints that can range anywhere from 24 to 120 hours, it generally uses an inoculum high enough to study resistance, and it allows for several growth and regrowth cycles. Using this model, it has been shown that the AUC/MIC best predicts the effects of fluconazole against strains with a wide range of susceptibilities to this drug (24,43,44).

Peritonitis/Septicemia Models

The septicemia model described below is a mouse model (26,45); however, there is a similar model in neutropenic rats (25). Organisms used in peritonitis/sepsis models include, but are not limited to, *F. coli*, *S. aureus*, *Staphylococcus epidermidis*, *P. aeruginosa*, *K. pneumoniae*, *S. pneumoniae*, *F. faecalis*, *Serratia marcescens*, *P. mirabilis*, group B Streptococcus, *C. albicans*, and *A. fumigatus*.

For this model, almost any mouse species can be used. Mice may or may not need to be rendered neutropenic. Inocula should be prepared in a suitable medium and allowed to grow into log phase. After dilution, the animals are infected by intraperitoneal injection of 0.1 to 0.5 mL of bacterial suspension ($\sim$1–10^9 CFU/mouse). Treatment is given in multiple dosing regimens for up to 72 hours. After treatment, animals still alive are considered long-term survivors. Deaths recorded throughout the experiment can be compared by Kaplan–Meier or probit analysis. Percent survival [(number alive/number treated) × 100] determined for each dosing regimen can then be fit to a suitable PD model.

Using this model, Knudsen et al. (45) found that the peak/MIC ratio and time above MIC were associated with survival for vancomycin and teicoplanin against *S. pneumoniae*. In a study using peritoneal washings and survival in mice for F_{max} modeling, der Hollander et al. (26) found that the peak/MIC ratio was associated with maximum bactericidal effects for azithromycin against *S. pneumoniae* and was achieved at a ratio of 4. In a similar model in rats, Drusano et al. (25) found that for lomefloxacin, a peak/MIC ratio that was 20:1 or 10:1 was associated with survival against *P. aeruginosa*. At peak/MIC ratios of <10:1, they found that the AUC/MIC ratio was associated with survival.

Meningitis Models

The meningitis model described below is a rabbit model (9,10). There are also models in the rat, guinea pig, cat, dog, goat, and monkey, but studies in the higher species are discouraged. Organisms used in meningitis include *Neisseria meningitidis*, *S. pneumoniae*, *Haemophilus influenzae*, *S. aureus*, *Streptococcus pyogenes*, *Streptococcus agalactiae*, *F. aerogenes*, *F. coli*, *K. pneumoniae*, *P. mirabilis*, *P. aeruginosa*, and *Listeria monocytogenes*.

For this model, male or female rabbits can be used. A dental acrylic helmet is attached to the skull of each rabbit by screws while it is under anesthesia. The helmet allows the animal to be secured to a stereotactic frame made for the puncture of the cisterna magna. Meningitis is induced by injection of a bacterial suspension directly into the cisterna magna. Treatment is started 18 hours after infection and is given in multiple dosing regimens for up to 72 hours. The CFU/mL CSF fluid or cure rates (sterile CSF cultures) determined for each dosing regimen can then be fit to a suitable PD model.

In this model, Tauber et al. (27) found that the peak CSF drug level/minimal bactericidal concentration (MBC) ratio emerged as an important factor contributing to the efficacy of ampicillin against *S. pneumoniae*. In earlier studies, these investigators also found that there was a linear correlation between β-lactam CSF peak concentrations and bactericidal activity in rabbits with *S. pneumoniae* meningitis (9). Antibiotic concentrations in the CSF in the range of the MBC produced static effects, and concentrations of 10 to 30 times the MBC produced a maximal bactericidal effect.

In the same model of pneumococcal meningitis, Lutsar et al. (31) found that the bactericidal activity of gatifloxacin in the CSF was closely related to the AUC/MBC ratio but that maximal activity was achieved only when drug concentrations exceeded the MBC for the entire dosing interval. In another study of pneumococcal meningitis, Lutsar et al. (10) found that for ceftriaxone the time above MBC predicted the bacterial killing rate and that there was a linear correlation between time above MBC and bacterial killing rate during the first 24 hours of therapy. Sterilization of the CSF was achieved only with $T >$ MBC of 95% to 100%. Ahmed et al. (46) found that concentrations of vancomycin needed to be at least four- to eightfold the MBC of *S. pneumoniae* in order to obtain adequate bacterial clearance. The suggestion was also that time above the MBC in the CSF was required for efficacy.

Osteomyelitis Models

The osteomyelitis model described below is both a rat model (32) and a rabbit model (47). Organisms used in osteomyelitis models include *S. aureus* and *P. aeruginosa*. For this model, male or female rats or rabbits can be used. Animals are anesthetized, then the tibia is exposed, and a 1 mm hole is bored with a dental drill into the medullary cavity of the proximal tibia. Bones are infected by first injecting 5% sodium morrhuate followed by an injection of bacterial suspension. The hole is plugged with dental gypsum, and the wound is closed. Treatment is initiated 10 days after surgery and is given in multiple dosing regimens for up to 21 days for each compound.

In this model, O'Reilly et al. (32) chose single daily doses of 50 mg/kg for azithromycin, single daily doses of 20 mg/kg for rifampin, and three 90 mg/kg doses of clindamycin daily for treatment regimens against *S. aureus*. Despite having large bone peak/MIC and trough/MIC ratios, azithromycin failed to sterilize a single bone. Clindamycin sterilized 20% of the animals treated, and rifampin sterilized 53% of the animals treated. These results underscore the difficulties in predicting the efficacies of antibiotics in osteomyelitis on the basis of in vitro studies and antibiotic levels in bone.

Endocarditis Models

The endocarditis model has been used extensively to evaluate antimicrobial regimens. Its advantages are that it produces an infection comparable to that observed in humans. The endocarditis model described below is a rabbit model (48); however, there is also a rat model (49). Organisms used in osteomyelitis models include, but are not limited to, *S. aureus, P. aeruginosa, S. epidermidis, F. aerogenes, F. faecalis, F. faecium, Streptococcus sanguis, S. mitis, C. albicans,* and *A. fumigatus.*

For this model, New Zealand rabbits are usually used. Rabbits are anesthetized; then a catheter is placed across the heart valve and left in place for the duration of the study. Animals are infected after 24 hours by an intravenous injection of bacterial inoculum. Treatment begins 24 hours after infection and is given as multiple dosing regimens for 3 to 10 days. Animals are killed; then heart valve vegetations are collected, homogenized, serially 10-fold diluted, and then cultured on a suitable medium to perform colony counts. Terminal blood samples are taken to check for sepsis. The CFU/g vegetation determined for each dosing regimen can be fit to a suitable PD model.

In the rabbit model, Powell et al. (50) found that single daily doses or continuous infusion of tobramycin provided equally efficacious results, suggesting that the AUC/MIC ratio may be the variable that best describes the efficacy of this compound. In another rabbit model of endocarditis, Fantin et al. (51) found that RP 59500 was efficacious against *S. aureus* despite the fact that it was above the MIC for only 33% of the time. However, it was found that RP 59500 penetrated vegetations well with a vegetation/blood ratio of 4:1 and a prolonged in vitro PAB.

In a rat model of endocarditis, Entenza et al. (49) found that despite having concentrations of the quinolone Y-688 that were the same as human concentrations, the compound failed to be efficacious against quinolone-resistant *S. aureus.* This failure may be due to insufficient vegetation penetration (52). In separate in vitro time-kill studies with *S. aureus,* low concentrations of Y-688 selected for drug resistance. Poor drug penetration into vegetations could have provided ideal conditions for selection of resistance in vivo, and hence the failure of Y-688 in this model.

In an analysis of 19 publications on the treatment of experimentally induced endocarditis caused by *S. aureus, S. epidermidis,* viridians streptococci, *F. aerogenes,* and *P. aeruginosa* in rabbit or rat models, Andes and Craig (35) found that for fluoroquinolones, a 24-hour AUC/MIC of ≥100, a peak/MIC of >8, and a time above the MIC of 100% were all associated with significant bactericidal activity after three to six days of therapy. However, they determined that the 24-hour AUC:MIC ratio exhibited the best linear correlation. A conclusion was that the results found in endocarditis with quinolones were similar to those found in the neutropenic thigh model.

EXAMPLES OF THE USE OF ANIMAL MODELS
TO STUDY PK–PD ISSUES
In Vivo PABs

PABs have been recognized since the very early studies of penicillin. Although in vitro studies can describe the persistent effects of a drug following its complete removal from a growing culture, they often have little relevance to the in vivo setting when drug concentrations fall more slowly over time. In contrast, demonstration of persistent antibiotic effects following a single dose can be helpful in determining the optimal dosage interval. Single dose experiments in animal models can be used to define the in vivo PAB [in vivo post antibiotic effect (PAE)] (53,54). In these experiments a single dose of drug is given and the serum concentration

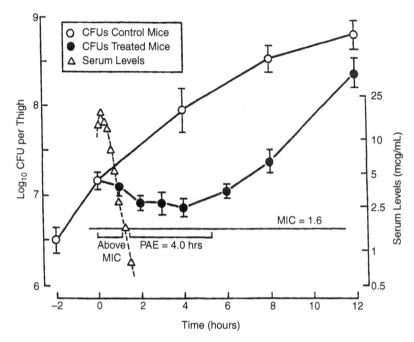

FIGURE 6 Example of an in vivo PAE for cefazolin. Growth curves of control (O) and antibiotic-exposed (●) *Staphylococcus aureus* ATCC 25923 in mouse thighs after a single dose of cefazolin 12.5 mg/kg. Serum levels of cefazolin after a single dose at 12.5 mg/kg (Δ). *Abbreviations*: CFU, colony forming units; PAE, post antibiotic effect. *Source*: From Ref. 38.

and tissue burden of the test organisms are measured over time. The in vivo PAE is calculated by the equation

$$PAE = T - C - M$$

where M the time for which plasma levels exceeded the MIC, T is the time required for the bacterial counts of treated mice to increase by 1 $\log_{10}$ CFU/thigh above the count at time M; and C is the time necessary for the counts in control animals to increase by 1 $\log_{10}$ CFU/thigh. An example is shown in Figure 6.

There are some important differences between the in vitro and in vivo PABs. Most notably, the in vivo PAE tends to be longer than the in vitro PAB. A possible explanation for the differences between the in vitro and in vivo PABs is the effect of subinhibitory drug concentrations on bacterial growth and regrowth. Sub-MIC concentrations can increase the length of the PAE (55), as reflected in the in vitro measurement, PAB-sub-MIC effects (SME). Measurements of the in vivo PAE from single-dose experiments can then be used to better define the duration of antimicrobial effects in vivo with declining concentrations.

Serum Protein Binding

Unfortunately, the effect of serum protein binding on antimicrobial activity still attracts considerable controversy and confusion among many clinicians and researchers. The proportional reduction of antimicrobial activity in the presence of

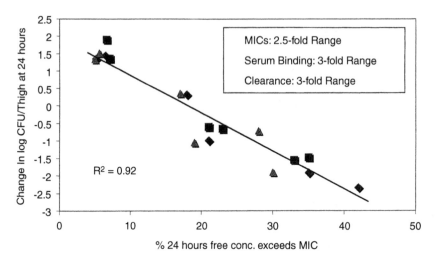

FIGURE 7 Relationship between exposure in vivo to free drug concentrations and bacterial killing in a neutropenic mouse thigh infected model due to two strains of methicilin-resistant *Staphylococcus aureus* and three cephalosporin analogs. *Abbreviations:* CFU, colony forming units; MIC, minimum inhibitory concentration.

serum or binding proteins has been thoroughly demonstrated for several anti-infectives in susceptibility testing.

Although of high interest, there are too few examples of well-controlled experiments that demonstrate the importance of serum protein binding on efficacy in vivo. The difficulty in showing the importance of protein binding in animal models largely lies in the fact that the class of anti-infectives with the greatest variability in serum protein binding is the β-lactam antibiotics. Since the in vivo efficacy of these agents is dependent upon $T > MIC$ and they have relatively short half-lives in small animals, large differences in serum protein binding are required to produce significant differences in free-drug $T > MIC$. Merriken et al. (56) demonstrated the importance of serum protein binding on the efficacy of several structurally related analogs of penicillin in a mouse model of sepsis due to *S. aureus*. All of the agents had similar in vitro potency against the test organism (MIC between 0.25 and 0.5 mg/L) and PK properties, but the percent bound to serum proteins ranged between 36% and 98%. Although the differences in PK properties of total drug were small (2.5-fold range), there was a 70-fold difference among agents in dose required for survival in 50% of animals (ED-50 ranged from 0.7 to 49.7). In a neutropenic mouse thigh model, we also compared the efficacy of three cephalosporin analogs with varying MICs to two strains of methicillin-resistant *S. aureus*, PKs, and protein binding. As shown in Figure 7, bacterial killing was best described by the number of hours that free-drug concentrations exceeded the MIC.

Drug Resistance

The increasing problem of resistance to anti-infective drugs has required critical analysis of the significance of novel resistance mechanisms. The study of drug PDs in animal models of infection either using isogenic strains with or without a resistance factor or using relevant clinical isolates can assess the importance of resistance under in vivo conditions. This information can be important in establishing resistance breakpoints for in vitro MIC testing (see below) as well as for formulating optimal dosage regimens to prevent or overcome established resistance. Generally, if resistance in vivo is less than that observed in vitro, this will be reflected by the disruption of expected relationships between drug exposure (e.g., AUC) and MIC.

The clinical relevance of resistance to extended spectrum cephalosporins by novel plasmid-mediated β-lactamases was studied by Craig et al. in the neutropenic mouse thigh model against several isogenic strains producing extended-spectrum β-lactamases. Mice were pretreated with uranyl nitrate to simulate human exposures to the drug. The results showed that although MICs were elevated at a high inoculum, in vivo results were best correlated with MICs obtained at the lower inoculum (Craig W, personal communication).

Reduced susceptibility to vancomycin in enterococci, and more recently *S. aureus*, is of increasing clinical concern because of the absence of alternative therapies. In experiments in the neutropenic mouse thigh model with *S. aureus* strains that were vancomycin-susceptible or intermediate (VISA), the efficacy was best described by the C_{max}/MIC or AUC/MIC ratio. When results for the vancomycin susceptible strains were compared with those for VISA, only slightly higher vancomycin exposures (C_{max} or AUC) were required for the same level of efficacy despite the higher MICs. Although further studies are required, this suggests that these strains have a reduced level of susceptibility in vivo to vancomycin that is less than that predicted by the in vitro MIC (57). Optimization of vancomycin dosage regimens could be a successful strategy for the clinical management of strains with reduced susceptibility to vancomycin.

Drug efflux is increasingly recognized as an important mechanism of resistance in bacteria and fungi. However, little is known concerning the efficiency of these pumps to produce resistance under in vivo conditions. Using isogenic strains of *P. aeruginosa* with varying levels of expression of the multicomponent mexAB-oprM efflux pump, a reduced response to levofloxacin and ciprofloxacin in the neutropenic mouse thigh model and mouse sepsis model was observed; the reduction in efficacy due to efflux was proportional to the change in AUC/MIC (58). In contrast, Andes and Craig (59) reported that AUC:MIC ratios associated with response in the neutropenic mouse thigh model for NOR-A efflux-related resistance to fluoroquinolones in *S. pnuemoniae* were lower than that in susceptible strains or strains with reduced susceptibility due to nonefflux (e.g., gyrA) mechanisms. These data suggest that efflux mechanisms of resistance may be significant in some bacteria but not in others.

Resistance to fluconazole due to target modifications and/or efflux has also been shown to be significant in vivo using PD modeling. Sorensen et al. studied several strains of *C. albicans* with over a 2000-fold range in MIC in a mouse model of disseminated candidiasis. The reduction in counts in kidneys at 24 hours following fluconazole was found to be described by the AUC/MIC ratio for all doses and strains tested (44), suggesting that elevated fluconzole MICs due to target or efflux-based mechanisms correspond to similar levels of reduced activity in vivo.

Establishing Susceptibility Breakpoints

Given the usefulness of PK–PD relationships for predicting efficacy in vivo, animal models using these analyses have been used for establishing breakpoints for in vitro

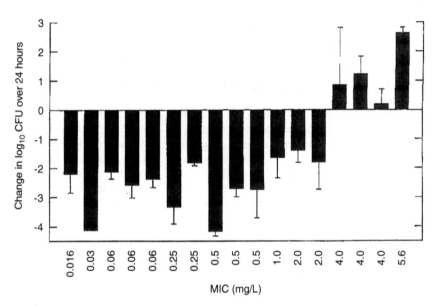

FIGURE 8 Effect of simulated human dosage regimens of amoxicillin on recovery of bacteria from the thighs of neutropenic mice according to amoxicilin MIC. Mice were pretreated with uranyl nitrate to enable simulation of amoxicillin levels corresponding to that observed in humans with usual doses as shown in Figure 2. Organisms with an MIC 2 mg/L all showed a reduction in CFU/ thigh, thus supporting a susceptibility breakpoint of this value. *Abbreviations*: CFU, colony forming units; MIC, minimum inhibitory concentration. *Source*: From Ref. 8.

susceptibility testing. The Antimicrobial Susceptibility Testing Subcommittee of the NCCLS has used such analyses for consideration of breakpoints for certain classes of drugs where clinical data are scant as well as for new agents. A combined approach where data from human clinical trials involving treatment with organisms with varying MICs along with animal model PK–PD data provides a rational basis for establishing susceptibility breakpoints.

PK–PD data derived in animal models of infection are used to establish susceptibility/resistance breakpoints for MIC testing in several ways. Studies in a relevant model of infection can be used to characterize the parameter that best describes efficacy in the model. The "target" PK–PD parameter (e.g., 24-hour AUC/ MIC ratio) for producing a bacteristatic drug, 50% or even 90% of the maximum response can be derived from the experiments. Based on the PK properties in humans at safe doses, PK–PD parameters for various dosage regimens can be calculated for various MIC values. The highest MIC value that still achieves the target PK–PD parameter for the drug would correspond to a suitable MIC break-point for susceptibility testing. For example, for a drug whose efficacy in an animal model of infection is maximal when the 24-hour free-drug AUC/MIC ratio exceeds 30 and safe regimens in humans produce a 24-hour AUC of 60 mg hr/L, a suscept-ibility breakpoint of 2 mg/L could be recommended. Of note is that although the exact serum concentration versus time curve, dosing frequency, and protein binding may differ between humans and small animals, these differences are considered in reducing the exposure relative to the MIC by using PK measures for free drug (e.g., free-drug AUC, free C_{max}).

The rigor of the selected breakpoint can be further evaluated using simulation. The availability of population PK parameters and their variability in target patient groups can be used to determine the probability of individual patients attaining a PK–PD target parameter using a selected breakpoint. The simulation can be even further developed by incorporating the distribution of MICs in organisms of interest and serum concentration data for individual patients for the population means and variances (60). This corresponds to the equivalent of a simulated clinical trial.

An alternative approach employs direct simulation of human PKs in the animal and testing of strains with varying MICs to the test agent. One would expect to see a graded response according to MIC and ultimately a "no effect" at a threshold MIC value. This approach is shown in Figure 8.

Human dosage regimens of amoxicillin were simulated in neutropenic mice, whose thighs were then infected with several strains of *S. pneumoniae* with varying levels of susceptibility to the drug. For strains with MICs exceeding 2 mg/L, little or no effect was seen on bacterial counts recovered from mice at 24 hours, thus supporting a susceptibility breakpoint of 2 mg/L or less.

CONCLUSIONS

Animal models of infection are a pivotal tool in the study of PK–PD properties of anti-infectives. Consideration of PK–PD issues in the design and interpretation of experiments in animals have strengthened the usefulness of these models for the study of human infection. Animal experimentation has been greatly improved because of recognition of the importance of these issues. In addition, many of the recognized limitations of animal models for application to treatment of human infections have been overcome by recognition of the importance of PKs in the outcome of infection. These principles are routinely applied in the study of new drugs in all phases of drug discovery and development as well as in the optimization of dosage in the pre- and postmarketing evaluation of agents.

ACKNOWLEDGMENTS

We acknowledge the excellent assistance of scientists and animal care personnel in Discovery and Preclinical Pharmacology at Microcide Pharmaceuticals, who contributed to generating data for some of the examples provided.

REFERENCES

1. Eagle H, Fleishman R, Levy M. "Continuous" vs. "discontinuous" therapy with penicillin. N Engl J Med 1953; 248:481–488.
2. Eagle H, Fleishman R, Levy M. On the duration of penicillin action in relation to its concentration in the serum. J Lab Clin Med 1953; 40:122–132.
3. Eagle H, Fleishman R, Musselman A. Effect of schedule of administration on the therapeutic efficacy of penicillin. Am J Med 1950; 9:280–299.
4. Collins H, Cross A, Dobek A, Opal S, McClain J, Sadoff J. Oral ciprofloxacin and a monoclonal antibody to lipopolysaccharide protect leukopenic rats from lethal infection with *Pseudomonas aeruginosa*. J Infect Dis 1989; 159:1073–1082.
5. Tisdale J, Pasko M, Mylotte J. Antipseudomonal activity of simulated infusion of gentamycin alone or with piperacillin assessed by serum bactericidal rate and area under the killing curve. Antimicrob Agents Chemother 1989; 33(9):1500–1505.

6. Firsov A, Vostrov S, Shevchenko A, Cornaglia G. Parameters of bacterial killing and regrowth kinetics and antimicrobial effect examined in terms of area under the concentration-time curve relationships: action of ciprofloxacin against *Escherichia coli* in an in vitro model. Antimicrob Agents Chemother 1997; 41(6):1281–1287.

7. Dudley M, Gilbert D, Longest T, Zinner S. Dose ranging of the pharmacodynamics of cefoperazone plus sulbactam in an in vitro model of infection: comparison using a new parameter—area under the survival fraction versus time curve. Pharmacotherapy 1989; 9:189.

8. Andes D, Craig W. In vivo activities of amoxicillin and amoxicillin-clavulanate against *Streptococcus pneumoniae*: application to breakpoint determinations. Antimicrob Agents Chemother 1998; 42:2375–2379.

9. Tauber M, Doroshow C, Hackbarth C, Rusnak M, Drake T, Sande M. Antibacterial activity of beta-lactam antibiotics in experimental meningitis due to *Streptococcus pneumoniae*. J Infect Dis 1984; 149:568–574.

10. Lutsar I, Ahmed A, Friedland I, et al. Pharmacodynamics and bactericidal activity of ceftriaxone therapy in experimental cephalosporin-resistant pneumococcal-meningitis. Antimicrob Agents Chemother 1997; 41:2414–2417.

11. Madaras-Kelly K, Larsson A, Rotschafer J. A pharmacodynamic evaluation of ciprofloxacin and ofloxacin against two strains of *Pseudomonas aeruginosa*. J Antimicrob Agents Chemother 1996; 37:703–710.

12. Hanberger H. Pharmacodynamic effects of antibiotics. Studies on bacterial morphology, initial killing, postantibiotic effect and effective regrowth time. Scand J Infect Dis Suppl 1992; 81:1–52.

13. Mattie H. A predictive parameter of antibacterial efficacy in vivo, based on efficacy in vitro and pharmacokinetics. Scand J Infect Dis Suppl 1990; 74:133–136.

14. Zhi J, Nightingale C, Quintiliani R. A pharmacodynamic model for the activity of antibiotics against microorganisms under measurable conditions. J Pharm Sci 1986; 75:1063–1067.

15. Dudley M, Zinner S. Simultaneous mathematical modeling of the pharmacokinetic and pharmacodynamic properties of cefoperazone. 89th Meeting of the American Society of Clinical Pharmacology and Therapeutics, San Diego, CA, 1988.

16. D'Argenio D, Schumitsky A. ADAPT II: a program for simulation, identification, and optimal experimental design: user manual. Biomedical Simulations Resource, Univ. Southern California, Los Angeles, 1992.

17. Dudley M, Shyu W, Nightingale C, Quintiliani R. Effects of saturable protein binding on the pharmacokinetics of unbound cefonicid. Antimicrob Agents Chemother 1986; 30:565–569.

18. Mordenti J. Pharmacokinetic scale-up: accurate prediction of human pharmacokinetic profiles from animal data. J Pharm Sci 1985; 74:1097–1099.

19. Tsuji M, Ishii Y, Ohno A, Miyazaki S, Yamaguchi K. In vitro and in vivo antibacterial activities of S-4661, a new carbapenem. Antimicrob Agents Chemother 1998; 42: 94–99.

20. Gerber A, Craig W. Aminoglycoside-selected subpopulations *of Pseudomonas aeruginosa*: characterization and virulence in normal and leukopenic mice. J Lab Clin Med 1982; 100:671–681.

21. Woodnutt G, Berry V. Two pharmacodynamic models for assessing the efficacy of amoxicillin-clavulanate against experimental respiratory tract infections caused by strains of *Streptococcus pneumoniae*. Antimicrob Agents Chemother 1999; 43:29–34.

22. Roosendaal R, Bakker-Woudenber I, Bergh JVD, Michel M. Therapeutic efficacy of continuous versus intermittent administration of ceftazidime in an experimental *Klebsiella pneumoniae* pneumonia in rats. J Infect Dis 1985; 152:373–378.

23. Azoulay-Dupuis E, Vallee E, Bedos J, Muffat-Joly M, Pocidalo J. Prophylactic and therapeutic efficacies of azithromycin in a mouse model of pneumococcal pneumonia. Antimicrob Agents Chemother 1991; 35:1024–1028.

24. Andes D, Ogtrop MV. Characterization and quantitation of the pharmacodynamics of fluconazole in a neutropenic murine disseminated candidiasis infection model. Antimicrob Agents Chemother 1999; 43:2116–2120.

25. Drusano G, Johnson D, Rosen M, Standiford M. Pharmacodynamics of a fluoroquinolone antimicrobial agent in a neutropenic rat model of *Pseudomonas sepsis*. Antimicrob Agents Chemother 1993; 37:483–490.
26. der Hollander J, Knudsen J, Mouton J, et al. Comparison of pharmacodynamics of azithromycin and erythromycin in vitro and in vivo. Antimicrob Agents Chemother 1998; 42:377–382.
27. Tauber M, Kunz S, Zak O, Sande M. Influence of antibiotic dose, dosing interval and duration of therapy on outcome in experimental pneumococcal meningitis in rabbits. Antimicrob Agents Chemother 1989; 33:418–423.
28. Gerber A, Brugger H, Feller C, Stritzko T, Stalder B. Antibiotic therapy of infections due to *Pseudomonas aeruginosa* in normal and granulocytopenic mice: comparison of murine and human pharmacokinetics. J Infect Dis 1986; 153:90–97.
29. Mizen L. Methods for obtaining human-like pharmacokinetic patterns in experimental animals. In: Zak O, Sande M, eds. Handbook of Animal Models of Infection. New York: Academic Press, 1999:93–103.
30. Lin J, Lin T. Renal handling of drugs in renal failure. I: Differential effects of uranyl nitrate- and glycerol-induced acute renal failure on renal excretion of TEAB and PAH in rats. J Pharmacol Exp Ther 1988; 246:896–901.
31. Lutsar I, Friedland I, Wubbel L, et al. Pharmacodynamics of gatifloxacin in cerebrospinal fluid in experimental cephalosporin-resistant pneumococcal meningitis. Antimicrob Agents Chemother 1998; 42:2650–2655.
32. O'Reilly T, Kunz S, Sande E, Zak O, Sande M, Tauber M. Relationship between antibiotic concentration in bone and efficacy of treatment of *Staphylococcal osteomyelitis* in rats: azithromycin compared with clindamycin and rifampin. Antimicrob Agents Chemother 1992; 36:2693–2697.
33. Genko F, Mannion T, Nightingale C, Schentag J. Integration of pharmacokinetics and pharmacodynamics of methicillin in curative treatment of experimental endocarditis. J Antimicrob Chemother 1984; 14:619–631.
34. Carpenter T, Hackbarth C, Chembers H, Sande M. Efficacy of ciprofloxacin for experimental endocarditis caused by methicillin-susceptible or -resistant strains of *Staphylococcus aureus*. Antimicrob Agents Chemother 1986; 30:382–384.
35. Andes D, Craig W. Pharmacodynamics of fluoroquinolones in experimental models of endocarditis. Clin Infect Dis 1998; 27:47–50.
36. Gerber A, Craig W, Brugger H-P, Feller C, Vastola A, Brandel J. Impact of dosing intervals on activity of gentamicin and ticarcillin against *Pseudomonas aeruginosa* in granulocytopenic mice. J Infect Dis 1983; 145:296–329.
37. Andes D, Ogtrop MV, Craig W. Impact of neutrophils on the in-vivo activity of fluoroquinolones. 37th Annual Meeting of the Infectious Disease Society of America, Philadelphia, PA, 1999.
38. Craig W, Gudmundsson S. The postantibiotic effect. In: Lorian V, ed. Antibiotics in Laboratory Medicine. Baltimore, MD: Williams and Wilkins, 1996:296–329.
39. Smith G, Abbott K. Development of experimental respiratory infections in neutropenic rats with either penicillin-resistant *Streptococcus pneumoniae* or beta-lactamase-producing *Haemophilus influenzae*. Antimicrob Agents Chemother 1994; 38:608–610.
40. Arai S, Gohara Y, Akashi A, et al. Effects of new quinolones on *Mycoplasma pneumoniae* infected hamsters. Antimicrob Agents Chemother 1993; 37:287–292.
41. Piroth L, Martin L, Coulon A, et al. Development of an experimental model of penicillin-resistant *Streptococcus pneumoniae* pneumonia and amoxicillin treatment by reproducing human pharmacokinetics. Antimicrob Agents Chemother 1999; 43:2484–2492.
42. Rogers T, Galgiani J. Activity of fluconazole (UK 49,858) and ketoconazole against *Candida albicans* in vitro and in vivo. Antimicrob Agents Chemother 1986; 30:418–422.
43. Louie A, Drusano G, Banerjee P, et al. Pharmacodynamics of fluconazole in a murine model of systemic candidiasis. Antimicrob Agents Chemother 1998; 42:1105–1109.
44. Sorensen K, Corcoran E, Chen S, et al. Pharmacodynamic assessment of efflux- and target-based resistance to fluconazole (FLU) on efficacy against *C. albicans* in a mouse kidney infection model. 39th Interscience Conference on Antimicrobial Agents and Chemotherapy, San Francisco, CA, 1999.

45. Knudsen J, Fuursted K, Esperson F, Frimodt-Moller N. Activities of vancomycin and teicoplanin against penicillin-resistant pneumococci in vitro and in vivo and correlations to pharmacokinetic parameters in the mouse peritonitis model. Antimicrob Agents Chemother 1997; 41:1910–1915.

46. Ahmed A, Jafri H, Lutsar I, et al. Pharmacodynamics of vancomycin for the treatment of experimental penicillin- and cephalosporin-resistant pneumococcal meningitis. Antimicrob Agents Chemother 1999; 43:876–881.

47. Smeltzer M, Thomas J, Hickmon S, et al. Characterization of a rabbit model of staphylococcal osteomyelitis. J Orthop Res 1997; 15:414–421.

48. Sande M, Johnson M. Antimicrobial therapy of experimental endocarditis caused by *Staphylococcus aureus*. J Infect Dis 1975; 131:367–375.

49. Entenza J, Marchetti O, Glauser M, Moreillon P. Y-688, a new quinolone active against quinolone-resistant *Staphylococcus aureus*: lack of in vivo efficacy in experimental endocarditis. Antimicrob Agents Chemother 1998; 42:1889–1894.

50. Powell S, Thompson W, Luthe M, et al. Once-daily vs. continuous aminoglycoside dosing: efficacy and toxicity in animal and clinical studies of gentamicin, netilmicin, and tobramicin. J Infect Dis 1983; 147:918–932.

51. Fantin B, Leclercq R, Ottaviani M, et al. In vivo activities and penetration of the two components of the streptogramin RP 59500 in cardiac vegetations of experimental endocarditis. Antimicrob Agents Chemother 1994; 38:432–437.

52. Cremieux A, Maziere B, Vallois J, et al. Evaluation of antibiotic diffusion into cardiac vegetations by quantitative autoradiography. J Infect Dis 1989; 159:938–944.

53. Craig W. Post-antibiotic effects in experimental infection models: relationship to in-vitro phenomena and to treatment of infections in man. J Antimicrob Chemother 1993; 31:149–158.

54. Vogelman B, Gudmundsson S, Tumidge J, Leggett J, Craig W. In vivo postantibiotic effect in a thigh infection in neutropenic mice. J Infect Dis 1988; 157:287–298.

55. Cars O, Odenholt-Tomqvist I. The post-antibiotic sub-MIC effect in vitro and in vivo. J Antimicrob Chemother 1993; 31:159–166.

56. Merrikin D, Briant J, Rolinson G. Effect of protein binding on antibiotic activity in vivo. J Antimicrob Chemother 1983; 11:233–238.

57. Dudley M, Griffith D, Corcoran E, et al. Pharmacokinetic-pharmacodynamic (PK–PD) indices for vancomycin (V) treatment of susceptible (VSSA) and intermediate (VISA) *S. aureus* in the neutropenic mouse thigh model. 39th Interscience Conference on Antimicrobial Agents and Chemotherapy, San Francisco, CA, 1999.

58. Griffith D, Corcoran E, Lofland D, et al. Pharmacodynamics of levofloxacin against *Pseudomonas aeruginosa* with reduced susceptibility due to different efflux pumps: do elevated MICs always predict reduced in vivo efficacy? Antimicro Agents Chemo 2006; 50:1628–1632.

59. Andes D, Craig W. Pharmacodynamics of gemifloxacin (GEM) against quinolone-resistant strains of *S. pneumoniae* with known resistance mechanisms. 39th Interscience Conference on Antimicrobial Agents and Chemotherapy, San Francisco, CA, 1999.

60. Dudley M, Ambrose PG. Pharmacodynamics in the study of drug resistance and establishing in vitro susceptibility breakpoints: ready for prime time. Curr Opin Microbiol 2000; 3:515–521.

5 The Predictive Value of Laboratory Tests for Efficacy of Antibiotic Combination Therapy

Jan G. den Hollander
Department of Internal Medicine and Infectious Diseases, Medical Center Rotterdam Zuid, Rotterdam, The Netherlands

Johan W. Mouton
Department of Medical Microbiology and Infectious Diseases, Canisius Wilhelmina Hospital, Nijmegen, The Netherlands

INTRODUCTION

The use of combinations of antimicrobial agents is a common practice during clinical therapy. There are several reasons why combination therapy is used or should be used, but all have the intention of increasing efficacy. Although one of the major reasons is the supposedly synergistic action between two antimicrobial agents, it is not always clear whether certain classes of antimicrobial agents can or do act synergistically in vivo. Indeed, in two recent meta-analyses, no significant difference in outcome was found between patients who received combination therapy and those who received monotherapy (1,2), except for perhaps infections caused by *Pseudomonas aeruginosa*. One of the reasons that no significant difference in outcome was found between the groups receiving monotherapy and combination therapy might have been diversity of patients and indications in the trials studied, and the finding that there is no difference between the groups a result of the relative low power of the analysis. If combination therapy is advantageous in certain settings and not in others, the ability to predict that synergistic effects exist would help to make a rational choice.

A number of in vitro tests have been developed and used over the years that were thought to predict the degree of interaction between antimicrobial agents, and furthermore, if they do, whether a particular combination is synergistic against the specific strain causing the infection in a particular patient. We here focus on the methods and its relevance to clinical practice to predict the effect of antimicrobial combinations. To that purpose, we discuss the correlation of results of various in vitro methods, and correlation of these in vitro results with clinical efficacy as determined in in vitro pharmacokinetic models (IVPMs), animal models, and clinical trials. The efficacy of combination therapy for endocarditis is not taken into consideration because this is a specific disease entity.

LABORATORY METHODS TO TEST ANTIBIOTIC INTERACTIONS

The susceptibility of micro-organisms to antibiotics is usually expressed as a minimal inhibitory concentration (MIC). Further insight in the killing kinetics of an antibiotic can be gained from time-kill experiments. In both methods, bacteria are exposed to various antibiotic concentrations and the result, i.e., growth and killing at certain concentrations, is somehow interpreted in a way that is considered to predict efficacy in vivo (3–6). Similarly, the laboratory tests most frequently used for investigation of the interaction between two antimicrobial agents are combination tests of MICs and interpretation thereof [e.g., checkerboard titrations, fractional inhibitory concentrations (FICs), isobolograms, and surface response], time-kill experiments, and agar diffusion tests.

Checkerboard Titrations

The method used most frequently to study antibiotic interactions is the checkerboard titration. In this method, serial dilutions of two antibiotics in concentrations equal to, below, and above the individual MIC of the micro-organism are tested. The checkerboard consists of columns in which each well contains the same amount of antibiotic A being over i dilutions along the x-axis, and in rows in which each well contains the same amount of antibiotic B over j dilutions on the y-axis (Fig. 1). The result is that each well contains a different combination of concentration of the two antibiotics. The total number of wells is $i \times j$ wells.

Although the results obtained from a checkerboard experiment seem to be quiet obvious, the interpretation of these results is less clear. Three problems can be distinguished. The first is compilation of $i \times j$ results in one single number. The method most commonly used is the FIC. For each well, the fractional concentrations of these drugs can be calculated [FIC = concentration drug A/MIC A) + (concentration drug B/MIC B)] (7,8). These FICs are calculated for all wells with the lowest concentrations that show no visible growth after 24 hours [or other time period, depending on the micro-organism–antimicrobial combination (9)] incubation with the target micro-organism. The mean FIC or fractional inhibitory index for the complete checkerboard is then calculated as the sum of these FICs divided by the total number of wells (FICi = Σ FIC/n). Although this approach seems to be clear enough, several methods have been described to calculate a FICi but the standardization of the FIC index as introduced by Hallander is now most commonly applied (10). Especially in earlier years, the methods used to determine the FICi varied widely and the results reported should be viewed with some

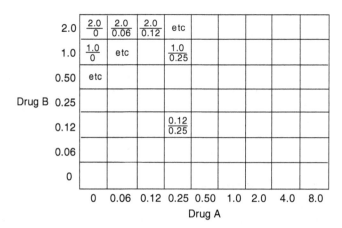

FIGURE 1 In the checkerboard, serial dilutions of two drugs are combined, usually employing ranges of concentrations including the MICs of the drugs being tested. The concentrations of the drugs are expressed in mg/L. *Abbreviation*: MICs, minimal inhibitory concentrations.

caution. In a survey among 11 publications examining the relation between FIC and outcome of combination therapy, none of the authors described exactly how the FIC index was calculated (11). Of the four authors who responded to a questionnaire, three had used different methods to calculate the FIC and none of these were as originally described by Hallander.

The second problem is the interpretation of the FICi itself, that is which values should be interpreted as synergistic and which as antagonistic. The interpretation of the American Society of Microbiology as well as the British Society of Antimicrobial Chemotherapy use the following interpretation: A combination of the drugs is defined as synergistic if FICi ≤ 0.5, as additive or indifferent if $0.5 <$ FICi < 4.0, and as antagonistic if FICi ≥ 4.0 (12,13). However, there is no universal consensus regarding the criteria that define additivity and antagonism and several other definitions exist (14). In various other guidelines and studies, the definition for synergism varies between FICi < 0.5 and FICi ≤ 1 (14–17). One of the reasons is that there are different interpretations of the method used. The error of measurement of an MIC is one dilution. There is therefore also an error of measurement in the determination of the FICi, which is one of the main reasons that FICi values between 0.5 and 4 are designated as indifferent. If the experiments were repeated a number of times and a confidence interval (CI) determined, strict cutoff values and the interpretation thereof would be less controversial. Values with 95% CI including 1 would be indifferent, otherwise there would be some interaction. Yet, in a study comparing the results of repeat experiments, Rand et al. (18) found that 25% of replicate sets gave discordant classification results despite the excellent reproducibility of the individual MICs. They concluded that experiments should be performed at least in fivefold with more than 80% agreement between replicates required for classification. Alternatively, Te Dorsthorst showed that, for antifungals, the conclusion with respect to the interpretation—that is synergism, indifference or antagonism—was largely dependent on the underlying—often wrong—assumptions made and concluded that the use of the FIC index for antifungals should be viewed with the greatest caution (19).

Another approach that has been used is smaller concentration intervals, known as modified dilution checkerboard titrations, because such titrations are thought to result in greater precision (16,17,20). In tests using twofold dilutions, the precision of the test diminishes especially at higher concentrations of the drugs than happens in the modified dilution system. Using this latter dilution scheme will result in more precise FICi values and consequently the modified checkerboard titration should predict synergism more precisely (17). However, the modified checkerboard titration is very laborious and has received little attention.

The third and perhaps most important problem is the fact that FICis are determined at static concentrations and that concentrations in vivo vary over time due to dosing and elimination of the drug. This is further discussed in one of the next paragraphs, pharmacodynamic studies.

Although the method used most commonly to perform a checkerboard titration is the microdilution broth assay, several other methods exist that are being used. The interpretation of these methods is more or less similar. These methods include the macrodilution broth assay and the agar dilution method. The latter is comparable to the MIC agardilution assay, except that the semisolid agar bases contain combinations of antimicrobials instead of the single drug. The advantage of this latter method is that a large number of strains can be tested simultaneously on single series of plates (15).

Time-Kill Curves

One of the major disadvantages of the checkerboard methods described earlier is that killing kinetics are not taken into account, i.e., the rate of killing over time during an exposure of a bacterial culture to an antibiotic. One of the approaches to this problem is the use of time-kill experiments. This method is more laborious as it depends on repeated sampling over 24 hours, but compared to the checkerboard titrations it results in more information. For example, it gives an insight into both the rate and extent of killing.

Two different approaches can be distinguished. The first is to perform time-kill curve experiments and taking samples after a fixed period of time, e.g., 6 hours or 24 hours (Fig. 2). This results in a certain number of CFU per regimen and the results of CFU count of the combination is compared to that found after exposure to the most active drug alone.

Similar to the interpretation of the FICi, the interpretation of this test is less obvious than one might wish for. Three problems can be distinguished. The first is the interpretation of the end point itself. Synergism is generally defined as a 2 $\log_{10}$ smaller CFU/mL count remaining after 24 hours exposure to the combination compared to that found after exposure to the most active drug alone (Fig. 2) (15). However, this definition is valid only if at least one of the two tested drugs produces no inhibition or killing when given alone. Thus, if both agents are bactericidal, there is no generally accepted definition of synergy in time-kill experiments. Other definitions which have been used are the time needed to kill the inoculum below 10^3 CFU/mL and the time to regrowth above 10^3 CFU/mL (21) or a 3 $\log_{10}$ CFU/mL difference in bacterial activity for the combination to qualify as synergistic (22). These interpretations are however somewhat subjective. Apart from the subjective interpretation that a 2 $\log_{10}$ kill is clinically relevant (for which no data exist), time-kill curves are usually not performed in duplicate and one of the reasons that a 2 or 3 $\log_{10}$ kill is being used is the certainty that indeed there is a difference between the effects of the drugs alone and in combination if that difference is found. One could argue that a significant difference between the drug given alone and in combination should in itself be enough reason to argue

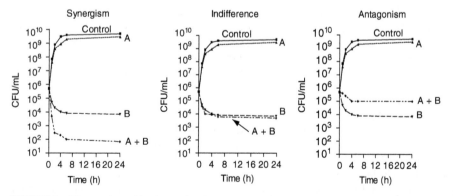

FIGURE 2 Effects of combinations of antimicrobial agents as determined with a time-kill experiment. The different panels show how synergism, indifference, and antagonism may become evident during time-kill experiments.

that interaction between two drugs exists. Ideally, the difference between single drug and combination should be expressed as a difference with 95% CI.

The second problem is that time-kill curves are performed at one, static concentration, while concentrations in vivo vary over time due to dosing and elimination of the drug. An interaction between two drugs found at one concentration may not be valid at other concentrations. For example, a combination may appear to be synergistic at a concentration of $0.5 \times$ MIC but not at $1 \times$ MIC. The approach usually taken is to perform time-kill experiments at different combinations, usually at or around the MICs for the two drugs. However, concentrations in vivo vary much more and it is therefore unclear what the conclusions from time-kill curves should be with respect to the in vivo interaction of the two drugs. Another approach would be to use the outcome of the time-kill experiments in a surface response plot and use a modeling approach to interpret the data. The advantage of having a quantitative measure of response in a checkerboard type of experimental design offers that opportunity, and, now extensively being used for antifungals (23), has been explored for antibacterial agents as well (24).

The third interpretation problem resembles that for the MIC: the outcome is determined after exposure to static concentrations after a certain period of time, while concentrations in vivo fluctuate over time. IVPMs may overcome this (see below) but some of the drawbacks mentioned here still remain.

In any case, the use of time-kill curves provides the advantage over the MIC of having a continuous outcome parameter (CFU) as compared to a dichotomous one (visible growth vs. no visible growth). The use of a continuous outcome parameter offers major advantages. For antifungals, response surface analysis has shown to be a very promising approach to analyzing interactions. An extensive review comparing these various methods can be found in Ref. (23). Alternatively, Li et al. (25,26) have introduced the fractional maximum effect method that in the evaluation of the interaction takes the nonlinear nature of the concentration–effect relationship into account.

The second approach using kill curves is taking multiple samples during the time-kill curve. This offers not only insight in the extent of killing, but also in the rate of killing. Recently, this approach has been taken up by a number of investigators for single drug exposures but has not been used for drug interaction studies yet.

Diffusion Techniques
During the last decades, many other tests for synergy/antagonism have been described in literature. Most of these tests were simplifications of the established synergy tests, including diffusion around antibiotic containing discs (27–31) or from antibiotic containing paper strips (32). However, none of these methods has become generally accepted nor did they find their way to the routine laboratory. One of the reasons is perhaps that quantitation of the results found is almost impossible.

E-Test
A more promising approach using diffusion techniques is the E-test (AB-Biodisk, Solna, Sweden). E-tests are plastic strips coated with a continuous gradient of antibiotic concentrations on one side and a concentration scale of antimicrobial agent on the other side. Two methods have been described to use these tests. The first method is that described by White et al. (33) and the second by Manno et al. (34), although various variations had been presented before. The methods used enable one to determine a value, which is thought to be comparable to the FICi in

A

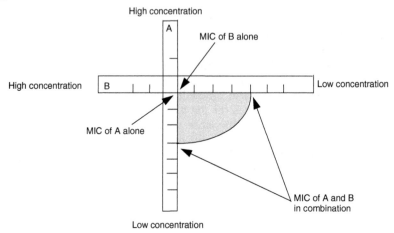

B

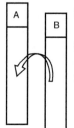

1. Perform standard MIC either in parallel with combination setup or separately first.

2. Place strip A on the inoculated agar surface and leave for 1h at room temperature.
3. Mark the strip's position on the back of the plate.
4. Remove strip A, clean with alcohol and save as a MIC reading scale.

5. Position strip B on top of the imprint of A e.g. ratio 1:1(256/256)1:8 (32/256).
6. Leave strip B on the agar plate and incubate.
7. Use the respective MIC strips/scales to read the MIC of each drug in the combination.

FIGURE 3 Schematic diagram of the E-test method to determine synergism as described by White et al. (33) (**A**, upper panel) and Manno et al. (35) (**B**, lower panel). *Abbreviation*: MIC, minimal inhibitory concentration. *Source*: Adapted from the EAS-023; Courtesy of AB-Biodisk, Solna, Sweden.

checkerboard titrations. White et al. (33) uses two E-test strips that are applied on the agar in a cross form (Fig. 3A), where the MIC for monotherapy is the cross point of both strips. The method described by Manno et al. (35) involves using two E-test strips, which are applied consecutively on the agar at exactly the same spot. The first one is put on the agar for one hour, and thereafter the second is put on the same place (Fig. 3B). The advantages of this test are its ease of application and its high reproducibility. These two properties allow routine laboratories to test for synergy of combinations of antibiotics at short notice. The clinical importance has, however, not yet been established.

Correlation Between Checkerboard Titrations, Time-Kill Curves, and E-Test

As mentioned during the description of each of the laboratory tests, several problems arise as to how to interpret the results of these tests. Irrespective of these problems, one might expect that, if these tests do describe the interaction between antimicrobial agents in a valid manner, the outcome of these tests correlate with each other. However, there are relatively few studies that do attempt to validate the results of interaction studies by using more than one method and comparing the results obtained. Table 1 shows studies that included both checkerboard

TABLE 1 Studies Describing the Percentage Synergy as Obtained by Checkerboard and Time-Kill Method[a]

References	Strain (N)	Percentage synergism		Percentage agreement
		Checkerboard	Time-kill	
36	*Acinetobacter baumannii* (2 for both methods)	100	100	100
37	*Salmonella enterica* (1)	100	100	100
38	*Streptococcus pneumoniae* (11)	0	0	100
39	*A. baumannii* (10)	40	57	17
40	*Burkholderia cepacia* (14), *Staphylococcus aureus* (2), *Klebsiella pneumoniae* (9)	59	59	100
41	*Enterococcus faecium* (12)	43	52	45
42	*Stenotrophomonas maltophilia* (20 for both methods)	11–58	80–85	?[b]
43	*Pseudomonas aeruginosa* (12 for both methods)	0.8–7.3	33–83	?[b]
44	Nonfermentative rods (20 for both methods)	14–28	10–30	?[b]
45	*Acinetobacter* sp (15 for both methods)	1–3	~50	0
46	*P. aeruginosa* (1)	100	0	0
33	*P. aeruginosa, S. aureus, Escherichia coli, Enterobacter cloacae* (4)	0–50	0–50	25–100[c]
47	*S. pneumoniae* (1)	0	100	0
48	*S. pneumoniae* (9 for both methods)	18–31	100	11–66
49	*P. aeruginosa* (7 for both methods)	0	83	17
50	*P. aeruginosa* (5)	0	100	0
51	*P. aeruginosa* (3)	44	67	39
52	*P. aeruginosa* (30)	47	7	53
53	*P. aeruginosa* (1)	100	100	100
22	*S. marcescens, S. aureus, P. aeruginosa* (18)	50–57, 67, 40–56	100	44–88
54	*P. aeruginosa* (1)	100	100	100
10	*P. aeruginosa* (1), *E. coli* (1), *P. maltophilia* (1)	19–70	50	42
55	*K. pneumoniae* (22)	23	77	18
56	*E. coli* (1), *S. marcescens* (1)	0	50–100[c]	0
57	*E. coli* (1), *C. freundii* (1), *Proteus* (2), *P. rettgeri* (1)	80	40–100[c]	20–80[c]
58	*S. aureus* (1)	0	100	0
59	*S. aureus* (2)	0	100	0
20	*P. aeruginosa* (123)	14	58	49
60	*P. aeruginosa* (2)	100	100	100
61	*P. aeruginosa* (2)	100	100	100

[a]Percentage of agreement was calculated as the number of strains yielding the same outcome in both methods.
[b]Cannot be calculated because of strain selection for one of the two methods.
[c]Dependent on strain and/or combination used.

titrations and time-kill techniques. The overall correlation between the two tests varied between 0% and 100%. Furthermore, in all these studies time-kill curves are performed on a limited number of strains, which were, for the major part, selected based on the results of the checkerboard. Thus, the correlations found are highly biased. The wide range of the correlations is further explained by the differences in definitions for synergy for each method as described above.

Studies correlating E-test with other techniques are still limited for antimicrobials. Manno et al. (35) compared 10 *Burkholderia cepacia* strains and found 90% concordance; however, the definitions used for interaction were not the ones as described above. Pankey et al. (62) compared 31 *P. aeruginosa* isolates using E-test and time-kill curves and found 65% concordance.

These studies show that differences in methodology will result in different answers and thus make it almost impossible to predict clinical efficacy.

PHARMACODYNAMIC STUDIES

The methods described above show one important difficulty with respect to the applicability for the clinical setting, i.e., they are performed at fixed or static antibiotic concentrations. In contrast, in patients, the antibiotic concentrations vary over time due to administration and elimination. To complicate matters even more, the pharmacokinetic profiles of the drugs in the combination will not be similar and will result in varying concentration ratio's over time. These ratios will vary not only in the individual patient due to changing clinical conditions, but also between patients due to individual elimination characteristics.

Another problem is that efficacy has been shown to be not only dependent on the concentrations reached in general, but also on the concentration profile and thus the dosing regimen itself, such as four times daily versus continuous infusion. This applies for the individual agents but has also been shown to apply during combination therapy. For instance, in a few earlier studies (63,64), it was remarked that the degree of synergism was dependent on the dosing schedule of the β-lactam in the combination. It was rather the efficacy of dosing regimens that determined outcome than a supposed presence of synergism. Recently, it was shown that efficacy of the combination is dependent on the dosing schedule of the individual drugs, while the degree of synergism seemed to be independent of the dosing regimen, at least over the dosing range studied (65). Finally, two drugs may or may not be given simultaneously and this has been shown to influence outcome. This was demonstrated by Konig et al. (66), who found that with susceptible strains ($n = 7$), nonsimultaneous administration was superior to simultaneous administration. In a more recent publication, a similar observation was made for gentamicin and ceftazidime (21), while in another study using ceftazidime and tobramycin, no difference was found between various modes of administration (46).

In spite of all these uncertainties, attempts have been made to correlate in vitro laboratory synergy tests with outcome. In general, two types of experiments have addressed these problems, IVPMs and animal model experiments.

The same problem as previously described for time-kill curves arises, since there is no unequivocal definition for synergism in these models. The most frequently used definitions for synergism are a 2 $\log_{10}$ larger decrease in CFU for the combination therapy as compared to the results of monotherapy of the most active drug, comparable to the definition used for time-kill experiments. Other less frequently used methods are to determine the time needed to kill the inoculum below 10^3 CFU/mL (21) and a decrease of 3 $\log_{10}$ CFU/mL of the combination therapy compared to monotherapy (22). Finally, another variable is the time of the end point of the experiment, which varies between 4 hours (67) and 96 hours (68) in various publications. Apart from the change in CFU as an end point, the animal model has the advantage that, in addition to CFUs, survival can be measured. However, since the half-life of most drugs in animals is shorter than in humans, the exposures of the antimicrobials to the micro-organisms is different from that encountered in the humans, and the results should be viewed with caution if this has not been taken into account in the evaluations.

IN VITRO PHARMACOKINETIC MODELS

Several investigators have studied the efficacy of combination therapy in an IVPM and correlated these results with checkerboard titrations and/or conventional time-kill experiments. While combination therapy was, in general, more efficacious than monotherapy (Table 2), results of checkerboard titrations and/or time-kill curves showed a poor relationship with outcome. For example, Zinner et al. (69) found that a *P. aeruginosa* strain with FICi = 13 (being extremely antagonistic) for the combination of piperacillin and thienamycin was slightly killed by this combination in the IVPM and the combination thus seemed to have at least an additive effect on this strain. If the definition for synergy as an FICi $\leq$ 0.5 and antagonism as an FIC $\geq$ 4 (15) is used, 15 out of 31 strains showed a concordant result for both in vitro synergy test and synergy tested in vitro models. However, 7 strains out of 31 showed a major error, which indicates that the in vitro test results in a better FICi than the efficacy of the combination in the in vitro model.

Taking an alternative approach, den Hollander et al. (77) defined the MIC_{combi} and the FIC_{combi}. The first was the MIC obtained in vitro using the E-test method described by White et al. (see above), while the FIC_{combi} describes the theoretical interaction between two drugs at each point in time based on the actual concentrations at that time point, and thereby took the effect of fluctuating concentrations on interaction into account. This index was subsequently used to construct pharmacodynamic indices, whereby the MIC_{combi} and the FIC_{combi} were used instead of the MIC. Finally, these indices were correlated with effect during combination therapy. Figure 4 shows the results of these studies.

The general conclusion from the studies in IVPMs is twofold. The first is that the relationship between the outcome of laboratory tests and outcome in the model is marginal at best and that there are no studies which unequivocally demonstrate the predictive value of laboratory tests. The second conclusion is that interaction studies performed in IVPMs provide much more information and are probably more relevant, than do checkerboards and in vitro time-kill curve experiments. The use of IVPM however, although providing insight in the interaction between antimicrobial agents in general, is unsuitable to use in the clinical setting.

ANIMAL MODELS

The study of infections in experimental animals provides the opportunity to study the antimicrobial response in vivo. Again the difference between in vitro synergy data and treatment outcome in animal models is striking. Several review articles in the past two decades have presented this incongruity (64,78–80).

TABLE 2 Studies Comparing Outcome of Killing Curves in In Vitro Pharmacokinetic Models with In Vitro Techniques to Determine Synergism

References	Strain	Combination	dCFU[a]	In vitro technique[b]	FIC or dCFU[c]	Predictive[d]
69	Pseudomonas aeruginosa ATCC 27853	Piperacillin + thienamycin	0.3	C	13	N
	P. aeruginosa ATCC 27853	Piperacillin + amikacin	0.5	C	0.7	Y
70	P. aeruginosa ATCC 27853	Netilmicin + ceftazidime	4.3–5.0	C	1	N
	P. aeruginosa 14974	Netilmicin + ceftazidime	0.4–1.2	C	0.7	Y
	P. aeruginosa A-10	Netilmicin + ceftazidime	0.3–0.6	C	0.5	N
	P. aeruginosa J-3	Netilmicin + ceftazidime	–0.3–0.6	C	0.8	?
	P. aeruginosa E29/2	Netilmicin + ceftazidime	0.5–0.7	C	0.7	Y
71	P. aeruginosa ATCC 27853	Amikacin + aztreonam	2.5	C	0.5	Y
		Amikacin + cefepime	1.3	C	0.38	N
		Amikacin + ceftazidime	1.2	C	0.25	N
		Aztreonam + cefepime	2.2	C	1.5	N
		Aztreonam + ceftazidime	–0.1	C	2.5	Y
	P. aeruginosa 16690	Amikacin + aztreonam	3	C	0.31	Y
		Amikacin + cefepime	1.6	C	0.31	N
		Amikacin + ceftazidime	3.8	C	0.31	Y
		Aztreonam + cefepime	–0.4	C	0.5	N
		Aztreonam + ceftazidime	1.2	C	0.75	Y
72	P. aeruginosa ATCC 27853	Amikacin + imipenem	0–2.5[e]	C	0.19	?
	P. aeruginosa ATCC 27853R	Amikacin + imipenem	4.2–6.0	C	0.25	Y
73	P. aeruginosa ATCC 27853	Amikacin + ceftazidime	>0	C	1.25	Y
	P. aeruginosa ATCC 27853CR	Amikacin + ceftazidime	1–5[e]	C	1	?
46	P. aeruginosa CF 133	Ceftazidime + tobramycin	2.6–4.0	C	0.37	Y

74	Enterococcus faecium 4162 (Van A)	Ampicillin/sulbactam + trovafloxacin	2.6–4.0	T	−0.5 log CFU/mL	N
	E. faecium 5924 (Van B)	Ampicillin/sulbactam + trovafloxacin	0	T	−4 log CFU/mL	N
58	Staphylococcus aureus MRSA 494	Vancomycin + gentamicin	0.5	T	+1.5 log CFU/mL	N
			−0.7–4.9[e]	T/C	FICi = 0.75, log 3 CFU/mL	?
68	E. faecium 1231 + 12366	Quinupristin/dalfopristin + doxycycline	0.5–0.9[e]	T	Additive or none	Y
59	S. aureus ATCC 25923	Quinupristin/dalfopristin + vancomycin	1.1	T/C	1.9, −log 4 CFU/mL	?
	S. aureus MRSA 67	Quinupristin/dalfopristin + vancomycin	0.9	T/C	2.5, −log 4 CFU/mL	?
75	S. aureus MSSA 1199	Levofloxacin/ofloxacin/ciprofloxacin + rifampicin	0.2–0.4	C	2.0	Y
	S. aureus MRSA 494		0.3–2.0[e]	C	<0.5	?
76	P. aeruginosa CsAs	Azlocillin + ciprofloxacin	2.2–3.5	C	0.19	Y
	P. aeruginosa CsAr		5.0–6.5	C	0.31	Y
	P. aeruginosa CrAs		−4.0– 2.0	C	0.38	N
	P. aeruginosa CrAr		3.6–4.0	C	0.38	Y

[a]Decrease in 10 log CFU [(s): $P < 0.05$] of combination vs. most active single agent.

[b]M, MICs; C, checkerboard; T, time-kill curves.

[c]Data provided are dependent on technique shown in previous column. MICx: factor MIC decrease in combination vs. single agent; dCFU given in $\log_{10}$ decrease vs. agent most active when used a single agent; S, I, A: synergistic, indifferent or antagonistic as stated by the author. In case of time-kill curves synergism is always defined as larger than 2 $\log_{10}$ CFU decrease of the combination vs. the most active agent.

[d]Y, yes; N, no; ?, uncertain.

[e]Dependent on dosing regimen used in experiment.

Abbreviations: FIC, fractional inhibitory concentration; MIC, minimal inhibitory concentration.

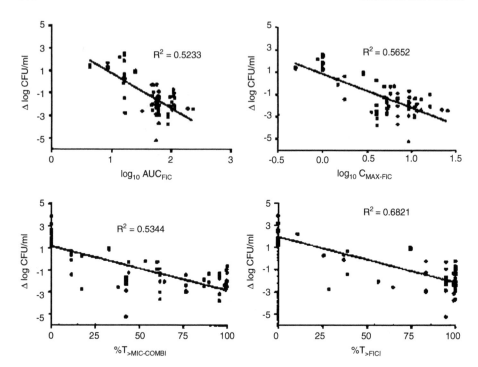

FIGURE 4 Correlation between PDI based on FIC$_{combi}$ and MIC$_{combi}$ and efficacy, expressed as dCFU/mL. *Abbreviations*: FIC, fractional inhibitory concentration; MIC, minimal inhibitory concentration.

The efficacy in vivo itself poses even more problems. Basically, two outcome parameters are being used. One is survival analysis and the other is CFU counts. The problem in the studies using survival analysis is that usually only one or at most two strains are tested. The advantage is that some form of statistical analysis is usually performed on outcome, for instance, Fisher tests or, in more recent studies, survival analysis. Table 3 shows a summary of the results obtained by survival analysis. Although the majority of the in vitro results are concordant with the in vivo outcome, it must be borne in mind that in most cases, strains were selected to perform animal experiments, usually based on in vitro synergism. Conclusions drawn from these studies may thus be highly biased. Studies using CFU counts as an end point usually yield much more information, mainly because of the quantity of strains tested, but basically suffer from the same drawbacks as time-kill curves in vitro in that different definitions are being used and thus may lead to conflicting interpretations (95). The major conclusions from the survival studies is that although interaction itself is usually described by the survival analysis, this is more mechanism based and the prediction from the in vitro test is at most indicative because of the small number of strains tested. The results of in vivo time-kill curves should be viewed with some caution (Table 4). Ideally the same strain should be tested in more than one experimental system.

A pharmacodynamic analysis on the interaction between drugs in vivo was performed by Mouton et al. (65). Although the primary objective of that study was to show that the effect of the antimicrobials during combination therapy correlated

TABLE 3 Studies Comparing Outcome of Survival Studies in Animal Models with Various In Vitro Techniques to Determine Synergism

References	Organism	Combination	Outcome of[a] combination vs. single agent	In vitro[b] technique	MICx, FIC or dCFU[c]	Predictive[d]
81	Pseudomonas 540	Carbenicillin + gentamicin	↑	M	0.25 + 0.25	Y
	Pseudomonas 407	Carbenicillin + gentamicin	↑	M	0.25 + 0.25	Y
	Pseudomonas 692	Carbenicillin + gentamicin	↑	M	0.25 + 0.5	?
82	Pseudomonas 407	Carbenicillin + gentamicin	~,↑[e]	M	0.25 + 0.25	N,Y
60	Pseudomonas aeruginosa BA	Ceftriaxon + tobramycin or gentamicin or amikacin	↑,↑,↑	C,T	<0.5,≥3	Y,Y
	P. aeruginosa 1438111	Ceftriaxon + tobramycin or gentamicin or amikacin	↑,↑,↑	C,T	<0.5,≥3	Y,Y
83	P. aeruginosa	Ciprofloxacin + azlocillin	↑	C	S	Y
84	P. aeruginosa 3757	Amikacin + imipenem	↑	T	≥3	Y
	P. aeruginosa 5075	Amikacin + imipenem	↑	T	≥3	Y
	Klebsiella pneumoniae 5166	Amikacin + imipenem	?	T	≥3	N
	K. pneumoniae 3823	Amikacin + imipenem	↑	T	≥3	Y
	S. marcescens 7744	Amikacin + imipenem	↑	T	≥3	Y
47	Streptococcus pneumoniae P15986	Amoxicillin + gentamicin	↑[f]	C,T	I, ≥2	N, Y
85	P. aeruginosa YED	Ciprofloxacin + cefpiramide	↑	M	0.025 + 0.25	Y
	P. aeruginosa YED	Tobramycin + ticarcillin	↑	M	0.25 + 0.03	Y
86	P. aeruginosa YED	Gentamicin + piperacillin or apalcillin	↑	C	S	Y
	P. aeruginosa Ps4	Gentamicin + piperacillin or apalcillin	↑	C	S	Y
53	P. aeruginosa	Tobramycin + ceftazidime	↑	M	0.25 + 0.25	Y
87	P. aeruginosa	Amikacin + azlocillin or ticarcillin	↑,↑	C	S, S	Y, Y
88	P. aeruginosa 25	Amikacin + piperacillin or ceftazidime	↑,↑	C	S, S	Y, Y
89	P. aeruginosa P323	Gentamicin + carbenicillin	↑	M	0.25 + 0.25	Y
90	Klebsiella sp. 8	Gentamicin + cephalothin or chloramphenicol	?,?	C	S, S	N, N
	Klebsiella sp. 60	Gentamicin + cephalothin or chloramphenicol	?,?	C	S, A	N, N
	Klebsiella sp. 71	Gentamicin + cephalothin or chloramphenicol	?,?	C	S, A	N, N

(Continued)

Table 3 Studies Comparing Outcome of Survival Studies in Animal Models with Various In Vitro Techniques to Determine Synergism (*Continued*)

References	Organism	Combination	Outcome of combination vs. single agent	In vitro[a] technique	MICx, FIC or dCFU	Predictive[d]
91	P. aeruginosa P-4	Carbenicillin + gentamicin or tobramycin	~	M	1 + 1, 0.5 + 1	Y
92	P. aeruginosa P-4	Tobramycin + cefsulodin	~	M	1 + 1, 0.5 + 1	Y
54	P. aeruginosa	Tobramycin + ticarcillin	~	M	S	N
	P. aeruginosa	Tobramycin + ceftazidime	~	M, T	S,0–4	N, ?
	P. aeruginosa	Tobramycin + azlocillin	~	M	S	N, ?
	P. aeruginosa	Netilmicin + ceftazidime	~	M, T	S,0–4	N, ?
93	Pseudomonas 407	Gentamicin + carbenicillin	↑	T	S	Y
50	Pseudomonas	L-658310 + tobramycin	↑	C,T	S,≥3	y
		L-658310 + gentamicin	↑	C,T	S,≥3	y
		L-658310 + amikacin	↑	C,T	S,≥3	y
94	K. pneumoniae 3823	Amikacin + cefazolin	~	M	0.25 + 0.25	N
	K. pneumoniae 5973	Amikacin + cefazolin	↑	M	0.125 + 0.06	Y
	K. pneumoniae 5166	Amikacin + cefazolin	↑	M	0.125 + 0.25	Y

a↑, Increased survival of combination vs. single agent; ~, no effect of combination vs. single agent.

bM, MICs; C, checkerboard; T, time-kill curves.

cData provided are dependent on technique shown in previous column. MICx: factor MIC decrease in combination vs. single agent; dCFU given in $\log_{10}$ decrease vs. single agent. In case of time-kill curves, synergism is always defined as larger than 2 $\log_{10}$ CFU decrease of the combination vs. the most active agent.

dY, yes; N, no; ?, uncertain.

e3 of 4 experiments showed increased survival and one no effect.

fResults of two experiments using different dosing regimens.

Abbreviations: FIC, fractional inhibitory concentration; MIC, minimal inhibitory concentration.

TABLE 4 Studies Comparing Outcome of In Vivo Killing Curves in Animal Models with In Vitro Techniques to Determine Synergism

References	Organism	Combination	dCFU[a]	in vitro technique[b]	MICx, FIC or dCFU[c]	Predictive[d]
96–98	S. marcescens 157	Ciprofloxacin + amikacin	3	C	0.5–2	N
		Ciprofloxacin + ceftizoxim	7	C	0.5–2	N
	Pseudomonas aeruginosa 864	Azlocillin + amikacin	≥4.3	C	0.5–2	N
		Ciprofloxacin + azlocillin	4.3	C	0.5–2	N
	P. aeruginosa 876	Azlocillin + amikacin	0.7	C	<0.5	N
		Ciprofloxacin + azlocillin	1.5	C	<0.5	N
	P. aeruginosa 913	Azlocillin + amikacin	≥4.7	C	0.5–2	N
		Ciprofloxacin + azlocillin	3	C	0.5–2	N
	P. aeruginosa 915	Azlocillin + amikacin	≥5.5	C	0.5	?
		Ciprofloxacin + azlocillin	2	C	<0.5	Y
99	P. aeruginosa 107	Tobramycin + ticarcillin	0	M,T	0.25 + 0.25, ≥7	N
47	Streptococcus pneumoniae	Amoxicillin + gentamicin	≥2	C,T	1, ≥2	N,Y
56,100	Escherichia coli 4672	Ciprofloxacin + gentamicin	1.2	C,T	Add, Add[e,g]	?[e]
57	E. coli 4672	Ciprofloxacin + mezlocillin	1.1 (s)	T	Add[g]	Y
	P. aeruginosa 12149	Ciprofloxacin + azlocillin	1.8 (s)	C,T	Add[g],1–2	Y
	C. freundii 13039	Ciprofloxacin + mezlocillin	1	C,T	S,≥2	N,N
	P. rettgeri 10020	Ciprofloxacin+ cefotaxime	1	C,T	S,≥2	N,N
67	P. aeruginosa 220	Ciprofloxacin + azlocillin	0[f]	C	0.5	?
		Ciprofloxacin + ceftazidime	0[f]	C	1	Y
		Ciprofloxacin + tobramycin	0[f]	C	1.5	Y
101	Enterobacter cloacae 474s	Cefepime + amikacin	1 (s)	T	0–4[e]	?
	E. cloacae 474r	Cefepime + amikacin	1 (s)	T	0–4[e]	?
102	E. cloacae	Cefepime + amikacin	1 (s)	T	0–4[e]	?

(Continued)

Table 4 Studies Comparing Outcome of In Vivo Killing Curves in Animal Models with In Vitro Techniques to Determine Synergism (Continued)

References	Organism	Combination	dCFU[a]	in vitro technique[b]	MICx, FIC or dCFU[c]	Predictive[d]
65	P. aeruginosa 27853	Tobramycin + ticarcillin	S	E	S	Y
		Netilmicin + ceftazidime	S	E	S	Y
		Ciprofloxacin + netilmicin	S	E	S	Y
		Ciprofloxacin + ceftazidime	S	E	S	Y
91	P. aeruginosa P-4	Carbenicillin + aminoglycoside	0.2	M	I	Y
103	Staphylococcus aureus 25923	Cloxacillin + netilmicin	0.6	C	0.2	N
		Clindamycin + rifampicin	1.9	C	2	N
	S. aureus 8906	Cloxacillin + netilmicin	0	C	0.2	N
		Clindamycin + rifampicin	1.4	C	0.3	N
54	P.aeruginosa	Tobramycin + ticarcillin	1.5	M	S[e]	N
		Ceftazidime + tobramycin	2.1	M,T	S, 0–4[e]	Y,?
		Azlocillin + tobramycin	2.1	M	S[e]	Y
		Ceftazidime + netilmicin	1.2	M,T	S, 0–4[e]	N,?
94	Klebsiella pneumoniae 5166	Amikacin + cefazolin	1.7 (s)	M	S	Y

[a]Decrease in log$_{10}$ CFU [(s): $P < 0.05$] of combination vs. most active single agent.
[b]M, MICs; C, checkerboard; T, time-kill curves; E, E-test.
[c]Data provided are dependent on technique shown in previous column. MICx: factor MIC decrease in combination vs. single agent; dCFU given in log$_{10}$ decrease vs. agent most active when used a single agent; S, I, A: synergistic, indifferent, or antagonistic as stated by the author. In case of time-kill curves synergism is always defined as larger than 2 log$_{10}$ CFU decrease of the combination vs. the most active agent.
[d]Y, yes; N, no; ?, uncertain.
[e]Dependent on concentrations used in experiment.
[f]Additive as described in text.
[g]After only 4 hr of therapy.
Abbreviations: FIC, fractional inhibitory concentration; MIC, minimal inhibitory concentration.

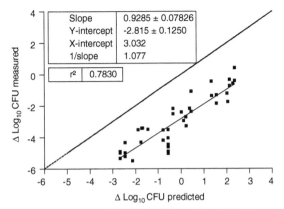

Predicted: t > 0.25* MIC ticarcillin and log AUC tobramycin

FIGURE 5 Observed versus predicted values of various dosing regimens for ticarcillin and tobramycin in an experimental model of infection. Predicted values based on the PDI, which best explains the efficacy of each antibiotic. The regression line deviates significantly from the line of symmetry. *Abbreviation*: MIC, minimal inhibitory concentration.

with the same pharmacodynamic index as during single drug therapy, the experimental setup also allowed for conclusions with respect to interaction of the drugs. An example is shown in Figure 5.

The figure shows the prediction of the effect of various single-agent dosing regimens of tobramycin and ticarcillin versus the effect measured during combination therapy based on the effect correlation with AUC/MIC and $T_{>0.25*MIC}$, respectively. The correlation between the predicted effects and the effect actually measured is excellent, thereby showing that the pharmacodynamic index that correlates with effect during single drug exposure is the same one during combination therapy. The second conclusion is that these agents act synergistically. If there were no interaction at all, the regression line would equal the line of symmetry. The regression line is clearly shifted to the right, and the intercept of the regression line of the predicted and observed responses is significantly different from zero. In this model, two antimicrobial agents show synergy. Similar results were found for the combination of netilmicin and ceftazidime. Conclusions with respect to interaction and a combination of a quinolone and a β-lactam or an aminoglycoside were less unequivocal, however, although antagonism did not seem to be present.

COMBINATION THERAPY IN CLINICAL INFECTIONS

As stated earlier, there are several reasons to use combination therapy in the clinical setting. During recent years, most clinical studies compare the efficacy of combination therapy to that of monotherapy, especially when a single new drug is promoted to be as efficacious as the combination of two agents used so far. Unfortunately, few studies have investigated the correlation between clinical outcome of combination therapy and in vitro data, such as MICs, FICs, or time-kill curves of the clinical isolates. Studies in which in vitro data were compared with clinical outcomes are discussed next.

Klastersky et al. (104) showed that combinations of antibiotics that were synergistic in vitro (as measured by FICs) were associated with favorable clinical outcome in 75% of cases, while infections caused by bacteria that showed no synergism to the combination therapy responded in only 41% of cases ($P < 0.01$). The difference was especially striking in severe infections, i.e., those associated with bacteremia and patients that were granulocytopenic. Young (105) in a prospective randomized study comparing the combinations of gentamicin + carbenicillin versus amikacin + carbenicillin in neutropenic patients observed that there was a positive association between in vitro synergism (determined as decrease in MIC to one-fourth for the combination) and favorable clinical outcome. Anderson et al. (106) likewise demonstrated a significantly higher response rate in patients with gram-negative rod bacteremia whose infecting organisms were synergistically inhibited (synergy was determined as the MIC of the combination being one-fourth of the drug alone). Klastersky and Zinner (107) showed a favorable clinical response in 79% of cancer or neutropenic patients when treated with synergistic combinations (FICs) of antibiotics against severe infections. When the combination of the antibiotics appeared to be nonsynergistic, the clinical outcome was much less favorable (45%). Fainstein et al. (108) described a clinical trial in which 253 patients (321 febrile episodes) were treated with ceftazidime alone or combined with tobramycin. Of all patients treated with monotherapy from whom an infecting strain could be isolated, the strains were susceptible and they showed a good clinical response in 88%. For patients treated with combination therapy, the clinical response was 100%. All strains isolated were susceptible to both antibiotics. Hilf et al. (20) studied 200 patients with *P. aeruginosa* bacteremia and could neither establish a correlation between the MICs and the clinical outcome nor find a correlation between in vitro synergy and clinical outcome. The most striking finding was that the mortality in the patients receiving combination therapy was 27% compared to 47% in the group receiving monotherapy ($P = 0.023$). Although combination therapy showed a favorable outcome compared to monotherapy, there was no correlation with in vitro susceptibility data in terms of synergism as determined by FICs. Thus, the FIC was not predictive for clinical outcome.

In conclusion, most studies described in Table 5 show a better outcome when the antibiotic combination used is synergistic in vitro. This seems to be the case in particular when synergism is tested by a decrease in MIC and not by checkerboard or time-kill method.

CONCLUSION

Considering the studies described above, there seems to be no in vitro test, which has been sufficiently evaluated and is predictive enough (although popular in many laboratories) to warrant its use in a routine laboratory with the possible exception of endocarditis. A predictive, easy to apply test is definitely needed, and, as has been indicated above, may be based more on decrease in MICs than on the use of checkerboard analysis. Alternatively, if a checkerboard analysis is performed, ways should be sought to analyze these data in a more meaningful way, such as a surface response analysis. The use of E-test may hold some promise for the future, but an unbiased correlation between synergy in vitro and efficacy in vivo still has to be shown.

TABLE 5 Studies Describing Outcome of Clinical Trials Where Combination Therapy was Evaluated and In Vitro Techniques were Used to Determine Synergism

References	Strain (N)	Combination	In vitro technique[a]	Percent in vitro synergy	Percent clinical cure		Predictive[c]
					Total	When synergistic	
52	*P. aeruginosa* (30)	Cefsulodin or ticarcillin + tobramycin or amikacin	C,T	47, 7%	62%	70%	N[b]
20	*P. aeruginosa* (200)	Numerous	C,T	14, 58%	59%	11%, 46%	N
104	Gram-negative rods (52)	Amikacin + penicillin or carbenicillin	M				
	bacteremic			52%	59%	75%	Y
	non-bacteremic			64%	58%	60%	Y
	total			59%	58%	66%	Y
109	Gram negatives (75)	Numerous	C				
	persistent granulopenia (<100/L)			58%	44%	26%	Y
	improving granulopenia (>100/L)			ND	82%	75%	NE
110	Gram-negative rods (38)	Moxalactam + piperacillin or moxalactam + amikacin	C	56%	68%	ND	ND
				93%	72%	ND	ND
111	Gram-negative rods(148)	Five combinations	M	48%	66%	58%	Y
106	Gram-negative rods (173)	Numerous	M	48%	72%	80%	Y
105	Gram-negative rods (49)	Carbenicillin + amikacin or gentamicin	M	59%	67%	87%	Y
107			C				

[a]M, MICs; C, checkerboard; T, time-kill curves.
[b]Y, yes; N, no; ND, not done; NE, not evaluated because the improvement of the granulocytes was a more important confounding factor.
[c]Antagonism in vitro was predictive for failure.
Abbreviations: FIC, fractional inhibitory concentration; MICs, minimal inhibitory concentrations.

The use of IVPMs and animal models does provide an insight into the interaction between antimicrobials in general, but application is possible in research settings only.

REFERENCES

1. Paul M, et al. Beta lactam monotherapy versus beta lactam-aminoglycoside combination therapy for sepsis in immunocompetent patients: systematic review and meta-analysis of randomised trials. BMJ 2004; 328(7441):668.
2. Safdar N, et al. Does combination antimicrobial therapy reduce mortality in Gram-negative bacteraemia? A meta-analysis. Lancet Infect Dis 2004; 4(8):519–527.
3. Ambrose PG, et al. Pharmacodynamics of fluoroquinolones against *Streptococcus pneumoniae* in patients with community-acquired respiratory tract infections. Antimicrob Agents Chemother 2001; 45(10):2793–2797.
4. Ambrose PG, et al. Clinical pharmacodynamics of quinolones. Infect Dis Clin North Am 2003; 17(3):529–543.
5. Mouton JW. Impact of pharmacodynamics on breakpoint selection for susceptibility testing. Infect Dis Clin North Am 2003; 17(3):579–598.
6. Mouton JW. Impact of Pharmacodynamics on Dosing Schedules: Optimizing Efficacy, Reducing Resistance, and Detection of Emergence of Resistance. In: Gould I, Van deer Meer J, eds. Antibiotic Policies: Theory and Practice. New York: Plenum Publishers, 2004:2004–2387.
7. Elion GB, et al. Antagonists of nucleic acid derivatives. VIII. Synergism in combinations of biochemically related antimetabolites. J Biol Chem 1954; 208(2):477–488.
8. Berenbaum MC. A method for testing for synergy with any number of agents. J Infect Dis 1978; 137(2):122–130.
9. NCCLS. Performance Standards for Antimicrobial Susceptibility Testing; 14th International Supplement. M100-S14 (M7). Wayne, NCCLS, 2004.
10. Hallander HO, et al. Synergism between aminoglycosides and cephalosporins with antipseudomonal activity: interaction index and killing curve method. Antimicrob Agents Chemother 1982; 22(5):743–752.
11. Hsieh MH, et al. Synergy assessed by checkerboard. A critical analysis. Diagn Microbiol Infect Dis 1993; 16(4):343–349.
12. Odds FC. Synergy, antagonism, and what the chequerboard puts between them. J Antimicrob Chemother 2003; 52(1):1.
13. American Society for Microbiology. Instructions to authors. Antimicrob Agents Chemother 2005; 49:i–xv.
14. Eliopoulos GM, Eliopoulos CT. Antibiotic combinations: should they be tested? Clin Microbiol Rev 1988; 1(2):139–156.
15. Eliopoulos GM, Moellering RC. Antimicrobial Combinations. In: Lorian V, ed. Antibiotics in Laboratory Medicine. Baltimore: Williams & Wilkins, 1986:1986–1432.
16. Horrevorts AM, et al. Chequerboard titrations: the influence of the composition of serial dilutions of antibiotics on the fractional inhibitory concentration index and fractional bactericidal concentration index. J Antimicrob Chemother 1987; 19(1):119–125.
17. Horrevorts AM, et al. Antibiotic interaction: interpretation of fractional inhibitory and fractional bactericidal concentration indices [letter]. Eur J Clin Microbiol 1987; 6(4):502–503.
18. Rand KH, et al. Reproducibility of the microdilution checkerboard method for antibiotic synergy. Antimicrob Agents Chemother 1993; 37(3):613–615.
19. Te Dorsthorst DT, et al. In vitro interactions between amphotericin B, itraconazole, and flucytosine against 21 clinical Aspergillus isolates determined by two drug interaction models. Antimicrob Agents Chemother 2004; 48(6):2007–2013.
20. Hilf M, et al. Antibiotic therapy for *Pseudomonas aeruginosa* bacteremia: outcome correlations in a prospective study of 200 patients [see comments]. Am J Med 1989; 87(5):540–546.

21. Barclay ML, et al. Improved efficacy with nonsimultaneous administration of first doses of gentamicin and ceftazidime in vitro. Antimicrob Agents Chemother 1995; 39(1):132–136.
22. Moody JA, et al. Evaluation of ciprofloxacin's synergism with other agents by multiple in vitro methods. Am J Med 1987; 82(4A):44–54.
23. Meletiadis J, et al. Assessing in vitro combinations of antifungal drugs against yeasts and filamentous fungi: comparison of different drug interaction models. Med Mycol 2005; 43(2):133–152.
24. Tam VH, et al. Novel approach to characterization of combined pharmacodynamic effects of antimicrobial agents. Antimicrob Agents Chemother 2004; 48(11):4315–4321.
25. Li RC, et al. The fractional maximal effect method: a new way to characterize the effect of antibiotic combinations and other nonlinear pharmacodynamic interactions. Antimicrob Agents Chemother 1993; 37(3):523–531.
26. Li RC, et al. Performance of the fractional maximal effect method: comparative interaction studies of ciprofloxacin and protein synthesis inhibitors. J Chemother 1996; 8(1):25–32.
27. Yourassowsky E, et al. A simple test to demonstrate antibiotic synergism. J Clin Pathol 1975; 28(12):1005–1006.
28. Lee WS, Komarmy L. New method for detecting in vitro inactivation of penicillins by *Haemophilus influenzae* and *Staphylococcus aureus*. Antimicrob Agents Chemother 1976; 10(3):564–566.
29. Meyer M, Hofherr L. A broth-disc technique for the assay of antibiotic synergism. Can J Microbiol 1979; 25(11):1232–1238.
30. Thabaut A, Meyran M. Methods for in vitro studies of antibiotic combinations. Indications and limits. Presse Med 1987; 16(43):2148–2152.
31. Chinwuba ZG, et al. Determination of the synergy of antibiotic combinations by an overlay inoculum susceptibility disc method. Arzneimittelforschung 1991; 41(2):148–150.
32. Anand CM, Paull A. A modified technique for the detection of antibiotic synergism. J Clin Pathol 1976; 29(12):1130–1131.
33. White RL, et al. Comparison of three different in vitro methods of detecting synergy: time-kill, checkerboard, and E test. Antimicrob Agents Chemother 1996; 40(8):1914–1918.
34. Poupard J, et al. Use of the AB Biodisk E-test as a screen for ticarcillin clavulanate (T/C)-amikacin (Ak) synergy with isolates of Xanthomonas maltophilia. 18th International Congress of Chemotherapy, Stockholm, 1993.
35. Manno G, et al. Use of the E test to assess synergy of antibiotic combinations against isolates of Burkholderia cepacia-complex from patients with cystic fibrosis. Eur J Clin Microbiol Infect Dis 2003; 22(1):28–34.
36. Yoon J, et al. In vitro double and triple synergistic activities of Polymyxin B, imipenem, and rifampin against multidrug-resistant *Acinetobacter baumannii*. Antimicrob Agents Chemother 2004; 48(3):753–757.
37. Mandal S, et al. Combination effect of ciprofloxacin and gentamicin against clinical isolates of *Salmonella enterica* serovar typhi with reduced susceptibility to ciprofloxacin. Jpn J Infect Dis 2003; 56(4):156–157.
38. Lin E, et al. Lack of synergy of erythromycin combined with penicillin or cefotaxime against *Streptococcus pneumoniae* in vitro. Antimicrob Agents Chemother 2003; 47(3):1151–1153.
39. Bonapace CR, et al. Evaluation of antibiotic synergy against Acinetobacter baumannii: a comparison with E-test, time-kill, and checkerboard methods. Diagn Microbiol Infect Dis 2000; 38(1):43–50.
40. Mackay ML, et al. Comparison of methods for assessing synergic antibiotic interactions. Int J Antimicrob Agents 2000; 15(2):125–129.
41. Matsumura SO, et al. Synergy testing of vancomycin-resistant *Enterococcus faecium* against quinupristin-dalfopristin in combination with other antimicrobial agents. Antimicrob Agents Chemother 1999; 43(11):2776–2779.

42. Visalli MA, et al. Determination of activities of levofloxacin, alone and combined with gentamicin, ceftazidime, cefpirome, and meropenem, against 124 strains of *Pseudomonas aeruginosa* by checkerboard and time-kill methodology. Antimicrob Agents Chemother 1998; 42(4):953–955.

43. Visalli MA, et al. Activities of three quinolones, alone and in combination with extended-spectrum cephalosporins or gentamicin, against *Stenotrophomonas maltophilia*. Antimicrob Agents Chemother 1998; 42(8):2002–2005.

44. Visalli MA, et al. Comparative activity of trovafloxacin, alone and in combination with other agents, against gram-negative nonfermentative rods. Antimicrob Agents Chemother 1997; 41(7):1475–1481.

45. Bajaksouzian S, et al. Activities of levofloxacin, ofloxacin, and ciprofloxacin, alone and in combination with amikacin, against acinetobacters as determined by checkerboard and time-kill studies. Antimicrob Agents Chemother 1997; 41(5):1073–1076.

46. den Hollander JG, et al. Synergism between tobramycin and ceftazidime against a resistant *Pseudomonas aeruginosa* strain, tested in an in vitro pharmacokinetic model. Antimicrob Agents Chemother 1997; 41(1):95–100.

47. Darras-Joly C, et al. Synergy between amoxicillin and gentamicin in combination against a highly penicillin-resistant and -tolerant strain of *Streptococcus pneumoniae* in a mouse pneumonia model. Antimicrob Agents Chemother 1996; 40(9):2147–2151.

48. Bajaksouzian S, et al. Antipneumococcal activities of cefpirome and cefotaxime, alone and in combination with vancomycin and teicoplanin, determined by checkerboard and time-kill methods. Antimicrob Agents Chemother 1996; 40(9):1973–1976.

49. Cappelletty DM, Rybak MJ. Comparison of methodologies for synergism testing of drug combinations against resistant strains of *Pseudomonas aeruginosa*. Antimicrob Agents Chemother 1996; 40(3):677–683.

50. Valiant ME, et al. L-658,310, a new injectable cephalosporin. II. In vitro and in vivo interactions between L-658,310 and various aminoglycosides or ciprofloxacin versus clinical isolates of *Pseudomonas aeruginosa*. J Antibiot (Tokyo) 1989; 42(5):807–814.

51. Perea EJ, et al. Interaction of aminoglycosides and cephalosporins against *Pseudomonas aeruginosa*. Correlation between interaction index and killing curve. J Antimicrob Chemother 1988; 22(2):175–183.

52. Chandrasekar PH, et al. Comparison of the activity of antibiotic combinations in vitro with clinical outcome and resistance emergence in serious infection by *Pseudomonas aeruginosa* in non-neutropenic patients. J Antimicrob Chemother 1987; 19(3):321–329.

53. Gordin FM, et al. Evaluation of combination chemotherapy in a lightly anesthetized animal model of *Pseudomonas pneumonia*. Antimicrob Agents Chemother 1987; 31(3):398–403.

54. Rusnak MG, et al. Single versus combination antibiotic therapy for pneumonia due to *Pseudomonas aeruginosa* in neutropenic guinea pigs. J Infect Dis 1984; 149(6):980–985.

55. Norden CW, et al. Comparison of techniques for measurement of in vitro antibiotic synergism. J Infect Dis 1979; 140(4):629–633.

56. Haller I. Comprehensive evaluation of ciprofloxacin-aminoglycoside combinations against *Enterobacteriaceae* and *Pseudomonas aeruginosa* strains. Antimicrob Agents Chemother 1985; 28(5):663–666.

57. Haller I. Comprehensive evaluation of ciprofloxacin in combination with beta-lactam antibiotics against *Enterobacteriaceae* and *Pseudomonas aeruginosa*. Arzneimittelforschung 1986; 36(2):226–229.

58. Houlihan HH, et al. Pharmacodynamics of vancomycin alone and in combination with gentamicin at various dosing intervals against methicillin-resistant Staphylococcus aureus-infected fibrin-platelet clots in an in vitro infection model. Antimicrob Agents Chemother 1997; 41(11):2497–2501.

59. Kang SL, Rybak MJ. Pharmacodynamics of RP 59500 alone and in combination with vancomycin against *Staphylococcus aureus* in an in vitro-infected fibrin clot model. Antimicrob Agents Chemother 1995; 39(7):1505–1511.

60. Angehrn P. In vitro and in vivo synergy between ceftriaxone and aminoglycosides against *Pseudomonas aeruginosa*. Eur J Clin Microbiol 1983; 2(5):489–495.

61. Sonne M, Jawetz E. Combined action of carbenicillin and gentamicin on *Pseudomonas aeruginosa* in vitro. Appl Microbiol 1969; 17(6):893–896.
62. Pankey GA. Antimicrob Agents Chemother 2005. In press.
63. Mordenti JJ, et al. Combination antibiotic therapy: comparison of constant infusion and intermittent bolus dosing in an experimental animal model. J Antimicrob Chemother 1985; 15(Suppl A):313–321.
64. Renneberg J. Definitions of antibacterial interactions in animal infection models. J Antimicrob Chemother 1993; 31(Suppl D):167–175.
65. Mouton JW, et al. Use of pharmacodynamic indices to predict efficacy of combination therapy in vivo. Antimicrob Agents Chemother 1999; 43(10):2473–2478.
66. Konig P, et al. Kill kinetics of bacteria under fluctuating concentrations of various antibiotics. I. Description of the model. Chemotherapy 1986; 32(1):37–43.
67. Kemmerich B, et al. Comparative evaluation of ciprofloxacin, enoxacin, and ofloxacin in experimental *Pseudomonas aeruginosa pneumonia*. Antimicrob Agents Chemother 1986; 29(3):395–399.
68. Aeschlimann JR, et al. Treatment of vancomycin-resistant Enterococcus faecium with RP 59500 (Quinupristin-dalfopristin) administered by intermittent or continuous infusion, alone or in combination with doxycycline, in an in vitro pharmacodynamic infection model with simulated endocardial vegetations [In Process Citation]. Antimicrob Agents Chemother 1998; 42(10):2710–2717.
69. Zinner SH, et al. Use of an in-vitro kinetic model to study antibiotic combinations. J Antimicrob Chemother 1985; 15(Suppl A):221–226.
70. Blaser J, et al. Impact of netilmicin regimens on the activities of ceftazidime-netilmicin combinations against *Pseudomonas aeruginosa* in an in vitro pharmacokinetic model. Antimicrob Agents Chemother 1985; 28(1):64–68.
71. McGrath BJ, et al. Pharmacodynamics of once-daily amikacin in various combinations with cefepime, aztreonam, and ceftazidime against *Pseudomonas aeruginosa* in an in vitro infection model [published erratum appears in Antimicrob Agents Chemother 1993 Apr; 37(4):930]. Antimicrob Agents Chemother 1992; 36(12):2741–2746.
72. McGrath BJ, et al. Pharmacodynamic effects of extended dosing intervals of imipenem alone and in combination with amikacin against *Pseudomonas aeruginosa* in an in vitro model. Antimicrob Agents Chemother 1993; 37(9):1931–1937.
73. Cappelletty DM, et al. Pharmacodynamics of ceftazidime administered as continuous infusion or intermittent bolus alone and in combination with single daily-dose amikacin against *Pseudomonas aeruginosa* in an in vitro infection model. Antimicrob Agents Chemother 1995; 39(8):1797–1801.
74. Zinner SH, et al. Activity of trovafloxacin (with or without ampicillin-sulbactam) against enterococci in an in vitro dynamic model of infection. Antimicrob Agents Chemother 1998; 42(1):72–77.
75. Kang SL, et al. Pharmacodynamics of levofloxacin, ofloxacin, and ciprofloxacin, alone and in combination with rifampin, against methicillin-susceptible and -resistant *Staphylococcus aureus* in an in vitro infection model. Antimicrob Agents Chemother 1994; 38(12):2702–2709.
76. Dudley MN, et al. Combination therapy with ciprofloxacin plus azlocillin against *Pseudomonas aeruginosa*: effect of simultaneous versus staggered administration in an in vitro model of infection. J Infect Dis 1991; 164(3):499–506.
77. den Hollander JG, et al. Use of pharmacodynamic parameters to predict efficacy of combination therapy by using fractional inhibitory concentration kinetics. Antimicrob Agents Chemother 1998; 42(4):744–748.
78. Ernst JD, Sande MA. Antibiotic combinations in experimental infections in animals. Rev Infect Dis 1982; 4(2):302–310.
79. Calandra T, Glauser MP. Immunocompromised animal models for the study of antibiotic combinations. Am J Med 1986; 80(5C):45–52.
80. Fantin B, Carbon C. In vivo antibiotic synergism: contribution of animal models. Antimicrob Agents Chemother 1992; 36(5):907–912.
81. Andriole VT. Synergy of carbenicillin and gentamicin in experimental infection with Pseudomonas. J Infect Dis 1971; 124(Suppl 124):46.

82. Andriole VT. Antibiotic synergy in experimental infection with Pseudomonas. II. The effect of carbenicillin, cephalothin, or cephanone combined with tobramycin or gentamicin. J Infect Dis 1974; 129(2):124–133.
83. Chin NX, et al. Synergy of ciprofloxacin and azlocillin in vitro and in a neutropenic mouse model of infection. Eur J Clin Microbiol 1986; 5(1):23–28.
84. Chadwick EG, et al. Correlation of antibiotic synergy in vitro and in vivo: use of an animal model of neutropenic gram-negative sepsis. J Infect Dis 1986; 154(4):670–675.
85. Fu KP, et al. Therapeutic efficacy of cefpiramide-ciprofloxacin combination in experimental Pseudomonas infections in neutropenic mice. J Antimicrob Chemother 1987; 20(4):541–546.
86. Fu KP, et al. Synergistic activity of apalcillin and gentamicin in a combination therapy in experimental *Pseudomonas bacteraemia* of neutropenic mice. J Antimicrob Chemother 1986; 17(4):499–503.
87. Johnson DE, Thompson B. Efficacy of single-agent therapy with azlocillin, ticarcillin, and amikacin and beta-lactam/amikacin combinations for treatment of *Pseudomonas aeruginosa* bacteremia in granulocytopenic rats. Am J Med 1986; 80(5C):53–58.
88. Johnson DE, et al. Comparative activities of piperacillin, ceftazidime, and amikacin, alone and in all possible combinations, against experimental *Pseudomonas aeruginosa* infections in neutropenic rats. Antimicrob Agents Chemother 1985; 28(6):735–739.
89. Lumish RM, Norden CW. Therapy of neutropenic rats infected with *Pseudomonas aeruginosa*. J Infect Dis 1976; 133(5):538–547.
90. McNeely DJ, et al. In vitro antimicrobial synergy and antagonism against Klebsiella species-what does it all mean? Clin Res 1976; 24:636A.
91. Pennington JE, Stone RM. Comparison of antibiotic regimens for treatment of experimental pneumonia due to Pseudomonas. J Infect Dis 1979; 140(6):881–889.
92. Pennington JE, Johnson CE. Comparative activities of N-formimidoyl thienamycin, ticarcillin, and tobramycin against experimental *Pseudomonas aeruginosa* pneumonia. Antimicrob Agents Chemother 1982; 22(3):406–408.
93. Scott RE, Robson HG. Synergistic activity of carbenicillin and gentamicin in experimental *Pseudomonas bacteremia* in neutropenic rats. Antimicrob Agents Chemother 1976; 10(4):646–651.
94. Winston DJ, et al. Antimicrobial therapy of septicemia due to *Klebsiella pneumoniae* in neutropenic rats. J Infect Dis 1979; 139(4):377–388.
95. Frimodt-Moller N, Frolund Thomsen V. Interaction between beta-lactam antibiotics and gentamicin against *Streptococcus pneumoniae* in vitro and in vivo. Acta Pathol Microbiol Immunol Scand [B] 1987; 95(5):269–275.
96. Peterson LR, et al. Comparison of azlocillin, ceftizoxime, cefoxitin, and amikacin alone and in combination against *Pseudomonas aeruginosa* in a neutropenic-site rabbit model. Antimicrob Agents Chemother 1984; 25(5):545–552.
97. Bamberger DM, et al. Ciprofloxacin, azlocillin, ceftizoxime and amikacin alone and in combination against gram-negative bacilli in an infected chamber model. J Antimicrob Chemother 1986; 18(1):51–63.
98. Fasching CE, et al. Treatment of ciprofloxacin- and ceftizoxime-induced resistant gram- negative bacilli. Am J Med 1987; 82(4A):80–86.
99. Chusid MJ, et al. Experimental *Pseudomonas aeruginosa* sepsis: absence of synergy between ticarcillin and tobramycin. J Lab Clin Med 1983; 101(3):441–449.
100. Haller I. Evaluation of ciprofloxacin alone and in combination with other antibiotics in a murine model of thigh muscle infection. Am J Med 1987; 82(4A):76–79.
101. Mimoz O, et al. Cefepime and amikacin synergy in vitro and in vivo against a ceftazidime-resistant strain of *Enterobacter cloacae*. J Antimicrob Chemother 1998; 41(3):367–372.
102. Mimoz O, et al. Cefepime and amikacin synergy against a cefotaxime-susceptible strain of *Enterobacter cloacae* in vitro and in vivo. J Antimicrob Chemother 1997; 39(3):363–369.
103. Renneberg J, et al. Interactions of drugs acting against *Staphylococcus aureus* in vitro and in a mouse model. J Infect 1993; 26(3):265–277.

104. Klastersky J, et al. Significance of antimicrobial synergism for the outcome of gram negative sepsis. Am J Med Sci 1977; 273(2):157–167.
105. Young LS. Review of clinical significance of synergy in Gram-negative infections at the University of California Los Angeles Hospital. Infection 1978; 6(Suppl. 1): s47–s52.
106. Anderson ET, et al. Antimicrobial synergism in the therapy of gram-negative rod bacteremia. Chemotherapy 1978; 24(1):45–54.
107. Klastersky J, Zinner SH. Combinations of antibiotics for therapy of severe infections in cancer patients. Infection 1980; 8(Suppl 1):229–233.
108. Fainstein V, et al. A randomized study of ceftazidime compared to ceftazidime and tobramycin for the treatment of infections in cancer patients. J Antimicrob Chemother 1983; 12(Suppl A):101–110.
109. De Jongh CA, et al. Antibiotic synergism and response in gram-negative bacteremia in granulocytopenic cancer patients. Am J Med 1986; 80(5C):96–100.
110. De Jongh CA, et al. A double beta-lactam combination versus an aminoglycoside-containing regimen as empiric antibiotic therapy for febrile granulocytopenic cancer patients. Am J Med 1986; 80(5C):101–111.
111. Klastersky J, et al. Clinical significance of in vitro synergism between antibiotics in gram-negative infections. Antimicrob Agents Chemother 1972; 2(6):470–475.
112. Aeschlimann JR, Rybak MJ. Pharmacodynamic analysis of the activity of quinupristin-dalfopristin against vancomycin-resistant *Enterococcus faecium* with differing MBCs via time-kill-curve and postantibiotic effect methods. Antimicrob Agents Chemother 1998; 42(9):2188–2192.

Section III: Antibacterial Agents

6 β-Lactam Pharmacodynamics

Zenzaburo Tozuka
Exploratory ADME, JCL Bioassay Corporation, Hyogo, Japan

Takeo Murakawa
Graduate School of Pharmaceutical Sciences, Kyoto University, Kyoto, Japan

INTRODUCTION

The target of antibiotic chemotherapy is bacteria. An in vitro study using the isolated bacterium predicts the efficacy of minimum inhibitory concentration (MIC), minimum bacterial concentration (MBC), post antibiotic effect (PAE), mutant prevention concentration (MPC), and time kill curve. The pharmacokinetics (PK) of clinical chemotherapy after administration of antibiotics in patients shows parameters such as maximum concentration (C_{max}) (peak), area under concentration-time curve (AUC), half-life, mean residence time, distribution volume (Vd) and clearance (CL). The relationship between PK and pharmacodynamics (PD) (PK/PD) shows parameters such as peak/MIC, AUC/MIC, and time above MIC ($T > MIC$). The β-lactam antibiotics effect by $T > MIC$ need the best-suited dosage (dose, dosing interval, and period). PK/PD with populationkinetics of β-lactam chemotherapy is a current clinical study to evaluate efficacy and solve ethical problems and damage by statistical treatment of two or three times point data at steady-state for drugs for which it is difficult to collect blood samples. This includes infants, the elderly, and serious illness-infected neonates, burn patients, and so on, on the base of population among participant patients in the mother group of pharmacokinetics data. Antibiotic chemotherapy has the important problems regarding the efficacy, safety, economy, and resistance of antibiotics. β-lactam chemotherapy is useful because of high efficacy and low toxicity.

History

After penicillin G, the original β-lactam antibiotic, was discovered by Fleming in 1928 and used for the first time in 1941 to treat a staphylococcal infection (1), many efforts were made to develop new antibiotics having better chemical and physical properties, better antimicrobial activity, a broader spectrum, better pharmacokinetic and pharmacodynamic properties, and less resistance by β-lactamase. The chemical modification of 6-aminopenicillanic acid produced new antibiotics. The structure deriver of 6-acyl is the most important modification to solve the above problem. The ester deriver of 1-carboxylic acid produces a prodrug for an oral formulation. Giuseppe Brotzu discovered cephalosporin at Sardinia in 1945 (2). The chemical modification of 7-aminocephalospollanic acid (7-ACA) produced new antibiotics. In addition to 7- and 3-derivatives, the modification of the cephalosporin nucleus produced new β-lactams.

The pharmacokinetic property β-lactam is their rapid elimination in urine as unchanged β-lactam after intravenous (i.v.) injection. The pharmacodynamic characteristic of β-lactam is a time-dependent killing above MIC or MBC. So the clinical application of β-lactam needs a slow distribution into the blood after intramuscular (i.m.) injection, a slow absorption from the intestinal tract after oral (p.o.) administration, or i.v. infusion for long duration of the antibiotic level above MIC. Initial success was obtained through the development of procaine penicillin G, benzathine penicillin G, penicillin V, phenethicillin, or propicillin. However, the antimicrobial activities of these compounds were limited mainly to gram-positive bacteria. Their antimicrobial spectrum was improved to include gram-negative bacteria. Ampicillin was the first derivative in this category, followed by hetacillin, ciclacillin, and amoxycillin. Later, ester derivatives of ampicillin were developed for increased oral absorption, such as bacampicillin, talampicillin, and pivampicillin. In addition, the antimicrobial spectrum was expanded to *Pseudomonas aeruginosa*. The first such synthesized compound was carbenicillin, followed by ticarcillin and mezlocillin. For oral use, esters of carbenicillin were developed. Continuous infusion (CI) of carbenicillin plus various antibiotics was effective in the treatment of febrile episodes in cancer patients (3).

After more than 60 years have passed since the development of penicillin G, more than 50 cephalosporins have been developed because of two major reasons: (*i*) many possible chemical modifications of the cephalosporin nuclei such as cephem, oxacephem, cephamycin, and so on; and (*ii*) strong bactericidal activity. On the basis of its antimicrobial profile, i.e., its activity and spectrum, the classification of cephalosporins by generations is very popular. The first generation is active against most common organisms such as Staphylococcus spp., Streptococcus spp., *Escherichia coli, and* Klebsiella spp. The second generation is similar to the first generation but has better activity against indole-positive Proteus and some rare organisms. The third generations are highly active against common gram-negatives and active against opportunistic pathogens. However, the activity against gram-positive bacteria is rather less.

PK/PD Parameters

The efficacy of antibiotics for the isolated bacterium is evaluated using such parameters as MIC, MBC, PAC, MPC, and time–kill curve by an in vitro study. The clinical chemotherapy of antibiotics shows such PK parameters as C_{max} (peak), AUC, half-life, mean residence time, V_d, and CL. PD shows the relationship between the pharmacological (or toxicological) effects and drug exposure. About two decades earlier, PK/PD parameters such as peak/MIC, AUC/MIC, and $T>$MIC were shown to evaluate the in vivo efficacy of antibiotics.

An animal PK/PD study is possible for searching the best-suited dosage (dose, dosing interval, and period), controlling peak/MIC, AUC/MIC, and $T>$MIC (β-lactam antibiotics) (4). The clinical antibiotic chemotherapy should be improved in accordance with the result of the animal PK/PD study considering the problems of efficacy, safety, economy, and resistance of antibiotics.

β-Lactam chemotherapy is useful because of its high efficacy and low toxicity. The β-lactam antibiotics effect by $T>$MIC that is possible to be changed by the best-suited dosage such as the divided administration (5), i.v. drip injection, and infusion. Especially, the CI of β-lactam antibiotics maintains the serum level above MIC, as shown by Craig in 1992 (6). In the case of patients with acute otitis media, $T>$MIC for multiple β-lactam antibiotics is effective for evaluating bacteriological cure (7–9). The relationship between $T>$MIC and the bacteriological cure rate for many β-lactam antibiotics against *Streptococcus pneumoniae* and *Haemophilus* in patients was shown by Craig et al. (10).

MECHANISM OF ACTION

The peptidoglycan layer of the bacterial cell wall plays an important role in maintaining the structural integrity of the cell wall. The mechanism of action of β-lactam antibiotics was well investigated (11) and it was found that the antibiotic causes damage following synthesis of the peptidoglycan layer by inhibition of murein-transpeptidase, which performs the final cross-linking of the nascent peptidoglycan layer. Inhibition of the antibiotic to the enzyme damages the peptidoglycan layer and then destroys the cell wall integrity (bacterial death). The β-lactam ring of the antibiotics is structurally similar to D-alanyl-D-alanine—the terminal amino acid residues on the precursor peptide subunits of the nascent peptidoglycan layer. Inhibition to murein-transpeptidase may lead to the activation of bacterial autolytic enzyme in the bacterial cell wall among gram-positive bacteria.

Antibacterial activity varies from β-lactam to β-lactam according to the following parameters:

Inhibitory activity (affinity) of β-lactam to the murein transpeptidase.
The activity of β-lactam antibiotics is limited to bacterial cells that have a peptidoglycan layer and there is no inhibitory activity to cell wall synthesis in human cells. This selectivity in the action mechanism accounts for the safety of β-lactam antibiotics.
Permeability of the antibiotic to the outer membrane of cell wall in gram-positive bacteria.
In gram-positive bacteria, the cell wall is usually simple (thick peptidoglycan layer), but in gram-negative bacteria, the cell wall is more complex and the peptidoglycan layer is surrounded by an outer layer consisting of a hydrophobic lipopolysaccharide and lipoprotein-phospholipid. The β-lactam antibiotic must have access to the inner layer, the peptidoglycan layer. The accessibility (permeability) of β-lactam antibiotics depends on molecule size, hydrophobicity, and electrical charge.
Resistance to enzymatic hydrolysis by bacterial β-lactamase that hydrolyzes the β-lactam ring of the antibiotic.
In gram-negative bacteria, β-lactam antibiotic must pass through the outer membrane and then reach the peptidoglycan layer without inactivation by the β-lactamase located at periplasmic space (space between outer membrane + peptidoglycan layer and cytoplasmic membrane).

ANTIMICROBIAL ACTIVITY

β-lactam antibiotics are active only to bacteria. In the development of β-lactam antibiotics, most of the effort is focused on the following points:

Broad spectrum
High activity (low MIC)
Activity against resistant bacterial strains

The antibacterial spectrum differs from drug to drug. Penicillin G is active mainly against gram-positive bacteria. After penicillin G, many kinds of β-lactam antibiotics such as penicillin (penam, oxapenam, and carbapenem), cephalosporin (cephem, oxacephem, and carbacephem), and monobactam are synthesized. More than 100 compounds are marketed. There may be a suitable drug among β-lactam antibiotics susceptible to bacterial pathogens.

MECHANISM OF RESISTANCE

Bacterial resistance to β-lactam antibiotics is caused by the following three mechanisms (12):

1. *Inactivation of β-lactam antibiotics by β-lactamase produced by the bacteria (13).* Genes for β-lactamase production are located in the chromosome and/or plasmid, which is an extra chromosomal element of DNA that replicates within bacterial cell. Genes for β-lactamase are commonly carried on plasmids. The plasmids are transferred from one bacterial cell to the other bacterial cell when they come in contact. The production of β-lactamase is inducible, and induction may result in a highly increased level of β-lactamase. The ability to induct varies from one β-lactam antibiotic to another:
2. *Alteration of target site (14).* The alteration results in reduced affinity between the drug and murein transpeptidase, or the so-called penicillin-binding proteins.
3. *Change of outer membrane to reduce permeability of the drug (15).*

Susceptibility of bacteria to antibiotics is not homogeneous among bacterial cells. To determine MIC, about 10^6 cells/mL (broth dilution method) or 10^4 to 10^5 cells of bacteria is used, and then MIC means concentration that kills highest resistant cell among the tested cell population. The susceptibility distribution of cell population may be stable at a bacterial growth circumstance. However, exposure to antibiotics may lead to change of susceptibility distribution of bacterial cells. If a resistant mutant appears, with exposure, the drug leads to increase of the mutant cells, and the susceptibility distribution of cell population will be altered to resistance. In treating patients, inadequate dosage of antibiotics (long-term treatment at insufficient concentrations to kill bacteria) may cause appearance of a resistant mutant, increase of bacterial resistance, and spread of resistant bacteria.

SUMMARY OF PHARMACOKINETIC PROPERTIES

The pharmacokinetic property of β-lactam antibiotics is their rapid elimination in urine as unchanged β-lactam antibiotics after i.v. injection. A one-compartmental i.v. model provided the best fit to the observed serum concentration data after i.v. injection of penicillin antibiotics. The model was defined by the following equation:

$$C_t = Ae^{-\alpha t}$$

C_t is the serum concentration at time t, e represents the base of the natural logarithm, A is the coefficient of the elimination phase, and α is the elimination phase rate constant. The calculated pharmacokinetic parameter of half-life $t_{1/2\alpha}$ is very short: 0.6 to 1.0 hour (16).

After i.v. injection of cephalosporin antibiotics, a two-compartmental i.v. model provided the best fit to the observed serum concentration data. The model was defined by the following equation:

$$C_t = Ae^{-\alpha(t-T)} + Be^{-\beta(t-T)}$$

C_t is the serum concentration at time t, T is the lag time, e represents the base of the natural logarithm, and A and B are the coefficient of the distribution phase and elimination phase, respectively. α and β are the rate constant of the distribution phase and elimination phase, respectively. The calculated pharmacokinetic parameter of the distribution phase half-life $t_{1/2} \times t_{1/2\alpha}$ and the elimination phase half-life $t_{1/2\beta}$ ranged from 0.01 to 2.13 hours and from 0.5 to 5.1 hours, respectively (17).

If the protein binding of β-lactam antibiotics is larger than 90%, then they have longer half-life. The serum protein binding of cefotetan, cefpiramide, cefonicid, and ceftriaxone is larger than 90% and their half-life is more than three hours. The protein binding of many β-lactam antibiotics is less than 90% and their half-life is less than three hours (18). The longer half-life of β-lactam antibiotics is related to the duration of β-lactam antibiotic level in the biological fluid and affects the antibiotic activity. However, the free fraction of β-lactam antibiotics is related to antibiotic activity and distribution volume.

The slow release formulations and slow absorption after p.o. administration, the slow distribution after i.m. injection, and the i.v. infusion for long periods affect the duration of β-lactam antibiotics in biological fluid in comparison with i.v. injection of β-lactam antibiotics. After i.m. injection or p.o. administration of penicillin and cephalosporin antibiotics, a one-compartmental i.m. or p.o. model provided the best fit to the observed serum concentration data. The model was defined by the following equation:

$$C_t = A(e^{-\alpha(t-T)} - e^{-\beta(t-T)})$$

T is the lag time, and α and β are the rate constants of the absorption phase and elimination phase, respectively. In the case of p.o. administration of penicillin antibiotics, the calculated pharmacokinetic parameter of the absorption phase half-life $t_{1/2\alpha}$ and the elimination phase half-life $t_{1/2\beta}$ are very short: 0.4 to 2.0 hours and 0.02 to 0.8 hour, respectively (16). In the case of i.m. injection of penicillin antibiotics, the calculated pharmacokinetic parameter of the absorption phase half-life $t_{1/2\alpha}$ and the elimination phase half-life $t_{1/2\beta}$ are very short: 0.7 to 2.5 hours and 0.03 to 0.29 hours, respectively (16). In the case of p.o. administration of cephalosporin antibiotics, the calculated pharmacokinetic parameter of the absorption phase half-life $t_{1/2\alpha}$ and the elimination phase half-life $t_{1/2\beta}$ ranged from 0.6 to 8.0 hours and from 0.3 to 1.9 hours, respectively (17). In the case of i.m. injection of cephalosporin antibiotics, the calculated pharmacokinetic parameter of the absorption phase half-life $t_{1/2\alpha}$ and the elimination phase half-life $t_{1/2\beta}$ are very short: 0.7 to 2.7 hours and 0.04 to 0.86 hours, respectively (17).

The putative molecular functions of 26,383 human genes contain 533 (1.7%) transporters (19). Recently many transporters were discovered. The influx transporter PEPT1 in the gut was discovered (20) and was found to mediate the transport of peptide-like drugs such as β-lactam antibiotics (21), especially after the timely discovery of third-generation oral cephalosporins (22). The efflux organic anion transporters OAT1 and OAT3 in the kidneys were discovered (23) and were found to mediate the transport of β-lactam antibiotics (24). Cephalosporins and β-lactam antibiotics accumulate extensively at OAT1 via proximal tubule–induced nephrotoxicity (25). Multidrug resistance–associated protein 2 on the bile canalicular membrane mediated the biliary excretion of cefodizime and ceftriaxone (β-lactam antibiotics) (26).

The clinically useful dosing of β-lactam antibiotics is by i.v. infusion. After i.v. infusion of penicillin antibiotics, the one-compartmental i.v. infusion model provided the best fit to the observed serum concentration data. The model was defined by the following equation:

Infusion phase ($0 < t < T$):

$$C_t = A((1 - e^{-\alpha t})/(1 - e^{-\alpha T}))$$

Postinfusion phase ($t > T$):

$$C_t = Ae^{-\alpha(t-T)}$$

T is the infusion time in hours, and it is possible to set a long-enough duration of penicillin antibiotics level above MIC.

After i.v. infusion of cephalosporin antibiotics, the two-compartmental i.v. infusion model provided the best fit to the observed serum concentration data. The model was defined by the following equation:

Infusion phase ($0 < t < T$):

$$C_t = A(1 - e^{-\alpha t})/(1 - e^{-\alpha T}) + B(1 - e^{-\beta T})/(1 - e^{-\beta T})$$

Postinfusion phase ($t > T$):

$$C_t = Ae^{-\alpha(t-T)} + Be^{-\beta(t-T)}$$

T is the infusion time in hours, such as 9.15 to 130.4 hours (17), and it is possible to set a long-enough duration of antibiotics level above MIC.

PK–PD OF THE CLASS (NONCLINICAL MODELS)

The in vivo data from animals correlate with observations from in vitro experiments. Using a mouse model, Gerber et al. (27) correlated bacterial regrowth to the $T > \text{MIC}$. When ticarcillin was administered as either a single bolus or a fractionated dose (more closely simulating human PK, resulting in similar AUCs), neutropenic animals were found to have reduced bacterial growth, with more frequent dosing extending the $T > \text{MIC}$. The extremely large peak/MIC ratio produced from a bolus dose (20:1) did not reduce the bacterial count more than the fractionated doses (5:1). This same study evaluated the effect of these two dosing schemes in normal mice. Unlike previous results, the bacterial regrowth with the bolus dose was not demonstrated, suggesting that the effect of the combination with antibiotic and white blood cells was significant in suppressing *P. aeruginosa*. No attempt was made to determine the optimal $T > \text{MIC}$ required to produce a good outcome, but an examination of these data shows that concentrations during bolus dosing were above the MIC for only 25% of the dosing interval. The required $T > \text{MIC}$ for obtaining the best outcome may vary for individual antibiotic–pathogen combinations. A univariate analysis of ticarcillin, cefazolin, and penicillin all demonstrated that the $T > \text{MIC}$ was the most important parameter in determining outcome as opposed to the AUC or peak concentrations (27). To produce a bactericidal effect, *E. coli* required a longer exposure to cefazolin (>60% vs. 20%) compared to *Streptococcus aureus*. This large difference is the result of a significant post antibiotic effect (PAE) of cefazolin for *S. aureus*. This evidence supports earlier work with cephalosporins and *S. pneumoniae* that found a relationship between $T > \text{MIC}$ and the effective dose required for protection of 50% of the animals (28). Studies comparing normal and neutropenic animals can show the pharmacodynamic influence of host defenses. As described previously, the work of Gerber et al. (29) demonstrated that normal mice could suppress bacterial growth regardless of the optimal dosing scheme. Roosendaal et al. (30) also discovered a striking difference between normal and leukopenic rats. For normal rats, the intermittent (every six hours) and continuous administration of ceftazidime required equivalent doses (0.35 and 0.36 mg/kg, respectively) to reach the protective dose for 50% (PD_{50}) of the animals. For leukopenic rats, the CI required 3.75 mg/kg to produce the PD_{50}, whereas 30 mg/kg of ceftazidime was needed with the intermittent dosing (II). This study highlights two important differences between normal and neutropenic animals: (*i*) Host defenses can overcome the deficiencies when the PD is not maximized and (*ii*) neutropenic hosts require larger doses than normal hosts to eradicate organisms.

PK–PD OF THE CLASS (CLINICAL)

The pharmacodynamic characteristic of β-lactam antibiotics is their time-dependent killing. A plasma concentration of β-lactam antibiotics above MIC or MBC relates with their interstitial and intracellular level to kill the bacterial pathogens at the site of many infections. The slow absorption and distribution phase of antibiotics after p.o. administration expand the duration time of antibiotics in the plasma in comparison with the rapid elimination phase of antibiotics after i.v. administration. Distribution of β-lactam antibiotics to lymph, cerebrospinal fluid (CSF), tear, sweat, and cellular fluid affects the efficacy against infection. The lipophilicity of synthetic β-lactam antibiotics relates with distribution volume. Urinary or biliary excretion of β-lactam antibiotics affects urinary or biliary infection respectively. The CI prolongs the duration time of antibiotics level above the MIC in the plasma.

Absorption of antibiotics after p.o. administration affects the plasma concentration of antibiotics. The substitution group of synthetic β-lactam antibiotics relates with oral absorption. Esterification of carboxylic acid of β-lactam antibiotics (prodrug) increases oral absorption. The slow-release formulations and slow absorption after p.o. administration and the slow distribution after i.m. injection or the i.v. infusion for a long period affect the duration of β-lactam antibiotics in biological fluid in comparison with i.v. injection of β-lactam antibiotics.

X: non	Y: H	Penicillin G
O	H	Penicillin V
O	CH_3	Phenethicillin
O	CH_2CH_3	Propicillin

After i.m. injection of 500 mg of penicillin G in healthy young men, a mean maximum serum level of 5.03 µg/mL was observed one hour after dosing and it rapidly decreased to 0.75 µg/mL after five hours. About 47.6% of the unchanged drug was excreted in the urine within five hours. After p.o. of 600 mg, the serum levels were less than 1 µg/mL, and the urinary recovery rate was 9% in the first five hours (31). Renal tubular excretion of the drug is partially blocked by probenecid, and the serum levels of penicillin G approximately doubled with the concomitant use of probenecid. The levels in lymph after i.v. injection were reported. The concentrations in both lymph and plasma were approximately equal by four hours (32). Penicillin G diffuses poorly into the CSF with uninflamed meninges. Procaine penicillin G was developed as one of the slow-release formulations of penicillin G using the less water-soluble nature of this compound. After an i.m. dose of 300,000 units of an aqueous suspension, peak serum levels were maintained for 48 hours at the level of 0.05 µg/mL. Benzathine penicillin G (benzylpenicillin) is also one of the slow-release formulations of penicillin G. Following an i.m. injection of 300,000 units, the serum levels of 0.03 to 0.05 unit/mL were maintained for several days. After an oral dose of 400,000 units, 0.02 to 1.25 unit/mL in serum was detected at one hour and decreased to 0.02 to 0.11 unit/mL at six hours after dosing (33). After p.o. administration of 250 mg of penicillin V (phenoxymethyl penicillin), the mean serum concentrations were 1.46, 0.38, and 0.03 µg/mL at one, two, and four hours after dosing, respectively. A mean of 50% of the dose of this drug was excreted in the zero-to-six hour urine samples (31). Another study was reported with the serum levels and urine excretion of 500 mg of penicillin V before and after breakfast. A mean peak serum concentration of 3.78 µg/mL at half an hour and urinary excretion of 32% in the first five hours were observed before the meal. After the meal, a mean peak serum concentration of 4.63 µg/mL occurred at one hour after dosing, and urinary excretion of 24% was observed (34).

After p.o. of 250 mg of phenethicillin (phenoxyethyl penicillin) solution to healthy volunteers, a mean peak serum concentration of 4.06 µg/mL was obtained at one hour after dosing followed by 1.16 and 0.11 µg/mL at two and four hours, respectively. A mean of 64% of the dose was excreted in the urine over a zero- to six-hour period. After 500 mg of the dose to the same subjects, a mean peak serum level of 8.7 µg/mL was obtained one hour after dosing. It was concluded that doubling the dose produced at least double the serum concentration (31). There were no differences in PK between the serum levels before and after meals (34). One of the major advantages with this compound is the production of serum concentration twice as high as that of penicillin V after an equivalent oral dose. After p.o. of 250 mg of propicillin (phenoxypropyl penicillin), the levels of penicillin in serum reached 3.58 µg/mL at one hour followed by 2.15, 0.49, and 0.11 µg/mL at two, four, and six hours, respectively. About 71.9% of the dosed drug was recovered in urine within eight hours (35).

Oral cephalosporins, cephaloglycine, with a phenylglycine moiety at the seventh position and methyl group at the third position of the nuclei, have been developed as well as different chemical structures such as cefaclor and cefatrizine. Several prodrugs of parenteral cephalosporins such as cefuroxime axetil and cefotiam hexetil have become available for clinical use. Nonprodrug oral cephalosporins with high antibacterial activity such as third-generation cephalosporins have also become available.

X: H	Y: CH3	Cephalexin
H	Cl	Cefaclor
H	CH2S-(triazole)	Cefatrizine

Cephalexin type (phenylglycine or phenylglycine-like moiety at the seventh position and methyl or methoxy group at the third position) derivatives are well absorbed orally, but their antibacterial activity is generally less than that of ampicillin, and their spectrum is limited mainly by activity against common pathogens. Cefatrizine and cefaclor are oral cephalosporins with similar antimicrobial activity to ampicillin or amoxicillin. They are not stable in solutions of neutral pH. To determine drug concentrations in serum, it is necessary that the sample should be adjusted to an acidic pH. The half-lives of oral cephalosporins vary from 0.4 to 2.6 hours. Regarding dosage regimen, most of the compounds are applied two or three times a day.

| X : H | Y: CH=C$_2$H | Cefdinir |
| CH2COOH | CH=CH$_2$ | Cefixime |

The kinetic profile of cefixime makes it possible to use this agent once a day. At present, there is no oral cephalosporin that can be used to treat severe infections in place of parenteral cephalosporins. A breakthrough would be an oral cephalosporin with high bioavailability and a rather long half-life so that it would be possible to achieve high serum concentrations (for example 10–20 pg/mL for one to two hours). This ensures that sufficient tissue and fluid penetration will be achieved. These drugs are mainly excreted into urine. The compounds with a similar chemical structure to cephalexin are highly absorbed and rapidly excreted into urine. Cefaclor and cefatridine are also well absorbed. Newer oral cephalosporins have intermediate or moderate oral absorption, i.e., 20% to 50% of urinary recovery. The biliary excretion rate is low, but biliary concentrations for most compounds are higher than serum concentrations.

Pharmacodynamic characteristic of β-lactam antibiotics is time-dependent killing. Plasma concentration of β-lactam antibiotics above MIC or MBC relates with their interstitial and intracellular level to kill the bacterial pathogens at the site of many infections. The slow absorption and distribution phase of antibiotics after p.o. administration expands the duration time of antibiotics in the plasma in comparison with the rapid elimination phase of antibiotics after i.v. administration. The CI prolongs the duration time of antibiotics level above the MIC in the plasma.

Ampicillin was developed as a new type of broad-spectrum penicillin. After p.o. administration of ampicillin, a peak serum concentration is obtained between one and two hours. There are many reports regarding the bioavailability of ampicillin, but there is much difference on serum levels in each report. After the p.o. administration of 500 mg, serum levels reached 4.1 µg/mL one hour after dosing and the drug was still detectable in serum after eight hours. About 50.4% of the drug was recovered in urine within 24 hours (36). Another study reported that after a 250 mg oral dose, a serum concentration of 1.7 µg/mL was observed at the peak and around 20% of the drug was recovered in urine within six hours (37). It was reported that when doubling the dose, the serum concentration doubles (38). The pharmacokinetic parameters after an oral dose of 500 mg in the tasted state were k_a 0.58, C_{max} 5.9 µg/mL, t_{max} 1.49 hours, AUC 19.8 h.µg/mL, and k_e 0.6. After a meal, the parameters were k_a 0.57, C_{max} 4.6 µg/mL, t_{max} 2.48 hours, AUC 13.7 h.µg/mL, and k_e 0.75. Urinary recovery over eight hours was 37.1% at tasting and 26.8% after eating. The results showed that ampicillin absorption was significantly decreased in the nontasting state (39). The oral absorption is limited in this compound and urinary excretion rate varied between 25% and 50%. After i.v. administration of 500 mg, a mean serum concentration of 25.8 and 16.8 µg/mL was observed at half an hour and one hour after dosing, respectively. And the drug cannot be detected in serum after four hours (40). Another author reported that ampicillin penetrated blister fluid rapidly after i.v. dosing and the ampicillin levels in serum and blister fluid were approximately equal at one hour. The rate of elimination from blister fluid was similar to that from serum (41). After an oral dose of 500 mg, saliva and sweat contained no detectable ampicillin, and tears contained a small amount of drugs (0.07–0.65 µg/mL) (36). The levels in both lymph and plasma were approximately equal by four hours after i.v. administration. The concentration of ampicillin in lymph was higher than that of benzylpenicillin at all times after injection (32). Based on the knowledge of ampicillin, several derivates were synthesized such as hetacillin, ciclacillin, amoxicillin, mecillinam, and epicillin. Also, many ester forms of ampicillin were developed as prodrugs to improve the oral absorption of ampicillin. These compounds are split off in the body to ampicillin and have ampicillin's antimicrobial activity. The merit of these compounds is good absorption through the intestinal tract compared with that of ampicillin.

POPULATION PK–PD

Population PK–PD study of antibiotic was investigated in a clinical study using nonlinear mixed-effects modeling (NONMEM), a mathematical/statistical analysis that considers the population study sample as a unit of analysis for the estimation of the distribution of parameters and their relationships with covariates within the population. NONMEM provides estimates of population characteristics that define the population distribution of the PK-PD parameter. Serum and plasma samples are collected from all or a special part of patients two or three times at steady state for drugs and are evaluated on the basis of population to predict quantitatively the individual variant of efficacy/dose among participant patients. Decreasing the collection times of blood samples solves ethical problems and the damage of many-times collection of blood from infected neonates burn patients and so on. The population PK-PD study gives a good result.

A multicenter study investigated the population PK-PD of piperacillin and tazobactam injected by either CI (13.5 g over 24 hours, $n = 130$) or intermittent infusion (3.375 g every six hours, $n = 132$) in hospitalized patients with complicated intra-abdominal infection (42). NONMEM was used to perform population pharmacokinetic analysis in a subset of patients ($n = 56$) who had serum samples obtained at steady state (mean steady-state concentration 35.31 ± 12.15 mg/L for piperacillin and 7.29 ± 3.28 mg/L for tazobactam). Classification and regression tree analysis was used to identify the breakpoints of piperacillin PK-PD indexes in 94 patients with causative pathogen's MIC. Creatine clearance and body weight (BW) were the most significant variables to explain patient variability in piperacillin and tazobactam clearance and distribution volume (42). Population PK-PD of cefepime infused (2 g every eight hours, 2 g every 12 hours, CI 4 g over 24 hours) against 1000 patients with *P. aeruginosa* by medical and surgical intensive care unit in the United States (43). CI over 24 hours offered the most promising pharmacodynamic target, attaining 65% to 81% ($p < 0.001$) (43). Population PK-PD of isepamicin and broad-spectrum β-lactam did not provide useful information in 196 intensive-care-unit patients with nosocomial pneumonia (44). Population PK of cefozopran in neonatal infections using NONMEM indicates that the elimination of cefozopran depends on the postnatal age and is approximately 38% lower in the younger group than in the older group. Dosing of cefozopran is necessary, particularly in prolongation of intervals of administration, in cases of postnatal age of one day or less (45). Population PK of cefepime dosed at 30 mg/kg/dose every 12 hours in neonates with infections less than 14 days of age using NONMEM provide antibiotic exposure equivalent to or greater than 50 mg/kg every eight hours in older infant and children (46). Population PK of panipenem dosed at 10 to 20 mg/kg every 12 hours in neonates should yield a concentration within the accepted therapeutic range (47). Population PK of arbekacin ($n = 41$), vancomycin ($n = 19$), or papipenem ($n = 23$) in neonates of the postconceptional ages (PCAs) from 24.1 to 48.4 weeks, and the BWs from 458 to 5200 g using NONMEM found that the mean clearance for subjects with PCAs of <33 to 34 weeks was significantly smaller than those with PCAs of ≥33 to 34 weeks, and clearance showed an exponential increase with PCA. Many antibiotics are excreted by glomerular filtration, and maturation of glomerular filtration is the most important factor for estimation of antibiotic clearance. Clinicians should consider PCA, serum creatinine level, BW, and chemical feature in determining the initial antibiotic dosing regimen for neonates (48). Population PK of imipenem in patients with burns ($n = 47,118$ samples) using NONMEM integrated a linear–inverse relationship between imipenem clearance and creatinine plasma level. The estimates of imipenem clearance (16.37 ± 0.204 L/hr) and of the distribution volume of the central compartment (0.376 ± 0.039 L/kg) are higher in the population of patients with burns than the estimates in healthy subjects (49). Population PK of ceftazidime in patients with burns ($n = 41$ patient, 94 samples) using NONMEM shows no relationship between covariates and PK parameters was established with the exception of a linear–inverse relationship between ceftazidime total clearance and creatinine plasma level. The lower ceftazidime clearance could be explained by the relative decrease in ceftazidime elimination in relation to the burn area, and the higher ceftazidime distribution volume in the presence of interstitial edema, which could act as a reservoir from which ceftazidime returns slowly to the circulation (50).

ECONOMY OF ANTIBIOTIC CHEMOTHERAPY

The goal of antibiotic therapy is to achieve the best possible clinical outcomes while consuming the least amount of hospital resources. Health-care systems are under intense pressure to increase quality of care and at the same time reduce costs. Pressure to reduce the cost of antimicrobial therapy is especially intense because these drugs may account for up to 50% of a hospital pharmacy budget. Although β-lactam antibiotics have traditionally been given by intermittent infusion, administration by CI is gaining popularity because it takes full advantage of the known PD of the β-lactams and potentially consumes the least amount of hospital resources. Other options that will also maintain the $T > MIC$ for the entire dosing interval are the use of antibiotics with longer half-lives or more frequent, larger doses. An advantage of CI, however, is that it can be given at a lower total daily dose than standard dosing schedules. Although the i.v. drip method used in the past was cumbersome and inaccurate, today, because of lightweight, portable, and accurate infusion pumps, a CI is relatively easy and cost-effective to administer. Giving a loading dose prior to starting the CI will bring the antibiotic concentration into the therapeutic range immediately and minimize any lag time in drug tissue equilibration. Constant infusion of β-lactam antibiotics has the potential for appreciable cost reductions because it may represent the best method to maintain levels above the MIC during the entire dosing interval using the least amounts of drug, labor, and supplies.

A number of in vitro and animal models substantiate the equivalent or superior efficacy of CI of β-lactams compared to standard dosing. These models have been used extensively to elucidate the PD of β-lactams because they hold a significant advantage over studies in humans. The doses and dosing intervals of antibiotics are easily varied, reducing the interdependence of the pharmacodynamic parameters. Using an in vitro model, investigators have demonstrated that a CI at $5 \times$ MIC of P. aeruginosa is as efficacious as II using less total daily drug (51,52). Experiments in animal models have confirmed these in vitro results and demonstrated further that a CI may be the better dosing strategy, if the same total daily dose is used (53,54).

The mouse thigh model has been used to evaluate the difference between neutropenic and nonneutropenic host response (55). The real difference in efficacy between CI and II shows up in the neutropenic host: Even with a lower total daily dose, CI provides much better efficacy than bolus dosing (56,57). In neutropenic rats with gram-negative infection, a ceftazidime bolus dose that protects 50% of the animals from death (PD_{50}) had to be 65-fold higher than the PD_{50} dose for a CI of the same drug. Once again, this experiment confirms that $T > $MIC correlates to efficacy for the β-lactams, and when host defenses are not present, the β-lactam concentration should exceed the MIC of the pathogen for the entire dosing interval (58).

CI is a practical way to maintain 100% $T >$MIC with less total daily drug (e.g., 3 g/24 hr CI of ceftazidime vs. 1–2 g every eight hours). Although clinical efficacy data are sparse, the PD of CI is well characterized in both normal volunteers and critically ill patients, and several small clinical trials have shown the equivalence of CI and standard dosing (30,59–62). In one of the first clinical trials of CI, Bodey et al. (67) compared the efficacy of cefamandole dosed as either a CI or an intermittent dose given in combination with carbenicillin. There was no significant difference in clinical cure between the two regimens. By subgroup analysis, patients with persisting, severe neutropenia had a better clinical outcome (65% vs. 21%, $p = 0.03$) with CI (63). In a study of CI benzylpenicillin versus daily i.m. procaine penicillin G in 123 patients with pneumococcal pneumonia, there was also no difference in clinical cure rates (64). A nonrandomized trial of continuous versus II of cefuroxime showed that the CI results in a lower total antibiotic dose. The CI results in equivalent efficacy, shorter length of hospital stay, and overall cost savings to the institution (65).

At Hartford Hospital, there have been two CI β-lactam clinical studies. The investigators compared cefuroxime given either as a CI of 1500 mg or intermittently as 750 mg every eight hours ($n = 25$ in each group) to treat hospitalized community-acquired pneumonia (CAP) (66). Steady-state cefuroxime serum concentrations were 13.25 ± 6.29 µg/mL, more than two to four times the MIC_{90} of typical CAP pathogens. There was no difference in clinical cure rates, but the CI regimen was associated with a shorter length of treatment, decreased length of stay, lower total cefuroxime dose, and overall cost savings. The average amount of i.v. cefuroxime per patient decreased significantly ($p = 0.04$) from 8.0 ± 3.4 g for II to 5.9 ± 3.2 g for the CI. The average daily costs (including antibiotics, labor, and supplies) decreased significantly ($p = 0.04$), $\$63.64 \pm 30.95$ for the CI compared with $\$83.85 \pm 34.82$ for II.

The second Hartford Hospital study was a prospective, randomized trial of the efficacy and economic impact of ceftazidime given as either a CI (3 g/day) or an II (2 g q8hr) plus once-daily tobramycin for the treatment of nosocomial pneumonia in the ICU (59). The investigators evaluated 35 patients, 17 in the CI group and 18 in the II group. Clinical efficacy did not vary significantly between groups (94% success in CI vs. 83% in II). Number of adverse events, duration of treatment, and total length of hospital stay also did not vary significantly. The CI regimen used half of the intermittent dose and maintained concentrations above the MIC of the pathogen for 100% of the dosing interval. The II regimen maintained concentrations above the MIC for 76% of the dosing interval. The costs (including drug acquisition, antibiotic preparation and administration, adverse events, and treatment failures) associated with the CI of ceftazidime, $\$625.69 \pm 387.84$, were significantly lower ($p = 0.001$) than with the II, $\$1004.64 \pm 429.95$.

There are some potential disadvantages to giving antibiotics by CI. For patients with limited i.v. access, the CI may require their only i.v. line. Drug compatibility will be an issue if two drugs must be infused simultaneously through one line. Although the cost of infusion pumps should be considered in any cost-effectiveness analysis of drug delivery by CI, since most manufacturers will provide the use of the infusion pumps with a supply contract, the cost to the hospital is usually limited to the price of the administration sets. Overall, the advantages of the CI, both pharmacodynamically and economically, far outweigh the disadvantages. The CI of $β$-lactams in place of frequent II is a good example of how the knowledge and application of pharmacodynamic concepts can lead to cost-effective antibiotic therapy.

CONCLUSION

Infectious diseases are now estimated to account for 15 million (>25%) of 57 million annual deaths worldwide by the World Health Organization. This value does not include the additional millions of deaths that occur as a consequence of past infections (for example, streptococcal rheumatic heart disease), or because of complications associated with chronic infections, such as liver failure and hepato-cellular carcinoma in people infected with hepatitis B or C viruses (67). Emerging infections can be defined as infections that have newly appeared in a population or have existed previously but are rapidly increasing in incidence or geographic range, such as severe acute respiratory syndrome in 2003, acquired immune deficiency syndrome in 1981, the 1918–1920 influenza pandemics, and the small-pox. The identification of specific microbes of infectious diseases and the development of vaccines and antibiotics led to enormous progress. By the 1950s, the widespread use of penicillin, the development of polio vaccines, and the discovery of drugs for tuberculosis had set in (68), and in 1967, the USA Surgeon General stated that the war against infectious disease has been won (69).

Intestinal β-lactamase metabolize oral β-lactam agents such as cefixime and the metabolites are absorbed into the body. The metabolite was studied by mass spectrometry as exemplified by the metabolism of cefixime (70). The structures of M4 and M5 were detected in biological samples such as serum, urine, and feces and identified by comparison to synthesized authentic samples. M3 was detected only in feces and is a key compound to study on the metabolism of cephalosporins, but it was difficult to determine the chemical structure because of a mixture between aldehyde and acetal. NMR spectra of the mixture showed a small amount of the aldehyde proton and the counter amount of the acetal proton at the field of the specific chemical shift. It was determined as dinitrophenylhydrazone of which secondary ion mass spectrometry SIMS showed m/z 467 as the quasi-molecular ion.

Ethnic differences of valid biomarkers such as CYP2C9, 2C19 and 2D6 affect plasma concentration of drugs (71) and their metabolites that are related with FDA Guidance 2005 "Safety testing of metabolites" (72). SNIP of CYP and transporter genes need genotype diagnosis for tailor-made chemotherapy. Genomics, proteomics and metabolomics of bacterium are important to develop new antibiotics to be studied by microdose clinical studies (73) or exploratory IND studies (74).

β-Lactamases inactivate beta-lactam antibiotics and are a major cause of antibiotic resistance. The recent outbreaks of *Klebsiella pneumoniae* carbapenem-resistant (KPC) infections mediated by KPC type β-lactamases are creating a serious threat to our "last resort" antibiotics, the carbapenems. KPC β-lactamases are serum carbapenemases and are a subclass of class A β-lactamases that have evolved to efficiently hydrolyze carbapenems and cephamycins which contain substitutions at the alpha-position proximal to the carbonyl group that normally render these β-lactams resistant to hydrolysis. To investigate the molecular basis of this carbapenemase activity, we have determined the structure of KPC-2 at 1.85 Å resolution (75).

REFERENCES

1. Fleming A. On the antibacterial action of cultures of a penicillium, with special reference to their use in the isolation of B influenzae. Br J Exp Pathol 1929; 10:226.
2. Abraham EP, Loder PB. Isolation of the cephalosporium sp. in Sardinia. In: Flynn EH, ed. Cephalosporins and Penicillins. Chemistry and Biology. New York & London: Academic Press, 1972:3–5.
3. Bodey GP, Ketchel SJ, Rodrigues N. A randomized study of carbenicillin plus cefamandole or tobramycin in the treatment of febrile episodes in cancer patients. Am J Med 1979; 67:608–616.
4. Craig WA, Interrelationship between pharmacokinetics and pharmacodynamics in determining dosage regimens for broad spectrum cephalosporins. Diagn Microbiol Infect Dis 1995; 21:1–8.
5. Leggett JE, Fantin B, Ebert S, et al. Comparative antibiotics dose-effect relations at several dosing intervals in murine pneumonitis and thigh-infection models. J Infect Dis 1989; 159:281–292.
6. Craig WA, Ebert SC. Continuous infusion of β-lactam antibiotics. Antimicrob Agents Chemother 1992; 36:2577–2583.
7. Dagan R, Abramason O, Leibovitz E, et al. Bacteriologic response to oral cephalosporins. J Infect Dis 1997; 176:1253–1259.
8. Dagan R, Leibovitz E, Fliss DM, et al. Bacteriologic efficacies of oral cefaclor in treatment of acute otitis media in infant and young children. Antimicrob Agents Chemother 2000; 44:43–50.
9. Dangan R et al. Evidence to support the rationale that bacterial eradication in respiratory tract infection is an important aim of antimicrobial therapy. J Antimicrob Chemother 2001; 47:129–140.
10. Craig WA, Andes D. Pharmacokinetics and pharmacodynamics of antibiotics in otitis media. Pediatr Infect Dis J 1996; 15:255–259.
11. Georgopapadakou NH. Penicillin-binding proteins and bacterial resistance to β-lactams. Antimicrob Agents Chemother 1993; 37:2045–2053.
12. Danziger LH, Pendland SL. Bacterial resistance to β-lactam antibiotics. Am J Health-Syst Pharm 1995; 52(suppl 2):S3–S8.
13. Pitout JD, Sanders CC, Sanders WE Jr. Antimicrobial resistance with focus on β-lactam resistance in gram-negative bacilli. Am J Med 1997; 103:51–59.
14. Neuwirth C, Siebor E, Duez JM. Imipenem resistance in clinical isolates of Proteus mirabilis with alteration in penicillin-binding proteins. J Antimicrob Chemother 1995; 36:335–342.
15. Nikaido H. Prevention of drug access to bacterial targes: permeability barieres and actives efflux. Science 1994; 264:382–388.

16. Tozuka Z. Penicillins-parameters. In: Kuemmere HP, Murakawa T, Nightingale CH, eds. Pharmacokinetics of Antimicrobial Agents. Principle. Mthod. Application, Ecomed, Gmbh & Co.KG, 1993:65–67.
17. Tozuka Z. Cephalosporin-parameters. In: Kuemmere HP, Murakawa T, Nightingale CH, eds. Pharmacokinetics of Antimicrobial Agents. Principle. Mthod. Application, Ecomed, Gmbh & Co.KG 1993:93–98.
18. Murakawa T, Tozuka Z. Cephalosporins and other β-lactam antibiotics. In: Kuemmere HP, Murakawa T, Nightingale CH, eds. Pharmacokinetics of Antimicrobial Agents: Principle. Mthod. Application, Ecomed, Gmbh & Co.KG, 1993:69–86.
19. Venter JC, Adams MD, Myers EW, et al. The sequence of the human genome. Science 2001; 291:1304–1351.
20. Tsuji A, Tamai I. Carrier-mediated intestinal transport of drugs. Pharm Res (NY) 1996; 13:963–977.
21. Hori R, Tomita Y, Katsura M, Inui K, Takano M. Transport of bestatin in rat renal brush-border membrane vesicles. Biochem Pharmacol 1993; 45:1763–1768.
22. Tsuji A, Terasaki T, Tamai I, Takeda K. In vivo evidence for carrier-mediated uptake of beta-lactam antibiotics through organic anion transport systems in rat kidney and liver. J Pharmacol Exp Ther 1990; 253:315–320.
23. Sekine T, Watanabe N, Hosoyamada M, Kanai Y, Endou H. Expression cloning and characterization of a novel multispecific organic anion transporter. J Biol Chem 1997; 272:18526–18529.
24. Jariyawat S, Sekine T, Takeda M, et al. The interaction and transport of beta-lactam antibiotics with the cloned rat renal organic anion transporter 1. J Pharmacol Exp Ther 1999; 290:672–677.
25. Takeda M, Tojo A, Sekine T, Hosoyamada M, Kanai Y, Endou H. Role of organic anion transporter 1 (OAT1) in cephaloridine (CER)-induced nephrotoxicity. Kidney Int 1999; 56:2128–2136.
26. Sathirakul K, Suzuki H, Yasuda K, et al. Kinetic analysis of hepatobiliary transport of organic anions in Eisai hyperbilirubinemic mutant rats. J Pharmacol Exp Ther 1993; 265:1301–1312.
27. Vogelman B, Gudmundsson S, Leggett J, Tumidge J, Ebert S, Craig WA. Correlation of antimicrobial pharmacokinetic parameters with therapeutic efficacy in an animal model. J Infect Dis 1988; 158:831–847.
28. Frimodt-Moller, Bentzon MW, Thomsen VF. Experimental infection with *Streptococcus pneumoniae* in mice: correlation of in vitro activity and pharmacokinetic parameters with in vivo effect for 14 cephalosporins. J Infect Dis 1986; 154:511–517.
29. Gerber AU, Brugger H-P, Fell C, Stritzko T, Stadler B. Antibiotic therapy of infections due to Pseudomonas aeruginosa in normal and granulocytopenic mice: comparison of murine and human pharmacokinetics J Infect Dis 1986; 153:90–97.
30. Roosendaal R, Bakker-Woudenbert IA, van den Berghe-van Raffe M, Michel MF. Continuous versus intermittent administration of ceftazidime in experimental *Klebsiella pneumoniae* pneumonia in normal and leukopenic rats. Antimicrob Agents Chemother 1986; 30:403–408.
31. McCarthy CG, Finland M. Absorption and excretion of four penicillins, penicillin G, penicillin V, phenethicillin and phenylmercaptomethyl penicillin. New Engl J Med 1960; 263 (7):315–326.
32. Daschner FD et al. Ticarcillin concentration in serum, muscle and fat after a single intravenous injection. Antimicrob Ag Chemother 1980; 738–739.
33. Bunn PA, Phenoxymethyl Penicillin (V): pharmacologic observations. I Lab Clin Med 1956; 48(3):392–398.
34. Knudsen ET, Rolingson GN. Absorption and excretion of the potassium salt of 6(alpha-phenoxypropionamido)penicillanic acid. Lancet 1959; 1105–1109.
35. Bond JM, Lightbrown JW, Barber M, Waterworth PM, A comparison of four phenoxypenicillin. Med J 1;2(5363):956–961 p63.
36. Phlipson A, Sabath LD, Rosner B. Sequence effect on ampicillin blood levels noted in an amoxicillin, ampicillin and epicillin triple crossover study. Antimicrob Ag Chemother 1975; 3:311–320.
37. Rotholt K, Nielsen B, Kristensen E. Clinical pharmacology of Pivampicillin. 1974; 6:563–571.

38. Knudsen ET, Rolinson GN. Absorption and excretion of Penbritin. Br Med J 1961; 2:198–200.
39. Eshelman FN, Spyker DA. Pharmacokinetic of amoxicillin and ampicillin: crossover study of the effect of food. Antimicrob AG Chemother 1978; 14:539–543.
40. Kunst MW, Mattie H. Absorption of pivampicillin in postoperative patients. Antimicrob AG Chemother 1975; 8:11–14.
41. Brown RM et al. Comparative pharmacokinetics and tissue penetration of sulbactam and ampicillin after concurrent intravenous administration. Antimicrob Agents Chemother 1982; 21(4):738–739.
42. Li C, Kuti JL, Nightingale CH, Mansfield DL, Dana A, Nicolau DP. Population pharmacokinetics and pharmacodynamics of piperacillin/tazobactam in patients with complicated intra-abdominal infection. J Antimicrob Chemother 2005; 56:388–395.
43. Tam VH, Louie A, Lomaestro BM, Drusano GL. Integration of population pharmacokinetics, a pharmacodynamic target, and microbiologic surveillance data to generate a rational empiric dosing strategy for cefepime against Pseudomonas aeruginosa. Pharmacotherapy 2003; 23:291–295.
44. Tod M, Minozzi C, Beaucaire G, Ponsonnet D, Cougnard J, Petitjean O. Isepamicin in intensive care unit patients with nosocomial pneumonia: population pharmacokinetic-pharmacodynamic study. J Antimicrob Chemother 1999; 44:99–108.
45. Sakurai Y, Hishikawa T, Hiramatsu N, et al. Pharmacokinetic analysis of cefozopran in neonatal infections—population pharmacokinetics using nonmem. Jpn J Antibiot 1999; 52:16–23.
46. Capparelli E, Hochwald C, Rasmussen M, Parham A, Bradley J, Moya F. Population pharmacokinetics of cefepime in the neonate. Antimicrob Agents Chemother 2005; 49:2760–2766.
47. Kimura T, Kokubun H, Nowatari M, Matsuura N, Sunakawa K, Kubo H. Population pharmacokinetics of panipenem in neonates and retrospective evaluation of dosage. J Antimicrob Chemother 2001; 47:51–59.
48. Kimura T, Sunakawa K, Matsuura N, Kubo H, Shimada S, Yago K. Population pharmacokinetics of arbekacin, vancomycin, and panipenem in neonates. Antimicrob Agents Chemother 2004; 48:1159–1167.
49. Dailly E, Kergueris MF, Pannier M, Jolliet P, Bourin M. Population pharmacokinetics of imipenem in burn patients. Fundam Clin Pharmacol 2003; 17:645–650.
50. Dailly E, Pannier M, Jolliet P, Bourin M. Population pharmacokinetics of ceftazidime in burn patients. Br J Clin Pharmacol 2003; 56:629–634.
51. Mouton JW, Hollander JG. Killing of Pseudomonas aeruginosa during continuous and intermittent infusion of ceftazidime in an in vitro pharmacokinetic model. Antimicrob Agents Chemother 1994; 38:931–936.
52. Cappelletty DM, Kang SL, Palmer SM, Rybak MJ. Pharmacodynamics of ceftazidime administered as continuous infusion or intermittent bolus alone and in combination with single daily-dose amikacin against Pseudomonas aeruginosa in an in vitro infection model. Antimicrob Agents Chemother 1995; 33:1797–1801.
53. Mordenti JJ, Quintiliani R, Nightingale CH. Combination antibiotic therapy: comparison of constant infusion and intermittent bolus dosing in an experimental animal model. J Antimicrob Chemother 1985; 15:313–321.
54. Livingston DH, Wang MT. Continuous infusion of cefazolin is superior to intermittent dosing in decreasing infection after hemorrhagic shock. Am J Surg 1993; 165:203–207.
55. Gerber AU. Impact of the antibiotic dosage schedule on efficacy in experimental soft tissue infections. Scand J Infect Dis 1991; (suppl 74):147–154.
56. Bakker-Woudenberg IA, van den Berg JC, Fontijne P, Michel MF. Efficacy of continuous versus intermittent administration of penicillin G in Streptococcus pneumoniae pneumonia in normal and immunodeficient rats. Eur J Clin Microbiol 1984; 3:131–135.
57. Totsuka K, Shimizu K, Leggett J, Vogelman B, Craig WA. Correlation between the pharmacokinetic parameters and the efficacy of combination therapy with an aminoglycoside and a β-lactam.
58. Ueda Y ed. Proceedings of the International Symposium on Netilmicin. Professional Postgraduate Services, Japan, 1985:41–48.

59. Nicolau DP, McNabb JC, Lacy MK, Li J, Quintiliani R, Nightingale CH. Pharmacokinetics and pharmacodynamics of continuous and intermittent ceftazidime during the treatment of nosocomial pneumonia. In press.

60. Nicolau DP, Nightingale CH, Banevicius MA, Fu Q, Quintiliani R. Serum bactericidal activity of ceftazidime: continuous infusion versus intermittent injections. Antimicrob Agents Chemother 1996; 40:61–64.

61. Daenen S, Erjavec Z, Uges DRA, Vries HG-Hospers, Jonge P, Halie MR. Continuous infusion of ceftazidime in febrile neutropenic patients with acute myeloid leukemia. Eur J Clin Microbiol Infect Dis 1995; 14:188–192.

62. Benko AS, Cappelletty DM, Kruse JA, Rybak MJ. Continuous infusion versus intermittent administration of ceftazidime in critically ill patients with suspected gram-negative infections. Antimicrob Agents Chemother 1996; 40:691–695.

63. Lagast H, Meunier F-Carpenter, Klastersky J. Treatment of gram-negative bacillary septicemia with cefoperazone. Eur J Clin Microbiol 1983; 2:554–558.

64. Bodey GP, Ketchel SJ, Rodriguez V. A randomized study of carbenicillin plus cefamandole or tobramycin in the treatment of febrile episodes in cancer patients. Am J Med 1979; 67:608–611.

65. Zeisler JA, McCarthy JD, Richelieu WA, Nichol MB. Cefuroxime by continuous infusion: a new standard of care? Infect Med 1992; 9:54–60.

66. Ambrose PG, Quintiliani R, Nightingale CH, Nicolau DP. Continuous vs. intermittent infusion of cefuroxime for the treatment of community-acquired pneumonia. Infect Dis Clin Prac 1998; 7:463–470.

67. Woodnutt Berry V, Mizen L. Effect of protein binding on penetration of β-lactams into rabbit peripheral lymph. Antimicrob Agents Chemother 1995; 39:2678–2683.

68. Ryan DM, Hodges B, Spencer GR, Harding SM. Simultaneous comparison of three methods for assessing ceftazidime penetration into extravascular nuid. Antimicrob Agents Chemother 1982; 22:995–998.

69. Kalager T, Digranes A, Bergan T, Solberg CO. The pharmacokinetics of ceftriaxone in serum, skin blister and thread nuid. J Antimicrob Chemother 1984; 13:479–485.

70. Tozuka, Z. Isolation and identification of decomposed products of cefixime with gastrointestinal contents or feces. Internal Report CRR860144 in Fujisawa Pharmaceutical Co. Ltd., J Mass Spectrom 2003; 38:793–808.

71. FDA Guidance 2005, Pharmacogenomics data submission.

72. FDA Guidance 2005, Safety testing of drug metabolites.

73. EMEA Position paper on non-clinical safety studies to support clinical trials with a single-dose microdose.

74. FDA Guidance 2006, Exploratory IND study.

75. Ke W, Bethel CR, Thomson JM, Bonomo RA, Akker FV. Crystal structure of KPC-2: Insights into carbapenemase activity in class A beta-lactamases. Biochemistry 2007 Apr 19.

76. Brewin A, Arango L, Hadley WK, Murray JF. High-dose penicillin therapy and pneumococcal pneumonia. JAMA 1974; 230:409A13.

7 Aminoglycosides

Myo-Kyoung Kim
Pharmacy Practice, University of the Pacific Thomas J. Long School of Pharmacy, Stockton, California, U.S.A.

David P. Nicolau
Center for Anti-Infective Research and Development, Department of Medicine, Division of Infectious Diseases and Pharmacy, Hartford Hospital, Hartford, Connecticut, U.S.A.

INTRODUCTION

Aminoglycosides are highly potent, broad-spectrum antibiotics, which have remained an important therapeutic option for the treatment of life-threatening infections. Since the introduction of this class of agents into clinical practice some five decades ago, the major obstacle to curtail their use is the potential for drug-related toxicity. However, over the last decade, new information concerning the pharmacodynamic profile of these agents has been revealed, which not only leads to the potential for improved antibacterial effectiveness, but also leads to the minimization of their toxicodynamic profile. As a result of our contemporary understanding of these principles, parenteral dosing techniques for the aminoglycosides have been modified from the administration of frequent small intermittent dosages to once-daily regimens, which not only optimize the pharmacodynamic and toxicodynamic profiles but also substantially reduce expenditure associated with this therapeutic option. Application of these new principles together with the aminoglycosides in vitro activity, proven clinical effectiveness, and synergistic potential are the rationale behind their continued use in the management of serious infections.

HISTORY AND MECHANISM OF ACTION OF AMINOGLYCOSIDES

The aminoglycosides include an important group of natural and semisynthetic compounds. The first parenterally administered aminoglycoside, streptomycin, was introduced in 1944 and was followed by a number of other naturally occurring compounds, which include neomycin, kanamycin, tobramycin, gentamicin, sisomicin, and paromomycin. Amikacin and netilmicin are semisynthetic derivatives of kanamycin and sisomicin, respectively, while isepamicin is a semisynthetic derivative of gentamicin. Arbekacin, known as habekacin, is also a semisynthetic derivative obtained by acylation of dibekacin in a reaction analogous to that used to produce amikacin (1,2).

Like many antibiotics (i.e., macrolides, tetracyclines, and streptogramins), the bactericidal activity of the aminoglycosides is thought to be ribosomally mediated. Existing data suggest that their antibacterial activity results from inhibition of protein biosynthesis by irreversible binding of the aminoglycoside to the bacterial ribosome. The intact bacterial ribosome is a 70S particle that consists of two subunits (50S and 30S) that are assembled from three species of rRNA (5S, 16S, and 23S) and from 52 ribosomal proteins. The smaller 30S ribosomal submit, which contains the 16S rRNA, has been identified as a primary target for aminoglycoside, which ultimately induces mistranslation on prokaryotic ribosomes (3,4).

In order to reach their cytoplasmic ribosomal target, aminoglycosides must initially cross the outer membrane (in gram-negative organisms) and the cytoplasmic membrane (in gram-negative and gram-positive bacteria). In gram-negative bacteria, the initial step involves ionic binding of the highly positively charged aminoglycosides to negatively charged phosphates mainly in lipopolysaccharides on the outer membrane surface, while uptake across this membrane is likely due to a "self-promoted uptake" mechanism (5,6). The cationic aminoglycosides may act by competitively displacing the divalent cations in the membrane resulting in the entry of the antibiotic (3). The rapid initial binding of the aminoglycosides to the cell accounts for the rapid bactericidal activity, which appears to increase with increasing aminoglycoside concentration. This characteristic of concentration- or dose-dependent killing in part explains the recent attention given to giving the entire dose of the aminoglycoside on a once-daily basis in order to maximize bacterial killing.

Aminoglycoside uptake across the cytoplasmic membrane is the result of their electrostatic binding to the polar heads of phospholipids while the driving force for aminoglycoside entry is provided by a cellular transmembrane electrical potential. The combination of these effects is characterized by rapid binding to the ribosome and an acceleration of aminoglycoside uptake across the cytoplasmic membrane.

Aminoglycosides of the gentamicin, kanamycin, and neomycin families induce misreading of mRNA codons during translation as well as inhibit translocation (7). Streptomycin induces misreading of the genetic code in addition to inhibiting translational initiation. By contrast, spectinomycin, an agent with only bacteriostatic activity, does not cause translation errors but inhibits translocation. These findings support the notion that translational misreading is at least partly responsible for the bactericidal activity characteristic of aminoglycosides (7–10).

While the ribosome has been identified as a primary target for these agents, the precise mechanism by which aminoglycosides exert their bactericidal activity has remained elusive, since these drugs manifest pleiotropic effects on bacterial cells. These effects include, but are not limited to, disruption of the outer membrane, irreversible uptake of the antibiotic, blockade of initiation of DNA replication, and "stabilization of DNA–RNA hybrid duplexes or triple helices" (11,12).

MICROBIOLOGIC SPECTRUM

While the aminoglycosides are highly potent, broad-spectrum antibiotics, their in vitro activity is considered to be most notable against a variety of gram-negative pathogens. These pathogens include common clinical isolates of Acinetobacter spp., Citrobacter spp., Enterobacter spp., *Escherichia coli*, Klebsiella spp., Serratia spp., Proteus spp., Morganella spp., and *Pseudomonas aeruginosa*. However, while these agents are generally considered active against these microbes, substantial differences in antimicrobial potency exist among the various aminoglycosides. For example, even though the antimicrobial spectra of gentamicin and tobramycin are quite similar, tobramycin is generally more active in vitro against *P. aeruginosa*, whereas gentamicin is more active against *Serratia*.

Although streptomycin has been used extensively for many years, the emergence of resistance in against *Mycobacterium tuberculosis* and aerobic gram-negative bacilli, as well as the relatively frequent occurrence of vestibular toxicity combined with the availability of less toxic antibiotics has greatly diminished its clinical utility. The introduction of kanamycin provided a broader spectrum of activity against gram-negative bacilli, including streptomycin-resistant strains, but it was not active against *P. aeruginosa*. As with streptomycin, extensive use of kanamycin quickly led to the emergence and widespread dissemination of kana-mycin resistance among Enterobacteriaceae. The development of other agents within the class (i.e., gentamicin, tobramycin, netilmicin, and amikacin) further expanded the spectrum of antimicrobial activity of this class to cover many kanamycin-resistant strains, including *P. aeruginosa*.

While the aminoglycosides are also active against Salmonella spp., Shigella spp., *Neisseria gonorrhea*, and *Haemophilus influenzae*, this class of agents is not recommended for infections caused by these species because of the wide avail-ability of effective and less toxic drugs.

Aminoglycosides are generally active against Staphylococci. Several reports demonstrated that arbekacin, marketed in Asian countries such as Japan, has a superior in vitro activity against methicillin-resistant *Staphylococcus aureus* (MRSA) compared to other aminoglycosides, although outbreaks of arbekacin-resistant MRSA have been recently reported in Japan (13–15). A study using an in vitro infection model also demonstrated that arbekacin had better in vitro activity than vancomycin against MRSA and one strain of glycopeptide intermediate-resistant *Staphylococcus aureus* (GISA), which is a New Jersey strain (*HIP5836*). However, both arbekacin and vancomycin were not effective against the other tested GISA strain, which is a Japanese strain (*Mu-50*) (16).

Aminoglycosides are not generally advocated as single agents for infections due to gram-positive pathogens. However, an aminoglycoside, usually gentamicin, is frequently administered in combination with a cell wall active agent to provide synergy in the treatment of serious infections due to Staphylococci, Enterococci, and *Viridans streptococci*. (Refer to the section entitled Resistance and Synergy.)

PHARMACOKINETIC CHARACTERIZATION

Aminoglycosides are poorly absorbed after oral administration due to their hydro-philicity and poor membrane permeability. Recently, there are many attempts and studies carried to enhance oral absorption and thus to invent an oral formulation (17–20). These potential absorption enhancers include a derivative of taurocholic acid (i.e., TC002) (17), various nonionic, anionic, and cationic surfactants (18), a medium length alkyl chain surfactant (i.e., labrasol) (19), and a polysaccharide (i.e., 11) (20). Some of these agents statistically enhanced bioavailability in animal models.

Due to similar reasons for poor oral absorption, these agents penetrate poorly through intact skin. In spite of the poor skin penetration, their use as a topical antibacterial for large areas of denuded skin (i.e., thermal injury) may cause substantial systemic absorption, especially in patients with altered renal function. Furthermore, the use of aminoglycosides for local irrigation of closed body cavities may result in considerable systemic accumulation and potential toxicity.

Like concerns for potential systemic toxicity from their topical usage, the swallowed portions of inhaled or nebulized aminoglycosides may cause detectable aminoglycoside serum concentrations and the development of toxicity. Pharmacokinetics and pharmacodynamics of inhaled or nebulized aminoglycosides will be further discussed in subsequent sections of this chapter.

As a consequence of poor oral bioavailability, the aminoglycosides must be given parenterally in order to achieve a consistent systemic serum concentration profile. While the intramuscular route is well tolerated and results in essentially complete absorption, intravenous administration is generally preferred because of the rapid and predictable serum profile. The importance of the rapid and reliable attainment of sufficient peak concentrations will be further discussed in subsequent sections of this chapter.

The aminoglycosides are weakly bound to serum proteins and therefore freely distribute into the interstitial or extracellular fluid. The apparent volume of distribution of this class of agents is approximately 25% of the total body weight, which corresponds to the estimated extracellular fluid volume. While the volume of distribution is generally approximated at 0.25 to 0.3 L/kg, patients who are malnourished, obese, pregnant, or in the intensive care unit or have ascites may have substantial alterations in this parameter, which require dosage and/or schedule modifications to maintain the desired serum profile. In general, the concentrations of the aminoglycosides attained in tissue and body fluids are less than that obtained in serum, with the notable exceptions of the kidney, perilymph of the inner ear, and urine. Approximately 20% to 50% of the serum concentration can be achieved in bronchial, sputum, pleural, and synovial fluid and unobstructed bile. Although low aminoglycoside bronchial fluid concentrations have been reported, the administration of large single daily doses versus conventional dosing substantially improves drug penetration into this fluid, while reducing drug accumulation in the renal tissue (21–23). As compared with other body sites, the penetration into prostate tissue and bone penetration is poor. Penetration of aminoglycosides into cerebral spinal fluid in the presence of inflammation or in the fluid of the eye is inadequate and variable, and therefore direct instillation is often required to provide sufficient concentrations at these sites. Additionally, these agents cross the placenta; therefore, the potential risk to the fetus and mother must be considered prior to use.

The kidneys, via glomerular filtration, are responsible for essentially all aminoglycoside elimination from the body. As a result, there is a proportional relationship between drug clearance and glomerular filtration rate, which is routinely utilized to assist with aminoglycoside dosage modification (24). In adults and children older than six months with normal renal function, the elimination half-life is approximately two to three hours. In premature, low-birth weight and infants less than one week old, the half-life is 8 to 12 hours, whereas the half-life decreases to five hours for neonates whose birth weight exceeds 2 kg. Finally, it should be expected that substantial increases in the half-life would be observed in patients with renal dysfunction as a result of the primary renal elimination.

PHARMACODYNAMIC OVERVIEW

Over the last several decades, new data have emerged which extended our understanding of the complex interactions, which take place among the pathogen-drug-host during the infection process. Much of this focus has revolved around a more complete understanding of the influence of drug concentration on bacterial cell death. The pharmacodynamic properties or the correlation of drug concentration and the clinical effect (e.g., bacterial killing) of a specific antibiotic class are therefore an integration of two related areas, one being microbiologic activity and the other pharmacokinetics (refer to Chapter 2 for a more complete review of this topic). A distinct pharmacodynamic profile exists for all antimicrobials, since the influence of drug concentration on the rate and extent of bactericidal activity is different among the various classes of drugs.

The pharmacodynamic profile of the aminoglycosides has been characterized both in vitro and in vivo. Utilization of both static (i.e., time-kill studies) and dynamic (i.e., pharmacokinetic modeling) in vitro techniques have provided fundamental information concerning the pharmacodynamic profile of the aminoglycosides. Based on these data, a general pharmacodynamic division among antimicrobials occurs between agents, in which the rate and extent of bactericidal activity is dependent upon drug concentration (aminoglycosides and fluoroquinolones) and agents such as the β-lactams, which have bactericidal activity independent of drug concentration when their concentration exceeds four times the minimum inhibitory concentration (MIC) (25–28).

Figure 1 depicts these principles as illustrated with ciprofloxacin, tobramycin, and ticarcillin. In this experiment, bacteria are exposed to various multiples of the MIC in vitro and as shown with ticarcillin, little difference in the rate of bactericidal activity is noted when its concentration exceeds four time the MIC. Therefore, this type of killing, which is characteristic of β-lactams, is termed nonconcentration or dose-independent bactericidal activity (refer to Chapter 7 for a more complete discussion of β-lactam pharmacodynamics).

By contrast, when the same multiples of the MIC are studied with tobramycin and ciprofloxacin, the number of organisms is seen to decrease more rapidly with each rising MIC interval. Since these agents eliminate bacteria more rapidly when their concentrations are appreciably above the MIC of the organism, their killing activity is referred to as concentration or dose-dependent bactericidal

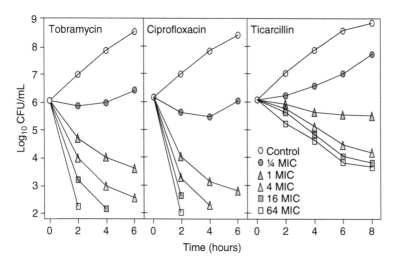

FIGURE 1 Influence of drug concentration on the bactericidal activity of tobramycin, ciprofloxacin, and ticarcillin against *Pseudomonas aeruginosa*. *Abbreviation*: MIC, minimum inhibitory concentration. *Source*: From Ref. 26.

activity (26). Based on these in vitro data, optimum bactericidal activity for the aminoglycosides is achieved when the exposure concentration is approximately 8 to 10 times the MIC (25,26,29,30). In addition to maximal bactericidal activity, Blaser et al. (31) have demonstrated that the peak/MIC ratio of 8:1 is correlated with a decrease in the selection and regrowth of resistant subpopulations occurring during treatment with netilmicin.

Antimicrobial activity in vivo is a complex and multifactorial process. As described elsewhere in this text, to be effective, an antimicrobial must reach and maintain adequate concentrations at the target site and interact with the target site for a period of time so as to interrupt the normal functions of the cell in vivo. This description of the interaction between the pathogen and drug (i.e., microbiologic activity) is influenced by the drug disposition or pharmacokinetic profile of the host species. As a result of the complex interactions occurring among the pathogen, drug, and host triad, in vivo pharmacodynamic characterization of compounds is required.

Since we are not yet able to measure drug concentrations at the site of action (i.e., ribosome for the aminoglycosides), we commonly employ a microbiologic parameter (i.e., MIC) as the critical value in the interpretation of these in vivo pharmacodynamic relationships. When integrating the microbiologic activity and pharmacokinetics, several parameters appear to be significant constituents of drug efficacy. The pharmacokinetic parameter of area under the concentration–time curve (AUC), maximum observed concentration (C_{max} or peak), and half-life are often integrated with the MIC of the pathogen to produce pharmacodynamic parameters such as the AUC/MIC, peak/MIC ratio, and the time, which the drug concentration remains above the MIC (time > MIC). For the aminoglycosides the AUC/MIC, peak/MIC ratio, and time > MIC have all been shown to be pharmacodynamic correlates of efficacy (25,31,32). However, it is not surprising that several pharmacodynamic parameters have been related to efficacy with these agents since all of these parameters are correlated (Fig. 2).

Therefore, since the amount of drug delivered against the pathogen is proportional to the amount of drug delivered to the host (AUC), the AUC is the primary pharmacokinetic parameter associated with efficacy. However, since the AUC is a product of concentration and time under certain conditions, the influence of concentration will appear to be a predominant factor, whereas under a different set of conditions, the exposure to the drug or the time > MIC may assume a larger role in bacterial eradication. In the case of the aminoglycosides, which display concentration-dependent killing and a relatively long postantibiotic effect (PAE), the influence of the time > MIC is small when compared to the influence of peak concentration. As a result, the pharmacodynamic parameter, which is believed to best characterize the profile of the aminoglycosides in vivo, is the peak/MIC ratio.

In support of the concentration-dependent bactericidal activity of the aminoglycosides displayed in in vitro studies and animal models of infection, several studies in man have also demonstrated the importance of achieving sufficient peak/MIC ratios as related to defined treatment success. Aminoglycoside concentrations have been associated with treatment success in a triad of studies reported by Moore et al. (33–35). In the first report by this group, higher C_{max} concentrations were associated with improved outcomes in gram-negative pneumonia, while in the second report higher concentrations were associated with improved survival in gram-negative bacteremia (33,34). Results of these studies suggest the importance of achieving adequate and early aminoglycoside concentrations in

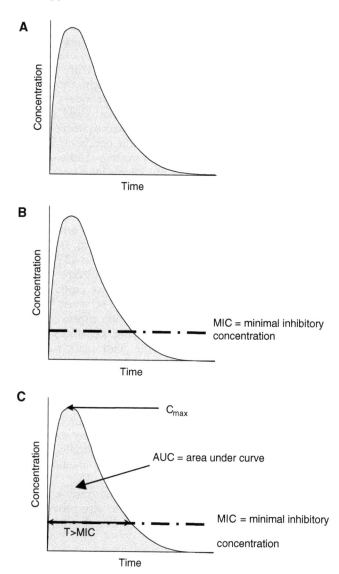

FIGURE 2 Antimicrobial pharmacodynamics: integration of microbiologic potency and selected pharmacokinetic parameters.

severely ill patients with gram-negative infection. Their last study published in 1987 evaluated the relationship between the ratio of the peak concentration to MIC and clinical outcome through data collection from four randomized, double-blind, controlled clinical trials, which utilized gentamicin, tobramycin, or amikacin for the treatment of gram-negative bacterial infections (33). For the purposes of the study, the maximal peak concentration (C_{max}) was defined as the highest concentration determined during therapy, while the mean peak concentration was calculated as the average of all C_{max} values during the course of treatment. The investigators demonstrated that high maximal and mean peak aminoglycoside concentration (8.5 ± 5.0 and $6.6 \pm 3.9 \, \mu g/mL$, respectively) to MIC ratios were significantly ($P < 0.00001$ and $P < 0.0001$, respectively) correlated with clinical response. Of the 188 patients who had a clinical response to therapy, the $C_{max}/$ MIC average value was $8.5 \pm 5.0 \, \mu g/mL$, whereas the 48 nonresponders had a ratio of $5.5 \pm 4.6 \, \mu g/mL$ ($P < 0.00001$). Although these studies used fixed dosing intervals and were not designed to assess the in vivo pharmacodynamic parameters and their relationship to outcome, these data provide the backbone for our commitment to the value of the peak/MIC ratio in clinical practice. Deziel-Evans et al. demonstrated that a 91% cure rate was observed in patients with peak/MIC ratios greater than 8, while only a 12.5% cure rate was observed for patients with ratios between ≤ 4 in a retrospective study with 45 patients (36). In another study by Keating et al., response rates of 57%, 67%, and 85% were observed in neutropenic patients with mean serum aminoglycoside concentration/MIC ratios of 1 to 4, 4 to 10, and greater than 10, respectively (37). Williams et al. (38) have also reported that C_{max}/MIC ratios correlated significantly with cure in 42 patients undergoing amikacin treatment who were evaluable for clinical outcome. Additionally, Fiala and Chatterjee (39) also noted that infection was cured more frequently in patients with severe gram-negative infections who achieved higher peak/MIC ratios. The association of high peak/MIC ratios with improved outcomes has also been noted in orthopedic patients receiving gentamicin and recently a similar relationship has been noted for 61 febrile neutropenic patients with hematologic malignancies (40,41). Other investigators have also observed beneficial correlations between serum concentrations or pharmacodynamic parameters and therapeutic outcomes in patients treated with aminoglycosides (42–44).

More recently, Kashuba et al. (45,46) reported that achieving an aminoglycoside peak/MIC of ≥ 10 within 48 hours of initiation of therapy for gram-negative pneumonia resulted in a 90% probability of therapeutic response by day 7 of therapy. They also note that aggressive aminoglycoside dosing (initial dose of $7 \, mg/kg$) followed by individualized pharmacokinetic monitoring should maximize the rate and extent of response in this patient population.

Additionally many once-daily aminoglycoside trials and their meta-analysis studies (which will be described in a later section in this chapter) also proved the importance of the correlation between the peak/MIC ratio and clinical outcome.

POSTANTIBIOTIC EFFECT

The PAE is defined as the persisting suppression activity against bacterial growth after limited exposure of bacteria to an antibiotic. In order to measure the in vitro PAE of aminoglycosides, after a one- to two-hour exposure of an antibiotic or antibiotics to bacteria, drug is removed rapidly by dilution, drug inactivation [e.g., cellulose phosphate powder or tobramycin-acetylating enzyme AAC (3)-II and acetyl coenzyme A], or filtration ($0.45 \, \mu m$ pore-size filters). After this drug elimination process, viable counts [colony-forming unit (CFU)/mL] at each time point are required to develop viability curves. The in vitro PAE is calculated by the following equation: $PAE = T - C$ where T is the time required for the count of CFU in the test culture to increase one $\log_{10}$ above the count observed immediately after drug removal and C is the time needed for the count of CFU in an untreated control culture to increase one $\log_{10}$ above the count observed immediately after the same procedure used on the test culture for drug removal (47).

Like quinolones, aminoglycosides represent another antibiotic class, which has a clinically meaningful PAE; however, several factors may affect the presence and duration of the PAE. A range of 0.5 to 7.5 hours has been reported for the PAE of aminoglycosides (47,48). Major factors influencing PAE include organism, concentration of antibiotic, duration of antimicrobial exposure, and antimicrobial combinations. Minor factors include size of the inoculum, growth phase of the organism at the time of exposure, mechanical shaking of the culture, type of medium, pH and temperature of the medium, and the effect of reexposure (47).

The examples below illustrate how the above factors affect the presence and duration of the PAE. Unlike β-lactam antibiotics with PAEs against only gram-positive organisms, aminoglycosides exhibit a PAE on both gram-positive and gram-negative organisms (48,49). However, the PAE duration differs depending on the types of bacteria. For example, the durations of the PAE following exposure of *P. aeruginosa* to gentamicin and tobramycin are 2.2 hours and 2.1 hours, respectively, while those of *E. coli* are 1.8 hours and 1.2 hours, respectively (47). Additionally, it has been shown that there is a positive correlation between subsequent dose and concentration of aminoglycosides and duration of the PAE (50–53). A maximum concentration to exert a maximal PAE effect of aminoglycosides is difficult to determine because most bacteria are completely and rapidly killed at high drug concentrations. In contrast, the PAE of penicillin G gradually increases up to a point of maximal effect at a concentration 8 to 16 times the MIC (53–55). The same theory also applies to duration of exposure.

The duration of PAE also varies depending on concurrently applied antibiotics when it is tested as a combination therapy. The combined effect of aminoglycosides and cell wall inhibitors on the duration of the PAE was studied by several researchers (56–59). In general, these combinations produced additive effects (i.e., similar to the sum of PAEs for individual drugs) or synergistic effects (i.e., at least one hour longer than the sum of PAEs for individual drugs) in *S. aureus* and various Streptococci. The effects of antibiotic combinations against gram-negative bacilli were mainly additive or indifferent (i.e., no different from the longest of the individual PAEs). As an exception, the addition of tobramycin to rifampin, which can achieve prolonged PAEs in gram-negative bacilli, showed synergism of the PAE in *P. aeruginosa*, *E. coli*, and *Klebsiella pneumonia* (59).

However, there are unavoidable limitations in this in vitro determination of the PAE duration. One major drawback is that bacteria undergo a single exposure for a short period of time to a fixed concentration of a testing antimicrobial agent. However, in a clinical setting, the antimicrobial agent should be used multiple times. In addition, it should maintain the concentration above the MIC for a relatively longer time period than that which occurs during PAE testing, and the concentration should decline continuously throughout the dosing interval. Karlowsky et al. (60,61) demonstrated that multiple exposures of *E. coli* and *P. aeruginosa* to aminoglycosides significantly decreased the duration of PAE, along with an attenuation in bacterial killing activity. McGrath et al. (52) suggested that the reasons for this phenomenon may be because of adaptive resistance or the selection of drug-resistant variants. Li et al. (62) demonstrated that *P. aeruginosa* exposed to constant tobramycin concentrations have longer PAE than those exposed to exponentially decreasing tobramycin concentrations at similar AUC above the MIC. These studies suggest that conventional testing yields an overestimate of the PAE in comparison to the PAE presented in a clinical situation with continuously changing concentrations.

Furthermore, the duration of PAE significantly varies depending on the testing environment; important factors influencing the testing environment include inoculum concentration, temperature, pH, oxygen tension, and free cation (Ca^{2+}, Mg^{2+}) content (63–67). The other obstacle to apply this in vitro PAE duration to clinical practice is that the PAE does not consider host immunity. However, some effort has been undertaken to include host immunity by using other terminologies such as postantibiotic leukocyte enhancement (PALE) and postantibiotic sub-MIC effect. PALE is a phenomenon that pathogens in the PAE phase are more susceptible to the antimicrobial effect of human leukocytes than non-PAE controls. Postantibiotic sub-MIC effect illustrates the joining of the PAE and the additive effects of exposure to sub-MIC levels (68).

Despite some of the limitations involved in predicting the exact duration of PAE, the general consensus is that PAE is an important factor to be considered when developing a drug regimen. The precise mechanisms of the PAE are largely unknown. However, several hypotheses have been suggested. They include limited persistence of antibiotic at the site of action, recovery from nonlethal damage to cell structures, and the time required for synthesis of new proteins or enzymes before growth. Drug-induced nonlethal damage due to the irreversible binding to bacterial ribosomes represents a feasible mechanism of the PAE of aminoglycoside (53,69). In a study measuring the rate of [³H] adenosine incorporation, Gottfredsson et al. (70) showed that DNA synthesis by *P. aeruginosa* after exposure to tobramycin was markedly affected during the PAE phase. However, in a study, which utilized cumulative radio-labeled nucleoside precursor uptake in a clinical strain of *E. coli*, Barmada et al. showed that DNA and RNA synthesis resumed almost immediately following exposure to tobramycin, whereas protein synthesis did not recover until four hours later. Therefore the duration of PAE produced by aminoglycosides against *E. coli* seems to be better correlated with inhibition of protein synthesis than inhibition of DNA or RNA synthesis (71). Even if the rationale of this difference is unknown, it may be from discrepancy in the mechanism of action between two species. Theoretically, provided that a certain threshold of growth suppression to restrain DNA synthesis is attained, greater accumulation or entrapment of intracellular tobramycin in *P. aeruginosa* may account for this disagreement (70,71).

Although numerous data are available on the in vitro PAE, there is less in vivo information. Six animal models have been developed to evaluate the in vivo PAE: thigh infection in mice, pneumonia in mice, infected subcutaneous threads in mice, meningitis in rabbits, infected tissue-cages in rabbits, and endocarditis in rats (47). Among these models, the mice thigh infection model is commonly used to evaluate the PAE of aminoglycosides, although the pneumonia model is also adopted for aminoglycosides (50,72–74). The endocarditis rat model has been used to evaluate the in vivo PAEs of aminoglycosides when they are added to penicillin or imipenem (51,75). In these above models, antibiotic is administered to achieve a concentration, which exceeds the MIC during the first one to two hours. Next, bacterial loads from tissue are counted at various time points while drug concentrations of plasma are measured simultaneously. After graphing the bacterial growth curve, in vivo PAE can be calculated by the following equation: PAE = $T \_ C \_ M$. M is the time serum concentration exceeds the MIC, T is the time required for the counts of CFU in tissue to increase one

$\log_{10}$ above the count at the time closest to but not less than time M, and C is the time required for the counts of CFU in tissue of untreated control to increase one $\log_{10}$ above the count at time zero (47).

Major factors affecting the in vivo PAE include the infection site, type of organism, type of antimicrobial agent, the drug dose, simulation of human pharmacokinetics, and the presence of leukocytes (47,76). For example, the in vivo PAEs for 15 clinical isolates of Enterobacteriaceae following administration of gentamicin (8 mg/kg) ranged from 1.4 to 7.3 hours (76). Like the in vitro PAE, higher doses of drugs are also correlated with longer in vivo PAEs (47). The PAEs of single doses of 4, 12, and 20 mg/kg tobramycin in the thighs of neutropenic mice infected with *P. aeruginosa* were 2.2, 4.8, and 7.3 hours, respectively. Generally, the combinations of aminoglycoside and β-lactam lengthened the PAE for *S. aureus* and *P. aeruginosa* by 1.0 to 3.3 hours, compared to the longest PAE of the individual drugs. However, no difference was observed against *E. coli* and *K. pneumoniae* (74). The adoption of a different infection model also may influence the duration of the in vivo PAE. PAEs with amikacin against *K. pneumoniae* in the mouse pneumonia model were roughly 1.5 to 2.5 times longer than that observed in the mouse thigh model at the corresponding dose (74). Furthermore, other environmental conditions also influence the in vivo PAE (76).

The in vivo PAE can be utilized to incorporate the effect of host immunity in conjunction with the PAE. The duration of the in vivo PAE of aminoglycosides was prolonged 1.9- to 2.7-fold by the presence of leukocytes (76). In addition, neutrophils are also proven to prolong the in vivo PAEs for aminoglycosides against a standard strain of *K. pneumoniae* (77). The other benefit of the in vivo PAE is that the half-life of some antimicrobials can be prolonged to simulate human pharmacokinetics by inducing transient renal impairment in mice with uranyl nitrate. The in vivo PAE in the renally impaired mice was approximately seven hours longer than that observed in normal mice with large doses inducing a similar effect. This difference is likely due to sub-MIC levels that persist for a longer time with renal impairment than with normal renal function (47).

In spite of the lack of an ideal method to apply the PAE to clinical practice, the PAE has a major impact on antimicrobial dosing regimens. For antibiotics with longer PAE, dosing frequency may be less frequent than that of antibiotics with shorter PAE. Therefore, PAE may be one of the rationales for the implementation of once-daily aminoglycoside dosing.

RESISTANCE AND SYNERGY

While the focus of this chapter concerns the pharmacodynamic profile of the aminoglycosides and its implications for clinical practice, it is important to realize that the development of antimicrobial resistance is often the rate-limiting step for a compound's clinical utility. Not unlike other antimicrobials, the aminoglycosides face similar issues regarding resistance. While this topic is beyond the scope of this chapter, it should be noted that at least three mechanisms confer resistance to the aminoglycosides: impaired drug uptake, mutations of the ribosome, and enzymatic modification of the drug. Intrinsic resistance is often due to impaired uptake while acquired resistance usually results from acquisition of transposon- and plasmid-encoded modifying enzymes (78). To this end, the pharmacodynamic implications regarding resistance are that one should select regimen, which maximizes the rate and extent of killing. If this approach is universally endorsed, it will likely minimize the development of resistance in vivo, since this pharmaco-dynamic optimization of aminoglycosides has been shown to have this effect in vitro (31,79). Additionally, adaptive resistance and refractoriness to aminoglyco-sides have been demonstrated in vitro and in a neutropenic murine model by exposing *P. aeruginosa* to concentrations below or at the MIC of the organism (80–82). Exposure to an aminoglycoside without a drug-free period leads to decreased bacterial killing. Therefore, longer dosing intervals, as can be achieved with the pharmacodynamically based once-daily aminoglycoside dosing approach, allow for a drug-free period in which the bacteria are not exposed to an aminoglycoside and should further preserve the antibacterial activity of these agents after multiple doses.

Aminoglycosides exhibit synergistic bactericidal activity when given in combination with cell wall active agents such as β-lactams and vancomycin (83,84). For example, enterococcal endocarditis should be treated with a combina-tion of an aminoglycoside plus a penicillin or vancomycin because by themselves neither of the agent is sufficiently bactericidal. However, when combination therapy is advocated to achieve synergy for gram-negative organisms, maximally effective doses of both agents should be maintained because synergy does not occur universally for all pathogens to all β-lactam plus aminoglycoside combina-tions (83,85). Additionally, it should also be noted that combination exposure may also prolong the in vitro and in vivo PAE observed with the aminoglycosides (see section "Postantibiotic Effect" above), although the clinical relevance of this effect is not fully understood.

TOXICODYNAMICS

Since their introduction into clinical practice, a variety of adverse events have been reported during aminoglycoside therapy. Although the precise cellular mechanism of toxicities remains elusive, aberrant vesicle fusion, mitochondrial toxicity/free radical generation, and decreased protein synthesis either by reduced transcription or by translation after aminoglycoside exposure are suggested (86).

Most (i.e., gastrointestinal) adverse reactions of aminoglycosides are mild and resolve with drug discontinuation. The aminoglycosides rarely produces hypersensitivity reactions and despite direct injection into the central nervous system and the eye, local adverse events (i.e., seizures and hypersensitivity reaction) are generally not observed. Although infrequent in contemporary clinical practice, the aminoglycosides have the potential to cause or exacerbate neuromus-cular blockade. Despite the concern for increased risk with the administration of the high doses routinely used in once-daily dosing protocols, this adverse event has not been observed (87,88). However, while generally well tolerated, the major obstacle, which has curtailed the use of aminoglycosides is the potential for ototoxicity and nephrotoxicity.

Although ototoxicity has long been recognized as a potential complication of aminoglycoside therapy, questions still remain regarding the full delineation of risk factors and a universally accepted definition. As a result of discrepancies in both definition and sensitivity of testing, the reported incidence of ototoxicity has spanned a wide range (2–25%). "Two distinct forms" of ototoxicity, cochleotoxicity and vestibulotoxicity, have been reported and may occur alone or simultaneously (89). While the precise mechanism of injury remains elusive, cochleotoxicity is believed to result from injury to the outer hair cells (mostly) and/or inner hair cells (in severe cases) (90,91). Pathology of vestibulotoxicity includes the injury to hair cells in the crista ampullaris of the semicircular canals (90,91).

Auditory toxicity often occurs at frequencies, which are higher than that required for conversation and thus patient complaints which usually manifest as tinnitus or a feeling of fullness in the ear are usually rendered once considerable auditory damage has already been done (92). Like the progression of auditory loss, the initial symptoms of vestibular toxicity often go unrecognized due to the nonspecific nature of its initial presentation (i.e., nausea, vomiting, cold sweats, nystagmus, vertigo, and dizziness) (93). While considered to be less frequent than auditory toxicity, these vestibular effects are by and large irreversible and therefore may have a profound impact on the daily function status. Although there is lack of well-controlled comparative trials with sufficient power to detect differences in ototoxicity among aminoglycosides after systemic administration, several groups of researchers reported relative comparisons among aminoglycoside groups; gentamicin, kanamycin, and tobramycin tend to be more cochleotoxic than amikacin (89,90). In contrast, streptomycin is thought to be more vestibulotoxic than gentamicin, which is more toxic than tobramycin (90). From the perspective of ear drops, the use of neomycin/polymyxin B is well documented for its safe usage for the short-term treatment of otorrhea (94–96), although the safety profile of the long-term use remains elusive. In contrast, gentamicin ear drops are documented to be ototoxic when it is used for the treatment of Meniere's diseases (97–99). Interestingly, there are many attempts reported to protect patients on aminoglycosides from its ototoxicity; administration of free-radical scavengers, iron chelators, caspase inhibitors may reduce intensity or incidence of ototoxicity (100–102). However, these protective agents should be proven for their efficacy and safety in a well-controlled human study before their clinical usage.

Besides these protective agents, the application of pharmacodynamic principles may be another method to reduce ototoxicity of aminoglycosides. Although serum concentration data may be useful to ensure an adequate pharmacodynamic profile, these data cannot accurately predict the development of ototoxicity. Recent data have suggested that toxicity is related to drug accumulation within the ear, not peak concentrations, and have supported the concept of saturable transport and reinforced the belief that higher peak concentrations should not result in increased ototoxicity (103). For these reasons, the once-daily administration techniques may minimize drug accumulation and therefore drug-related toxicity (104,105).

While nephrotoxicity has been reported in more than half of patients receiving aminoglycoside therapy, the broad range of definitions and the poor risk factor assessment of the affected patient population often make the true incidence difficult, if not impossible, to determine. While considered by many to be a noteworthy event, toxicity is generally mild and reversible as few patients have progressive toxicity severe enough to warrant dialysis (106). At present it is thought that this toxicity is due to aminoglycoside accumulation in the lysosomes of the renal proximal tubule cells. which results in necrosis of the tubular cells and the clinical presentation of acute tubular necrosis manifested by nonoliguric renal failure within a week (107).

Several investigators have reported that advanced age, preexisting renal dysfunction, hypovolemia, shock, liver dysfunction, obesity, duration of therapy, use of concurrent nephrotoxic agents, and elevated peak/trough aminoglycoside concentrations are risk factors for development of nephrotoxicity (108–111). Additionally in the last study, multiple logistic regression analysis also revealed that trough concentration, duration of therapy, advanced age, leukemia, male gender, decreased albumin, ascites, and concurrent clindamycin, vancomycin, or piperacillin were independent risk factors for nephrotoxicity (111). Similar risk factors were identified in patients receiving once-daily aminoglycosides (112).

There are many methods reported to reduce nephrotoxicity of aminoglycosides, but they may be, by and large, categorized into three main methods. The first method is to modify cellular structures, which has been shown to statistically reduce nephrotoxicity in animal models (78,107). Secondly, protective agents may ameliorate nephrotoxicity. They include antioxidants (i.e., desferrioxamine, methimazole, vitamin E, vitamin C, or selenium), certain antibiotics (i.e., ceftriaxone or fleroxacin), and alternative agents (i.e., melatonin, *Rhazya stricta*, *Ginkgo biloba*, Garlic, *Nigella sativa* oil, and glycyrrhizin) (113–116). Similar to protective agents against ototoxicity, these agents are only tested in animal models, and thus they should be proven to be effective and safe in well-controlled human trials before their clinical application. Finally, dosing methods to enhance principles of pharmacokinetics and pharmacodynamics may be adopted. Similar to that previously described in discussion of ototoxicity, a saturable aminoglycoside transport system has been used to describe the uptake of drug in the kidney. Therefore, less frequent single-daily dose administration may minimize accumulation and nephrotoxicity (21,117). In this regard, once-daily regimens have been reported to reduce the incidence of nephrotoxicity (88,105,118).

Recently, modeling approaches to reduce toxicity using pharmacodynamic principles have been performed (119–121). These modeling approaches allow researchers to consider multivariables such as times of day of administrations, nonlinear accumulation process, and inter and intraindividual variabilities as well as administration methods. In a recent study, Rougier et al. took into consideration for nonlinear processes such as the amount of aminoglycosides taken up in the renal cortex or the tubuloglomerular feedback. This study demonstrated that nephrotoxicity associated with a thrice-daily administration occurs more rapidly, with greater intensity and for a longer duration, as compared to once-daily aminoglycosides (121).

In summary, application of pharmacodynamic principles such as implementing once-daily aminoglycosides may ameliorate toxicity of aminoglycosides. However, further explanations on the correlated relationship of systemic serum concentrations and local cellular concentrations, and well-accepted description of toxicodynamic surrogate markers reflecting cellular concentrations remained to be delineated.

CLINICAL USAGE AND APPLICATION OF PHARMACODYNAMICS

The parenteral aminoglycosides, particularly gentamicin, tobramycin, and amikacin, have long been used empirically for treatment of the febrile neutropenic patient or of patients with serious nosocomial infection. While aminoglycoside utilization has generally been declining due to the introduction of parenteral fluoroquinolones, emergence of fluoroquinolone-resistant *P. aeruginosa* will likely result in resurgence in clinical use of the aminoglycosides. To this point it is also apparent that the antipseudomonal β-lactams should not be given alone to treat systemic pseudomonal infections since this organism often develops resistance under therapy, thus the aminoglycosides plan an important role in the combination therapy for gram-negative infections.

As discussed earlier, the aminoglycosides are also commonly utilized with a cell wall active agent for synergistic purposes for gram-positive infections. In this situation, gentamicin is frequently administered to provide synergy in the treatment of serious infections due to Staphylococci, Enterococci, and *V. streptococci*.

In the current era of aminoglycoside utilization, two predominate intravenous administration techniques are employed in clinical practice. The older of the two approaches is the administration of multiple doses usually 1.7 to 2 mg/kg every eight hours for gentamicin and tobramycin, while amikacin was frequently dosed using regimens of 5 mg/kg every eight hours or 7.5 mg/kg every 12 hours (Fig. 3).

Using this technique, maintenance of concentrations within the therapeutic range for patient with alterations in elimination or volume of distribution was achieved with the use of a nomogram or by individualized pharmacokinetic dosing methods based on the patient-specific aminoglycoside disposition. Of the nomogram-based methods, the scheme of Sarubbi and Hull appears to have gained the widest acceptance (122). By this method, a loading dose of gentamicin or tobramycin of 1 to 2 mg/kg and of amikacin 5 to 7.5 mg/kg based on ideal body weight was given to adults with renal impairment. After the loading dose, subsequent dose were selected as a percentage of the chosen loading dose according to the desired dosing interval and the estimated creatinine clearance of the patient (Table 1).

Alternatively, one-half of loading dose may be given at intervals equal to that of the estimated half-life. While the nomogram approach was utilized frequently, the preferred method of dosage adjustment is to individualize the

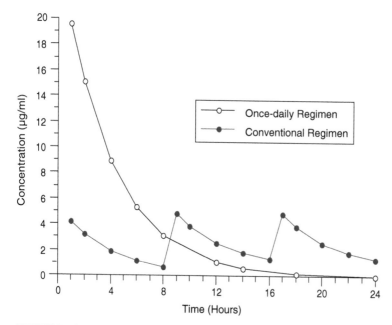

FIGURE 3 Concentration–time profile comparison of conventional q8h intermittent dosing versus the once-daily administration technique.

TABLE 1 Selection of Aminoglycoside Maintenance Dosing Using the Method of Sarubbi and Hull

Creatinine clearance (mL/min)	Half-life (hr)	Dose interval (8 hr)	Dose interval (12 hr)	Dose interval (24 hr)
90	3.1	84%	–	–
80	3.4	80	91%	–
70	3.9	76	88	–
60	4.5	71	84	–
50	5.3	65	79	–
40	6.5	57	72	92%
30	8.4	48	63	86
25	9.9	43	57	81
20	11.9	37	50	75
17	13.6	33	46	70
15	15.1	31	42	67
12	17.9	27	37	61
10	20.4	24	34	56
7	25.9	19	28	47
5	31.5	16	23	41
2	46.8	11	16	30
0	69.3	8	11	21

Source: From Ref. 122.

regimen using the standard pharmacokinetic dosing principles when aminoglycoside concentrations are available (123,124).

The second method has been referred to as the once-daily, single-daily, or the extended interval dosing method (Fig. 3). While the potential benefits of this second method were not well described until the early portion of the 1990s, this administration technique has now become the standard of practice in the United States as three of every four hospitals survey in 1998 utilized this administration technique (125,126).

For this reason and the wide availability of tertiary text references concerning the dosing of aminoglycosides using the more frequent intermittent approach, the remainder of this section will focus on the once-daily dosing methodology.

When considering the pharmacodynamic profile of aminoglycosides as described earlier in this chapter, four distinct advantages of using extended dosing intervals are readily apparent (87). As stated previously, giving aminoglycosides as a single daily dose, as opposed to conventional strategies, provides the opportunity to maximize the peak concentration/MIC ratio and the resultant bactericidal activity (Fig. 3). Second, this administration technique should minimize drug accumulation within the inner ear and kidney and therefore minimize the potential for toxic effects to these organs. Third, the PAE may also allow for longer periods of bacterial suppression during the dosing interval. Lastly, this aminoglycoside dosing approach may prevent the development of bacterial resistance.

Once-daily aminoglycoside therapy has been evaluated in several large clinical studies with a total study population of 100 or more patients (127–136). When compared with multidose aminoglycoside regimens, the once-daily regimen was shown to be as efficacious or superior to traditional dosing for the treatment of a wide variety of infections. Toxicity evaluations showed that there were no differences between the two dosing methods for either nephrotoxicity or ototoxicity. These toxicity data have been supported by other recent observations from investigators in Detroit (105,118). In addition, clinical experience at our own institution in a large patient population, which received either 7 mg/kg of gentamicin or tobramycin, indicates a reduced potential for nephrotoxicity (88). Lastly, several recently published meta-analyses evaluating once-daily dosing with standard dosing regimens also demonstrated that increased bacterial killing and trends for decreased toxicity are actually borne out in clinical practice when the extended interval dosing is used (137–145).

Studies have also been conducted in special populations such as pediatrics (146–157) and pregnant populations (158) for determination of serum concentrations, as well as comparisons for efficacy and safety between conventional administration and extended interval regimens. Considerable numbers of researches published in many countries suggest the international acceptance of extended interval regimens for infants and neonates including preterm and full-term babies, although multifactors such as postnatal age, gestational age, methods of ventilations, and other physiological status should be considered to decide patient-oriented extended interval regimen. Recently, Contopoulos-Ioannidis et al. (159) reported a meta-analysis on extended interval administration of aminoglycoside in pediatric populations. The study showed that efficacy data measured in the clinical failure rate, microbiologic failure rate, and combined effects favored once-daily dosing over multidaily dosing, although statistical difference was not achieved. Similarly, safety profiles measured for ototoxicity or nephrotoxicity tend to be better with once-daily dosing, in spite of insignificant statistical difference.

At present, the strategy for once-daily dosing has not been consistent in the literature as doses for gentamicin, tobramycin, and netilmicin have ranged from 3 to 7 mg/kg, whereas, the usually amikacin doses is 15 to 20 mg/kg. Dosing regimens, which use doses of less than 6 mg/kg for gentamicin, tobramycin, and netilmicin have arrived at this dose based on a conversion of the conventional milligram/kilogram dose, which is then administered once-daily. At present there appears to be four commonly advocated methods for the administration of once-daily aminoglycosides. While each of these approaches differs somewhat with regard to dose and/or interval, all reflect the need for dosage modification in the patient with renal disease. As of yet, no method has been shown to be superior over another. While concerns about the extended intervals and possible risk of increased toxicity should be mentioned in patients with reduced drug clearance, it should be no greater than that encountered with conventional dosing based on our current understanding of aminoglycoside-induced toxicity.

The first method of once-daily dosage determination has been proposed and implemented based on the pharmacokinetic and pharmacodynamic profile of these agents. This method, which was developed at our institution, intends to optimize the peak/MIC ratio in the majority of clinical situations by administering a dose of 7 mg/kg of either gentamicin or tobramycin (88). Similar to that of conventional regimens, once-daily protocols also require modification for patients with renal dysfunction in order to minimize drug accumulation. In the Hartford Hospital program, this is accomplished by administering a fixed dose with dosing interval adjustments for patients with impaired renal function (88). Due to the high peak concentrations obtained and the drug-free period at the end of the dosing interval, it is no longer necessary to draw standard peak and trough samples, rather a single random blood sample is obtained between 6 and 14 hours after the start of the aminoglycoside infusion. This serum concentration is used to determine the dosing interval based on a nomogram for once-daily dosing (Fig. 4) (88).

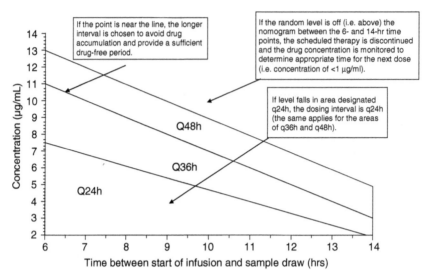

FIGURE 4 Once-daily aminoglycoside nomogram for the assessment of dosing interval using a 7 mg/kg dose of gentamicin or tobramycin. *Source*: From Ref. 88.

While Demczar et al. (160) have suggested that the nomogram may be inappropriate for the monitoring of therapy based on their assessment of aminoglycoside distribution in 11 healthy subjects, a subsequent population pharmacokinetic analysis using data derived from more than 300 patients receiving 7 mg/kg of tobramycin further supports the clinical utility of the original nomogram (161).

As a result of low toxicity, the short duration of therapy, and the excellent renal function of most patients, criteria have been developed to withhold the initial random concentration (which is obtained after the first or second dose) in patients: (*i*) receiving 24 hour dosing, (*ii*) without concurrently administered nephrotoxic agents (e.g., amphotericin, cyclosporine, and vancomycin), (*iii*) without exposure to contrast media, (*iv*) not quadriplegic nor amputee, (*v*) not in the intensive care unit, and (*vi*) less than 60 years of age (88). Even though the initial random concentration may be withheld in eligible patients, monitoring of the serum creatinine should continue to occur at two- to three-day intervals throughout the course of therapy. For patients who continue on the once-daily regimen for five days or more, a random concentration is obtained on the fifth day and weekly thereafter. Even though an initial random concentration may no longer be necessary in many patients, for those experiencing rapidly changing creatinine clearances or those in whom the creatinine clearance is significantly reduced (i.e., ≤30 mL/min) it may be necessary to obtain several samples to adequately structure the administration schedule to maximize efficacy and minimize toxicity. The 7 mg/kg dosage regimen has also been advocated by other investigators to rapidly obtain sufficient aminoglycoside exposures (45,162).

The second method proposed by Gilbert and coworker utilizes a 5 mg/kg gentamicin or tobramycin dose in patients without renal dysfunction (87,163). If dosage adjustment is required to compensate for diminished renal function, the

TABLE 2 Suggested Once-Daily Dosage Requirements for Patients with Altered Renal Function

Aminoglycoside	Creatinine clearance (mL/min)	Dosage interval (hr)	Dose (mg/kg)
Gentamicin/tobramycin	>80	24	5.0
	70	24	4.0
	60	24	4.0
	50	24	3.5
	40	24	2.5
	30	24	2.5
	20	48	4.0
	10	48	3.0
	Hemodialysis[a]	48	2.0
Amikacin	>80	24	15
	70	24	12
	50	24	7.5
	30	24	4.0
	20	48	7.5
	10	48	4.0
	Hemodialysis[a]	48	5.0

[a]Administer posthemodialysis.
Source: From Ref. 163.

dose and/or dosing interval may be modified to optimize therapy and minimize drug accumulation (Table 2).

A similar scheme for dosage modification has also been advocated by Prins et al. (164) for patients with renal dysfunction. Lastly, Begg et al. (165) have suggested two methods to optimize once-daily dosing. The first suggested for patients with normal renal function uses a graphical approach with target AUC values. The second method for patients with renal dysfunction uses two aminoglycoside serum concentrations and a target AUC value based on the 24-hour AUC that would result with multiple-dose regimens for dosage modifications.

While all the above noted once-daily methodologies have used fixed doses, subsequent dosage adjustments may be guided by individualized pharmacokinetic methods similar to that used for conventional multiple dose approach. On the other hand, while the individualization of therapy can be accomplished, no data are available to support that these manipulations will improve outcomes or minimize toxicity further than the fixed dose methodologies.

At the time of once-daily implementation, the methodology was introduced into clinical practice to further optimize the clinical outcomes of patients receiving these agents for serious infections. However, in addition to meeting this goal and reducing the incidence of drug-induced adverse events, this approach has also substantially reduced expenditures associated with the initiation of aminoglycoside therapy as compared to traditional dosing techniques (166–168).

APPLICATION OF PHARMACODYNAMICS AND DIRECT DELIVERY TO LOCAL SITES

To optimize efficacy, agents should arrive and remain at the site of infection for adequate duration of time to disrupt the life cycle of the target pathogens.

Adoption of contemporary pharmacodynamic principles described above suggests that higher concentrations, especially peak concentration, at target sites should be correlated with improved efficacy because the adequate pharmacodynamic surrogate marker is the C_{max}/MIC (169,170).

However, due in part to hydrophilicity of aminoglycosides, aminoglycosides may not achieve adequate or consistent local concentrations in certain sites such as bronchial fluids or bones and connective tissues. Therefore, a method of direct delivery to local sites may enhance concentrations at infection sites and potentially reduce systemic toxicity such as nephrotoxicity or ototoxicity. Utilization of aerolized tobramycin is the most prominent example to enhance efficient delivery to the intended site via direct local administration (171,172).

Clinicians used to utilize injectable solutions of tobramycin for this purpose. However, quite recently, tobramycin solution for inhalation (TOBI®) has launched on the market. Since types of compressors and/or nebulizers significantly affect its delivery and therefore pharmacokinetic profile, the manufacturer of TOBI suggests to administer with the reusable hand-held PARI LC PLUS® nebulizer and the DeVilbiss Pulmo-Aide® compressor (173–176). Pharmacokinetic studies of TOBI (300 mg bid) demonstrated significantly higher drug concentrations in the respiratory tract such as lung epithelial lining fluid concentrations or sputum concentrations, in spite of wide variation. As expected from its pharmacodynamics, TOBI significantly improved lung function such as forced expiratory (FEV1) volume per second, and reduced sputum bacterial density, hospitalization, or intravenous antibacterial treatment in randomized, double-blind, placebo-controlled studies performed in cystic fibrosis patients (177,178). Recipients demonstrated fewer systemic toxicities such as ototoxicity and nephrotoxicity after administration of TOBI, although it may cause local toxicities such as bronchospasms. In addition to its proven efficacy for cystic fibrosis patients, TOBI treatment significantly reduced bacterial loads in patients with bronchiectasis and *P. aeruginosa* (179). Another investigational inhaled tobramycin formulation, PulmoSphere formulation (PStob), by a passive dry powder completed a pharmacokinetics study in healthy human volunteers (180). It produced seven times higher whole-lung deposition and two times higher serum concentrations, as compared with TOBI. Although serum concentrations of PStob were higher than those of TOBI, they were low enough to avoid systemic toxicity. However, its efficacy and safety in patients should be tested before clinical application.

Besides aerosolized aminoglycosides, unique investigational products such as gentamicin-Eudragit microspheres (intended for intraocular formulation) (181), tobramycin- or gentamicin-bone cement (intended to prevent infection associated with prostheses) (182,183), and amikacin-treated fibrin glue (intended to prevent local graft infection) (184) have been invented. Although their efficacy and safety should be investigated before clinical utilization, efforts to invent and investigate novel products to enhance drug concentrations at the sites of infections should be valued to enhance clinical utilization of aminoglycosides by optimizing pharmacodynamics of aminoglycosides.

SUMMARY

The pharmacodynamic profile of aminoglycosides is maximized when high-dose, extended interval aminoglycoside therapy is employed. The use of this aminoglycoside administration technique has considerable in vitro and in vivo scientific support, which justifies its wide scale use within this country. The implementation of such programs should maximize the probability of clinical cure, minimize toxicity, and may help to avoid the development of resistance. Although such dosing is not appropriate for all patients, this strategy appears to be useful in the majority of patients requiring aminoglycoside therapy and can be successfully employed as a hospital-wide program.

REFERENCES

1. Kondo S, Iinuma K, Yamamoto H, Ikeda Y, Maeda K. Letter: synthesis of (S)-4-amino-2-hydroxybutyryl derivatives of 3',4'-dideoxykanamycin B and their antibacterial activities. J Antibiot (Tokyo) 1973; 26(11):705–707.
2. Fillastre JP, Leroy A, Humbert G, Moulin B, Bernadet P, Josse S. Pharmacokinetics of habekacin in patients with chronic renal insufficiency. Pathol Biol (Paris) 1987; 35(5 Pt 2):739–741.
3. Cundliffe E. Antibiotics and prokaryotic ribosomes: action, interaction and resistance. In: Hill WE, Dahlberg A, Garrett RA, Moore PB, Schlessinger D, Warner JR, eds. Ribosomes. Washington, D.C.: American Society for Microbiology, 1990:479–490.
4. Davies JE. Resistance to aminoglycosides: mechanisms and frequency. Rev Infect Dis 1983; 5:S261–S266.
5. Hancock REW. Aminoglycoside uptake and mode of action with special reference to streptomycin and gentamicin. J Antimicrob Chemother 1981; 8:249–276.
6. Hancock REW, Farmer SW, Li Z, Poole K. Interaction of aminoglycosides with the outer membranes and purified lipopolysaccharide and OmpF porin of *Escherichia coli.* Antimicrob Agents Chemother 1991; 35:1309–1314.
7. Gale EF, Cundliffe E, Reynolds PE, Richmond MH, Waring MJ. The Molecular Basis of Antibiotic Action. London: John Wiley & Sons, 1981.
8. Davis BD, Tai PC, Wallace BJ. Complex interactions of antibiotics with the ribosome. In: Nomura M, Tissieres A, Lengyel P, eds. Ribosomes. New York: Cold Spring Harbor Laboratory, 1974:771–789.
9. Gorini L. Streptomycin and misreading of the genetic code. In: Nomura M, Tissieres A, Lengyel P, eds. Ribosomes. New York: Cold Spring Harbor Laboratory, 1974: 791–803.
10. Ruusala T, Kurland CG. Streptomycin preferentially perturbs ribosomal proofreading. Mol Gen 1984; 198:100–104.
11. Matsunaga K, Yamaki H, Nishimura T, Tanaka N. Inhibition of DNA replication initiation by aminoglycoside antibiotics. Antimicrob Agents Chemother 1986; 30: 468–474.
12. Arya DP, Xue L, Willis B. Aminoglycoside (neomycin) preference is for A-form nucleic acids, not just RNA: results from a competition dialysis study. J Am Chem Soc 2003; 125(34):10148–10149.
13. Fujimura S, Tokue Y, Takahashi H, et al. Novel arbekacin- and amikacin-modifying enzyme of methicillin-resistant *Staphylococcus aureus.* FEMS Microbiol Lett 2000; 190(2):299–303.
14. Matsuo H, Kobayashi M, Kumagai T, Kuwabara M, Sugiyama M. Molecular mechanism for the enhancement of arbekacin resistance in a methicillin-resistant *Staphylococcus aureus.* FEBS Lett 2003; 546(2–3):401–406.
15. Tanaka N, Matsunaga K, Hirata A, Matsuhisa Y, Nishimura T. Mechanism of action of habekacin, a novel amino acid-containing aminoglycoside antibiotic. Antimicrob Agents Chemother 1983; 24(5):797–802.
16. Akins RL, Rybak MJ. in vitro activities of daptomycin, arbekacin, vancomycin, and gentamicin alone and/or in combination against glycopeptide intermediate-resistant *Staphylococcus aureus* in an infection model. Antimicrob Agents Chemother 2000; 44(7):1925–1929.
17. Axelrod HR, Kim JS, Longley CB, et al. Intestinal transport of gentamicin with a novel, glycosteroid drug transport agent. Pharm Res 1998; 15(12):1876–1881.

18. Swenson C, Curatolo WJ. Means to enhance penetration. Adv Drug Deliv Rev 1992; 8:39–92.
19. Rama Prasad YV, Eaimtrakarn S, Ishida M, et al. Evaluation of oral formulations of gentamicin containing labrasol in beagle dogs. Int J Pharm 2003; 268(1–2):13–21.
20. Ross BP, DeCruz SE, Lynch TB, Davis-Goff K, Toth I. Design, synthesis, and evaluation of a liposaccharide drug delivery agent: application to the gastrointestinal absorption of gentamicin. J Med Chem 2004; 47(5):1251–1258.
21. De Broe ME, Verbist L, Verpooten GA. Influence of dosage schedule on renal cortical accumulation of amikacin and tobramycin in man. J Antimicrob Chemother 1991; 27(suppl C):41–47.
22. Santre C, Georges H, Jacquier JM, et al. Amikacin levels in bronchial secretions of 10 pneumonia patients with respiratory support treated once daily versus twice daily. Antimicrob Agents Chemother 1995; 39:264–267.
23. Valcke YJ, Vogelaers DP, Colardyn FA, Pauwels RA. Penetration of netilmicin in the lower respiratory tract after once-daily dosing. Chest 1992; 101:1028–1032.
24. Zarowitz BJ, Robert S, Peterson EL. Prediction of glomerular filtration rate using aminoglycoside clearance in critically ill medical patients. Ann Pharmacother 1992; 26:1205–1210.
25. Begg EJ, Peddie BA, Chambers ST, Boswell DR. Comparison of gentamicin dosing regimens using an in-vitro model. J Antimicrob Chemother 1992; 29:427–433.
26. Craig WA, Ebert SC. Killing and regrowth of bacteria in vitro: a review. Scand J Infect Dis 1991; (suppl 74):63–70.
27. Drusano GL, Johnson DE, Rosen M, Standiford HC. Pharmacodynamics of a fluoroquinolone antimicrobial agent in a neutropenic rat model of *Pseudomonas sepsis*. Antimicrob Agents Chemother 1993; 37:483–490.
28. Dudley MN. Pharmacodynamics and pharmacokinetics of antibiotics with special reference to the fluoroquinolones. Am J Med 1991; 91(suppl 6A):45S–50S.
29. Davis BD. Mechanism of the bactericidal action of the aminoglycosides. Microbiol Rev 1987; 51:341–350.
30. Ebert SC, Craig WA. Pharmacodynamic properties of antibiotic: application to drug monitoring and dosage regimen design. Infect Control Hosp Epidemiol 1990; 11: 319–326.
31. Blaser J, Stone B, Groner M, et al. Comparative study with enoxacin and netilmicin in a pharmacodynamic model to determine importance of ratio of antibiotic peak concentration to MIC for bactericidal activity and emergence of resistance. Antimicrob Agent Chemother 1987; 31:1054–1060.
32. Leggett JE, Ebert S, Fantin B, Craig WA. Comparative dose-effect relationships at several dosing intervals for beta-lactam, aminoglycoside and quinolone antibiotics against gram-negative bacilli in murine thigh-infection and pneumonitis models. Scan J Infect Dis 1991; 74:179–184.
33. Moore RD, Smith CR, Lietman PS. Association of aminoglycoside plasma levels with therapeutic outcome in gram-negative pneumonia. Am J Med 1984; 77:657–662.
34. Moore RD, Smith CR, Lietman PS. The association of aminoglycoside plasma levels with mortality in patients with gram-negative bacteremia. J Infect Dis 1984; 149: 443–448.
35. Moore RD, Lietman PS, Smith CR. Clinical response to aminoglycoside therapy: importance of the ratio of peak concentration to minimal inhibitory concentration. J Infect Dis 1987; 155:93–99.
36. Deziel-Evans L, Murphy J, Job M. Correlation of pharmacokinetic indices with therapeutic outcomes in patients receiving aminoglycoside. Clin Pharm 1986; 5: 319–324.
37. Keating MF, Bodey GP, Valdivieso M, Rodriguez V. A randomized comparative trial of three aminoglycosides—comparison of continuous infusions of gentamicin, amikacin, and sisomicin combined with carbenicillin in the treatment of infections in neutropenic patients with malignancies. Medicine 1979; 58:159–170.
38. Williams PJ, Hull JH, Sarubbi FA, Rogers JF, Wargin WA. Factors associated with nephrotoxicity and clinical outcome in patients receiving amikacin. J Clin Pharmacol 1986; 26:79–86.

39. Fiala M, Chatterjee SN. Antibiotic blood concentrations in patients successfully treated with tobramycin. Postgrad Med 1981; 57:548–551.
40. Berirtzoglou E, Golegou S, Savvaidis I, Bezirtzoglou C, Beris A, Xenakis T. A relationship between serum gentamicin concentrations and minimal inhibitor concentration. Drugs Exp Clin Res 1996; 22:57–60.
41. Binder L, Schiel X, Binder C, et al. Clinical outcome and economic impact of aminoglycoside peak concentrations in febrile immunocompromised patients with hematologic malignancies. Clin Chem 1998; 44:408–414.
42. Anderson ET, Young LS, Hewitt WL. Simultaneous antibiotic levels in "breakthrough" gram-negative rod bacteremia. Am J Med 1976; 61:493–497.
43. Noone P, Parsons TMC, Pattison JR, Slack RCB, Garfield-Davies D, Hughes K. Experience in monitoring gentamicin therapy during treatment of serious gram-negative sepsis. Br Med J 1974; 1:477–481.
44. Reymann MT, Bradac JA, Cobbs CG, Dismukes WE. Correlation of aminoglycoside dosage with serum concentrations during therapy of serious gram-negative bacillary disease. Antimicrob Agent Chemother 1979; 16:13353–13361.
45. Kashuba ADM, Bertino JS Jr, Nafziger AN. Dosing of aminoglycosides to rapidly attain pharmacodynamic goals and hasten therapeutic response by using individualized pharmacokinetic monitoring of patients with pneumonia caused by gram-negative organism. Antimicrob Agent Chemother 1998; 42:1842–1844.
46. Kashuba ADM, Nafziger AN, Drusano GL, Bertino JS Jr. Optimizing aminoglycoside therapy for nosocomial pneumonia caused by gram-negative bacteria. Antimicrob Agent Chemother 1999; 43:1623–1629.
47. Craig W, Gunmundsson S. Postantibiotic effect. In: Lorian V, ed. Antibiotics in Laboratory Medicine. 4th ed. Baltimore, Philadelphia: Williams & Wilkins, 1987:296–329.
48. Zhanel G, Hoban D, Harding G. The postantibiotic effect: a review of in-vitro and in vivo data. Ann Pharmacother 1991; 25:153–163.
49. Bundtzen R, Gerber A, Cohn D, et al. Postantibiotic suppression of bacterial growth. Rev Infect Dis 1981; 3:28–37.
50. Vogelman B, Gudmundsson S, Turnidge J, et al. In vivo postantibiotic effect in a thigh infection in neutropenic mice. J Infect Dis 1988; 157:287–298.
51. Hessen M, Pitsakis P, Levison M. Postantibiotic effect of penicillin plus gentamicin versus *Enterococcus faecalis* in vitro and in vivo. Antimicrob Agent Chemother 1989; 33:608–611.
52. McGrath B, Marchbanks C, Gilbert D, et al. In vitro postantibiotic effect following exposure to imipenem, temafloxacin, and tobramycin. Antimicrob Agent Chemother 1993; 37:1723–1725.
53. Vogelman B, Craig W. Postantibiotic effects. J Antimicrob Chemother 1985; 15(suppl A):37–46.
54. Odenholt-Tornqvist I. Pharmacodynamics of beta-lactam antibiotics: studies on the paradoxical effect and postantibiotic effects in vitro and in an animal model. Scand J Infect Dis 1989; 58(S):1–55.
55. Craig W, Leggett K, Totsuka K, et al. Key pharmacokinetic parameters of antibiotic efficacy in experimental animal infections. J. Drug Dev 1998; 1(S3):7–15.
56. Dornbusch K, Henning C, Linden E. In-vitro activity of the new penems FCE 22101 and FCE 24362 alone or in combination with aminoglycosides against streptococci isolated from patients with endocarditis. J Antimicrob Chemother 1989; 23(suppl C): 109–117.
57. Fuursted K. Comparative killing activity and postantibiotic effect of streptomycin combined with ampicillin, ciprofloxacin, imipenem, piperacillin, or vancomycin against strains of *Streptococcus faecalis* and *Streptococcus faecium*. Chemother 1988; 34:229–234.
58. Winstanley T, Hastings J. Penicillin-aminoglycosides synergy and post-antibiotic effect for enterococci. JAC 1989; 23:189–199.
59. Gudmundsson S, Erlendsdottir H, Gottfredsson M, et al. The postantibiotic effect induced by antimicrobial combinations. Scand J Infect Dis 1991; 74:80–93.
60. Karlowsky J, Zhanel G, Davidson R, et al. Postantibiotic effect in *Pseudomonas aeruginosa* following single and multiple aminoglycoside exposures in vitro. J Antimicrob Chemother 1994; 33:937–947.

61. Karlowsky J, Zhanel G, Davidson R, et al. In vitro postantibiotic effects following multiple exposures of cefotaxime, ciprofloxacin, and gentamicin against *Escherichia coli* in pooled human cerebrospinal fluid and Mueller-Hinton broth. Antimicrob Agent Chemother 1993; 37:1154–1157.

62. Li R, Zhu Z, Lee S, et al. Antibiotic exposure and its relationship to postantibiotic effect and bactericidal activity: constant versus exponentially decreasing tobramycin concentrations against *Pseudomonas aeruginosa*. Antimicrob Agent Chemother 1997; 41:1808–1811.

63. Rescott D, Nix D, Holden P, et al. Comparison of two methods for determining in vitro postantibiotic effects of three antibiotics on *Escherichia coli*. Antimicrob Agent Chemother 1998; 32:450–453.

64. Fuursted K. Postexposure factors influencing the duration of postantibiotic effect: significance of temperature, pH, cations, and oxygen tension. Antimicrob Agent Chemother 1997; 41:1693–1696.

65. Gudmundsson A, Erlendsdottir H, Gottfredsson M, et al. The impact of pH and cationic supplementation on in vitro postantibiotic effect. Antimicrob Agent Chemother 1991; 35:2617–2624.

66. Park MK, Myers R, Marzella L. Hyperoxia and prolongation of aminoglycoside-induced postantibiotic effect in *Pseudomonas aeruginosa*: role of reactive oxygen species. Antimicrob Agent Chemother 1993; 37:120–122.

67. Hanberger H, Nilsson L, Maller R, et al. Pharmacodynamics of daptomycin and vancomycin on *Enterococcus faecalis* and *Staphylococcus aureus* demonstrated by studies of initial killing and postantibiotic effect and influence of Ca^{2+} and albumin on these drugs. Antimicrob Agent Chemother 1991; 35:1710–1716.

68. Cars O, Odenholt-Tornqvist I. The postantibiotic sub-MIC effect in vitro and in vivo. J Antimicrob Chemother 1993; 31(suppl D):159–166.

69. Craig W, Vogelman B. The postantibiotic effect. Ann Intern Med 1987; 106:900–902.

70. Gottfredsson M, Erlendsdottir H, Gudmundsson, et al. Different patterns of bacterial DNA synthesis during the postantibiotic effect. Antimicrob Agent Chemother 1995; 39:1314–1319.

71. Barmada S, Kohlhepp S, Leggett J, et al. Correlation of tobramycin-induced inhibition of protein synthesis with postantibiotic effect in *Escherichia coli*. Antimicrob Agent Chemother 1993; 37:2678–2683.

72. Minguez F, Izquierdo J, Caminero M, et al. In vivo postantibiotic effect of isepamicin and other aminoglycosides in a thigh infection model in neutropenic mice. Chemother 1992; 38:179–184.

73. Gudmundsson S, Einarsson S, Erlendsdottir H. The post-antibiotic effect of antimicrobial combinations in a neutropenic murine thigh infection model. J Antimicrob Chemother 1993; 31(suppl D):177–191.

74. Craig W, Redington J, Ebert S. Pharmacodynamics of amikacin in vitro and in mouse thigh and lung infections. J Antimicrob Chemother 1991; 27(SC):29–40.

75. Hessen M, Pitsakis P, Levison M. Absence of a postantibiotic effect in experimental *Pseudomonas endocarditis* treated with imipenem, with or without gentamicin. J Infect Dis 1998; 158:542–548.

76. Craig W. Post-antibiotic effects in experimental infection model: relationship to in-vitro phenomena and to treatment of infection in man. J Antimicrob Chemother 1993; 31(suppl D):149–158.

77. Fantin B, Craig W. Factors affecting duration of in vivo postantibiotic effect for aminoglycosides against Gram-negative bacilli. Antimicrob Agents Chemother 1991; 27:829–836.

78. Mingeot-Leclercq MP, Glupczynski Y, Tulkens PM. Aminoglycosides: activity and resistance. Antimicrob Agents Chemother 1999; 43(4):727–737.

79. Karlowsky JA, Zhanel GG, Davidson RJ, Hoban DJ. Once-daily aminoglycoside dosing assessed by MIC reversion time with *Pseudomonas aeruginosa*. Antimicrob Agents Chemother 1994; 38:1165–1168.

80. Daikos GL, Jackson GG, Lolans V, Livermore DM. Adaptive resistance to aminoglycoside antibiotics from first-exposure down-regulation. J Infect Dis 1990; 162:414–420.

81. Daikos GL, Lolans VT, Jackson GG. First-exposure adaptive resistance to aminoglycoside antibiotics in vivo with meaning for optimal clinical use. Antimicrob Agents Chemother 1991; 35:117–123.
82. Gerber AU, Vastola AP, Brandel J, Craig WA. Selection of aminoglycoside-resistant variants of *Pseudomonas aeruginosa* in an in vivo model. J Infect Dis 1982; 146:691–697.
83. Owens RC Jr., Banevicius MA, Nicolau DP, Nightingale CH, Quintiliani R. In vitro synergistic activities of tobramycin and selected β-lactams against 75 gram-negative clinical isolates. Antimicrob Agents Chemother 1997; 41:2586–2588.
84. Marangos MN, Nicolau DP, Quintiliani R, Nightingale CH. Influence of gentamicin dosing interval on the efficacy of penicillin containing regimens in experimental *Enterococcus faecalis* endocarditis. J Antimicrob Chemother 1997; 39:519–522.
85. Hallander HO, Donrbusch K, Gezelius L, Jacobson K, Karlsson I. Synergism between aminoglycosides and cephalosporins with anti-pseudomonal activity: interaction index and killing curve method. Antimicrob Agents Chemother 1982; 22:743–752.
86. Sandoval RM, Molitoris BA. Gentamicin traffics retrograde through the secretory pathway and is released in the cytosol via the endoplasmic reticulum. Am J Physiol Renal Physiol 2004; 286(4):F617–F624.
87. Gilbert DN. Once-daily aminoglycoside therapy. Antimicrob Agents Chemother 1991:35:399–405.
88. Nicolau DP, Freeman CD, Belliveau PP, Nightingale CH, Ross JW, Quintiliani R. Experience with a once-daily aminoglycoside program administered to 2,184 adult patients. Antimicrob Agents Chemother 1995; 39:650–655.
89. Govaerts PJ, Claes J, van de Heyning PH, Jorens PG, Marquet J, De Broe ME. Aminoglycoside-induced ototoxicity. Toxicol Lett 1990; 52(3):227–251.
90. Matz G, Rybak L, Roland PS, et al. Ototoxicity of ototopical antibiotic drops in humans. Otolaryngol Head Neck Surg 2004; 130(3 suppl):S79–S82.
91. Hutchin T, Cortopassi G. Proposed molecular and cellular mechanism for aminoglycoside ototoxicity. Antimicrob Agents Chemother 1994; 38:2517–2520.
92. Fausti SA, Henry JA, Scheffer HI, Olson DJ, Frey RH, McDonald WJ. High-frequency audiometric monitoring for early detection of aminoglycoside ototoxicity. J Infect Dis 1992; 165:1026–1032.
93. Federspil P. Drug-induced sudden hearing loss and vestibular disturbances. Adv Otorhinolaryngol 1981; 27:144–158.
94. Rakover Y, Keywan K, Rosen G. Safety of topical ear drops containing ototoxic antibiotics. J Otolaryngol 1997; 26(3):194–196.
95. Merifield DO, Parker NJ, Nicholson NC. Therapeutic management of chronic suppurative otitis media with otic drops. Otolaryngol Head Neck Surg 1993; 109(1): 77–82.
96. Welling DB, Forrest LA, Goll F 3rd. Safety of ototopical antibiotics. Laryngoscope 1995; 105(5 Pt 1):472–474.
97. Kaplan DM, Nedzelski JM, Al-Abidi A, Chen JM, Shipp DB. Hearing loss following intratympanic instillation of gentamicin for the treatment of unilateral Meniere's disease. J Otolaryngol 2002; 31(2):106–111.
98. Kaplan DM, Hehar SS, Bance ML, Rutka JA. Intentional ablation of vestibular function using commercially available topical gentamicin-betamethasone eardrops in patients with Meniere's disease: further evidence for topical eardrop ototoxicity. Laryngoscope 2002; 112(4):689–695.
99. Lange G, Maurer J, Mann W. Long-term results after interval therapy with intratympanic gentamicin for Meniere's disease. Laryngoscope 2004; 114(1):102–105.
100. Rybak LP, Kelly T. Ototoxicity: bioprotective mechanisms. Curr Opin Otolaryngol Head Neck Surg 2003; 11(5):328–333.
101. Klemens JJ, Meech RP, Hughes LF, Somani S, Campbell KC. Antioxidant enzyme levels inversely covary with hearing loss after amikacin treatment. J Am Acad Audiol 2003; 14(3):134–143.
102. Matsui JI, Haque A, Huss D, et al. Caspase inhibitors promote vestibular hair cell survival and function after aminoglycoside treatment in vivo. J Neurosci 2003; 23(14):6111–6122.

103. Beaubien AR, Ormsby E, Bayne A, et al. Evidence that amikacin ototoxicity is related to total perilymph area under the concentration-time curve regardless of concentration. Antimicrob Agents Chemother 1991; 35:1070–1074.

104. Proctor L, Petty B, Lietman P, Thakor R, Glackin R, Shimizu H. A study of potential vestibulotoxicity effects of once daily versus thrice daily administration of tobramycin. Laryngoscope 1987; 97:1443–1449.

105. Rybak MJ, Abate BJ, Kang SL, Ruffing MJ, Lerner SA, Drusano GL. Prospective evaluation of the effect of an aminoglycoside dosing regimen on rates of observed nephrotoxicity and ototoxicity. Antimicrob Agents Chemother 1999; 43:1549–1555.

106. Garrison MW, Zaske DE, Rotschafer JC. Aminoglycosides: another perspective. DICP Ann Pharmacother 1990; 24:267–272.

107. Mingeot-Leclercq MP, Tulkens PM. Aminoglycosides: nephrotoxicity. Antimicrob Agents Chemother 1999; 43(5):1003–1012.

108. Moore RD, Smith CR, Lietman PS. Risk factors for nephrotoxicity in patients treated with aminoglycosides. Ann Intern Med 1984; 100:352–357.

109. Sawyers CL, Moore RD, Lerner SA, Smith CR. A model for predicting nephrotoxicity with aminoglycosides. J Infect Dis 1986; 153:1062–1068.

110. Whelton A. Therapeutic initiatives for avoidance of aminoglycoside toxicity. J Clin Pharmacol 1985; 25:67–81.

111. Bertino JS Jr, Booker LA, Franck PA, Jenkins PL, Nafziger AN. Incidence and significant risk factors for aminoglycoside-associated nephrotoxicity in patients dosed by using individualized pharmacokinetic monitoring. J Infect Dis 1993; 167:173–179.

112. Nodoushani M, Nicolau DP, Hitt CH, Quintiliani R, Nightingale CH. Evaluation of nephrotoxicity associated with once-daily aminoglycoside administration. J Pharm Tech 1997; 13:258–262.

113. Ali BH. Agents ameliorating or augmenting experimental gentamicin nephrotoxicity: some recent research. Food Chem Toxicol 2003; 41(11):1447–1452.

114. Maldonado PD, Barrera D, Medina-Campos ON, Hernandez-Pando R, Ibarra-Rubio ME, Pedraza-Chaverri J. Aged garlic extract attenuates gentamicin induced renal damage and oxidative stress in rats. Life Sci 2003; 73(20):2543–2556.

115. Parlakpinar H, Ozer MK, Sahna E, Vardi N, Cigremis Y, Acet A. Amikacin-induced acute renal injury in rats: protective role of melatonin. J Pineal Res 2003; 35(2):85–90.

116. Sohn EJ, Kang DG, Lee HS. Protective effects of glycyrrhizin on gentamicin-induced acute renal failure in rats. Pharmacol Toxicol 2003; 93(3):116–122.

117. Verpooten GA, Giuliano RA, Verbist L, Eestermans G, De Broe ME. Once daily dosing decreases the accumulation of gentamicin and netilmicin. Clin Pharmacol Ther 1989; 45:22–27.

118. Murray KR, McKinnon PS, Mitrzyk B, Rybak MJ. Pharmacodynamic characterization of nephrotoxicity associated with once-daily aminoglycoside. Pharmacother 1999; 19:1252–1260.

119. Matthews I, Kirkpatrick C, Holford N. Quantitative justification for target concentration intervention—parameter variability and predictive performance using population pharmacokinetic models for aminoglycosides. Br J Clin Pharmacol 2004; 58(1):8–19.

120. Xuan D, Nicolau DP, Nightingale CH. Population pharmacokinetics of gentamicin in hospitalized patients receiving once-daily dosing. Int J Antimicrob Agents 2004; 23(3):291–295.

121. Rougier F, Claude D, Maurin M, et al. Aminoglycoside nephrotoxicity: modeling, simulation, and control. Antimicrob Agents Chemother 2003; 47(3):1010–1016.

122. Sarubbi FA Jr, Hull JH. Amikacin serum concentrations: prediction of levels and dosage guidelines. Ann Intern Med 1978; 89:612–618.

123. Sawchuk RJ, Zaske DE. Pharmacokinetics of dosing regimens which utilize multiple intravenous infusions: gentamicin in burn patients. J Pharmacokinet Biopharm 1976; 4:183–195.

124. Sawchuk RJ, Zaske DE, Cipolle RJ, Wargin WA, Strate RG. Kinetic model for gentamicin dosing with the use of individual patient parameters. Clin Pharmacol Ther 1977; 21:362–369.

125. Schumock GT, Raber SR, Crawford SY, Naderer OJ, Rodvold KA. National survey of once-daily dosing of aminoglycoside antibiotics. Pharmacother 1995; 15:201–209.

126. Chuck SK, Raber SR, Rodvold KA, Areff D. National survey of extended-interval aminoglycoside dosing. Clin Infect Dis 2000; 30:433–439.

127. DeVries PJ, Verkooyen RP, Leguit P, Verbrugh HA. Prospective randomized study of once-daily versus thrice-daily netilmicin regimens in patients with intraabdominal infections. Eur J Clin Microbiol Infect Dis 1990; 9:161–168.

128. Mauracher EH, Lau WY, Kartowisastro H, et al. Comparison of once-daily and thrice-daily netilmicin regimens in serious systemic infections: a multicenter study in six Asian countries. Clin Ther 1989; 11:604–613.

129. Maller R, Ahrne H, Eilard T, et al. Efficacy and safety of amikacin in systemic infections when given as a single daily dose or in two divided doses. J Antimicrob Chemother 1991; 27(suppl C):121–128.

130. Maller R, Ahrne H, Holmen C, et al. Onceversus twice-daily amikacin regimen: efficacy and safety in systemic gram-negative infections. J Antimicrob Chemother 1993; 31:939–948.

131. Prins JM, Buller HR, Kuijper EJ, Tange RA, Speelman P. Once versus thrice daily gentamicin in patients with serious infections. Lancet 1993; 341:335–339.

132. Rozdzinski E, Kern WV, Reichle A, et al. Once-daily versus thrice-daily dosing of netilmicin in combination with beta-lactam antibiotics as empirical therapy for febrile neutropenic patients. J Antimicrob Chemother 1993; 31:585–598.

133. TerBraak EW, DeVries PJ, Bouter KP, et al. Once-daily dosing regimen for aminoglycoside plus beta-lactam combination therapy of serious bacterial infections: comparative trial with netilmicin plus ceftriaxone. Am J Med 1990; 89:58–66.

134. International Antimicrobial Therapy Cooperative Group of the EORTC. Efficacy and toxicity of single daily doses of amikacin and ceftriaxone versus multiple daily doses of amikacin and ceftazidime for infection in patients with cancer and granulocytopenia. Ann Intern Med 1993; 119:584–593.

135. Beaucaire G, Leroy O, Beuscart C, et al. Clinical and bacteriological efficacy, and practical aspects of amikacin given once daily for severe infections. J Antimicrob Chemother 1991; 27(suppl C):91–103.

136. Prins JM, Buller HR, Kuijper EJ, Tange RA, Speelman P. Once-daily gentamicin versus once-daily netilmicin in patients with serious infections-a randomized clinical trial. J Antimicrob Chemother 1994; 33:823–835.

137. Galoe AM, Graudal N, Christensen HR, Kampmann JP. Aminoglycosides: single or multiple daily dosing? A meta-analysis on efficacy and safety. Eur J Clin Pharmacol 1995; 48:39–43.

138. Freeman CD, Strayer AH. Mega-analysis of meta-analysis: an examination of meta-analysis with an emphasis on once-daily aminoglycoside comparative trials. Pharmacother 1996; 16:1093–1102.

139. Hatala R, Dinh T, Cook D. Once-daily aminoglycoside dosing in immunocompetent adults: a meta-analysis. Ann Intern Med 1996; 124:717–725.

140. Barza M, Ioannidis JPA, Cappelleri JC, Lau J. Single or multiple doses of aminoglycosides: a meta-analysis. BMJ 1996; 312:338–345.

141. Munckhof WJ, Grayson JL, Turnidge JD. A meta-analysis of studies on the safety and efficacy of aminoglycosides given with once daily or as divided doses. J Antimicrob Chemother 1996; 37:645–663.

142. Ferriols-Lisart R, Alos-Alminana M. Effectiveness and safety of once-daily aminoglycosides: a meta-analysis. Am J Health-Syst Pharm 1996; 53:1141–1150.

143. Bailey TC, Little JR, Littenberg B, Reichley RM, Dunagan WC. A meta-analysis of extended-interval dosing versus multiple daily dosing of aminoglycosides. Clin Infect Dis 1997; 24:786–795.

144. Ali MZ, Goetz MB. A meta-analysis of the relative efficacy and toxicity of single daily dosing versus multiple daily dosing of aminoglycosides. Clin Infect Dis 1997; 24:796–809.

145. Hatala R, Dinh TT, Cook DJ. Single daily dosing of aminoglycosides in immunocompromised adults: a systematic review. Clin Infect Dis 1997; 24:810–815.

146. Marik PE, Lipman J, Kobilski S, Scribante J. A prospective randomized study comparing once- versus twice-daily amikacin dosing in critically ill adult and paediatric patients. J Antimicrob Chemother 1991; 28:753–764.
147. Nicolau DP, Quintiliani R, Nightingale CH. Once-a-day aminoglycoside therapy. Report Ped Infect Dis 1997; 7:28.
148. Sung L, Dupuis LL, Bliss B, et al. Randomized controlled trial of once- versus thrice-daily tobramycin in febrile neutropenic children undergoing stem cell transplantation. J Natl Cancer Inst 2003; 95(24):1869–1877.
149. Mercado MC, Brodsky NL, McGuire MK, Hurt H. Extended interval dosing of gentamicin in preterm infants. Am J Perinatol 2004; 21(2):73–77.
150. Bhatt-Mehta V, Donn SM. Gentamicin pharmacokinetics in term newborn infants receiving high-frequency oscillatory ventilation or conventional mechanical ventilation: a case-controlled study. J Perinatol 2003; 23(7):559–562.
151. Dupuis LL, Sung L, Taylor T, et al. Tobramycin pharmacokinetics in children with febrile neutropenia undergoing stem cell transplantation: once-daily versus thrice-daily administration. Pharmacotherapy 2004; 24(5):564–573.
152. Piekarczyk A, Kaminska E, Taljanski W, Sosonowska K, Poszwinska B, Rutkowska, M. Pharmacokinetics of netilmicin in neonates. Med Wieku Rozwoj 2003; 7(4 Pt 2):547–555.
153. English M, Mohammed S, Ross A, et al. A randomised, controlled trial of once daily and multi-dose daily gentamicin in young Kenyan infants. Arch Dis Child 2004; 89(7):665–669.
154. Botha JH, du Preez MJ, Adhikari M. Population pharmacokinetics of gentamicin in South African newborns. Eur J Clin Pharmacol 2003; 59(10):755–759.
155. Kosalaraksa P, Janthep P, Jirapradittha J, Taksaphan S, Kiatchoosakun P. Once versus twice daily dose of gentamicin therapy in Thai neonates. J Med Assoc Thai 2004; 87(4):372–376.
156. Knight JA, Davis EM, Manouilov K, Hoie EB. The effect of postnatal age on gentamicin pharmacokinetics in neonates. Pharmacotherapy 2003; 23(8):992–996.
157. Hansen A, Forbes P, Arnold A, O'Rourke E. Once-daily gentamicin dosing for the preterm and term newborn: proposal for a simple regimen that achieves target levels. J Perinatol 2003; 23(8):635–639.
158. Bourget P, Fernandez H, Delouis C, Taburet AM. Pharmacokinetics of tobramycin in pregnant women, safety and efficacy of a once-daily dose regimen. J Clin Pharm Ther 1991; 16:167–176.
159. Contopoulos-Ioannidis DG, Giotis ND, Baliatsa DV, Ioannidis JP. Extended-interval aminoglycoside administration for children: a meta-analysis. Pediatrics 2004; 114(1): e111–e118.
160. Demczar DJ, Nafziger AN, Bertino JS Jr. Pharmacokinetics of gentamicin at traditional versus high doses: implications for once-daily aminoglycoside dosing. Antimicrob Agents Chemother 1997; 41:1115–1119.
161. Xuan D, Lu JF, Nicolau DP, Nightingale CH. Population pharmacokinetics study of tobramycin after once-daily dosing in hospitalized patients. Intern J Antimicrob Agents 2000; 15:185–191.
162. Konrad F, Wagner R, Neumeister B, Rommel H, Georgieff M. Studies on drug monitoring in thrice and once daily treatment with aminoglycosides. Intensive Care Med 1993; 19:215–220.
163. Gilbert DN, Bennett WM. Use of antimicrobial agents in renal failure. Infect Dis Clin North Am 1989; 3:517–531.
164. Prins JM, Koopmans RP, Buller HR, Kuijper EJ, Speelman P. Easier monitoring of aminoglycoside therapy with once-daily dosing schedules. Eur J Clin Microbiol Infect Dis 1995; 14:531–535.
165. Begg EJ, Barclay ML, Duffull SB. A suggested approach to once-daily aminoglycoside dosing. Br J Clin Pharmacol 1995; 39:605–609.
166. Nicolau DP, Wu AHB, Finocchiaro S, et al. Once-daily aminoglycoside dosing: impact on requests for therapeutic drug monitoring. Therapeutic Drug Mon 1996; 18:263–266.
167. Hitt CM, Klepser ME, Nightingale CH, Quintiliani R, Nicolau DP. Pharmacoeconomic impact of a once-daily aminoglycoside administration. Pharmacother 1997:17810–17814.

168. Parker SE, Davey PG. Once-daily aminoglycoside administration in gram-negative sepsis: economic and practical aspects. Pharmacoeconomics 1995; 7:393–402.
169. Klepser ME. Role of nebulized antibiotics for the treatment of respiratory infections. Curr Opin Infect Dis 2004; 17(2):109–112.
170. Flume P, Klepser ME. The rationale for aerosolized antibiotics. Pharmacotherapy 2002; 22(3 Pt 2):71S–79S.
171. Tiddens H. Inhaled antibiotics. Pediatr Pulmonol Suppl 2004; 26:92–94.
172. Cole PJ. The role of nebulized antibiotics in treating serious respiratory infections. J Chemother 2001; 13(4):354–362.
173. de Boer AH, Hagedoorn P, Frijlink HW. The choice of a compressor for the aerosolisation of tobramycin (TOBI) with the PARI LC PLUS reusable nebuliser. Int J Pharm 2003; 268(1–2):59–69.
174. Cheer SM, Waugh J, Noble S. Inhaled tobramycin (TOBI): a review of its use in the management of *Pseudomonas aeruginosa* infections in patients with cystic fibrosis. Drugs 2003; 63(22):2501–2520.
175. Eisenberg J, Pepe M, Williams-Warren J, et al. A comparison of peak sputum tobramycin concentration in patients with cystic fibrosis using jet and ultrasonic nebulizer systems. Aerosolized Tobramycin Study Group. Chest 1997; 111(4):955–962.
176. Geller DE, Pitlick WH, Nardella PA, Tracewell WG, Ramsey BW. Pharmacokinetics and bioavailability of aerosolized tobramycin in cystic fibrosis. Chest 2002; 122(1):219–226.
177. Moss RB. Administration of aerosolized antibiotics in cystic fibrosis patients. Chest 2001; 120(3 suppl):107S–113S.
178. Ramsey BW, Pepe MS, Quan JM, et al. Intermittent administration of inhaled tobramycin in patients with cystic fibrosis. Cystic Fibrosis Inhaled Tobramycin Study Group. N Engl J Med 1999; 340(1):23–30.
179. Barker AF, Couch L, Fiel SB, et al. Tobramycin solution for inhalation reduces sputum *Pseudomonas aeruginosa* density in bronchiectasis. Am J Respir Crit Care Med 2000; 162(2 Pt 1):481–485.
180. Newhouse MT, Hirst PH, Duddu SP, et al. Inhalation of a dry powder tobramycin PulmoSphere formulation in healthy volunteers. Chest 2003; 124(1):360–366.
181. Al-Kassas R. Design and in vitro evaluation of gentamicin-Eudragit microspheres intended for intra-ocular administration. J Microencapsul 2004; 21(1):71–81.
182. Sterling GJ, Crawford S, Potter JH, Koerbin G, Crawford R. The pharmacokinetics of simplex-tobramycin bone cement. J Bone Joint Surg Br 2003; 85(5):646–649.
183. Fletcher MD, Spencer RF, Langkamer VG, Lovering AM. Gentamicin concentrations in diagnostic aspirates from 25 patients with hip and knee arthroplasties. Acta Orthop Scand 2004; 75(2):173–176.
184. Nishimoto K, Yamamura K, Fukase F, Kobayashi M, Nishikimi N, Komori K. Subcutaneous tissue release of amikacin from a fibrin glue/polyurethane graft. J Infect Chemother 2004; 10(2):101–104.

8 Quinolones

Paul G. Ambrose
Institute for Clinical Pharmacodynamics, Ordway Research Institute, Albany, New York, and University of the Pacific School of Health Sciences, Stockton, California, U.S.A.

Sujata M. Bhavnani
Institute for Clinical Pharmacodynamics, Ordway Research Institute, Albany, New York, U.S.A.

Robert C. Owens, Jr.
Maine Medical Center, Portland, Maine, and University of Vermont, College of Medicine, Burlington, Vermont, U.S.A.

INTRODUCTION

Pharmacokinetics-pharmacodynamics (PK-PD) describes the science that relates drug concentration to an agent's pharmacological or toxicological effects. For antimicrobial agents, PK-PD describes not only the time-course of antimicrobial effect on microorganisms, but also the time-course of drug effect on patient signs and symptoms of infection. Over the past 25 years, animal and in vitro infection models have served to further our understanding of the PK-PD of antimicrobial agents. Over the last 15 years, clinical data have emerged and have demonstrated that the magnitude of the PK-PD measures associated with efficacy in animal and in vitro infection models is remarkably concordant with those required for efficacy in humans (1).

In this chapter, we concentrate on PK-PD first principles, as they apply to quinolones. The information provided herein should be of value to the practicing clinician, clinical pharmacologist, drug developer, and student.

FIRST PRINCIPLES OF ANTIMICROBIAL THERAPEUTICS
PK-PD Measures

The most common PK-PD measures that have been correlated with efficacy of antimicrobial agents are (*i*) duration of time that drug concentration exceeds the minimum inhibitory concentration (MIC) of the agent against the pathogen ($T >$ MIC), (*ii*) ratio of the maximal drug concentration of the agent to the MIC of the agent against the pathogen (C_{max}:MIC ratio), and (*iii*) ratio of the area under the concentration–time curve (AUC) at 24 hours of the agent to the MIC of the agent against the pathogen (AUC_{24}:MIC ratio) (2,3).

The PK-PD profile of most classes of antibacterial agents, including the quinolones, has been well characterized. Quinolones are classified as concentration-dependent" killing antibacterial agents, as they eradicate bacteria most rapidly when their concentrations are significantly above the MIC of the targeted microorganism. Additionally, quinolones display a moderate-to-prolonged persistent killing effect (2,3).

The free-drug (*f*) AUC:MIC ratio is the PK-PD measure that generally has correlated most strongly with efficacy in animal and in vitro infection models and in patients with a variety of infection types (4–9). The fC_{max}:MIC ratio is also important, especially relating to the suppression of drug-resistant subpopulations of bacteria (10,11).

QUINOLONE PK-PD AGAINST STREPTOCOCCI AND STAPHYLOCOCCI
Dynamic In Vitro Infection Models
There have been a variety of dynamic in vitro PK-PD infection models described. However, they all attempt to simulate the drug's human concentration–time profile in the presence of bacteria. Numerous in vitro PK-PD infection models have evaluated various quinolones against *Streptococcus pneumoniae*, and there has been remarkable concordance across quinolones (ciprofloxacin, garenoxacin, gatifloxacin, levofloxacin, moxifloxacin, and sparfloxacin) (12–15). For instance, the hollow-fiber infection model was used by Lister to determine the gatifloxacin $fAUC_{24}$:MIC ratio needed to eradicate *S. pneumoniae* (15). Log-phase cultures [5 × 10^7 colony-forming units (CFU)/mL] of pneumococci were exposed to clinically relevant gatifloxacin exposure (similar to that observed in humans). MIC values for the four pneumococcal strains evaluated ranged from 0.4 to 1 mg/L and $fAUC_{24}$:MIC ratios ranged from 9 to 48. While maintaining a fC_{max} of two- to three-fold the MIC value, the $fAUC_{24}$:MIC ratios were varied by altering gatifloxacin elimination from the model. Bacterial density was measured over a period of 30 hours. Regardless of MIC value, $fAUC_{24}$:MIC ratios 30 were associated with eradication of *S. pneumoniae* from the model; exposures with *low* $fAUC_{24}$:MIC ratios (i.e., $fAUC_{24}$:MIC ratios of 10–22) failed to eradicate *S. pneumoniae* from the model and in some instances regrowth occurred, with viable counts increased to that of drug-free controls by the end of the experiment (Fig. 1).

PK-PD Animal Infection Models
The murine-thigh infection model, which was initially developed by Dr. Harry Eagle and later refined by Craig, is perhaps the most commonly used infection model (2,3,16–19). Typically, two endpoints for efficacy are used: survival after four days of therapy and the change in bacterial density after 24 hours of therapy. The murine-thigh infection models have been used extensively to investigate the magnitude of the $fAUC_{24}$:MIC ratio needed for quinolones to eradicate gram-positive microorganisms, such as *S. pneumoniae* and *Staphylococcus aureus* (2,3).

Classically, mice are infected with 10^5 to 10^6 CFU per thigh muscle of the strain of interest. Subsequently, mice receive one to six or seven dosing regimens of the quinolone being studied. If the endpoint for efficacy is the change in bacterial density at 24 hours, thigh muscles are aseptically removed and CFU/thigh is determined using a standard plating technique. Efficacy is then calculated by subtracting the log_{10} CFU/thigh of each treated mouse at the end of therapy (24 hours) from the mean log_{10} CFU/thigh of control mice just prior to treatment (0 hour). If one uses mortality as the endpoint, efficacy is calculated by counting the number of mice that survive after four days of therapy for each exposure level.

In Figure 2, the relationship between the $fAUC_{24}$:MIC ratio and animal survival for non-neutropenic mice infected with *S. pneumoniae* that were treated

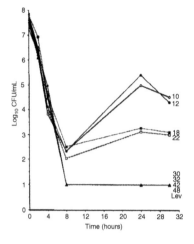

FIGURE 1 Time-kill pharmacokinetics-pharmacodynamics of gatifloxacin against *S. pneumoniae* in a hollow fiber model of infection. The numbers at the right side of each line represent the $fAUC_{24}$:MIC ratio for each experiment. $fAUC_{24}$:MIC ratios 30 or greater resulted in the elimination of *S. pneumoniae* from the model, while low ratios, ranging from 10 to 22, did not result in elimination of the strain from the model. *Abbreviations*: AUC, area under the concentration–time curve; MIC, minimum inhibitory concentration. *Source*: From Ref. 15.

for four days with ciprofloxacin, gatifloxacin, gemifloxacin, levofloxacin, moxifloxacin, or sitafloxacin is illustrated.

When the $fAUC_{24}$:MIC ratio was greater than approximately 25 to 34, regardless of the quinolone studied, survival was greater than 90% (20). Similarly, when one examines data derived from non-neutropenic mice and change in $\log_{10}$ CFU/thigh, $fAUC_{24}$:MIC ratios of 25 to 34 are associated with a 99% reduction in bacterial burden (20).

The magnitude of the $fAUC_{24}$:MIC ratio required for efficacy against *S. aureus* is greater compared with that of *S. pneumoniae*. $fAUC_{24}$:MIC ratios greater than 60 to 80 are generally associated with 90% animal survival when mortality is used as the efficacy endpoint and a 90% reduction in bacterial density when the net change in CFU is used as the efficacy endpoint in immunocompromised animals.

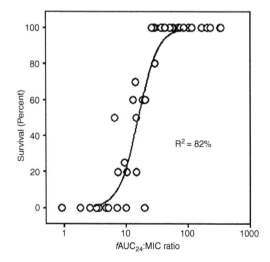

FIGURE 2 Relationship between the $fAUC_{24}$:MIC ratio for six quinolones (ciprofloxacin, gatifloxacin, gemifloxacin, levofloxacin, moxifloxacin, and sitafloxacin) and survival in non-neutropenic mice infected with *Streptococcus pneumoniae*. When the $fAUC_{24}$:MIC ratio was greater than approximately 25 to 34, survival was greater than 90%. *Abbreviations*: AUC, area under the concentration–time curve; MIC, minimum inhibitory concentration. *Source*: From Ref. 20.

Human PK-PD Data

In no other therapeutic area are there more clinical PK-PD data than that of the community-acquired respiratory tract (6–8). This should not be surprising, as these clinical indications comprise the most profitable market sector for antimicrobial agents. Given that S. pneumoniae is the most common pathogen associated with community-acquired respiratory tract infections, most analyses have focused on this organism.

There have been multiple analyses evaluating the relationship between the $fAUC_{24}$:MIC ratio and outcome for different quinolone agents involving patients with community-acquired pneumonia, acute maxillary sinusitis, or acute bacterial exacerbation of chronic bronchitis (6–8).

In one such study, Ambrose et al. evaluated the relationship between the $fAUC_{24}$:MIC ratio of gatifloxacin and levofloxacin against S. pneumoniae and microbiological response of patients enrolled in either of two Phase III, double-blinded, randomized studies (8). The analyses demonstrated that for quinolones, $fAUC_{24}$:MIC ratios of at least 33.7 correlated with the eradication of S. pneumoniae. $fAUC_{24}$:MIC ratios greater than 33.7 were associated with 100% of patients having a positive microbiological response to therapy, while those patients with $fAUC_{24}$: MIC ratios less than 33.7 had only a 64% response to therapy.

In Figure 3, the relationship between microbiological response and $fAUC_{24}$: MIC ratio in 121 patients with respiratory tract infections (pneumonia, acute bacterial exacerbation of chronic bronchitis, acute bacterial maxillary sinusitis) treated with various quinolones is shown (1). Patients in whom $fAUC_{24}$:MIC ratios of 34 or greater were attained had the highest probability (92.6% of a positive response to therapy), while those with low ratios had only a 66.7% probability of a favorable response (odds ratio = 6.3, $P = 0.01$). These data are very concordant with the data derived from the aforementioned in vitro and animal infection models involving quinolones and pneumococci.

QUINOLONE PK-PD AGAINST GRAM-NEGATIVE BACILLI
Dynamic In Vitro Infection Models

In vitro PK-PD infection models that have evaluated various quinolones against gram-negative bacilli are fewer when compared with that of gram-positive pathogens. One such study evaluated gatifloxacin against Salmonella typhi, the organism responsible for typhoid fever (also known as enteric fever) (9). Log-phase cultures (5×10^7 CFU/mL) of S. typhi were exposed to clinically relevant gatifloxacin exposure (similar to that observed in humans). Two strains were studied, one gatifloxacin-susceptible (MIC = 0.5 mg/L) and one -resistant (MIC = 5 mg/L). The gatifloxacin-susceptible strain had a GyrA mutation (Asp87 → Asn), while the resistant strain had two GryA (Ser83 → Try, Asp87 → Ile) and two ParC (Thr57 → Ser, Ser80 → Ile) mutations. Bacterial density was measured over a period of 24 hours. The $fAUC_{24}$:MIC ($r^2 = 0.96$) and fC_{max}:MIC ($r^2 = 0.93$) ratios were more predictive of bacterial killing than was $\%fT > MIC$ ($r^2 = 0.68$). The $fAUC_{24}$:MIC ratio associated with 90% E_{max} was 105. Perhaps the most important observation was that the $fAUC_{24}$:MIC ratio required for a given level of bactericidal activity did not differ by MIC value (Fig. 4). For instance, the $fAUC_{24}$:MIC ratio necessary 90% E_{max} was the same for the gatifloxacin-susceptible and -resistant strain despite difference in MIC value and the numbers of target site mutations.

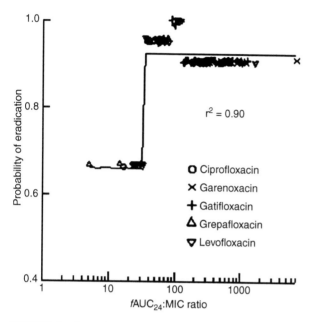

FIGURE 3 Jitter plot of the relationship between the $fAUC_{0-24}$:MIC ratio for five quinolones (ciprofloxacin, garenoxacin, gatifloxacin, grepafloxacin, and levofloxacin) and microbiological response in 121 patients with respiratory tract infection (pneumonia, acute exacerbation of chronic bronchitis, acute maxillary sinusitis) associated with *Streptococcus pneumoniae*. Microbiological eradication was higher in patients with $fAUC_{0-24}$:MIC ratio >34 (92.6%) and lower in patients with values <24 (66.7%) ($P = 0.01$). The *x*-axis is log transformed for graphical clarity. *Abbreviations*: AUC, area under the concentration–time curve; MIC, minimum inhibitory concentration. *Source*: From Ref. 1.

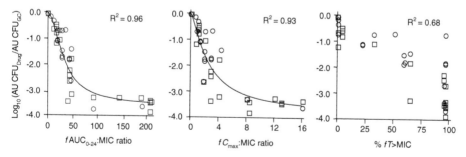

FIGURE 4 Relationships between gatifloxacin $fAUC_{0-24}$:MIC ratio (*left*), fC_{max}:MIC ratio (*middle*), and $fT > MIC$ (*right*) for two strains of *Salmonella enterica* Serotype Typhi with differing MIC values and changes in bacterial density. The square symbols represent a susceptible strain with a *GyrA* mutation (Asp87→Asn) and a gatifloxacin MIC of 0.5 mg/mL, while the circles represent a resistant strain with *GyrA* (Ser83→Try; Asp87→Gly) and *ParC* (Thr57→Ser; Ser80→Ile) mutations and a

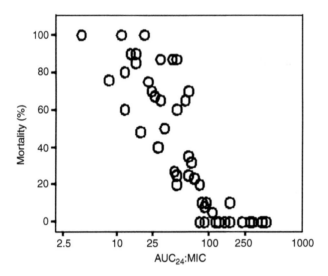

FIGURE 5 Relationship between the AUC_{24}:MIC ratio for quinolones and mortality in immuno-compromised animals infected with gram-negative bacilli and a few gram-positive cocci. When the AUC_{24}:MIC ratio is greater than approximately 100, survival is greater than 90%. *Abbreviations*: AUC, area under the concentration–time curve; MIC, minimum inhibitory concentration. *Source*: From Ref. 3.

PK-PD Animal Infection Models

The first study evaluating the PK-PD of quinolones against gram-negative bacilli was published in 1991. In neutropenic mouse-thigh and -lung infection models involving either *Pseudomonas aeruginosa* or *Klebsiella pneumoniae*, Leggett et al. demonstrated that the AUC_{24}:MIC ratio was the PK-PD measure most associated with bacterial killing (21). In Figure 5, the relationship between the AUC_{24}:MIC ratio for various quinolones and mortality in immunocompromised animals infected with gram-negative bacilli is shown (as well as few gram-positive cocci).

When the AUC_{24}:MIC ratio is greater than approximately 100, survival is greater than 90% (3). When efficacy is measured as the change in bacterial density (Log_{10} CFU) after 24 hours of therapy, rather than survival, AUC_{24}:MIC ratios of approximately 50 are associated with a net bacteriostatic effect (i.e., no net change in Log_{10} CFU) and ratios of 100 are generally associated with two Log-unit reduction (i.e., 99% reduction in bacterial density).

Human PK-PD Data

Forrest et al. published some of the first data to correlate PK-PD measures and response in humans (5). Ciprofloxacin was studied in critically ill patients with pneumonia involving predominantly gram-negative bacilli. The analyses demonstrated that AUC_{24}:MIC ratio was predictive of clinical and microbiological response ($P < 0.003$). As illustrated in Figure 6, a high probability of therapeutic response was observed when ciprofloxacin total-drug AUC_{24}:MIC ratios of 125 or greater against gram-negative bacilli were attained. As ciprofloxacin is approximately 40% bound to serum proteins, this value corresponds to a $fAUC_{24}$:MIC ratio of about 75.

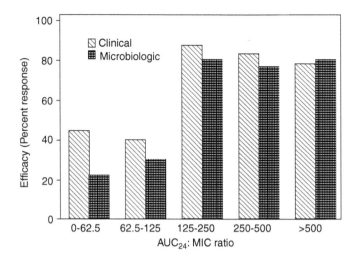

FIGURE 6 Relationship between AUC_{24}:MIC ratio and clinical and microbiological response to ciprofloxacin therapy in hospitalized patients infected with gram-negative microorganisms. *Abbreviations*: AUC, area under the concentration–time curve; MIC, minimum inhibitory concentration. *Source*: From Ref. 5.

In a similar analysis, involving different investigators, in patients with hospital-acquired pneumonia treated with levofloxacin, AUC_{24}:MIC ratio was also found to be predictive of response (22). Total-drug AUC_{24}:MIC ratios of greater than or equal to 87 correlated with the eradication of gram-negative bacilli ($P = 0.01$). Given that levofloxacin is approximately 29% bound to serum proteins, this value corresponds to a $fAUC_{24}$:MIC ratio of approximately 62. $fAUC_{24}$:MIC ratios greater than or equal to 62 were associated with 90% patients having a positive microbiological response to therapy, while those patients with $fAUC_{24}$:MIC ratios below 62 were associated with a 43% response to therapy.

PK-PD AND RESISTANCE COUNTER SELECTION

PK-PD and antimicrobial resistance represents a relative new and exciting area of research. Given that the era of multidrug resistance is upon us, perhaps there is no clinically more important area of research. There have been two main thrusts in this area: the dynamic in vitro PK-PD infection model and static in vitro experiments based upon the mutant selection window hypothesis.

Mutant Selection Window Hypothesis

The mutant selection window hypothesis was first put forth by Drlica and coworkers (23), a variant of which was studied clinically first by Kastner and Guggenbichler (24). Drlica and coworkers put forth the hypothesis that there was a mutant prevention concentration, which is defined as the drug concentration that prevents the amplification of single-step drug-resistant mutant (23). These static in vitro experiments are conducted on agar plates, rather than in a dynamic in vitro or in vivo system. Despite this limitation, this metric has been useful in providing a tool for comparing drug regimens. For instance, the mutant

prevention concentration was identified for five quinolones, moxifloxacin, trovafloxacin, gatifloxacin, grepafloxacin, and levofloxacin, against clinical isolates of
S. pneumoniae (25). The authors reported that moxifloxacin and gatifloxacin were
the most potent in this assay, while levofloxacin was the least potent (Fig. 7).

There are, however, a number of important limitations to the interpretation
derived from mutant prevention window experiments. First, it is critical to note
that any differences in protein binding between agents should be accounted for
when comparing the mutant prevention concentration with the concentration time
profile of agents in humans. Failure to do this may lead to a misleading comparison of drug regimens. Second, this metric may overestimate the dose regimen
required to prevent resistance, as it has been suggested by some to require 100%
time above the mutant prevention concentration. This, in essence, creates a time
above threshold metric, rather than an AUC metric, for a concentration-dependent
killing agent. As bacterial killing with quinolones is related to AUC_{24}:MIC ratio
(oftentimes drug concentrations remain below the MIC for a significant fraction of
the dosing interval), this requirement is not consistent with first principles and can
result in a recommendation of falsely high and potentially toxic drug regimens.
Finally, given that mutation prevention concentrations are derived in vitro, the
impact of host immune function cannot be evaluated. This final limitation is
shared with dynamic in vitro PK-PD infection models, which are discussed in the
following section.

Dynamic Infection Models

The hollow-fiber and the murine infection models have also been used to evaluate
the relationship between quinolone exposure, as measured by the 24 hour $fAUC_{24}$:
MIC ratio, and the suppression of preexistent resistant subpopulations of bacteria
(26–29). Levofloxacin against *P. aeruginosa* in a mouse-thigh infection model was
evaluated by Jumbe et al. (29). In this study, the magnitude of the PK-PD measure
predictive of resistance suppression was determined in initial experiments and
then validated in a second series of experiments. For levofloxacin and *P. aeruginosa*, a $fAUC_{24}$:MIC ratio greater than 110 was associated with the prevention of
amplification of preexisting mutant subpopulations of bacteria (29). Similarly, for
garenoxacin against *K. pneumoniae*, $fAUC_{24}$:MIC ratios of 88.3 or greater were
necessary to kill the susceptible population and prevent regrowth of the resistant
subpopulations in a hollow-fiber infection model (27).

A critical factor on resistance suppression is bacterial load at the infection
site. The impact of bacterial load on the $fAUC_{24}$:MIC ratio needed to suppress the
emergence of resistant subpopulations was evaluated by Fazili et al. The authors
demonstrated that for a pneumococcal strain with dual resistance mechanisms
(a *ParC* and efflux pump mutant), there was a 2.5-fold difference between 10^6 and
10^8 CFU/mL for $fAUC_{24}$:MIC ratio needed (200 vs. 500) to suppress the emergence
of resistant subpopulations (28).

These data demonstrate that it is possible to identify drug exposures that
suppress the emergence of resistant subpopulations, that resistance suppression
targets differ by genus and species, that resistance suppression targets are often of
greater magnitude compared with those associated with optimal outcomes in
traditional nonclinical infection models or those associated with positive outcomes
in infected patients, and, finally, that the bacterial load at the primary infection
site can make a difference in the drug exposure needed to prevent resistance
selection.

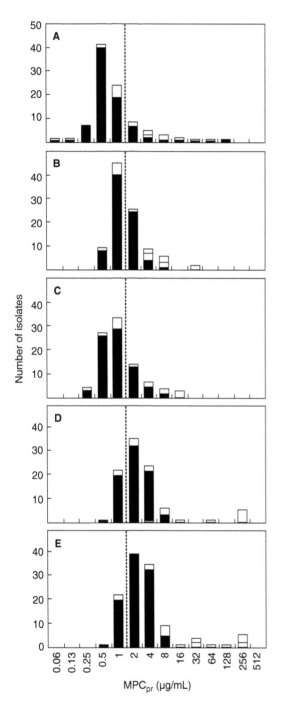

FIGURE 7 Distribution of *Streptococcus pneumoniae* isolates with respect to MPC. White areas of bars represent isolates containing *ParC* mutations known to confer resistance; shaded regions represent isolates containing *ParC* mutations that have not been demonstrated by genetic tests to confer quinolone resistance; solid regions represent unsequenced isolates. Dashed line is for alignment of panels. Panels: (**A**) moxifloxacin; (**B**) gatifloxacin; (**C**) trovafloxacin; (**D**) grepafloxacin; and (**E**) levofloxacin. *Abbreviation*: MPC, mutant prevention concentration. *Source*: From Ref. 25.

Human PK-PD Data

Clinical PK-PD studies evaluating the relationship between drug exposure and resistance are very few. Thomas et al. evaluated data from four clinical trials of acutely ill patients with hospital-acquired respiratory tract infections (30). PK-PD measures for five antimicrobial regimens and microbiological results for serial tracheal aspirates were examined across and by organism. Total-drug AUC_{24}:MIC ratios below 100 were associated with an 82.4% likelihood of emergence of resistance while ratios above 100 were only associated with a 9.3% emergence of resistance. Given that the greatest frequency of selected resistance was observed for Pseudomonas spp. treated with ciprofloxacin and that ratios exceeding 100 did not reduce the risk of emergence of resistance among patients with β-lactamase-producing gram-negative bacilli treated with β-lactam monotherapy, the reported PK-PD breakpoint was likely more reflective of that for ciprofloxacin for Pseudomonas spp. More data from clinical trials looking at such relationships by organism and drug are needed to further elucidate this question and confirm in vitro findings. While this analysis represented a step forward in the paradigm of using microbiological endpoints for evaluating the time-course of antimicrobial effect (i.e., by evaluating a serial microbiological endpoint which is more sensitive than dichotomous endpoints, like "success" and "failure" at test-of-cure), there were certain important limitations including the retrospective nature of the review and the controversy surrounding the relative value of tracheal aspirate specimens.

SUMMARY

An understanding of PK-PD concepts forms the basis for the rational use of antimicrobial agents. For quinolones against various bacterial genera, there has been good concordance among findings from in vitro infection models and animal infection models and data from well-controlled clinical trials. However, there remains much work to do, especially with regard to resistance prevention, where the mutation prevention concentration data and that from dynamic infection models are playing a critical role.

REFERENCES

1. Ambrose PG, Bhavnani SM, Rubino CM, et al. Pharmacokinetics-pharmacodynamics of antimicrobial therapy: it's not just for mice anymore. Clin Infect Dis 2007; 44:79–86.
2. Craig WA. Pharmacokinetic/pharmacodynamic parameters: rationale for antibacterial dosing in mice and men. Clin Infect Dis 1997; 26:1–12.
3. Craig WA. Pharmacodynamics of antimicrobials: general concepts and application. In: Nightingale CH, Murakawa T, Ambrose PG, eds. Antimicrobial Pharmacodynamics in Theory and Clinical Practice. New York: Marcel-Dekker, 2002:1–21.
4. Andes DR, Craig WA. Pharmacodynamics of fluoroquinolones in experimental models of endocarditis. Clin Infect Dis 1998; 27:47–50.
5. Forrest A, Nix DE, Ballow CH, Goss TF, Birmingham MC, Schentag JJ. Pharmacodynamics of intravenous ciprofloxacin in seriously ill patients. Antimicrob Agents Chemother 1993; 37:1073–1081.
6. Forrest A, Chodash S, Amantea MA, Collins DA, Schentag JJ. Pharmacokinetics and pharmacodynamics of oral grepafloxacin in patients with acute bacterial exacerbations of chronic bronchitis. J Antimicrob Chermother 1997; 40(suppl A):45–57.
7. Preston SL, Drusano GL, Berman AL, et al. Pharmacodynamics of levofloxacin: a new paradigm for early clinical trials. JAMA 1997; 279:125–129.

8. Ambrose PG, Grasela DM, Grasela TH, Passarell J, Mayer HB, Pierce PF. Pharmacodynamics of fluoroquinolones against *Streptococcus pneumoniae* in patients with community-acquired respiratory tract infections. Antimicrob Agents Chemother 2001; 45: 2793–2797.
9. Booker BM, Smith PF, Forrest A, et al. Application of an in vitro infection model and simulation for re-evaluation of fluoroquinolone breakpoints for *Salmonella enterica* serotype typhi. Antimicrob Agents Chemother 2005; 49:1775–1781.
10. Blaser J, Stone BB, Groner MC, Zinner SH. Comparative study with enoxacin and netilmicin in a pharmacodynamic model to determine importance of ratio of antibiotic peak concentration to MIC for bactericidal activity and emergence of resistance. Antimicrob Agents Chemother 1987; 31:1054–1060.
11. Drusano GL, Johnson DE, Rosen M, et al. Pharmacodynamics of a fluoroquinolone antimicrobial agent in a neutropenic rat model of *Pseudomonas sepsis*. Antimicrob Agents Chemother 1993; 37:483–490.
12. Lacy MA, Lu W, Xu X, et al. Pharmacodynamic comparisons of levofloxacin, ciprofloxacin and ampicillin against *Streptococcus pneumoniae* in an in vitro model of infection. Antimicrob Agents Chemother 1999; 43:672–677.
13. Lister PD, Sanders CC. Pharmacodynamics of levofloxacin and ciprofloxacin against *Streptococcus pneumoniae*. J Antimicrob Chemother 1999; 43:79–86.
14. Lister PD, Sanders CC. Pharmacodynamics of moxifloxacin, levofloxacin, and sparfloxacin against *Streptococcus pneumoniae* in an in vitro pharmacodynamic model. J Antimicrob Chemother 2001; 47:811–818.
15. Lister PD. Pharmacodynamics of gatifloxacin against *Streptococcus pneumoniae* in an in vitro pharmacokinetic model: impact of area under the curve/MIC ratios on eradication. Antimicrob Agents Chemother 2002; 46:69–74.
16. Eagle H. Effect of schedule of administration on therapeutic efficacy of penicillin: importance of aggregate time penicillin remains at effectively bactericidal levels. Am J Med 1950; 9:280–299.
17. Eagle H, Fleischman R, Mussleman AD. Effective concentrations of penicillin in vitro and in vivo for streptococci and pneumococci and Treponema. J Bacteriol 1950; 59:625–643.
18. Eagle H, Fleischman R, Levy M. Continuous versus discontinuous therapy with penicillin: effect of interval between injections on therapeutic efficacy. N Engl J Med 1953; 248:481–488.
19. Craig WA. Does dose matter? Clin Infect Dis 2001; 33(suppl 3):S233–S237.
20. Craig WA, Andes DA. Correlation of the magnitude of the AUC24/MIC for 6 fluoroquinolones against *Streptococcus pneumoniae* with survival and bactericidal activity in an animal model [abstr 289]. Abstracts of the 40th Interscience Conference of Antimicrobial Chemotherapy. Toronto, Canada: 2000.
21. Leggett JE, Ebert S, Fantin B, et al. Comparative dose–effect relations several dosing intervals for beta-lactam, aminoglycoside, and quinolone antibiotics against Gram-negative bacilli in murine thigh-infection and pneumonitis models. Scand J Infect Dis 1991; 74:179–184.
22. Drusano GL, Preston SL, Fowler C, Corrado M, Weisinger B, Kahn J. Relationship between fluoroquinolone AUC: MIC ratio and the probability of eradication of the infecting pathogen in patients with nosocomial pneumonia. J Infect Dis 2004; 189: 1590–1597.
23. Dony Y, Zhao X, Kreiswirth BN, Drlica K. Mutant prevention concentration as a measure of antibiotic potency: studies with clinical isolates of Mycobacterium tuberculosis. Antimicrob Agents Chemother 2000; 44:2581–2584.
24. Kastner U, Guggenbichler JP. Influence of macrolide antibiotics on promotion of resistance in the oral flora of children. Infection 2001; 29:251–256.
25. Blondeau JM, Zhao X, Hansen G, Karl Drlica K. Mutant prevention concentrations of fluoroquinolones for clinical isolates of *Streptococcus pneumoniae*. Antimicrob Agents Chemother 2001; 45:433–438.
26. Tam VH, Louie A, Deziel MR, et al. Pharmacokinetics of BMS-284756 in counter-selecting resistance in a hollow-fiber system [abstr A-442]. Abstracts of the 41st Interscience Conference of Antimicrobial Chemotherapy. Chicago, Illinois: 2001.

27. Tam VH, Louie A, Deziel MR, Liu W, Grasela DM, Miller MH, Drusano GL. Pharmacokinetics of BMS-284756 and ciprofloxacin against *Pseudomonas aeruginosa* and *Klebsiella pneumoniae* in a hollow-fiber system. [abstract A-443] In: Abstracts of the 41st Interscience Conference of Antimicrobial Chemotherapy. Chicago, Illinois: 2001.
28. Fazili T, Louie A, Tam V, Deziel M, Liu W, Drusano GL. Effect of innoculum on the pharmacodynamic breakpoint dosage that prevents selection of gatifloxacin resistance in ciprofloxacin-susceptible and -resistant *Streptococcus pneumoniae* [abstr A-445]. Abstracts of the 41st Interscience Conference of Antimicrobial Chemotherapy. Chicago, Illinois: 2001.
29. Jumbe N, Louie A, Leary R, et al. Application of a mathematical model to prevent in vivo amplification of antibiotic-resistant bacterial populations during therapy. J Clin Invest 2003; 112:275–285.
30. Thomas JK, Forrest A, Bhavnani SM, et al. Pharmacodynamic evaluation of factors associated with the development of bacterial resistance in acutely ill patients during therapy. Antimicrob Agents Chemother 1998; 42:521–527.

9 Glycopeptide Pharmacodynamics

Elizabeth D. Hermsen
Department of Pharmaceutical and Nutrition Care, The Nebraska Medical Center, Omaha, Nebraska, U.S.A.

Gigi H. Ross
Ortho-McNeil Pharmaceutical, Inc., Raritan, New Jersey, U.S.A.

John C. Rotschafer
Department of Experimental and Clinical Pharmacology, University of Minnesota College of Pharmacy, Minneapolis, Minnesota, U.S.A.

INTRODUCTION

Pharmacodynamics represents a blending of pharmacokinetic parameters with a measure of bacterial susceptibility, the minimum inhibitory concentration (MIC). Thus, the pharmacokinetic parameters of the antibiotic must be adequately defined prior to exploring the drug's pharmacodynamic properties. This has not been an easy task for vancomycin because the drug has undergone several different formulation changes to remove impurities.

Measuring vancomycin concentrations by any other method than microbiological assay was not possible until the late 1970s when a radioimmunoassay was introduced. Microbiologic assays were technically challenging, accurate at best to ±10% (1) and often could not be performed if patients were receiving other antibiotics.

Pharmacokinetically, vancomycin has been characterized using one-, two-, and three-compartment models, as well as noncompartmental models. As a result, there is model-dependent variability in the reporting of vancomycin pharmacokinetic parameters. Therefore, identifying and quantifying clinically applicable pharmacodynamic parameters has not been easy. Even today, there are extremely limited in vitro, animal, and human data characterizing vancomycin's performance against but a few bacteria. The purpose of this review is to examine the microbiology, pharmacology, and pharmacokinetics of vancomycin and attempt to build on the data presently available describing the pharmacodynamics of the drug.

History of Compound

The only commercially available glycopeptide antibiotic in the United States, vancomycin, was first introduced in 1956, with widespread clinical use by 1958 (2). Originally, the drug was isolated from the actinomycete *Streptomyces orientalis*; however, the structure and molecular weight were not identified until 1978. The compound consists of a seven-membered peptide chain and two chlorinated β-hydroxytyrosine moieties with a molecular weight of 1449 (2). Clinical use of the drug was highly prevalent in the late 1950s due to the emergence of penicillinase-producing strains of staphylococcus, which, however, soon lost favor with the introduction of methicillin. Impurities in early vancomycin formulations led to an unacceptable incidence of infusion-related reactions. Subsequently, for 20 years, vancomycin was exclusively used for the treatment of serious staphylococcal infections in patients with severe penicillin allergies. The current Eli Lilly formulation, marketed in 1986, is estimated to be 93% pure factor B (vancomycin) and is the result of several production changes and improved separation techniques (2). With the enhancement in purity and the heightened frequency of methicillin-resistant (MR) staphylococcus and ampicillin-resistant enterococcus, clinical use of vancomycin has significantly increased. Unfortunately, resistance to vancomycin is emerging and it is a clinical concern. The first vancomycin-resistant enterococci (VRE) were reported in 1988 and have quickly become endemic in intensive care units (3,4). The first case of vancomycin-intermediate *Staphylococcus aureus* (VISA) was identified in 1996, and in 2002, vancomycin-resistant *S. aureus* (VRSA) became a reality (5,6).

Antimicrobial Spectrum

Vancomycin is primarily effective against gram-positive cocci, including staphylococcus, streptococcus, and enterococcus, and is considered to be bactericidal [minimum bacterial concentration (MBC)/minimum inhibitory concentration (MIC) < 4] against most gram-positive pathogens with the exception of *S. aureus*, enterococci, limited numbers of tolerant (MBC/MIC > 32) *Streptococcus pneumoniae*, and tolerant staphylococci. The Clinical Laboratory Standards Institute (CLSI) has MIC standards of susceptibility for vancomycin against staphylococci and enterococci (7). Sensitive strains of coagulase-negative staphylococci and enterococci have MICs of ≤ 4 mg/L; MICs for intermediate isolates are 8 to 16 mg/L, and resistance is determined by an MIC ≥ 32 mg/L. *S. aureus* and *Staphylococcus epidermidis*, including both methicillin-susceptible and -resistant strains, are usually sensitive with MIC_{90}s of ≤ 2 mg/L (8). All strains of Streptococcus are sensitive to vancomycin, regardless of penicillin susceptibility, with MIC_{90}s < 1 mg/L (7). A 1999 report, however, claims approximately 2% of *S. pneumoniae* isolates have developed tolerance to vancomycin (9). *Enterococcus faecalis* are typically susceptible to vancomycin with MIC_{50}s of ≤ 1 mg/L while *Enterococcus faecium* are generally nonsusceptible with MIC_{50}s of ≥ 16 mg/L (8). Vancomycin is also effective against other Streptococcus spp., *Listeria monocytogenes*, Bacillus spp., Corynebacteria, and anaerobes such as diphtheroids and clostridium species, including *Clostridium perfringens* and *Clostridium difficile*. Vancomycin has no activity against gram-negative organisms, atypical pathogens, fungi, or viruses.

PHARMACOLOGY

Vancomycin has multiple mechanisms of action: preventing the synthesis and assembly of a growing bacterial cell wall, altering the permeability of the bacterial cytoplasmic membrane, and selectively inhibiting bacterial RNA synthesis (10).

Vancomycin prevents polymerization of the phosphodisaccharide-pentapeptide-lipid complex of the growing cell wall at the D-alanyl-D-alanine end of the peptidoglycan precursor during the latter portion of biosynthesis (10,11). By tightly binding the free carboxyl end of the cross-linking peptide, vancomycin sterically prevents binding to the enzyme peptidoglycan synthetase. This activity occurs at an earlier point and at a separate site from that of penicillins and cephalosporins (11). Therefore, no cross-resistance or competition of binding sites occurs between the classes. Vancomycin, like β-lactams, does require actively growing bacteria in order to exert a bactericidal effect. However, vancomycin's bactericidal activity is restricted to gram-positive organisms because the molecule is too large to cross the outer cell membrane of gram-negative species.

Many factors appear to impede vancomycin's bactericidal activity: the absence of environmental oxygen, the size of the bacterial inoculum, and the phase of bacterial growth. The antibiotic appears to kill bacteria under aerobic conditions more effectively than under anaerobic conditions (12). Because many gram-positive pathogens, including streptococcus and staphylococcus, can grow under aerobic and anaerobic conditions, this fact could prove problematic in clinical situations. Vancomycin activity was reduced by 19% and 99% with increases in inoculum size from 10^6 colony-forming units (CFU)/mL to 10^7 and 10^8 CFU/mL, respectively (13,14). When vancomycin was evaluated against growing and non-growing *S. epidermidis* cells, the drug was found to be effective only against actively growing cultures (15). Finally, activity is relatively unaffected by extremes in pH, however is maximal at pH 6.5 to 8.0 (13,14,16).

PHARMACOKINETICS

The pharmacokinetics of vancomycin are highly dependent upon the modeling method utilized to characterize the parameters. Data can be found in the literature that characterize vancomycin using one-, two-, three-, and noncompartmental pharmacokinetic models that employ different serum sampling schemes and vary in the duration of study. Consequently, the literature varies in the reporting of vancomycin pharmacokinetic parameters.

Absorption is complete only when given intravenously. Oral absorption is poor and intramuscular administration is painful and absorption is erratic. Vancomycin is readily absorbed after intraperitoneal administration, as well (17).

The distribution of vancomycin is a complex process and is best characterized using a multicompartmental approach. Vancomycin has a large volume of distribution, varying from 0.4 to 0.6 L/kg in patients with normal renal function and up to 0.9 L/kg in patients with end-stage renal disease (16,18,19). Distribution includes ascetic, pericardial, synovial, and pleural fluids, as well as bone and kidney. Penetration into bile, however, is generally considered poor. Cerebral spinal fluid concentrations are minimal unless sufficient inflammation is present, where 10% to 15% of serum concentrations can be obtained (16,19). Approximately 10% to 50% of vancomycin is protein bound (~40–50% in healthy volunteers), primarily to albumin, providing a relatively high free fraction of active drug (16,20). Most patients, because of lower albumin concentrations as compared to volunteers, will not bind vancomycin as avidly. Studies attempting to measure the effect of other serum proteins have reported virtually no binding to the reactive protein, α-1 glycoprotein, but have noted binding to IgA (20).

Drug elimination is almost exclusively via glomerular filtration with 80% to 90% of the vancomycin dose appearing unchanged in the urine within 24 hours in patients with normal renal function (16,18,19). The remainder of the dose is eliminated via biliary and hepatic means. When taken orally, vancomycin is excreted primarily in the feces. Vancomycin is not significantly removed by conventional hemo- or peritoneal dialysis due to its large molecular weight ($\sim$2000); however high-flux dialyzers can remove vancomycin and other molecules with molecular weights <20,000 (21). Levels should be monitored in this case to assess appropriate dosing.

The elimination of vancomycin is multicompartmental, with an α, or distribution, half-life of 0.6 to 3 hours and a β, or elimination, half-life of 4 to 8 hours with normal renal function (18,19). Renal insufficiency can prolong the terminal half-life up to 7 to 12 days. Due to the complexity of this biexponential decay, attempting to utilize various modeling techniques is difficult. A one-compartment model inappropriately characterizes the distribution phase by forming a regression line that is a hybrid of the α and β phases. The pharmacokinetic parameters produced are, accordingly, mythical values that may or may not relate to the actual parameters. The extrapolated peak concentration and the half-life can be greatly underestimated depending upon the sampling scheme used. Generally, pairing a serum concentration obtained early in the distribution phase with a serum concentration late in the elimination phase results in the greatest error. Because one compartment modeling also underestimates the area-under-the-serum-concentration-time-curve (AUC), this error is passed along in the calculation of both distribution volume and drug clearance.

For a concentration-independent or time-dependent antibiotic, vancomycin has almost an ideal pharmacokinetic profile. The drug has a large volume of distribution, low serum protein binding, a long terminal half-life, and limited drug interactions due to modest hepatic metabolism. Thus, vancomycin can be used effectively and conveniently to treat infections in most body sites.

GLYCOPEPTIDE RESISTANCE

Vancomycin has been in clinical use for over 40 years with only recent emergence of resistance. The multiple modes of action of vancomycin necessitate significant alterations in bacterial wall synthesis to occur in order for the intrinsically susceptible organisms to develop resistance. Thus, the rarity of acquired vancomycin resistance led to predictions that such resistance was unlikely to occur on any significant scale (22,23).

The first reports of VRE, however, began to appear from Europe in the mid-1980s (22). VRE was first identified in the United States in 1988 (3). How the enterococci were able to develop resistance to vancomycin is unclear; however, several hypotheses have been elucidated ranging from the overuse of antibiotics to the incorporation of glycopeptide antibiotics into animal feed.

Enterococci are normal gut flora and the emergence of resistance has been linked to vancomycin overuse in the treatment of *C. difficile* enterocolitis (23). Additionally, the parenteral use of vancomycin has steadily increased since the late 1970s and may have played a role in the development of VRE (24). The possibility also exists that the agricultural use of avoparcin, a related glycopeptide, may have been important in Europe, but this drug has not been used in the United States. Regardless, the enterococci were the first class of organisms to acquire vancomycin resistance, and subsequently, VRE is now problematic in both Europe and the United States (23).

The genetic basis for glycopeptide resistance to the enterococci is complex and is characterized by several different phenotypes. Resistance-conferring genes encode for a group of enzymes that enable the enterococci to synthesize cell-wall precursors generally ending in D-alanine-D-lactate or D-alanine-D-serine, rather than the usual D-alanine-D-alanine vancomycin binding site (25,26). Affinity of vancomycin and teicoplanin for D-alanine-D-lactate is 1000-fold less than that for D-alanine-D-alanine (23).

The most frequently encountered resistance phenotype, *vanA*, consists of high-level vancomycin resistance (MIC ≥ 32 mg/L), which is accompanied by high-level resistance to teicoplanin (25). The resistance found on *vanA* strains is vancomycin and/or teicoplanin inducible. The genes encoding *vanA* resistance are relatively easily transferred to other enterococcal species via conjugation (25,26). Significant concern has been expressed in both lay and professional literature that this plasmid-mediated form of resistance could be passed on not only to other enterococci but also to gram-positive organisms such as staphylococci, which could lead to catastrophic consequences worldwide. In June 2002, this event materialized. The *vanA* gene was identified in a VRSA isolate found in a patient with a diabetic foot ulcer infection, which also included vancomycin-resistant *E. faecalis* and *Klebsiella oxytoca*. General agreement exists that the *vanA* gene was transferred to the *S. aureus* from the VRE (27).

Enterococci with *vanB* phenotypic resistance have variable levels of vancomycin resistance and are susceptible to teicoplanin. The *vanB* phenotype is inducible by vancomycin but not teicoplanin, and vancomycin exposure produces teicoplanin resistance. *VanB* encoding genes are more commonly chromosomal but can be transferred by conjugation (25,28).

The *vanC* resistance phenotype consists of relatively low levels of vancomycin resistance (MIC $= 8$–16 mg/L) and is devoid of teicoplanin resistance. *VanC* resistance is chromosomally produced by encoded genes found in all strains of *Enterococcus flavescens*, *Enterococcus casseliflavus*, and *Enterococcus gallinarum*. *VanC* encoded genes are nontransferable (23). Perichon et al. (29) have described a fourth phenotype, *vanD*, similar to *vanB*, but not transferable by conjugation, found in a rare strain of *E. faecium*. Lastly, Fines et al. (30) identified a fifth phenotype, *vanE*, similar to *vanC*, in a strain of *E. faecalis*.

Following a steady increase of VRE prevalence in the United States over the past 15 years, over 25% of enterococci in hospital intensive care units (participating in the National Nosocomial Infections Surveillance surveys) exhibit vancomycin resistance (31). Similarly, rapid increases in VRE prevalence outside the intensive care units in U.S. hospitals have also been observed (31). Eliopoulos et al. reported that 72% of VRE isolates in the United States, from participating centers, exhibit the *vanA* resistance phenotype and genotype with the remaining 28% constituted by the *vanB* gene (32). Of note, the isolates were not probed for the *vanC*, *vanD*, or *vanE* genes.

Low-level vancomycin resistance was reported in clinical isolates of coagulase-negative staphylococci in the late 1980s and early 1990s (33–35). Although troubling, these reports were not terribly feared due to the relative lack of virulence associated with the coagulase-negative staphylococci. In vitro studies, however, demonstrated that both coagulase-negative staphylococci and *S. aureus* isolates, when exposed to increasing levels of glycopeptides, demonstrated the ability to select for resistant subpopulations (36,37). Given these findings and the spread of VRE, for which excessive use of vancomycin was identified as an important control measure, the prudent use of vancomycin was suggested by the Centers for Disease Control (CDC) as critical to prevent the emergence of resistance among staphylococci (38).

In May 1996, a MR *S. aureus* (MRSA) clinical isolate was isolated with reduced susceptibility to vancomycin (MIC = 8 mg/L) from a four-month-old boy with a sternal surgical incision site (39,40). This isolate has been referred to as Mu50 by the investigators who isolated the organism. By current CLSI standards, this *S. aureus* clinical isolate is classified as intermediately resistant to vancomycin. In August of 1997, the first MRSA isolates intermediately resistant to vancomycin in the United States were reported in Michigan and New Jersey (5,41). A number of cases have occured in the United States and worldwide. The majority of these strains were isolated from patients who had received multiple, extended courses of vancomycin therapy. Moreover, most of the strains appear to have developed from MRSA strains previously infecting the patients (42).

The exact mechanism of resistance for these VISA strains remains largely unknown. None of the VISA strains isolated to date has carried any of the *van* determinants (42). Changes in the VISA cell-wall structure have been noted, however, and may be in part responsible for the decreased sensitivity to vancomycin. This is inferred from three findings: the cell wall appeared twice as thick as the wall of control strains on electron microscopy; there was a threefold increase in cell-wall murein precursor production as compared with vancomycin-susceptible MRSA strains; and there was a threefold increase in the production of penicillin-binding protein (PBP)2 and PBP2′ (39,40).

Heterogeneous VISA (hVISA) appears to occur immediately prior to the development of a uniform VISA population. The MICs for hVISA strains fall within the susceptible range according to CLSI, but subpopulations of VISA exist within the overall bacterial population (42). Continued exposure to vancomycin favors the selection of the VISA cells, eventually leading to a uniform VISA population. Although several reports demonstrate that hVISA strains may be responsible for clinical failures with vancomycin therapy in *S. aureus* isolates that appear to be vancomycin susceptible (43–46), no standard hVISA identification criteria exist, making a correlation with clinical outcome nearly impossible (42).

As of September 2006, six VRSA isolates have been identified in the United States (132). All of the patients had a history of VRE infection. The *vanA* gene was found in all six of the VRSA isolates, suggesting transfer of this gene from the VRE.

PHARMACODYNAMICS
Introduction to Basic Principles
Evaluations of serum peak/MIC, the AUC for 24 hours/MIC (AUC_{24}/MIC), and the time antibiotic concentration exceeds the MIC of the infecting organism ($T >$ MIC) have been employed as surrogate markers of the bactericidal effects of antibiotics. Pharmacodynamic indices for vancomycin have been poorly characterized, and therefore, most dosing strategies have been based upon extrapolations from aminoglycoside studies. By modifying aminoglycoside dosing models, specific peak and trough concentrations have been proposed with the assumption that similar clinical outcomes will be produced—high-peak concentrations being

essential for bacterial killing and definitive trough concentration ranges minimizing drug-related toxicity.

Based upon limited in vitro studies, $T > MIC$ appears to most closely predict efficacy of vancomycin, although select animal model studies have identified both AUC/MIC and Cp-max/MIC as important. Some also debate the significance of the latter two parameters in the era of hVISA. Because pharmacokinetic parameters such as Cp-max and AUC are extremely sensitive to underlying model assumptions, the corresponding pharmacodynamic outcome parameter would be affected as well. Nonetheless, most agree that the time the antibiotic concentration exceeds the MIC of the offending organism and not the height of the peak above the MIC, as in aminoglycosides, should be considered the goal of the dosing of vancomycin. Although higher serum concentrations of vancomycin may be helpful in driving the drug to relatively inaccessible sites of infection such as endocardial vegetation or cerebrospinal fluid, higher concentrations are unlikely to improve the rate of bacterial kill. Clinicians attempting to increase the dose of vancomycin for serious but relatively accessible infections likely only expose the patient to an increased risk of adverse reactions but unlikely change the bacterial response to the drug.

Investigations of other pharmacodynamic parameters, including postantibiotic effect (PAE), sub-MIC effect (SME), and postantibiotic sub-MIC effect (PA-SME), have also been undertaken to create a more informative depiction of vancomycin bactericidal activity than MICs allow alone. The PAE, or the continued suppression of microbial growth after limited antibiotic exposures of vancomycin against gram-positive bacteria can persist for several hours depending on the organism and initial antibiotic concentration (49,50). This effect may inhibit regrowth when antibiotic concentrations fall below the MIC of the infecting organism and may be important to consider when dosing vancomycin due to the extended half-life and prolonged dosing intervals. The PAE of vancomycin was evaluated against *S. epidermidis* by Svensson et al. (15). The PAE was dependent upon concentration—as drug concentration increased from 0.5 to 8 times the MIC of the organism, the PAE increased from 0.2 to 1.9 hours. Another study found PAEs ranging from 0.6 to 2.0 hours for *S. aureus* to 4.3 to 6.5 hours for *S. epidermidis* after single doses (51).

As patients receiving antibiotics will always have some amount of drug remaining in the body after dosing and elimination, PAEs are typically studied in vitro. SMEs and PA-SMEs are parameters studied in vivo. Generally all of these effects are longer when measured in vivo versus in vitro. SMEs characterize inhibition of bacterial regrowth following initial sub-MIC concentrations of antibiotic (51). PA-SMEs, on the other hand, illustrate microbial suppression following bacterial exposure to supra-MIC concentrations that have declined below the MIC. This phenomenon is important clinically where patients given intermittent boluses will experience gradually lowered serum and tissue levels that will expose bacteria to both supra- and sub-MICs during the dosing interval (51).

In Vitro Studies

In vitro investigations have demonstrated that, like β-lactam antibiotics, vancomycin is a concentration-independent or time-dependent killer of gram-positive organisms and exhibits minimal concentration-dependent killing. Although vancomycin is considered to be a bactericidal antibiotic, the rate of bacterial killing is slow in comparison to other concentration-independent killers such as β-lactams. In vitro studies, however, can be limiting for several reasons: (*i*) one compartment models only represent concentrations as they would exist in the central compartment and not necessarily at the site of infection; (*ii*) typically only bacteria in log phase growth at standard inocula (10^5 or 10^6 CFU/mL) are utilized; and (*iii*) the effects of the immune system or protein binding are generally not considered (52). Despite the limitations, in vitro studies appear to correlate well with animal and human studies and, therefore, provide useful information for optimal dosing strategies in clinical situations.

Several investigators demonstrated the concentration-independent killing of vancomycin by exposing various bacteria to increasing amounts of drug. Vancomycin's killing effect against *S. aureus* was investigated in vitro by Flandrois et al. (53). The early portion of the time-kill curve was the focus of the study to characterize the bactericidal activity in the initial phases of the dosing interval. A decrease in CFU of only 1 log was obtained at the end of the eight-hour study at concentrations of 1, 2.5, 5, and 10 times the MIC indicating a concentration-independent, slow rate of kill. The killing phase occurred between hours 2 and 4 with the CFU/mL being held constant for the remainder of the curve. Ackerman et al. generated mono- and biexponential killing curves for vancomycin over a 2 to 50 mg/L concentration range to evaluate the relationship between concentration and pharmacodynamic response against *S. aureus* and coagulase-negative Staphylococcus (CNS) species (54). For all organisms tested, killing rates did not change with increasing concentrations of vancomycin and maximum killing appears to be achieved once concentrations of four to five times the MIC of the pathogen are obtained.

Limited studies exist that characterize vancomycin pharmacodynamics for VISA isolates. One such study was conducted by Aeschlimann et al. (55). A vancomycin concentration of 15 mg/L was tested against three different VISA strains (MIC = 8 mg/L). The results showed that the rate of killing by vancomycin was decreased but the extent of killing, as well as the PAE, was unaffected.

Since the pharmacokinetics of vancomycin involve, at minimum, biexponential decay, further studies attempting to simulate this elimination and any effects on bacterial killing were investigated. Utilizing an in vitro model simulating mono- or biexponential decay, Larrson et al. (12) found no statistically significant difference in either the rate or the extent of bacterial killing of *S. aureus*. Again, varying concentrations did not induce change in bactericidal activity, thereby demonstrating that the high drug concentrations achieved during the distribution phase did not enhance the bactericidal activity attained during the elimination phase.

With the understanding that vancomycin killed staphylococci in a concentration-independent fashion, the need to select a pharmacodynamic indice to best predict efficacy was warranted. Duffell et al. (52) used four different vancomycin regimens against *S. aureus* in an in vitro dynamic model. Three dosing schedules with different peak concentrations but the same AUC and a fourth dosing regimen with a smaller AUC were compared for efficacy. The authors found that killing was independent of both peak concentrations and total exposure to drug (AUC). As well, maintaining a constant concentration above the MIC was equally effective even with an AUC that was half of that obtained by the other three dosing regimens. Thereby, this investigation supported $T > \text{MIC}$ as the optimal parameter for efficacy.

Greenberg et al. (56) produced time-kill curves from experiments performed in a static environment with 50% bovine serum and constant antibiotic concentrations. They reported a significantly increased rate and extent of killing of *S. aureus* when the concentration of vancomycin increased from 20 to 80 mg/L, even though free drug concentrations for all regimens exceeded the MIC by at least threefold. This in vitro experiment is one of a few to demonstrate significant concentration-dependent killing with vancomycin alone with concentrations beyond the MIC of the organism.

Vancomycin in combination with other antimicrobials has also been evaluated. Houlihan et al. (57) investigated the pharmacodynamics of vancomycin alone and in combination with gentamicin at various dosing intervals against *S. aureus*-infected fibrin-clots in an in vitro dynamic model. Vancomycin monotherapy simulations included continuous infusion, 500 mg every 6 hours, 1 g every 12 hours, and 2 g every 24 hours, all of which produced varying peaks and troughs. Although all regimens produced concentrations above the MIC for 100% of the dosing intervals, no difference in kill was seen with higher peak concentrations. The investigators also discovered that vancomycin killing was significantly enhanced by the addition of gentamicin whether given every 12 or 24 hours and, in fact, killed in a concentration-dependent fashion. The 2 g dosing scheme of vancomycin significantly reduced bacterial counts greater than any other combination regimen. Whether this finding is due to augmented penetration into the fibrin clots in the presence of gentamicin is unknown.

Since the vast majority of pharmacodynamic investigations with vancomycin include the use of the *S. aureus,* few studies involve other gram-positive or anaerobic organisms. Levett demonstrated time-dependent killing of *C. difficile* by vancomycin in vitro (58). Vancomycin was subinhibitory at concentrations below the MIC of the organism. Once concentrations at the MIC were obtained, no difference in kill was seen whether 4 (at the MIC) or 1000 mg/L (250 × MIC) was utilized. Therefore, as for other organisms, vancomycin kills *C. difficile* in a concentration-dependent manner until the MIC is achieved, beyond which, time-dependent killing is observed.

Hermsen et al. (59) evaluated the activity of vancomycin against methicillin-susceptible and -resistant *S. aureus,* methicillin-susceptible *S. epidermidis,* and *Streptococcus sanguis* in an in vitro peritoneal dialysate model. Interestingly, vancomycin did not decrease the starting inoculum of any of the organisms when studied using peritoneal dialysate fluid (PDF) as the growth medium. The authors suggest that this is due to the bacteriostatic effects of PDF and vancomycin's need for an actively growing bacterial population to exert an effect.

Odenholt-Tornqvist et al. (60) have been the primary source of investigations on the SMEs and PA-SMEs of vancomycin. In an initial study with *Streptococcus pyogenes* and *S. pneumoniae*, the investigators found that the PA-SME with concentrations as low as 0.3 × MIC prevented regrowth of both *Streptococcus* species for 24 hours. In a more recent in vitro investigation of the pharmacodynamic properties of vancomycin against *S. aureus* and *S. epidermidis*, the same authors detected no concentration-dependent killing (51). Low killing rates were demonstrated by time-to-3 log-kill (T3K) at 24 hours with all strains, the exception being a methicillin-sensitive strain of *S. epidermidis*, which obtained T3K at nine hours. Regrowth occurred between 12 and 24 hours when drug concentration had declined to the MIC. PA-SME, SME, and post-MIC effect (PME) were also evaluated in this study. Long PA-SMEs (2.3 to >20 hours) were found with all strains while SMEs were shorter (0.0–15.8 hours). Both PA-SMEs and SMEs increased with increasing multiples of the MIC. Interestingly, longer PMEs, "the difference in time for the numbers of CFU to increase 1 log/mL from the values obtained at the time when the antibiotic concentration has declined to the MIC compared with the corresponding time for an antibiotic-free growth control" (51), were found with shorter vancomycin half-lives. Other investigations have suggested that the regrowth of bacteria can occur if insufficiently inhibited bacteria are allowed to synthesize new peptidoglycan to overcome the antimicrobial's bactericidal effect (61). The authors assumed that the PAE, PA-SME, and PME would emulate the time that the amount of peptidoglycan is kept below a critical level needed for bacterial growth (51). Subsequently, the investigators postulated that longer PMEs may occur with shorter half-lives due to the fact that the MIC is attained faster, thereby not allowing adequate peptidoglycan production to initiate regrowth. Conversely, shorter PMEs were found with longer half-lives. With a slower decline to the MIC and a longer period of time at the MIC, sufficient peptidoglycan could be produced to allow regrowth. How PA-SMEs, SMEs, and PMEs will influence dosing schedules is unknown and further investigations are needed.

Animal Studies

Animal studies focusing on pharmacodynamic predictors of efficacy for vancomycin are quite limited. Peetermans et al. (13), with a granulocytopenic mouse thigh infection model, showed concentration-dependent killing of staphylococcus for concentrations at or below the MIC. Once concentrations exceeded that value however, no further kill was seen with increasing doses.

In a dose-fractionation study using a neutropenic murine-thigh infection model, Ebert et al. (62) observed that both the log AUC and the log Cp-max and the log AUC alone were important predictive parameters for efficacy against methicillin-sensitive *S. aureus* (MSSA) and MRSA, respectively. Time above MIC was not identified as an important indice for either organism. However, the likely impact of a PAE on the outcome of this study was noted, and questions surfaced regarding whether the PAE should be included or subtracted from the evaluation of this parameter. Dudley et al., again using a murine-thigh infection model, have also identified AUC/MIC and Cp-max/MIC as important pharmacodynamic parameters for predicting efficacy with vancomycin against glycopeptide-susceptible and glycopeptides intermediate resistant *S. aureus* (GISA) (63). Substantial variability in AUC/MIC (range 86–462) and Cp-max/MIC (range 15–61) ratios required to reach 50% maximum killing was reported for GISA strains. As opposed to the vancomycin-susceptible strains, for a given level of in vivo exposure to vancomycin, killing of GISA was more complete and required lower AUC/MIC (range 23–31) and Cp-max/MIC (range 4.4–9.5) ratios for 50% of E_{max}.

Cantoni et al. (64), in an attempt to compare the efficacy of amoxicillin/clavulanic acid against MSSA and MRSA, respectively, versus vancomycin in a rat model of infection, found vancomycin activity to be dependent upon strain. Against the MSSA strain, vancomycin at 30 mg/kg given every 6 hours was more effective than giving the same dose every 12 hours. Against the MRSA strain, the four-times-daily regimen only marginally improved outcome as compared to the twice-daily regimen. Because vancomycin concentrations were undetectable after six hours of therapy, the four-times-daily regimen was the only therapy that allowed concentrations to remain above the MIC for a majority of the dosing interval. This finding further supports the dependence of vancomycin activity upon the $T > MIC$.

The activity of vancomycin was evaluated against penicillin-resistant pneumococci utilizing a mouse peritonitis model by Knudsen et al. (65). In comparing various pharmacokinetic/pharmacodynamic parameters at the values of the ED-50 (effective dose for 50% of the population), investigators concluded that both $T > MIC$ and C_{max} were important predictors of efficacy in their model. These parameters were deemed best predictors because they varied the least. Of significance with this study also was the discovery that vancomycin activity was not influenced by the penicillin susceptibility of the organism. The same investigators, again using their peritonitis model with immunocompetent mice, studied the pharmacodynamics of vancomycin against *S. pneumoniae* and *S. aureus* (66). Survival after six days was best predicted by Cp-max/MIC for both organisms in experiments where treatments with total doses close to $ED_{50}s$ for single doses were given as one or two doses. Cp-max/MIC ratios ranged from 4.07 to 8.56. Further, in multidosing studies, $ED_{50}s$ increased as the number of doses given over the 48-hour treatment period increased, although the same total amount of vancomycin was administered. The authors concluded from their observations that Cp-max/MIC is of major importance; however, this ratio alone cannot predict outcome because highly significant correlations were found for all three parameters and effect. Likewise, Cp-max/MIC, AUC/MIC, and $T > MIC$ were also correlated to each other.

Human Studies

In vivo, serum bactericidal titers (SBT) have been evaluated to determine antimicrobial efficacy. A SBT of 1:8 with vancomycin has been associated with clinical cure in patients with staphylococcal infections (67,68). This SBT was associated with serum concentrations greater than 12 mg/L. James et al. (69) conducted a prospective, randomized, crossover study to compare conventional dosing of vancomycin versus continuous infusions in patients with suspected or documented gram-positive infections. Because the most effective concentration of vancomycin against staphylococcus is not known, the investigators chose a target concentration of 15 mg/L via continuous infusion and peak and trough concentrations of 25 to 35 and 5 to 10 mg/L, respectively, with conventional dosing of 1 g every 12 hours. Despite variability in actual concentrations obtained, continuous infusion produced SBTs of 1:16 while conventional dosing produced trough SBTs of 1:8, which was not found to be statistically significant. Concentrations remained above the MIC throughout the entire dosing intervals for all patients, whether receiving conventional dosing or continuous infusion. Therefore, the authors concluded that both methods of intravenous administration demonstrated equivalent pharmacodynamic activities. Although continuous infusion therapy was more likely than conventional dosing to produce SBTs of 1:8 or greater, this study did not attempt to evaluate clinical efficacy associated with such values and therefore, whether improved patient outcome was obtained is unknown.

Lacy et al. (70) evaluated a twice-daily and a once-daily regimen of vancomycin in 12 healthy volunteers to assess the duration of time that the serum concentrations remained above the pathogen MIC. The doses studied were 1 g every 12 hours and 1 g every 24 hours, and the isolates studied were MRSA and two MR-CNS isolates, *S. epidermidis* and *Staphylococcus hominis* (MIC = 2–4 mg/L). Bactericidal activity was maintained for >80% of the dosing interval for the twice-daily regimen for both the MRSA and the MR-CNS isolates, and serum concentrations remained above the MIC for 100% of the interval. Bactericidal activity was maintained for 50% to 66% of the interval for the once-daily regimen, and the serum concentrations remained above the MIC for 54% to 75% of the interval. The authors suggest that the twice-daily regimen be used in patients with normal renal function when MR-CNS is suspected or if the MIC approaches the vancomycin susceptibility breakpoint of 4 mg/L.

Klepser et al. (71), in a preliminary report of a multicenter study of patients with gram-positive infections receiving vancomycin therapy, found increased rates of bactericidal activity with vancomycin trough concentrations of greater than 10 mg/L. Bacterial eradication was also correlated with trough SBTs of 1:8 or greater. Patients who failed therapy had pathogen MICs of >1 mg/L. Hyatt et al. (72) suggest that the area under the inhibitory serum concentration-time curve (AUIC) as well as organism MIC were associated with clinical outcome. By performing a retrospective analysis of 84 patients receiving vancomycin therapy for gram-positive infections, investigators found that therapy that produced AUIC <125 and pathogens with MICs >1 mg/L had a higher likelihood of failure. Moise et al. suggest that both an AUIC and a pneumonia scoring system are predictive of clinical and microbiological outcomes of vancomycin treatment for lower respiratory tract infections caused by S. aureus. The authors performed a retrospective review of 70 patients and used classification-and-regression-tree modeling to determine which variables were correlated with clinical and microbiological outcomes (73). Therefore, these studies propose that, not only may $T > MIC$ but also AUIC and trough values may be important for maximum clinical efficacy.

In summary, vancomycin demonstrates concentration-independent killing of gram-positive bacteria, and peak concentrations do not appear to correlate with rate or extent of kill. Maximum killing is achieved at serum concentrations of four to five times the MIC of the infecting pathogen and sustaining concentrations at or above these levels for the entire dosing interval will likely produce the best antimicrobial effect. Dosing strategies should therefore be aimed at maximizing the time in which concentration at the site of infection remains above the MIC of the pathogen. Whether the most efficient killing is obtained by continuous infusion of vancomycin versus intermittent bolus is controversial. Several studies revealed that no difference in killing is seen between the two methods of administration (57,69,74). However, such benefits, as predictable serum concentrations and ease of administration, might be advantageous (74). Conversely, due to vancomycin's long half-life and the perceived better tolerability associated with intermittent bolus injections, continuous infusion of this drug may not be needed and is often discouraged (74).

CLINICAL APPLICATION
Clinical Uses
Vancomycin is available as vancomycin hydrochloride (Vancocin®, Lyphocin®, Vancoled®, others) for intravenous use, as powder for oral solution, and as capsules for oral use (Vancocin Pulvules®). The indications for vancomycin use are limited in relation to its strong gram-positive spectrum. Although vancomycin is bactericidal against most gram-positive cocci and bacilli, the intravenous preparation should be reserved for serious gram-positive infections not treatable with β-lactams or other traditional options. Use should not precede therapy with β-lactams for susceptible organisms. Clinical outcomes with vancomycin versus nafcillin or ampicillin against both staphylococci and enterococci show vancomycin inferiority when comparing bactericidal rate and rapidity of blood sterility (75–79).

Vancomycin is the drug of choice for serious staphylococcal infections that cannot be treated with β-lactams due to bacterial resistance (MRSA and MR *S. epidermidis*) or due to the patient's inability to receive these medications (80). Staphylococcal infections include bacteremia, endocarditis, skin and soft-tissue infections, pneumonia, and septic arthritis. Dialysis peritonitis due to staphylococci may also be treated with intravenous or intraperitoneal vancomycin. Although vancomycin is indicated for *S. aureus* osteomyelitis, bone penetration is extremely variable, especially between published studies, and treatment with other options could prove more effective (81–83). Vancomycin is also indicated for infections due to coagulase-negative staphylococci including catheter-associated bacteremia, prosthetic valve endocarditis, vascular graft infections, prosthetic joint infections, central nervous system shunt infections, and other indwelling medical device-associated infections (80,84). Complete cure of most medical-device related infections usually requires the removal of the device due to the biofilm secreted by *S. epidermidis*. Staphylococcal treatment with vancomycin may require up to one week or longer for clinical response in serious infections such as MRSA (84). Courses of vancomycin that fail to cure serious staphylococcal infections may require the addition of gentamicin, rifampin, or both (80,84,85).

Two significant clinical issues surround the use of vancomycin for the treatment of staphylococcal endocarditis and other serious infections. First, controversy exists as to whether the addition of rifampin is synergistic or antagonistic. Although certain studies have proven the combination to be more efficacious than single therapy with vancomycin (86–88), other more recent publications cite the combination as antagonistic (77). Additionally, clinical experience with the combination has been inconsistent (89).

The second issue that surrounds vancomycin use for serious staphylococcal infections is the potentially better outcome with β-lactams. In addition to the in vitro data that suggest vancomycin is less rapidly bactericidal than nafcillin, clinical data exist to support this conclusion (75–79). Although no large-scale comparison studies exist to evaluate the efficacy of vancomycin versus β-lactams in staphylococcal endocarditis, assumptions can be formulated from published studies. In a study by Korzeniowski and Sande, the duration of bacteremia due to *S. aureus* endocarditis lasted a median of 3.4 days after treatment with nafcillin while bacteremia lasted a median of seven days for patients treated with vancomycin in a study conducted by Levine (77,79). The patients in the Levine et al. study were infected with MRSA in comparison to the methicillin-sensitive organisms from the Korzeniowski study; yet, in general, morbidity and mortality of bacteremic infections due to MSSA and MRSA are comparable (78). In a small study that compared vancomycin to nafcillin in *S. aureus* endocarditis, the investigators found patients treated with nafcillin plus tobramycin had a cure rate of 94%, while only 33% of patients treated with vancomycin plus tobramycin were cured (76). Worth mentioning, however, is the fact that while the nafcillin-plus-tobramycin group consisted of 50 patients, only 3 patients received vancomycin plus tobramycin due to β-lactam allergy. Small and Chambers performed another study that evaluated the use of vancomycin in 13 patients with staphylococcal endocarditis, 5 of whom failed therapy (75). The reason for vancomycin ineffectiveness in these cases may be the need for prolonged high levels of a bactericidal antibiotic; however, with longer durations of bacteremia and poorer clinical outcomes, serious consideration needs to be given as to whether vancomycin should be considered at all in patients with MSSA endocarditis who can tolerate β-lactam therapy.

Streptococcal infections not treatable with β-lactams or other traditional options are also proper indications for vancomycin (80,84). Endocarditis due to β-lactam-resistant *Streptococcus viridans* or *Streptococcus bovis* is a common use of vancomycin, although organisms with elevated MIC values may require combination with an aminoglycoside. Vancomycin is the drug of choice for pneumococci infections showing high-level resistance to penicillin (80,84). Cefotaxime or ceftriaxone plus rifampin may be needed to adequately cover *S. pneumoniae* meningitis due to vancomycin's poor penetration in the central nervous system (90,91). Although penetration is enhanced while meninges are inflamed, as in meningitis and shunt infections, certain cases may require intrathecal or intraventricular administration to obtain therapeutic levels.

As for enterococcal infections, vancomycin represents the treatment of choice for ampicillin-resistant enterococcus (80,84). *Enterococcus endocarditis* and other infections may require the addition of an aminoglycoside such as gentamicin. Vancomycin is also the treatment of choice for corynebacterium (80,84).

Empirically, vancomycin should only be used in limited situations. Vancomycin can be considered for febrile neutropenic patients presenting with clinical signs and symptoms of gram-positive infections in areas of high MRSA prevalence (38). Other indications for empiric use of vancomycin in neutropenic patients with fever include the presence of severe mucositis, colonization with MRSA or penicillin-resistant *S. pneumoniae*, prophylaxis with quinolone antibiotics, or obvious catheter-related infection (92). Vancomycin should be discontinued after four to five days if no infection is identified or if initial cultures for gram-positive organisms are negative after 24 to 48 hours. For prophylaxis, vancomycin may be used perioperatively with prosthesis implantation only in severely β-lactam allergic patients (38). Vancomycin is also used for endocarditis prophylaxis for β-lactam allergic patients.

Orally, vancomycin is indicated for metronidazole-refractory antibiotic-associated colitis caused by *C. difficile* (38,80,84). Intravenous administration of vancomycin typically does not achieve adequate levels in the colon lumen to successfully treat antibiotic-associated colitis. However, there are rare reports of success with this route cited in the literature (93). Administration via nasogastric tube, enema, ileostomy, colostomy, or rectal catheter may be needed if the patient presents with severe ileus. Oral vancomycin has also been used prophylactically to prevent endogenous infections in cancer and leukemia patients. This regimen seems to decrease the *C. difficile* associated with the chemotherapy (94,95).

Inappropriate Uses

Although vancomycin is an effective option for most gram-positive infections, the drug needs to be judiciously used to prevent the emergence and spread of resistance. Vancomycin should not be used when other drug options such as β-lactams are viable. Microbial susceptibilities need to be performed to determine the appropriateness of vancomycin therapy, and the antibiotic needs to be changed if the organism is susceptible to a different agent.

The CDC has published guidelines for the appropriate use of vancomycin (Table 1) (38). However, vancomycin misuse around the nation is widespread. A retrospective study from May 1993 to April 1994 identified 61% of vancomycin

TABLE 1 Use of Vancomycin

Appropriate Use:
Treatment of serious infections due to β-lactam–resistant gram-positive pathogens
Treatment of gram-positive infections in patients with serious β-lactam allergies
Antibiotic-associated colitis failure to metronidazole
Endocarditis prophylaxis per American Heart Association recommendations
Antibiotic prophylaxis for implantation of prosthetic devices at institutions with a high rate of
  infections due to methicillin-resistant Staphylococci

Inappropriate Use:
Routine surgical prophylaxis
Empiric treatment for febrile neutropenic patients without strong evidence of gram-positive infection
  and high prevalence of β-lactam–resistant organisms in the institution
Treatment in response to a single positive blood culture for coagulase-negative staphylococci
  when other blood cultures taken appropriately in the same time frame are negative
Continued empiric use without positive culture for β-lactam–resistant gram-positive pathogen
Systemic or local prophylaxis for central or peripheral catheter
Selective gut decontamination
Eradication of methicillin-resistant *Staphylococcus aureus* colonization
Primary treatment of antibiotic-associated colitis
Routing prophylaxis for patients on chronic ambulatory peritoneal dialysis
Routine prophylaxis for very low birth weight infants
Topical application or irrigation

Source: From Ref. 38.

usage as inappropriate, according to the CDC criteria (96). A similar evaluation published in 1997 found only 47% of vancomycin orders prescribed for 7147 patients were appropriate (97). According to this study, inadequate use and inappropriate control patterns were similar whether large teaching centers or small rural hospitals were evaluated. Thus, alternate methods of vancomycin control need to be implemented to assure adequate use and limit resistance. A more recent study evaluated the appropriateness of vancomycin use in hospitals, using data from 1999 in comparison to data from 1994 (before any of the studied hospitals had a vancomycin restriction policy) (98). The investigators found that the hospitals using vancomycin restriction policies had a decrease in inappropriate therapeutic and prophylactic vancomycin use, but inappropriate empiric use was unaffected. The authors suggest the implementation of a hospital-wide vancomycin restriction policy to help control the inappropriate use of this antibiotic.

Toxicity and Adverse Drug Reactions
A variety of adverse reactions have been associated with vancomycin, including fever, rash, phlebitis, hematologic effects, nephrotoxicity, auditory toxicity, interstitial nephritis, and infusion-related reactions. Many of the infusion-related reactions were likely due to impurities in the initial formulations and have been significantly reduced with the newer formulations. The Red Man or Red Neck Syndrome is an anaphylactoid reaction related to rapid infusion of large doses, typically >12 mg/kg/hr (16,80,84). The reaction begins 10 minutes after infusion and generally resolves within 15 to 20 minutes after stopping the infusion. Patients may experience tachycardia, chest pain, dyspnea, urticaria, and swelling of the face, lips, and eyelids. Additionally, patients may experience a hypotensive episode with a 25% to 50% reduction in systolic blood pressure. Interestingly,

healthy volunteers receiving vancomycin infusions have a higher propensity toward the reaction than patients (74). The reason is unknown. Symptoms of Red Man Syndrome appear to be histamine mediated, but investigations are inconclusive. Extending the administration of vancomycin to one hour or a maximum of 15 mg/min should prevent most infusion-related reactions.

Vancomycin toxicity was retrospectively studied by Farber and Moellering in 98 patients (99). They noted a 13% incidence of phlebitis, a 3% incidence of fever and rash, and a 2% incidence of neutropenia. However, this report may overestimate true adverse reactions because of the inclusion of many potentially high-risk patients. Interestingly, although other studies have shown that concomitant aminoglycosides are not a risk factor for nephrotoxicity (100), patients receiving both vancomycin and an aminoglycoside experienced a 35% incidence of reversible nephrotoxicity, which is more than expected from either antibiotic alone. Only 5% of patients receiving vancomycin alone experienced nephrotoxicity. The authors also found that patients with nephrotoxicity had trough concentrations of 20 to 30 mg/L.

Vancomycin ototoxicity has been reported with peak serum concentrations of 80 to 100 mg/L (101). Geraci identified two patients with vancomycin-induced ototoxicity, one of which had a history of renal disease, elevated blood-urea nitrogen on admission, and a recorded diastolic blood pressure of zero. Serum concentrations determined three to six hours after the dose was administered ranged from 80 to 95 mg/L. Due to the biexponential nature of the vancomycin serum concentration time curve, the true vancomycin peak was likely near 200 to 300 mg/L. Farber and Moellering, as well, reported the occurrence of ototoxicity in a patient who, at one hour post infusion, had serum concentrations of <50 mg/L (99); however the true peak was likely in the toxic range as defined by Geraci (101).

A recent study evaluated the hematologic toxicity of vancomycin compared to linezolid in long-term therapy for osteomyelitis (102). The hematologic effects observed were thrombocytopenia, neutropenia, and anemia. These adverse hematologic effects appeared to be more common in the vancomycin group, although the difference was not statistically significant. The only significant difference identified by the authors was a higher incidence of thrombocytopenia in the linezolid group who had received vancomycin in the previous two weeks versus those who had not received vancomycin (71% vs. 15%, respectively).

In summary, the incidence of adverse reactions associated with vancomycin is relatively low. Only approximately 40 cases of oto- and nephrotoxicity have been reported in the medical literature between the years 1956 and 1984 despite incessant use. Most of these cases were complicated by concomitant aminoglycoside therapy and preexisting renal problems as well as investigator discrepancies in interpreting serum levels.

Dosing and Therapeutic Monitoring

Medical literature abounds, questioning the need to therapeutically monitor vancomycin concentrations. Cantu et al. (103) suggest that monitoring vancomycin concentrations is unnecessary because no correlation has been demonstrated between drug levels, toxicity, and clinical response. Opponents propose that vancomycin can be dosed using published nomograms based on the patient's age, weight, and estimated creatinine clearance. Conversely, Moellering argues that therapeutic vancomycin monitoring would, in fact, be prudent for optimal clinical response and restriction of toxicity in such situations as patients on hemodialysis, patients with rapidly changing renal function, and patients receiving high-dose vancomycin or concomitant aminoglycoside therapy (104). Numerous strategies do exist for empirically dosing vancomycin. Administering 500 mg every 6 hours, 1 g every 12 hours, or 20 to 40 mg/kg of body weight/day is commonly employed. Additionally, nomograms exist such as those established by Matzke et al., Moellering et al., Lake and Peterson, and Nielsen et al. (104–107). Serious faults lie in the dependence on these nomograms for efficacious use of vancomycin, however, because the authors assume, rather than prove, that their method of pharmacokinetically modeling the data is appropriate. Most empiric regimens were designed to provide peak concentrations of between 20 and 40 mg/L and trough concentrations of 5 to 10 mg/L (or approximately five times the MIC of the infecting pathogen). However, such practices only place 3% to 23% of patients in this therapeutic range, according to one published study (108). Further, some investigators have suggested targeting trough concentrations of 10 to 20 mg/L, to prevent hVISA, or in the presence of hVISA, to stave off selection of a uniform VISA population. Unfortunately, although such goals in serum levels are set, no solid data is available to support any specific therapeutic range, and accordingly, serum peak and trough concentrations have been selected somewhat arbitrarily, based upon speculations from retrospective studies, case reports, and personal opinions. Peak concentrations appear to play little to no role in the efficacy of the drug and appear to have limited involvement in toxicity unless exceedingly large peak values are obtained. On the other hand, trough concentrations may be useful monitoring parameters. Since vancomycin is a concentration-independent killer, the goal of therapy should be to maintain the unbound concentration above the microbial MIC for a significant portion of the dosing interval because regrowth of most organisms will begin shortly after drug concentrations fall below the MIC. A depiction of predicted vancomycin pharmacodynamic indices obtained from a typical intravenous dose using various pathogen MICs is presented (Table 2).

The role of vancomycin degradation products also needs to be considered when interpreting levels in patients with renal failure where half-lives are significantly extended (109,110). In vitro and in vivo, vancomycin breaks down over time to form crystalline degradation products. Antibodies in commercial assays, such as TDx® fluorescence polarization immunoassay, cross-react with major and minor degradation products, thereby overstating factor B (active drug) content in the level. This can result in an overstated vancomycin concentration of 20% to 50%.

In summary, trough concentrations of 5 to 10 mg/L appear to be reasonable goals for vancomycin therapy because MICs of most gram-positive pathogens are ≤1 mg/L. Such concentrations would allow the unbound concentrations to remain above the MIC of the organism for at least 50% of the dosing interval. In situations where the MIC for *S. aureus* is >1 mg/L, vancomycin should be used with caution. Recent guidelines for nosocomial pneumonia, endocarditis, and meningitis have suggested trough concentrations of 10–20 mg/L depending on the disease state and the guideline. Currently, there is no convincing data that maintaining higher vancomycin trough concentrations will improve clinical or microbiologic outcomes or prevent the emergence of hVISA or VISA. Administering 10 to 15 mg/kg per dose and adjusting the dosing interval per renal function based upon numerous published nomograms is not likely to produce "toxic" peak concentrations and should allow therapeutic concentrations throughout the dosing interval in the majority of patients with normal renal function. Loading doses are not typically

TABLE 2 Estimated Vancomycin Pharmacodynamic Ratios for Various MIC Values[a]

MIC (mg/L)	Cpmax/MIC	$T >$ MIC (hr)	AUC$_{24}$/MIC
0.25	140	12	784
0.5	70	12	392
1.0	35	12	196
2.0	17.5	12	98
4.0	8.75	12	49
8.0	4.38	11	24.5

[a]Calculations based on total concentrations achieved after administration of a 1 g dose given every 12 hours to a 70 kg patient with normal renal function.
Abbreviations: AUC, area under the serum concentration-time curve; MIC, minimum inhibitory concentration.

needed because transiently high distribution-phase concentrations are unlikely to enhance bacterial killing. However, loading doses may be reasonable in patients in whom the site of infection is distal to the central compartment or poorly accessible. Until a relationship between clinical efficacy, toxicity, and vancomycin concentration is established, vancomycin therapy will inevitably continue to be monitored in an attempt to improve patient outcome. A study using decision analysis to model the cost-effectiveness of pharmacokinetic dosage adjustment of vancomycin to prevent nephrotoxicity found that monitoring was not cost effective in all patients (111). The investigators suggested cost-effectiveness when monitoring was used in patients receiving concomitant nephrotoxins, intensive care patients, and, possibly, oncology patients. Whether therapeutic monitoring of vancomycin should be a standard of practice or is only necessary in patients receiving high-dose therapy, patients on concomitant aminoglycoside therapy, or patients with renal insufficiency or failure on dialysis is likely a personal preference until further studies establish guidelines. Of note, if the CDC guidelines for appropriate vancomycin usage were stringently followed, at least half of vancomycin use could be eliminated, leaving the remaining patients to be monitored.

OTHER GLYCOPEPTIDES AND GLYCOPEPTIDE-LIKE COMPOUNDS
Teicoplanin
Teicoplanin, like vancomycin, binds to the terminal D-alanyl-D-alanine portion of the peptidoglycan cell wall of actively growing gram-positive bacteria to exert its bactericidal activity (112). Currently, only available in Europe, teicoplanin can be used to treat infections caused by both methicillin-sensitive and -resistant strains of *S. aureus* and *S. epidermidis*, streptococci, and enterococci. Clinical trials have demonstrated teicoplanin to be a safe, well-tolerated agent with reports of side effects occurring in 6% to 13% of recipients (112). The most prevalent adverse reactions reported are pain at the injection site and skin rash. Nephro- and ototoxicity are uncommon even when used concomitantly with other nephro- and ototoxic drugs. Pharmacokinetically, teicoplanin differs from vancomycin. The half-life is considerably longer (~47 hours), and the percent protein bound nears 90% (112). Because of the long half-life, an every other day dosing regimen has been suggested as being equivalent to a daily dosing regimen (113). Moreover, teicoplanin can be administered by either the intravenous or the intramuscular route as opposed to vancomycin, which is limited parenterally to the intravenous route. Pharmacodynamic evaluations virtually duplicate those of vancomycin once the heightened protein binding of teicoplanin and subsequent lower active free

concentrations are accounted for (114). Further review of teicoplanin can be found elsewhere (112,115,116).

Dalbavancin

Dalbavancin is an experimental semisynthetic glycopeptide being developed for the treatment of serious gram-positive infections. Dalbavancin has the same mechanism of action as vancomycin, but dalbavancin retains activity against some VRE and VISA and possesses activity against *Clostridium* spp., *Peptostreptococcus* spp., *Actinomyces* spp., *Corynebacterium* spp., and *Bacillus subtilis* (117,118). Like teicoplanin, dalbavancin is active against the *vanB* phenotype of VRE but not the *vanA* phenotype. Dalbavancin is bactericidal against these microorganisms. The most prevalent adverse reactions reported in one study were pyrexia, headache, and nausea (117). The half-life of dalbavancin is approximately one week, allowing for the possibility of weekly dosing (117). This agent looks promising for the treatment of serious gram-positive infections, including those due to resistant organisms.

Oritavancin

Oritavancin is an investigational synthetic glycopeptide with a broad spectrum of activity against gram-positive cocci, including VRE, both *vanA* and *vanB* phenotypes, and VISA. Several factors distinguish oritavancin from vancomycin: concentration-dependent bactericidal activity versus *S. aureus* and enterococcus, activity versus intracellular gram-positive organisms, and a prolonged PAE (119). The drug acts on the same molecular target as vancomycin and other glycopeptide antibiotics (120), but a *p*-chlorophenylbenzyl side chain allows for enhanced interaction with the cytoplasmic membrane and affects the proteins involved in transglycosylation, resulting in improved activity over that of vancomycin (121). The half-life is long, approaching 10.5 days, which may allow for infrequent dosing (122). Pharmacodynamic investigations are lacking, but one study suggests that the C_{max} appears to correlate better with activity than the time above the MIC or the AUC (119). Oritavancin is in Phase III development and shows particular promise for skin and soft-tissue infections as well as respiratory infections due to gram-positive bacteria.

Daptomycin

Daptomycin (Cubicin, Cubist Pharmaceuticals, Lexington, MA, U.S.A.) is the first in a new class of antibiotics, lipopeptides. Daptomycin, available only as an injectable, was approved by the Food and Drug Administration in September 2003 for the treatment of skin and soft-tissue infections caused by MSSA, MRSA, *S. pyogenes, Streptococcus agalactiae, Streptococcus dysgalactiae* subsp *equisimilis,* and vancomycin-susceptible *E. faecalis* and more recently was approved for the treatment of bacteremia and endocarditis due to *S. aureus* (123). Daptomycin exerts rapid bactericidal activity by inserting itself into the cell membrane of gram-positive bacteria and causing depolarization of the membrane potential, leading to cell death by halting bacterial DNA, RNA, and protein synthesis (124). The half-life of daptomycin is approximately nine hours, and the drug is highly protein bound (90–94%) (124). The most prevalent adverse reactions reported with daptomycin are constipation, nausea, and injection-site reactions (123). Initially, the development of daptomycin by Eli Lilly was halted in 1991, when several patients developed skeletal muscle toxicity. However, on the basis of several pharmacodynamic studies, a dosing regimen change from multiple daily doses to a once-daily regimen significantly lowered the skeletal muscle effects (125). Daptomycin exhibits a prolonged PAE and concentration-dependent activity, although the bactericidal activity of the drug is dependent on the presence of calcium ions. These characteristics allow once-daily dosing, which also minimizes the adverse effects. Several pharmacodynamic studies have identified the AUC/MIC ratio as the pharmacodynamic parameter that correlates best with activity (126–128). Clinical trials evaluating daptomycin for the treatment of *S. aureus* bacteremia and endocarditis are ongoing. Further review of daptomycin can be found elsewhere (125).

Others

Several other novel glycopeptide antibiotics are currently under investigation. Mannopeptimycin (AC98-6446) is a semisynthetic cyclic glycopeptide with good in vitro activity versus susceptible and resistant gram-positive bacteria. Mannopeptimycin has been shown to be bactericidal against MRSA, MSSA, VRE, and penicillin-susceptible and -resistant *S. pneumoniae* in animal models (129). Like oritavancin, mannopeptimycin demonstrates concentration-dependent killing against staphylococci and enterococci (130). Televancin (TD-6424) is a novel lipidated glycopeptide in Phase III clinical trials. Televancin exhibits bactericidal activity versus gram-positive bacteria, including susceptible and resistant strains. Televancin has a short half-life and is highly protein bound. Lastly, semisynthetic, lipophilic aglycon glycopeptide derivatives have been shown to be active against retroviruses, including HIV-1 and HIV-2 (131). The derivatives appear to hinder viral entry into the cells. Limited data exist for all of these novel compounds, and more studies must be completed before the clinical significance will be apparent.

CONCLUSION

With years of clinical experience, vancomycin has proven to be a safe and efficacious agent against gram-positive pathogens, including many multidrug-resistant strains. Despite this history, to date, the therapeutic range has not been rigorously defined, but going beyond the currently suggested therapeutic range is not likely to improve antibiotic performance. The accumulation of in vitro and in vivo studies suggests that vancomycin is a concentration-independent killer of gram-positive organisms, with maximum killing occurring at serum concentrations of four to five times the MIC of the infecting organism. At the present time, insufficient data exist to suggest one pharmacodynamic outcome parameter is superior to another in predicting clinical or microbiologic success. Likewise, a specific targeted indice value cannot be assigned. However, high-peak concentrations appear not to be associated with an improved rate or extent of kill, and, therefore, therapy should be targeted toward sustaining serum concentrations above the MIC for a large portion of the dosing interval. With the high level of vancomycin usage, the development and spread of vancomycin-resistant organisms is a major clinical concern. At a time when we are attempting to be more prudent and judicious in the use of vancomycin, we also find ourselves more dependent on the drug. Unfortunately, this combination of factors may drive bacterial resistance and ultimately nullify a drug that has been a gold standard product for half a century. Further, with the availability and pending approval of newer agents with activity against staphylococcal, streptococcal, and enterococcal infections, the use of vancomycin as a contemporary first-line agent against these resistant pathogens may be limited.

REFERENCES

1. Crossley KB, Rotschafer JC, Chern MM, Mead KE, Zaske DE. Comparison of a radioimmunoassay and a microbiological assay for measurement of serum vancomycin concentrations. Antimicrob Agents Chemother 1980; 17:654–657.
2. Cooper GL, Given DB. The development of vancomycin. In: Cooper GL, Given DB, eds. Vancomycin: A Comprehensive Review of 30 Years of Clinical Research. Park Row Publishers, 1986:1–6.
3. Leclercq R, Derlot E, Duval J, Courvalin P. Plasmid-mediated resistance to vancomycin and teicoplanin in *Enterococcus faecium*. N Engl J Med 1988; 319:157–161.
4. Uttley AH, Collins CH, Naidoo J, George RC. Vancomycin-resistant enterococci. Lancet 1988; 1:57–58.
5. Centers for Disease Control and Prevention. *Staphylococcus aureus* with reduced susceptibility to vancomycin—United States, 1997. JAMA 1997; 278:891–892.
6. Chang S, Sievert DM, Hageman JC, et al. Infection with vancomycin-resistant *Staphylococcus aureus* containing the *vanA* resistance gene. N Engl J Med 2003; 348: 1342–1347.
7. National Committee for Clinical Laboratory Standards. Performance Standards for Antimicrobial Susceptibility Testing. Vol. 19. Wayne, PA: National Committee for Clinical Laboratory Standards, 1999.
8. Jones RN, Ballow CH, Biedenbach DJ, Deinhart JA, Schentag JJ. Antimicrobial activity of quinupristin-dalfopristin (RP 59500, Synercid) tested against over 28,000 recent clinical isolates from 200 medical centers in the United States and Canada. Diagn Microbiol Infect Dis 1998; 31:437–451.
9. Nowak R et al. Emergence of vancomycin tolerance in *S. pneumoniae*. Nature 1999; 399:524–526, 590-593.
10. Reynolds PE. Structure, biochemistry and mechanism of action of glycopeptide antibiotics. Eur J Clin Microbiol Infect Dis 1989; 8:943–950.
11. Reynolds PE, Somner EA. Comparison of the target sites and mechanisms of action of glycopeptide and lipoglycodepsipeptide antibiotics. Drugs Exp Clin Res 1990; 16:385–389.
12. Larsson AJ, Walker KJ, Raddatz JK, Rotschafer JC. The concentration-independent effect of monoexponential and biexponential decay in vancomycin concentrations on the killing of *Staphylococcus aureus* under aerobic and anaerobic conditions. J Antimicrob Chemother 1996; 38:589–597.
13. Peetermans WE, Hoogeterp JJ, Hazekamp-van Dokkum A-M, Van Den Broek P, Mattie H. Antistaphylococcal activities of teicoplanin and vancomycin in vitro and in an experimental infection. Antimicrob Agents Chemother 1990; 34:1869–1874.
14. Lamp KC, Rybak MJ, Bailey EM, Kaatz GW. In vitro pharmacodynamic effects of concentration, pH, and growth phase on serum bactericidal activities of daptomycin and vancomycin. Antimicrob Agents Chemother 1992; 36:2709–2714.
15. Svensson E, Hanberger H, Nilsson LE. Pharmacodynamic effects of antibiotics and antibiotic combinations on growing and nongrowing *Staphylococcus epidermidis* cells. Antimicrob Agents Chemother 1997; 41:107–111.
16. Lundstrom TS, Sobel JD. Vancomycin, trimethoprim-sulfamethoxazole, and rifampin. Infect Dis Clin North Am 1995; 9:747–767.
17. Morse GD, Apicella MA, Walshe JJ. Absorption of intraperitoneal antibiotics. Drug Intell Clin Pharm 1988; 22:58–61.
18. Rotschafer JC, Crossley K, Zaske DE, Mead K, Sawchuk RJ, Solem LD. Pharmacokinetics of vancomycin: observations in 28 patients and dosage recommendations. Antimicrob Agents Chemother 1982; 22:391–394.
19. Moellering RC Jr. Pharmacokinetics of vancomycin. J Antimicrob Chemother 1984; 14 (suppl D):43–52.
20. Sun H, Maderazo EG, Krusell AR. Serum protein-binding characteristics of vancomycin. Antimicrob Agents Chemother 1993; 37:1132–1136.
21. Matzke GR, Frye RF. Drug therapy individualization for patients with renal insufficiency. In: DiPiro JT, Talbert RL, Yee GC, Matzke GR, Wells BG, Posey LM, eds.

Pharmacotherapy: A Physiologic Approach. 3rd ed. Stamford, CT: Appleton & Lange, 1997:1083–1103.

22. Woodford N, Johnson AP, Morrison D, Speller DC. Current perspectives on glycopeptide resistance. Clin Microbiol Rev 1995; 8:585–615.

23. Moellering RC Jr. Vancomycin-resistant enterococci. Clin Infect Dis 1998; 26:1196–1199.

24. Ena J, Dick RW, Jones RN, Wenzel RP. The epidemiology of intravenous vancomycin usage in a university hospital. A 10-year study. JAMA 1993; 269:598–602.

25. Arthur M, Reynolds PE, Depardieu F, et al. Mechanisms of glycopeptide resistance in enterococci. J Infect 1996; 32:11–16.

26. Gold HS, Moellering RC Jr. Antimicrobial-drug resistance. N Engl J Med 1996; 335:1445–1453.

27. Centers for Disease Control. *Staphylococcus aureus* resistant to vancomycin—United States, 2002. JAMA 2002; 288:824–825.

28. Quintiliani R Jr, Evers S, Courvalin P. The *vanB* gene confers various levels of self-transferable resistance to vancomycin in enterococci. J Infect Dis 1993; 167:1220–1223.

29. Perichon B, Reynolds P, Courvalin P. VanD-type glycopeptide-resistant *Enterococcus faecium* BM4339. Antimicrob Agents Chemother 1997; 41:2016–2018.

30. Fines M, Perichon B, Reynolds P, Sahm DF, Courvalin P. VanE, a new type of acquired glycopeptide resistance in *Enterococcus faecalis* BM4405. Antimicrob Agents Chemother 1999; 43:2161–2164.

31. Murray BE. Vancomycin-resistant enterococcal infections. N Engl J Med 2000; 342:710–721.

32. Eliopoulos GM, Wennersten CB, Gold HS, et al. Characterization of vancomycin-resistant *Enterococcus faecium* isolates from the United States and their susceptibility in vitro to dalfopristin-quinupristin. Antimicrob Agents Chemother 1998; 42:1088–1092.

33. Schwalbe RS, Stapleton JT, Gilligan PH. Emergence of vancomycin resistance in coagulase-negative staphylococci. N Engl J Med 1987; 316:927–931.

34. Veach LA, Pfaller MA, Barrett M, Koontz FP, Wenzel RP. Vancomycin resistance in *Staphylococcus haemolyticus* causing colonization and bloodstream infection. J Clin Microbiol 1990; 28:2064–2068.

35. Sanyal D, Johnson AP, George RC, Cookson BD, Williams AJ. Peritonitis due to vancomycin-resistant *Staphylococcus epidermidis*. Lancet 1991; 337:54.

36. Schwalbe RS, Ritz WJ, Verma PR, Barranco EA, Gilligan PH. Selection for vancomycin resistance in clinical isolates of *Staphylococcus haemolyticus*. J Infect Dis 1990; 161:45–51.

37. Daum RS, Gupta S, Sabbagh R, Milewski WM. Characterization of *Staphylococcus aureus* isolates with decreased susceptibility to vancomycin and teicoplanin: isolation and purification of a constitutively produced protein associated with decreased susceptibility. J Infect Dis 1992; 166:1066–1072.

38. Hospital Infection Control Practices Advisory Committee. Recommendations for preventing the spread of vancomycin resistance: recommendation of the Hospital Infection Control Practices Advisory Committee (HICPAC). MMWR 1995; 44:1–13.

39. Hiramatsu K, Hanaki H, Ino T, Yabuta K, Oguri T, Tenover FC. Methicillin-resistant *Staphylococcus aureus* clinical strain with reduced vancomycin susceptibility. J Antimicrob Chemother 1997; 40:135–136.

40. Centers for Disease Control. Reduced susceptibility of *Staphylococcus aureus* to vancomycin—Japan. MMWR Morb Mortal Wkly 1996; 46:624–628.

41. Centers for Disease Control and Prevention. Update: *Staphylococcus aureus* with reduced susceptibility to vancomycin—United States, 1997. JAMA 1997; 278: 1145–1146.

42. Liu C, Chambers HF. *Staphylococcus aureus* with heterogeneous resistance to vancomycin: epidemiology, clinical significance, and critical assessment of diagnostic methods. Antimicrob Agents Chemother 2003; 47:3040–3045.

43. Ariza J, Pujol M, Cabo J, et al. Vancomycin in surgical infections due to methicillin-resistant *Staphylococcus aureus* with heterogeneous resistance to vancomycin. Lancet 1999; 353:1587–1588.

44. Moore MR, Perdreau-Remington F, Chambers HF. Vancomycin treatment failure associated with heterogeneous vancomycin-intermediate *Staphylococcus aureus* in a

patient with endocarditis and in the rabbit model of endocarditis. Antimicrob Agents Chemother 2003; 47:1262–1266.

45. Howden BP, Ward PB, Charles PG, et al. Treatment outcomes for serious infections caused by methicillin-resistant *Staphylococcus aureus* with reduced vancomycin susceptibility. Clin Infect Dis 2004; 38:521–528.

46. Woods CW, Cheng AC, Fowler VG Jr, et al. Endocarditis caused by *Staphylococcus aureus* with reduced susceptibility to vancomycin. Clin Infect Dis 2004; 38:1188–1191.

47. Whitener CJ, Park SY, Browne FA, et al. Vancomycin-resistant *Staphylococcus aureus* in the absence of vancomycin exposure. Clin Infect Dis 2004; 38:1049–1055.

48. Centers for Disease Control. Brief report: vancomycin-resistant *Staphylococcus aureus*— New York. MMWR Morb Mortal Wkly 2004; 53:322–323.

49. Craig WA, Vogelman B. The post-antibiotic effect. Ann Intern Med 1987; 106:900–902.

50. Cooper MA, Jin YF, Ashby JP, Andrews JM, Wise R. In-vitro comparison of the post-antibiotic effect of vancomycin and teicoplanin. J Antimicrob Chemother 1990; 26:203–207.

51. Lowdin E, Odenholt I, Cars O. In vitro studies of pharmacodynamic properties of vancomycin against *Staphylococcus aureus* and *Staphylococcus epidermidis*. Antimicrob Agents Chemother 1998; 42:2739–2744.

52. Duffull SB, Begg EJ, Chambers HF, Barclay ML. Efficacies of different vancomycin dosing regimens against *Staphylococcus aureus* determined with a dynamic in vitro model. Antimicrob Agents Chemother 1994; 38:2480–2482.

53. Flandrois JP, Fardel G, Carret G. Early stages of in vitro killing curve of LY146032 and vancomycin for *Staphylococcus aureus*. Antimicrob Agents Chemother 1988; 32:454–457.

54. Ackerman BH, Vannier AM, Eudy EB. Analysis of vancomycin time-kill studies with *Staphylococcus* species by using a curve stripping program to describe the relationship between concentration and pharmacodynamic response. Antimicrob Agents Chemother 1992; 36:1766–1769.

55. Aeschlimann JR, Hershberger E, Rybak MJ. Analysis of vancomycin population susceptibility profiles, killing activity, and postantibiotic effect against vancomycin-intermediate *Staphylococcus aureus*. Antimicrob Agents Chemother 1999; 43:1914–1918.

56. Greenberg RN, Benes CA. Time-kill studies with oxacillin, vancomycin, and teicoplanin versus *Staphylococcus aureus*. J Infect Dis 1990; 161:1036–1037.

57. Houlihan HH, Mercier RC, Rybak MJ. Pharmacodynamics of vancomycin alone and in combination with gentamicin at various dosing intervals against methicillin-resistant *Staphylococcus aureus*-infected fibrin-platelet in and in vitro infection model. Antimicrob Agents Chemother 1997; 41:2497–2501.

58. Levett PN. Time-dependent killing of *Clostridium difficile* by metronidazole and vancomycin. J Antimicrob Chemother 1991; 27:55–62.

59. Hermsen ED, Hovde LB, Hotchkiss JR, Rotschafer JC. Increased killing of staphylococci and streptococci by daptomycin compared with cefazolin and vancomycin in an in vitro peritoneal dialysate model. Antimicrob Agents Chemother 2003; 47:3764–3767.

60. Odenholt-Tornqvist I, Lowdin E, Cars O. Postantibiotic sub-MIC effects of vancomycin, roxitrhomycin, sparfloxacin, and amikacin. Antimicrob Agents Chemother 1992; 36:1852–1858.

61. Greenwood D, Bidgood K, Turner M. A comparison of the responses of staphylococci and streptococci to teicoplanin and vancomycin. J Antimicrob Chemother 1987; 20:155–164.

62. Ebert S, Legett J, Vogelman B. Presented at the 27th Interscience Conference on Antimicrobial Agents and Chemotherapy, NY, 1987.

63. Dudley M, Griffith D, Corcoran E, et al. Pharmacokinetic-pharmacodynamic Indices for vancomycin treatment of susceptible and intermediate *Staphylococcus aureus* in the neutropenic mouse thigh model. Presented at the 39th Interscience Conference on Antimicrobial Agents and Chemotherapy, San Francisco, CA, 1999.

64. Cantoni L, Wenger A, Glauser MP, Bille J. Comparative efficacy of amoxicillin-clavulanate, cloxacillin, and vancomycin against methicillin-sensitive and methicillin-resistant *Staphylococcus aureus* endocarditis in rats. J Infect Dis 1989; 159:989–993.

65. Knudsen JD, Fuursted K, Espersen F, Frimodt-Moller N. Activities of vancomycin and teicoplanin against penicillin-resistant pneumococci in vitro and in vivo and correlation to pharmacokinetic parameters in the mouse peritonitis model. Antimicrob Agents Chemother 1997; 41:1910–1915.

66. Knudsen JD, Fuursted K, Raber S, Espersen F, Frimodt-Moller N. Pharmacodynamics of glycopeptides in the mouse peritonitis model of *Streptococcus pneumoniae* or *Staphylococcus aureus* Infection. Antimicrob Agents Chemother 2000; 44:1247–1254.

67. Louria DB, Kaminski T, Buchman J. Vancomycin in severe staphylococcal infections. Arch Intern Med 1961; 107:225–240.

68. Schadd UB, McCracken GH, Nelson JD. Clinical pharmacology and efficacy of vancomycin in pediatric patients. J Pediatr 1980; 96:119–126.

69. James JK, Palmer SM, Levine DP, Rybak MJ. Comparison of conventional dosing versus continuous-infusion vancomycin therapy for patients with suspected or documented gram-positive infections. Antimicrob Agents Chemother 1996; 40:696–700.

70. Lacy MK, Tessier PR, Nicolau DP, Nightingale CH, Quintiliani R. Comparison of vancomycin pharmacodynamics (1 g every 12 or 24 h) against methicillin-resistant staphylococci. Int J Antimicrob Agents 2000; 15:25–30.

71. Klepser ME, Kang SL, McGrath BJ. Presented at the American College of Clinical Pharmacy Annual Winter Meeting, San Diego, CA, Feb 6–9, 1994.

72. Hyatt JM, McKinnon PS, Zimmer GS, Schentag JJ. The importance of pharmacokinetic/pharmacodynamic surrogate markers to outcome. Focus on antibacterial agents. Clin Pharmacokinet 1995; 28:143–160.

73. Moise PA, Forrest A, Bhavnani SM, Birmingham MC, Schentag JJ. Area under the inhibitory curve and a pneumonia scoring system for predicting outcomes of vancomycin therapy for respiratory infections by *Staphylococcus aureus*. Am J Health Syst Pharm 2000; 57:S4–S9.

74. Klepser ME, Patel KB, Nicolau DP. Comparison of bactericidal activities of intermittent and continuous infusion dosing of vancomycin against methicillin-resistant *Staphylococcus aureus* and *Enterococcus faecalis*. Pharmacotherapy 1998; 18:1069–1074.

75. Small PM, Chambers HF. Vancomycin for *Staphylococcus aureus* endocarditis in intravenous drug users. Antimicrob Agents Chemother 1990; 34:1227–1231.

76. Chambers HF, Miller RT, Newman MD. Right-sided *Staphylococcus aureus* endocarditis in intravenous drug abusers: two-week combination therapy. Ann Intern Med 1988; 109:619–624.

77. Levine DP, Fromm BS, Ramesh Reddy B. Slow response to vancomycin or vancomycin plus rifampin in methicillin-resistant *Staphylcoccus aureus* endocarditis. Ann Intern Med 1991; 115:674–680.

78. Karchmer AW. *Staphylococcus aureus* and vancomycin: the sequel. Ann Intern Med 1991; 115:739–741.

79. Korzeniowski O, Sande MA. Combination antimicrobial therapy for *Staphylococcus aureus* endocarditis in patients addicted to parenteral drugs and in nonaddicts: a prospective study. Ann Intern Med 1982; 97:496–503.

80. Glew RH, Keroack MA. Vancomycin and teicoplanin. In: Grobach SL, Bartlett JG, Blacklow NR, eds. Infectious Diseases. Philadelphia, PA: WB Saunders Company, 1998:260–269.

81. Norden CW, Shaffer M. Treatment of experimental chronic osteomyelitis due to *Staphylococcus aureus* with vancomycin and rifampin. J Infect Dis 1983; 147:352–357.

82. Martin C, Alaya M, Mallet MN, et al. Penetration of vancomycin into mediastinal and cardiac tissues in humans. Antimicrob Agents Chemother 1994; 38:396–399.

83. Massias L, Dubois C, de Lentdecker P, Brodaty O, Fischler M, Farinotti R. Penetration of vancomycin in uninfected sternal bone. Antimicrob Agents Chemother 1992; 36:2539–2541.

84. Fekety R. Vancomycin and teicoplanin. In: Mandell GL, Bennett JE, Dolin R, eds. Principles and Practice of Infectious Diseases. 4th ed. New York, NY: Churchill Livingstone, 1995:346–353.

85. Gopal V, Bisno AL, Silverblatt FJ. Failure of vancomycin treatment in *Staphylococcus aureus* endocarditis; in vivo and in vitro observations. JAMA 1976; 236:1604–1606.

86. Bayer AS, Lam K. Efficacy of vancomycin plus rifampin in experimental aortic-valve endocarditis due to methicillin-resistant *Staphylococcus aureus*: in vitro-in vivo correlations. J Infect Dis 1985; 151:157–165.

87. Massanari RM, Donta ST. The efficacy of rifampin as adjunctive therapy in selected cases of *Staphylococcal endocarditis*. Chest 1978; 73:371–375.

88. Faville RJ Jr, Zaske DE, Kaplan EL, Crossley K, Sabath LD, Quie PG. *Staphylococcus aureus* endocarditis. Combined therapy with vancomycin and rifampin. JAMA 1978; 240:1963–1965.

89. Levine DP, Cushing RD, Jui J, Brown WJ. Community-acquired methicillin-resistant *Staphylococcus aureus* endocarditis in the Detroit Medical Center. Ann Intern Med 1982; 97:330–338.

90. Bradley JS, Scheld WM. The challenge of penicillin-resistant *Streptococcus pneumoniae* meningitis: current antibiotic therapy in the 1990s. Clin Infect Dis 1997; 24(suppl 2): S213–S221.

91. Viladrich PF, Gudiol F, Linares J, et al. Evaluation of vancomycin for therapy of adult pneumococcal meningitis. Antimicrob Agents Chemother 1991; 35:2467–2472.

92. Hughes WT, Armstrong D, Bodey GP, et al. 1997 guidelines for the use of antimicrobial agents in neutropenic patients with unexplained fever. Infectious Diseases Society of America. Clin Infect Dis 1997; 25:551–573.

93. Donta ST, Lamps GM, Summers RW, Wilkins TD. Cephalosporin-associated colitis and *Clostridium difficile*. Arch Intern Med 1980; 140:574–576.

94. Cudamore MA, Silva J, Fekety R, Liepman MK, Kim KH. *Clostridium difficile* colitis associated with cancer chemotherapy. Arch Intern Med 1982; 142:333–335.

95. Miller SD, Koornhof HJ. *Clostridium difficile* colitis associated with the use of antineoplastic agents. Eur J Clin Microbiol 1984; 3:10–13.

96. Johnson SV, Hoey LL, Vance-Bryan K. Inappropriate vancomycin prescribing based on criteria from the Centers for Disease Control and Prevention. Pharmacotherapy 1995; 15:579–585.

97. Gentry C. Wide overuse of antibiotic cited in study. Wall Street J 1997:B-1.

98. Thomas AR, Cieslak PR, Strausbaugh LJ, Fleming DW. Effectiveness of pharmacy policies designed to limit inappropriate vancomycin use: a population-based assessment. Infect Control Hosp Epidemiol 2002; 23:683–688.

99. Farber BF, Moellering RC Jr. Retrospective study of the toxicity of preparations of vancomycin from 1974 to 1981. Antimicrob Agents Chemother 1983; 23:138–141.

100. Vance-Bryan K, Rotschafer JC, Gilliland SS, Rodvold KA, Fitzgerald CM, Guay DR. A comparative assessment of vancomycin-associated nephrotoxicity in the young versus the elderly hospitalized patient. J Antimicrob Chemother 1994; 33:811–821.

101. Geraci JE. Vancomycin. Mayo Clin Proc 1977; 52:631–634.

102. Rao N, Ziran BH, Wagener MM, Santa ER, Yu VL. Similar hematologic effects of long-term linezolid and vancomycin therapy in a prospective observational study of patients with orthopedic infections. Clin Infect Dis 2004; 38:1058–1064.

103. Cantu TG, Yamanaka-Yuen NA, Lietman PS. Serum vancomycin concentrations: reappraisal of their clinical value. Clin Infect Dis 1994; 18:533–543.

104. Moellering RC Jr, Krogstad DJ, Greenblatt DJ. Vancomycin therapy in patients with impaired renal function: a nomogram for dosage. Ann Intern Med 1981; 94:343–346.

105. Matzke GR, Kovarik JM, Rybak MJ, Boike SC. Evaluation of the vancomycin-clearance:creatinine-clearance relationship for predicting vancomycin dosage. Clin Pharm 1985; 4:311–315.

106. Lake KD, Peterson CD. A simplified dosing method for initiating vancomycin therapy. Pharmacotherapy 1985; 5:340–344.

107. Nielsen HE, Hansen HE, Korsager B, Skov PE. Renal excretion of vancomycin in kidney disease. Acta Med Scand 1975; 197:261–264.

108. Zokufa HZ, Rodvold KA, Blum RA, et al. Simulation of vancomycin peak and trough concentrations using five dosing methods in 37 patients. Pharmacotherapy 1989; 9: 10–16.

109. Saunders NJ, Want SV, Adams DJ. Assay of vancomycin by fluorescence polarisation immunoassay and EMIT in patients with renal failure. J Antimicrob Chemother 1995; 36:411–415.

110. Somerville AL, Wright DH, Rotschafer JC. Implications of vancomycin degradation products on therapeutic drug monitoring in patients with end-stage renal disease. Pharmacotherapy 1999; 19:702–707.   ·

111. Darko W, Medicis JJ, Smith A, Guharoy R, Lehmann DE. Mississippi mud no more: cost-effectiveness of pharmacokinetic dosage adjustment of vancomycin to prevent nephrotoxicity. Pharmacotherapy 2003; 23:643–650.

112. Shea KW, Cunha BA. Teicoplanin. Med Clin North Am 1995; 79:833–844.

113. Rouveix B, Garnier M, Pinta P, Kreutz M. Residual serum concentrations and safety of teicoplanin administered intravenously to healthy volunteers (HV) at a 15 mg/kg dose on alternate days, compared with a 6 mg/kg daily dose. Presented at the 41st Interscience Conference on Antimicrobial Agents and Chemotherapy, Chicago, IL, 2001.

114. Lagast H, Dodion P, Klastersky J. Comparison of pharmacokinetics and bactericidal activity of teicoplanin and vancomycin. J Antimicrob Chemother 1986; 18:513–520.

115. MacGowan AP. Pharmacodynamics, pharmacokinetics, and therapeutic drug monitoring of glycopeptides. Ther Drug Monit 1998; 20:473–477.

116. Harding I, Sorgel F. Comparative pharmacokinetics of teicoplanin and vancomycin. J Chemother 2000; 12(suppl 5):15–20.

117. Leighton A, Gottlieb AB, Dorr MB, et al. Tolerability, pharmacokinetics, and serum bactericidal activity of intravenous dalbavancin in healthy volunteers. Antimicrob Agents Chemother 2004; 48:940–945.

118. Lefort A, Pavie J, Garry L, Chau F, Fantin B. Activities of dalbavancin in vitro and in a rabbit model of experimental endocarditis due to *Staphylococcus aureus* with or without reduced susceptibility to vancomycin and teicoplanin. Antimicrob Agents Chemother 2004; 48:1061–1064.

119. Boylan CJ, Campanale K, Iversen PW, Phillips DL, Zeckel ML, Parr TR Jr. Pharmacodynamics of oritavancin (LY333328) in a neutropenic-mouse thigh model of *Staphylococcus aureus* infection. Antimicrob Agents Chemother 2003; 47:1700–1706.

120. Allen NE, LeTourneau DL, Hobbs JN Jr. Molecular interactions of a semisynthetic glycopeptide antibiotic with D-alanyl-D-alanine and D-alanyl-D-lactate residues. Antimicrob Agents Chemother 1997; 41:66–71.

121. Loutit JS. Presented at the 42nd Interscience Conference on Antimicrobial Agents and Chemotherapy, San Diego, CA, 2002.

122. Chien J, Allerheiligen S, Phillips DL, Cerimele B, Thomasson HR. Presented at the 38th Interscience Conference on Antimicrobial Agents and Chemotherapy, San Diego, CA, 1999.

123. Cubicin Prescribing Information. Cubist Pharmaceuticals, Lexington, MA, 2003.

124. Sun HK. Daptomycin: a novel lipopeptide antibiotic for the treatment of resistant gram-positive infections. Formulary 2003; 38:634–645.

125. Carpenter CF, Chambers HF. Daptomycin: another novel agent for treating infections due to drug-resistant gram-positive pathogens. Clin Infect Dis 2004; 38:994–1000.

126. Dandekar PK, Tessier PR, Williams P, Nightingale CH, Nicolau DP. Pharmacodynamic profile of daptomycin against *Enterococcus* species and methicillin-resistant *Staphylococcus aureus* in a murine thigh infection model. J Antimicrob Chemother 2003; 52:405–411.

127. Dandekar PK, Tessier PR, Williams P, Zhang C, Nightingale CH, Nicolau DP. Determination of the pharmacodynamic profile of daptomycin against *Streptococcus pneumoniae* isolates with varying susceptibility to penicillin in a murine thigh infection model. Chemotherapy 2004; 50:11–16.

128. Louie A, Kaw P, Liu W, Jumbe N, Miller MH, Drusano GL. Pharmacodynamics of daptomycin in a murine thigh model of *Staphylococcus aureus* infection. Antimicrob Agents Chemother 2001; 45:845–851.

129. Murphy TM, Lenoy EB, Young M, Weiss WJ. Presented at the 42nd Interscience Conference on Antimicrobial Agents and Chemotherapy, San Diego, CA, 2002.

130. Petersen PJ, Hartman H, Wang T, Dushin R, Bradford P. Presented at the 42nd Interscience Conference on Antimicrobial Agents and Chemotherapy, San Diego, CA, 2002.
131. Preobrazhenskaya M, Printsevskaya SS, Pavlov AY, et al. Presented at the 43rd Interscience Conference on Antimicrobial Agents and Chemotherapy, Chicago, IL, 2003.
132. Hageman JC, Patel JB, Carey RC, et al. Investigation and control of vancomycin-intermediate and -*resistant Staphyococcus aureus*: a guide for health departments and infection control personnel. Atlanta, GA 2006. Available at www.cdc.gov/ncidod/.

10 | Macrolide, Azalide, and Ketolides

Sanjay Jain
Center for Tuberculosis Research and Pediatric Infectious Diseases, Johns Hopkins University School of Medicine, Baltimore, Maryland, U.S.A.

William Bishai
Center for Tuberculosis Research, Johns Hopkins University School of Medicine, Baltimore, Maryland, U.S.A.

Charles H. Nightingale
Center for Anti-infective Research and Development, Hartford Hospital and University of Connecticut School of Pharmacy, Hartford, Connecticut, U.S.A.

BRIEF DESCRIPTION AND HISTORY OF DEVELOPMENT

Erythromycin, the first member of the macrolide class to enter clinical use, was introduced in 1952. Erythromycin remained a mainstay antibiotic for several decades both as an inpatient, infusible agent and as an outpatient, orally administered drug until the introduction of the newer macrolide and azalide agents, clarithromycin, azithromycin, and dirithromycin in the early 1990s and recently the ketolide agent telithromycin in 2004. These antibiotics have activity against each of the major pyogenic pathogens playing etiologic roles in respiratory tract infections: *Streptococcus pneumoniae*, *Haemophilus influenzae*, and *Moraxella catarrhalis*. Moreover, they are potent against atypical respiratory tract pathogens: *Legionella pneumophila*, *Mycoplasma pneumoniae*, and *Chlamydophila (Chlamydia) pneumoniae*. As such, they comprise a class of agents with focused activity against respiratory pathogens, though azithromycin has been increasingly used for enteric infections caused by *Salmonella* and *Shigella*. Importantly, they have little role in the treatment of skin and soft tissue infections for which staphylococci are common causes, or opportunistic gram-negative infections such as *Pseudomonas aeruginosa*. Hence, these antibiotics are considered excellent focused therapy agents for use against respiratory tract infections. Ketolides are a new class of macrolides and are semisynthetic derivates of erythromycin A. Their defining characteristic is the removal of the neutral sugar, L-cladinose, from the third position of the ring and subsequent oxidation of the 3-hydroxyl to a 3-keto functional group. Telithromycin is the first of this new class of drugs to be approved for clinical use while ABT-733 is a ketolide agent in development. Ketolides are designed to combat respiratory tract pathogens that have acquired resistance to macrolides and have excellent microbiologic activity against drug-resistant *S. pneumoniae*.

MECHANISM(S) OF ACTION AND IMMUNOMODULATORY EFFECTS

All the drugs in this class inhibit protein synthesis by blocking the translation of mRNA into protein at the ribosome. They bind to the 23S ribosomal RNA (rRNA) in the 50S subunit (1), which contains a highly conserved domain forming the peptidyl transferase site (2). Inhibition of peptidyl transferase activity blocks the translocation of the peptidyl transfer RNA (tRNA) from the amino acid (A) site to the polypeptide (P) site. However, some drugs in this family (14 and 16 membered structures) also bind several proteins in the 50S subunit (1) and this may also contribute to their antibacterial effect. Macrolides are also known to weaken the nonspecific binding between the ribosome and the peptidyl-tRNA promoting disassociation (3). The mechanism of action of ketolides is similar to macrolides and they also inhibit protein synthesis by interacting with the peptidyl transferase site of the 50S subunit (4,5). Both macrolides and ketolides interact within domains II and V of the 23S rRNA, binding in a 1:1 ratio. Within their binding site on the bacterial ribosome, ketolides interact with two regions of the 23S rRNA. Telithromycin binds to A752 in addition to A2058 bound by erythromycin and other macrolides and azalides (Fig. 1) (7,8). Blurry vision, unmasking and worsening of myasthenia gravis and acute hepatotoxicity have been noted with use of telithromycin. This has led to loss of its indication for AECB and acute bacterial sinusitis. Though it still has an indication for CAP, it is not to be used in patients with myasthenia gravis. In the latest CAP guidelines, final recommendations for telithromycin are pending safety evaluation by the FDA (93).

The ketolides display a higher affinity than macrolides for forming interactions with the ribosome (9,10). Ketolides also have a significant inhibitory effect on the formation of the 50S ribosomal subunits (4,11). Ketolides have been shown to

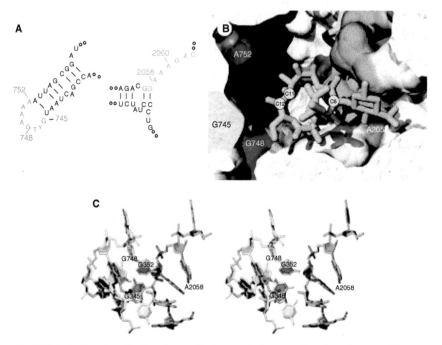

FIGURE 1 Within their binding site on the bacterial ribosome, ketolides interact with two regions of the 23S rRNA. The C11, 12 carbamate extension of telithromycin additionally spans the distance across the channel to bind with A752 in addition to A2058. (**A**) Secondary structures of the nucleotides that make up the target site, with the relevant portions of domain II in blue on the left and those of domain V in green on the right. (**B**) Cross-section of the ribosomal tunnel, showing erythromycin (*orange*) bound within the macrolide- and ketolide-binding site. (**C**) Stereo view showing the perspective of the domain II and V nucleotides that makes up the macrolide- and ketolide-binding site. *Source*: From Ref. 6.

accumulate at a greater rate than macrolides in bacterial cells (10). New guidelines for the treatment of CAP were released by the ATS, IDSA, and CDC in March 2007 (93). The guidelines have not been changed significantly from previous guidelines, Ertapenem was added as an acceptable β-lactam alternative for hospitalized patients with risk factors for gram-negative infections other than *Pseudomonas*.

The 14 and 15 membered ring structures in the macrolide, azalide, and ketolide classes have anti-inflammatory properties in addition to their antibacterial effects. Josamycin, which is a 16-membered agent available in Europe, appears to lack anti-inflammatory activity. Anti-inflammatory effects of macrolides are mediated via prevention of the production of proinflammatory mediators and cytokines (12,13) and appear to promote healing processes in chronic respiratory diseases (14,15). In a multicenter, randomized, double-blind, placebo-controlled trial in the United States, azithromycin was found to improve pulmonary function and reduce the number of pulmonary exacerbations in cystic fibrosis patients (16). Similarly, ketolides have been shown to decrease inflammation and neutrophil recruitment in response to heat-killed *S. pneumoniae* in animal models and in vitro (17–19); however, additional studies are required to determine if the benefit in this group of patients is due to the antibacterial or anti-inflammatory properties of these agents. On the other hand, there is concern for development of bacterial resistance with long-term use of these drugs and well-designed trials are needed to answer these questions more definitively.

MICROBIOLOGICAL ACTIVITY WITH SPECIAL REFERENCE TO RESPIRATORY TRACT PATHOGENS

While macrolides, azalides, and ketolides have in vitro activities against numerous pathogens, this section will focus on their activity against respiratory tract pathogens. In the United States, erythromycin, clarithromycin, dirithromycin, azithromycin, and telithromycin are approved for use against *S. pneumoniae, H. influenzae, Haemophilus parainfluenzae, M. catarrhalis, C. pneumoniae, and M. pneumoniae*. In the United States, a growing proportion of *S. pneumoniae* are multidrug resistant, manifesting resistance to both penicillins and macrolides. Approximately 28% of *S. pneumoniae* are therefore defined resistant to erythromycin, clarithromycin, dirithromycin, and azithromycin (20). Both clarithromycin and azithromycin have variable efficacy against *H. influenzae* (61–95% isolates susceptible to clarithromycin) though azithromycin was more active than clarithromycin, at least for isolates from North America (21). European surveillance shows that susceptibility patterns of *H. parainfluenzae* are similar to *H. influenzae* (22). These surveys however do not take into account the activity of 25-hydroxy clarithromycin, which is an active in vivo metabolite of clarithromycin and a potent inhibitor of *Haemophilus*. Nearly all strains of *M. catarrhalis* are susceptible to clarithromycin and azithromycin. However the breakpoints in these studies are extrapolated from *H. influenzae* (23,24). Finally, the macrolides and azalides are very active against atypical organisms such as *M. pneumoniae* and *C. pneumoniae* though a correlation between in vitro activity and clinical response is yet to be demonstrated (25,26). The ketolides have good activity against these same respiratory tract pathogens. They are also active against erythromycin-resistant *S. pneumoniae* regardless of resistance mechanism with minimum inhibitory concentration (MIC)$_{90}$ $\leq 0.12 \, \mu m/$ mL. In a recently published trial, 772 pediatric isolates from children with community-acquired respiratory tract infections (PROTEKT Global Surveillance

1999–2000), all *S. pneumoniae* were susceptible to telithromycin (27). Similarly, in a large U.S. study (PROTEKT US Year 2 2001–2002), though 27.9% of *S. pneumoniae* isolates were macrolide (erythromycin)-resistant and 96.4% or more of the macrolide-resistant isolates were susceptible to telithromycin, regardless of resistance mechanism (20). The ketolides also have good activity against nonpneumococcal Streptococcus spp. and may be more active than either clarithromycin or azithromycin. They are active against some erythromycin-resistant strains of *Streptococcus pyogenes* but not with *ermB* resistance genotype (28,29). Ketolides are active against *Corynebacterium diphtheriae* and *Listeria monocytogenes*, have variable activity against Enterococcus spp. (not active against *Enterococcus faecium*) and other Corynebacterium spp., and have poor activity against erythromycin-resistant *Staphylococcus aureus* and coagulase-negative *Staphylococcus* spp (29). Finally, ketolides have good activity against *H. influenzae, M. catarrhalis,* Neisseria spp., *Bordetella pertussis* (29–32) and atypical pathogens *M. pneumoniae, Ureaplasma urealyticum,* and *L. pneumophila* (33–37). The in vitro activities (MIC$_{50}$, MIC$_{90}$ and the range reported in the literature) of clarithromycin, azithromycin, telithromycin, and ABT-773 against *S. pneumoniae* are shown in Table 1 (38).

EPIDEMIOLOGY AND MECHANISM OF RESISTANCE WITH SPECIAL REFERENCE TO *STREPTOCOCCUS PNEUMONIAE*

Macrolide resistance is well documented for several pathogens. As with penicillin resistance, the prevalence of macrolide resistance has increased markedly in *S. pneumoniae* in the last decade. Twenty-six percent of *S. pneumoniae* isolates and 75% of penicillin-resistant *S. pneumoniae* isolates in the United States are resistant to macrolides (39,40). Similar rates are reported from countries in Europe, Asia, and South America (41). The most common resistance mechanism in gram-positive cocci is mediated by *mef*-encoded efflux or *erm*-encoded methylation of 23S rRNA. Efflux resistance in *S. pneumoniae* and *S. pyogenes* is encoded by *mefA* whereas ribosomal methylation is encoded by *ermB* in *S. pneumoniae* and *ermA* and *ermB* in *S. pyogenes* (42). Less common mechanisms include mutations in the ribosomal proteins or RNA (43). The *mefA* pump confers resistance to 14- and 15-member macrolides, but not to 16-member macrolides (e.g., spiramycin and josamycin), clindamycin, or streptogramins. This mechanism results in low- to mid-level resistance, with MICs for erythromycin between 1 and 32 µg/mL. The *ermB* mechanism may be constitutively or inducibly expressed in *S. pneumoniae* and results in high-level macrolide resistance (i.e., erythromycin MICs ≥64 µg/mL) and cross-resistance with lincosamides such as clindamycin and streptogramins (so-called "MLS" resistance) due to overlapping binding sites. The prevalence of the *S. pneumoniae* macrolide resistance mechanisms varies by geography. While the efflux mechanism accounts for more than two-thirds of resistant isolates in North America, it accounts for less than 20% in Europe and South Africa (41). The dramatic rise in *S. pneumoniae* macrolide resistance in the United States during the last decade is attributable primarily to the efflux mechanism (44,45). Nearly all strains of *H. influenzae* have an intrinsic macrolide efflux pump (46).

Ketolides were designed to overcome these resistance mechanisms. Efflux-mediated resistance is less effective against ketolides because they are poor substrates for the efflux pumps. Within their binding site on the bacterial ribosome, ketolides interact with two regions of the 23S rRNA. Telithromycin binds to A752 in addition to A2058 bound by erythromycin and other macrolides and

TABLE 1 The In Vitro Activities of Clarithromycin, Azithromycin, Telithromycin, and ABT-773 Against *Streptococcus pneumoniae*

Resistance mechanism	Clarithromycin			Azithromycin			Telithromycin			ABT-773		
	MIC_{50}	MIC_{90}	Range	MIC_{50}	MIC_{90}	Range	MIC_{50}	MIC_{90}	Range	MIC_{50}	MIC_{90}	Range
Streptococcus pneumoniae	0.03	≥64	0.001–≥64	0.12	≥64	0.008–≥64	0.015	0.06	≤0.001–≥64	0.004	0.03	≤0.001–≥64
Erythromycin susceptible	0.03	0.06	0.001–1	0.06	0.25	0.008–64	0.008	0.016	≤0.001–0.12	0.004	0.004	≤0.001–0.06
ermB resistance	≥64	≥64	0.25–≥64	≥64	≥64	1–≥64	0.06	0.12	≤0.001–≥64	0.004	0.06	0.002–≥64
mefA resistance	1	4	0.06–≥64	4	16	0.5–64	0.06	0.5	0.002–2	0.004	0.12	≤0.002–1

Abbreviation: MIC, minimum inhibitory concentration.
Source: From Ref. 38.

azalides (7,8). In addition, ketolides lack the capacity to induce *erm* expression and therefore are active against inducibly resistant macrolide strains (11). For these reasons, ketolides retain their activity against most *S. pneumoniae* regardless of their erythromycin susceptibility and are active against isolates constitutively expressing *ermB* (28,47). *S. pneumoniae* is less likely to develop resistance when exposed to ketolides than to macrolides (48,49). Clinically, this would mean that though ketolides will become inactive against other gram-positive organisms due to selection of constitutive *erm* expression, they may remain active against *S. pneumoniae* longer.

PHARMACOKINETIC AND PHARMACODYNAMIC PROPERTIES

While in vitro microbiological parameters are important, there are limitations to clinical inferences drawn about in vivo activity of an antimicrobial agent based upon its MIC against a pathogen. Effective in vivo activity can be achieved despite the pathogen being defined as nonsusceptible in vitro. As with β-lactams, the time above MIC correlates best with efficacy for erythromycin and clarithromycin and optimal efficacy is obtained when the time above MIC is greater than 40% of the dosing interval (50). However, for azithromycin, area under the curve (AUC)/MIC appears to be the most important parameter with the ratio exceeding 25 for optimal efficacy (51). A 500 mg oral dose of clarithromycin produces a C_{max} around 2.5 µg/mL, with a half-life of six hours (52) with expected serum levels to exceed 1 µg/mL [the National Committee for Clinical Laboratory Standards (NCCLS) susceptibility breakpoint for full resistance to erythromycin and clarithromycin)] for greater than 50% of the 12-hour dosing interval. On the other hand, a 500 mg oral dose of azithromycin achieves a C_{max} of only 0.4 µg/mL and an AUC of 4.5 mg h/mL (52). Therefore the target AUC/MIC ratio for optimal efficacy is obtainable in serum only as long as the MIC of the infecting isolate is less than 0.25 µg/mL, which is less than the current NCCLS breakpoint of 0.5 µg/mL or less for full resistance to azithromycin. However, through exceptional tissue penetration, macrolides may achieve concentrations at the site of infection, which are substantially greater than serum levels. This is especially relevant for extracellular pathogens such as *S. pneumoniae*. In pneumonia, antimicrobial levels in the alveolar epithelial lining fluid (ELF) are thought to be more important in determining therapeutic efficacy than serum levels (53–55). In normal volunteers, steady-state concentrations of both clarithromycin and azithromycin in the ELF are significantly higher than in serum (56,57). After repeated doses, mean clarithromycin concentrations may exceed 32 µg/mL in the ELF and may be another 10-fold higher inside alveolar macrophages (AM) six hours postdose. At 24 hours postdose, levels in ELF and AM are approximately 4.5 and 100 µg/mL, respectively. Although, azithromycin does not concentrate as well in the ELF (1–2 µg/mL in normal volunteers), it is heavily concentrated intracellularly (56). In addition, studies suggest that azithromycin is delivered to the site of infection by leukocytes leading to higher concentrations at the site of action in the presence of inflammation (58–60). Clarithromycin undergoes extensive first-pass metabolism involving the cytochrome P450 system. It is primarily excreted through the liver and kidneys in a nonlinear, dose-dependent manner accounting for disproportionate increases in C_{max} and AUC with increasing doses.

Data suggest that like azithromycin, AUC/MIC may be the most important parameter for telithromycin. Telithromycin is approximately 60% bioavailable and a single oral dose of 800 mg produces a C_{max} of 1.90 to 2.27 µg/mL in approximately one hour after ingestion with a half-life of about 7.2 hours (61,62). The plasma levels achieved for both telithromycin and ABT-773 are above the MICs for most common respiratory pathogens with the possible exception of *H. influenzae*. Ketolides are more lipophilic than macrolides and like macrolides penetrate extensively into tissue and fluids (9). They also have excellent uptake by leukocytes (63). The tissue/plasma or fluid/serum ratios for telithromycin are greater than 1 suggesting that like macrolides, their activity may not correlate with serum drug concentrations. Again, like macrolides, ketolides are primarily metabolized by the cytochrome P450 system. The terminal half-life of telithromycin after a single dose of 800 mg is 7.2 hours, allowing it to be dosed once-daily (61). It is eliminated via various pathways and only 13% is excreted unchanged in the urine. No dosage adjustment is therefore required in patients with mild to moderate renal impairment.

IN VIVO ACTIVITY, ANIMAL MODELS

Pulmonary infection of neutropenic mice with clarithromycin susceptible and low-level *mefA*-mediated resistant *S. pneumoniae* responded well to clarithromycin. However, clarithromycin failed against pulmonary and thigh infections in mice by *erm*-mediated resistant *S. pneumoniae* (64,65). Both telithromycin and ABT-773 achieved bactericidal activity against *S. pneumoniae* in animal models. This activity was higher than macrolides against erythromycin-resistant *S. pneumoniae* though like clarithromycin, telithromycin was less effective against the highly erythromycin-resistant *S. pneumoniae* (66).

PHARMACOKINETIC/PHARMACODYNAMIC RELATIONSHIP, IN VIVO IN VITRO PARADOX WITH PNEUMOCOCCAL RESPIRATORY TRACT INFECTIONS

Although breakthrough bacteremia with resistant isolates during or immediately following macrolide therapy have been reported, most occurred in out-patient settings with oral macrolide therapy (67–75). However, macrolide pharmacokinetics suggests that they are highly concentrated in the tissues and tissue fluid especially if they are inflamed. Therefore clarithromycin/azithromycin may be effective for pneumonia caused by *S. pneumoniae* with lower levels of resistance (MICs <32 µg/mL). In fact, there are inadequate data to determine whether in vitro macrolide resistance predicts adverse treatment outcomes. Mortality for macrolide nonsusceptible versus susceptible infections and for discordant versus concordant antimicrobial therapy for *S. pneumoniae* was similar in the only prospective study where such comparisons were made (76). In contrast to lower levels of resistance, MICs associated with *erm*B-mediated resistance are often 128 µg/mL or more. This level is far greater than levels routinely achieved in serum or ELF and would therefore be expected to be clinically relevant. However, this mechanism accounts only for a minority of macrolide resistance in North America, suggesting that macrolides may be effective for macrolide-resistant *S. pneumoniae* pulmonary infections in the United States. More information is needed from clinical trials and other studies to settle these questions. Therefore, as per the recent Infectious Disease Society of America (IDSA) guidelines, macrolides are recommended as monotherapy for uncomplicated outpatients, such as those who were previously healthy and not recently treated with antibiotics. However, a macrolide plus a β-lactam is recommended for initial empiric treatment of outpatients with underlying disease, age 65 years or more, and exposure to antibiotics in the previous three months and for hospitalized patients (77).

FOOD AND DRUG ADMINISTRATION APPROVED INDICATIONS, CLINICAL TRIALS FOR USE IN RESPIRATORY INFECTIONS, AND DOSING REGIMENS

Macrolides are effective against a number of upper and lower respiratory tract infections in adults and pediatrics. Clarithromycin, for example, is approved for pharyngitis/tonsillitis due to *S. pyogenes*, acute maxillary sinusitis, acute exacerbation of chronic bronchitis (AECB), and community-acquired pneumonia (CAP). Dirithromycin is approved for pharyngitis/tonsillitis, acute bronchitis (bacterial), AECB, and CAP in adults and children 12 or more years of age. Azithromycin is widely used for similar indications in both adults and pediatrics. Macrolides and azalides are known to have gastrointestinal adverse effects. Erythromycin is a risk factor for causing infantile hypertrophic pyloric stenosis after maternal postnatal use of macrolides (78). Another serious concern with oral erythromycin is the prolongation of cardiac repolarization associated with torsades de pointes. Because erythromycin is extensively metabolized by cytochrome P-450 3A (CYP3A) isozymes, commonly used medications that inhibit CYP3A (e.g., statins) may increase plasma erythromycin concentrations, thereby increasing the risk of ventricular arrhythmias and sudden death (79).

Telithromycin is the only ketolide currently approved for clinical use. Although it has been approved in several European countries for a few years, it was recently approved by the FDA in the United States for patients aged 18 years and older to treat AECB, acute bacterial sinusitis, and mild to moderate CAP, including those infections caused by multidrug-resistant *S. pneumoniae*. ABT-773 is still in Phase III development.

Telithromycin (800 mg/once-daily for five days) was found to be clinically and bacteriologically effective in treating AECB (80–82) and acute bacterial sinusitis (83,84). In noncomparative trials, 800 mg once-daily telithromycin for 7 to 10 days was 92.9% to 93.6% effective in achieving clinical cure in adults with mild to moderate or acute/hospitalized CAP (85–87). In comparative trials with the same regimen, telithromycin achieved clinical cure in 94.6%, 88.3%, and 90.0% compared to 90.1%, 88.5%, and 94.2% with high-dose amoxicillin, clarithromycin, and trovafloxacin for adults with CAP (88–90). Pooled analysis of data from eight clinical trials in patients with CAP showed high clinical cure rates in patients infected with *S. pneumoniae* (94%), *H. influenzae* (90%), *M. catarrhalis* (88%), and other pathogens such as *M. pneumoniae* (97%), *C. pneumoniae* (94%), and *L. pneumophila* (100%) (91,92). Data analyzed from 3935 patients who had participated in one Japanese Phase II study and 11 US/global Phase III (AECB, acute sinusitis, or CAP) found telithromycin to be clinically effective against *S. pneumoniae*, with clinical cure rates of 92.8% for all isolates, 91.7% for those with reduced susceptibility to penicillin G, and 86.0% for those with reduced susceptibility to erythromycin A (87). The most common adverse effects associated with telithromycin are gastrointestinal and include diarrhea, nausea, and vomiting. Telithromycin is also affected by agents altering CYP3A metabolism and care should be taken when using telithromycin with those agents. However, as per the manufacturer's briefing documentation to the FDA (Ketek, March 2001), telithromycin showed no significant effect on the QT interval when administered at therapeutic doses.

Guidelines for the treatment of CAP in adults: Four major classes of antimicrobials are recommended for out-patient treatment of CAP and are summarized

TABLE 2 Four Major Classes of Antimicrobials Are Recommended for Out-Patient Treatment of CAP

Antimicrobial class	Advantages	Disadvantages
Macrolides and azalides	Active against most common pathogens, including atypical agents; achieve high tissue and ELF concentrations; clinical trials have shown efficacy consistently; clarithromycin and azithromycin can be given once-daily	Macrolide resistance is high and increasing; breakthrough pneumococcal bacteria with macrolide-resistant strains more common than other classes
β-Lactams	Amoxicillin active against 90–95% of *Streptococcus pneumoniae* strains when used at a dosage of 3–4 g/day or 90–100 mg/kg/day; amoxicillin-clavulanate covers β-lactamase–producing organisms; oral II generation cephalosporins active against 75–85% of *S. pneumoniae* and virtually all *H. influenzae*	Lacks activity against atypical agents; high doses lead to more gastrointestinal intolerance; amoxicillin is more predictably active against *S. pneumoniae*
Doxycycline	Active against 90–95% of strains of *S. pneumoniae*; also active against *Haemophilus influenzae* and atypical agents	Limited use in pediatric patients; limited published clinical data on CAP
Fluoroquinolones (gatifloxacin, levofloxacin, moxifloxacin, and gemifloxacin)	Active against >98% of *S. pneumoniae* strains in the United States, including penicillin-resistant strains; active against *H. influenzae*, atypical agents; shown to be consistently efficient with significantly better outcomes than other classes; can be given once-daily	Limited use in pediatric patients; concern for abuse with risk of increasing resistance by *S. pneumoniae* including clinical failures attributed to emergence of resistance during therapy and selection of strains that are usually resistant to macrolides and β-lactams as well

Abbreviations: CAP, community-acquired pneumonia; ELF, epithelial lining fluid.

in Table 2. For a previously healthy patient with no recent antibiotic use, a macrolide or doxycycline may be used. However, if antibiotics have been used recently then a respiratory fluoroquinolone (moxifloxacin, gatifloxacin, levofloxacin, or gemifloxacin) alone, an advanced macrolide (clarithromycin or azithromycin) plus high-dose amoxicillin, or an advanced macrolide plus high-dose amoxicillin-clavulanate may be used. In patients with comorbidities (COPD, diabetes, renal or congestive heart failure, or malignancy) with no recent antibiotic therapy, an advanced macrolide or a respiratory fluoroquinolone may be used. However, if antibiotics have been used recently, then a respiratory fluoroquinolone alone or an advanced macrolide plus a β-lactam may be used. For suspected aspiration with infection, amoxicillin-clavulanate or clindamycin may be used while a β-lactam or a respiratory fluoroquinolone should be used where bacterial superinfection is suspected with influenza virus.

With the emergence of macrolide resistance, ketolides provide coverage against the most common causative organisms, including infections caused by multidrug-resistant *S. pneumoniae* and atypical pathogens. They will therefore become very useful for treatment of CAP and other respiratory tract infections. In addition, due to the relative protection for development of resistance and cross-resistance, both in vitro and in vivo, ketolides are an attractive option for empirical treatment of respiratory tract infections. Joint CAP guidelines from the IDSA and the American Thoracic Society are expected in 2005 and it is anticipated that this guideline will address the question of how telithromycin should optimally be used in the management of out-patient CAP.

REFERENCES

1. Mazzei T, Mini E, Novelli A, Periti P. Chemistry and mode of action of macrolides. J Antimicrob Chemother 1993; 31(suppl C):1–9.
2. Protein synthesis. In: Stryer L, ed. Biochemistry. 4th ed. 1995:875–910.
3. Brisson-Noel A, Trieu-Cuot P, Courvalin P. Mechanism of action of spiramycin and other macrolides. J Antimicrob Chemother 1988; 22(suppl B):13–23.
4. Champney WS, Tober CL. Structure-activity relationships for six ketolide antibiotics. Curr Microbiol 2001; 42(3):203–210.
5. Champney WS, Tober CL. Superiority of 11,12 carbonate macrolide antibiotics as inhibitors of translation and 50S ribosomal subunit formation in *Staphylococcus aureus* cells. Curr Microbiol 1999; 38(6):342–348.
6. Novotny GW, Jakobsen L, Andersen NM, Poehlsgaard J, Douthwaite S. Ketolide antimicrobial activity persists after disruption of interactions with domain II of 23S rRNA. Antimicrob Agents Chemother 2004; 48(10):3677–3683.
7. Hansen LH, Mauvais P, Douthwaite S. The macrolide-ketolide antibiotic binding site is formed by structures in domains II and V of 23S ribosomal RNA. Mol Microbiol 1999; 31(2):623–631.
8. Xiong L, Shah S, Mauvais P, Mankin AS. A ketolide resistance mutation in domain II of 23S rRNA reveals the proximity of hairpin 35 to the peptidyl transferase centre. Mol Microbiol 1999; 31(2):633–639.
9. Bertho G, Gharbi-Benarous J, Delaforge M, Lang C, Parent A, Girault JP. Conformational analysis of ketolide, conformations of RU 004 in solution and bound to bacterial ribosomes. J Med Chem 1998; 41(18):3373–3386.
10. Capobianco JO, Cao Z, Shortridge VD, Ma Z, Flamm RK, Zhong P. Studies of the novel ketolide ABT-773: transport, binding to ribosomes, and inhibition of protein synthesis in *Streptococcus pneumoniae*. Antimicrob Agents Chemother 2000; 44(6):1562–1567.
11. Champney WS, Tober CL. Inhibition of translation and 50S ribosomal subunit formation in *Staphylococcus aureus* cells by 11 different ketolide antibiotics. Curr Microbiol 1998; 37(6):418–425.
12. Ianaro A, Ialenti A, Maffia P, et al. Anti-inflammatory activity of macrolide antibiotics. J Pharmacol Exp Ther 2000; 292(1):156–163.
13. Jaffe A, Bush A. Anti-inflammatory effects of macrolides in lung disease. Pediatr Pulmonol 2001; 31(6):464–473.
14. Culic O, Erakovic V, Parnham MJ. Anti-inflammatory effects of macrolide antibiotics. Eur J Pharmacol 2001; 429(1–3):209–229.
15. Garey KW, Alwani A, Danziger LH, Rubinstein I. Tissue reparative effects of macrolide antibiotics in chronic inflammatory sinopulmonary diseases. Chest 2003; 123(1):261–265.
16. Saiman L, Marshall BC, Mayer-Hamblett N, et al; Macrolide Study Group. Azithromycin in patients with cystic fibrosis chronically infected with Pseudomonas aeruginosa: a randomized controlled trial. JAMA 2003; 290(13):1749–1756.
17. Duong M, Simard M, Bergeron Y, Ouellet N, Cote-Richer M, Bergeron MG. Immunomodulating effects of HMR 3004 on pulmonary inflammation caused by heat-killed *Streptococcus pneumoniae* in mice. Antimicrob Agents Chemother 1998; 42(12):3309–3312.
18. Duong M, Simard M, Bergeron Y, Bergeron MG. Kinetic study of the inflammatory response in *Streptococcus pneumoniae* experimental pneumonia treated with the ketolide HMR 3004. Antimicrob Agents Chemother 2001; 45(1):252–262.

19. Jung R, Bearden DT, Danziger LH. Effect of ABT-773 and other antimicrobial agents on the morphology and the release of interleukin-1B and interleukin-8 against *Haemophilus influenzae* and *Streptococcus pneumoniae* in whole blood (#2141). 39th Interscience Conference on Antimicrobial Agents and Chemotherapy, San Francisco, CA, Sep 26–29, 1999.

20. Farrell DJ, Jenkins SG. Distribution across the USA of macrolide resistance and macrolide resistance mechanisms among *Streptococcus pneumoniae* isolates collected from patients with respiratory tract infections: PROTEKT US 2001–2002. J Antimicrob Chemother 2004; 54(suppl 1):i17–i22.

21. Mathai D, Lewis MT, Kugler KC, Pfaller MA, Jones RN; SENTRY Participants Group (North America). Antibacterial activity of 41 antimicrobials tested against over 2773 bacterial isolates from hospitalized patients with pneumonia: I—results from the SENTRY Antimicrobial Surveillance Program (North America, 1998). Diagn Microbiol Infect Dis 2001; 39(2):105–116.

22. Felmingham D. Respiratory pathogens: assessing resistance patterns in Europe and the potential role of grepafloxacin as treatment of patients with infections caused by these organisms. J Antimicrob Chemother 2000; 45:1–8.

23. Doern GV, Pfaller MA, Kugler K, Freeman J, Jones RN. Prevalence of antimicrobial resistance among respiratory tract isolates of *Streptococcus pneumoniae* in North America: 1997 results from the SENTRY antimicrobial surveillance program. Clin Infect Dis 1998; 27(4):764–770.

24. Melo-Cristino J, Fernandes ML, Serrano N; Portuguese Surveillance Group for the Study of Respiratory Pathogens. A multicenter study of the antimicrobial susceptibility of *Haemophilus influenzae*, *Streptococcus pneumoniae*, and *Moraxella catarrhalis* isolated from patients with community-acquired lower respiratory tract infections in 1999 in Portugal. Microb Drug Resist 2001 (Spring); 7(1):33–38.

25. Critchley IA, Jones ME, Heinze PD, et al. In vitro activity of levofloxacin against contemporary clinical isolates of *Legionella pneumophila*, *Mycoplasma pneumoniae* and *Chlamydia pneumoniae* from North America and Europe. Clin Microbiol Infect 2002; 8(4):214–221.

26. McCracken GH Jr. Microbiologic activity of the newer macrolide antibiotics. Pediatr Infect Dis J 1997; 16(4):432–437.

27. Farrell DJ, Jenkins SG, Reinert RR. Global distribution of *Streptococcus pneumoniae* serotypes isolated from paediatric patients during 1999–2000 and the in vitro efficacy of telithromycin and comparators. J Med Microbiol 2004; 53(Pt 11):1109–1117.

28. Giovanetti E, Montanari MP, Marchetti F, Varaldo PE. In vitro activity of ketolides telithromycin and HMR 3004 against Italian isolates of *Streptococcus pyogenes* and *Streptococcus pneumoniae* with different erythromycin susceptibility. J Antimicrob Chemother 2000; 46(6):905–908.

29. Nilius AM, Bui MH, Almer L, et al. Comparative in vitro activity of ABT-773, a novel antibacterial ketolide. Antimicrob Agents Chemother 2001; 45(7):2163–2168.

30. Jones RN, Biedenbach DJ. Antimicrobial activity of RU-66647, a new ketolide. Diagn Microbiol Infect Dis 1997; 27(1–2):7–12.

31. Barry AL, Fuchs PC, Brown SD. In vitro activity of the ketolide ABT-773. Antimicrob Agents Chemother 2001; 45(10):2922–2924.

32. Goldstein EJ, Citron DM, Hunt Gerardo S, Hudspeth M, Merriam CV. Activities of HMR 3004 (RU 64004) and HMR 3647 (RU 66647) compared to those of erythromycin, azithromycin, clarithromycin, roxithromycin, and eight other antimicrobial agents against unusual aerobic and anaerobic human and animal bite pathogens isolated from skin and soft tissue infections in humans. Antimicrob Agents Chemother 1998; 42(5):1127–1132.

33. Bebear CM, Renaudin H, Bryskier A, Bebear C. Comparative activities of telithromycin (HMR 3647), levofloxacin, and other antimicrobial agents against human mycoplasmas. Antimicrob Agents Chemother 2000; 44(7):1980–1982.

34. Yamaguchi T, Hirakata Y, Izumikawa K, et al. In vitro activity of telithromycin (HMR3647), a new ketolide, against clinical isolates of *Mycoplasma pneumoniae* in Japan. Antimicrob Agents Chemother 2000; 44(5):1381–1382.

35. Strigl S, Roblin PM, Reznik T, Hammerschlag MR. In vitro activity of ABT 773, a new ketolide antibiotic, against *Chlamydia pneumoniae*. Antimicrob Agents Chemother 2000; 44(4):1112–1113.

36. Roblin PM, Hammerschlag MR. In vitro activity of a new ketolide antibiotic, HMR 3647, against *Chlamydia pneumoniae*. Antimicrob Agents Chemother 1998; 42(6): 1515–1516.

37. Hammerschlag MR, Roblin PM, Bebear CM. Activity of telithromycin, a new ketolide antibacterial, against atypical and intracellular respiratory tract pathogens. J Antimicrob Chemother 2001; 48(suppl T1):25–31.

38. Zhanel GG, Walters M, Noreddin A, et al. The ketolides: a critical review. Drugs 2002; 62(12):1771–1804.

39. Doern GV, Heilmann KP, Huynh HK, Rhomberg PR, Coffman SL, Brueggemann AB. Antimicrobial resistance among clinical isolates of *Streptococcus pneumoniae* in the United States during 1999–2000, including a comparison of resistance rates since 1994–1995. Antimicrob Agents Chemother 2001; 45(6):1721–1729.

40. Thornsberry C, Sahm DF, Kelly LJ, et al. Regional trends in antimicrobial resistance among clinical isolates of *Streptococcus pneumoniae*, *Haemophilus influenzae*, and *Moraxella catarrhalis* in the United States: results from the TRUST Surveillance Program, 1999–2000. Clin Infect Dis 2002; 34(suppl 1):S4–S16.

41. Lynch JP, Martinez FJ. Clinical relevance of macrolide-resistant *Streptococcus pneumoniae* for community-acquired pneumonia. Clin Infect Dis 2002; 34(suppl 1):S27–S46.

42. Roberts MC, Sutcliffe J, Courvalin P, Jensen LB, Rood J, Seppala H. Nomenclature for macrolide and macrolide-lincosamide-streptogramin B resistance determinants. Antimicrob Agents Chemother 1999; 43(12):2823–2830.

43. Tait-Kamradt A, Davies T, Cronan M, Jacobs MR, Appelbaum PC, Sutcliffe J. Mutations in 23S rRNA and ribosomal protein L4 account for resistance in pneumococcal strains selected in vitro by macrolide passage. Antimicrob Agents Chemother 2000; 44(8):2118–2125.

44. Gay K, Baughman W, Miller Y, et al. The emergence of *Streptococcus pneumoniae* resistant to macrolide antimicrobial agents: a 6-year population-based assessment. J Infect Dis 2000; 182(5):1417–1424.

45. Hyde TB, Gay K, Stephens DS, et al; Active Bacterial Core Surveillance/Emerging Infections Program Network. Macrolide resistance among invasive *Streptococcus pneumoniae* isolates. JAMA 2001; 286(15):1857–1862.

46. Sanchez L, Pan W, Vinas M, Nikaido H. The acrAB homolog of *Haemophilus influenzae* codes for a functional multidrug efflux pump. J Bacteriol 1997; 179(21):6855–6857.

47. Davies TA, Ednie LM, Hoellman DM, Pankuch GA, Jacobs MR, Appelbaum PC. Antipneumococcal activity of ABT-773 compared to those of 10 other agents. Antimicrob Agents Chemother 2000; 44(7):1894–1899.

48. Agouridas C, Bonnefoy A, Chantot JF. Antibacterial activity of RU 64004 (HMR 3004), a novel ketolide derivative active against respiratory pathogens. Antimicrob Agents Chemother 1997; 41(10):2149–2158.

49. Fernandez-Roblas R, Calvo R, Esteban J, Bryskier A, Soriano F. The bactericidal activities of HMR 3004, HMR 3647 and erythromycin against gram-positive bacilli and development of resistance. J Antimicrob Chemother 1999; 43(2):285–289.

50. Craig WA. Pharmacokinetic/pharmacodynamic parameters: rationale for antibacterial dosing of mice and men. Clin Infect Dis 1998; 26(1):1–10.

51. Craig WA. The hidden impact of antibacterial resistance in respiratory tract infection. Re-evaluating current antibiotic therapy. Respir Med 2001; 95(suppl A):S12–S19.

52. Carbon C. Pharmacodynamics of macrolides, azalides, and streptogramins: effect on extracellular pathogens. Clin Infect Dis 1998; 27(1):28–32.

53. Amsden GW. Pneumococcal macrolide resistance—myth or reality? J Antimicrob Chemother 1999; 44:1–6.

54. Baldwin DR, Honeybourne D, Wise R. Pulmonary disposition of antimicrobial agents: in vivo observations and clinical relevance. Antimicrob Agents Chemother 1992; 36(6): 1176–1180.

55. Baldwin DR, Honeybourne D, Wise R. Pulmonary disposition of antimicrobial agents: methodological considerations. Antimicrob Agents Chemother 1992; 36(6):1171–1175.

56. Rodvold KA, Gotfried MH, Danziger LH, Servi RJ. Intrapulmonary steady-state concentrations of clarithromycin and azithromycin in healthy adult volunteers. Antimicrob Agents Chemother 1997; 41(6):1399–402.

57. Patel KB, Xuan D, Tessier PR, Russomanno JH, Quintiliani R, Nightingale CH. Comparison of bronchopulmonary pharmacokinetics of clarithromycin and azithromycin. Antimicrob Agents Chemother 1996; 40(10):2375–2379.

58. Girard AE, Cimochowski CR, Faiella JA. Correlation of increased azithromycin concentrations with phagocyte infiltration into sites of localized infection. J Antimicrob Chemother 1996; 37(suppl C):9–19.

59. Ballow CH, Amsden GW, Highet VS, Forrest A. Healthy volunteer pharmacokinetics of oral azithromycin in serum, urine, polymorphonuclear leukocytes and inflammatory vs. non-inflammatory skin blisters. Clinical Drug Investigation 1998; 15:159–167.

60. Freeman CD, Nightingale CH, Nicolau DP, Belliveau PP, Banevicius MA, Quintiliani R. Intracellular and extracellular penetration of azithromycin into inflammatory and noninflammatory blister fluid. Antimicrob Agents Chemother 1994; 38(10):2449–2451.

61. Namour F, Wessels DH, Pascual MH, Reynolds D, Sultan E, Lenfant B. Pharmacokinetics of the new ketolide telithromycin (HMR 3647) administered in ascending single and multiple doses. Antimicrob Agents Chemother 2001; 45(1):170–175.

62. Perret P, Lenfant B, Weinling E, et al. Pharmacokinetics and absolute oral bioavailability of an 800-mg oral dose of telithromycin in healthy young and elderly volunteers. Chemotherapy 2002; 48(5):217–223.

63. Pascual A, Ballesta S, Garcia I, Perea EJ. Uptake and intracellular activity of ketolide HMR 3647 in human phagocytic and non-phagocytic cells. Clin Microbiol Infect 2001; 7(2):65–69.

64. Hoffman HL, Klepser ME, Ernst EJ, Petzold CR, Sa'adah LM, Doern GV. Influence of macrolide susceptibility on efficacies of clarithromycin and azithromycin against *Streptococcus pneumoniae* in a murine lung infection model. Antimicrob Agents Chemother 2003; 47(2):739–746.

65. Tessier PR, Kim MK, Zhou W, et al. Pharmacodynamic assessment of clarithromycin in a murine model of pneumococcal pneumonia. Antimicrob Agents Chemother 2002; 46(5):1425–1434.

66. Piroth L, Desbiolles N, Mateo-Ponce V, et al. HMR 3647 human-like treatment of experimental pneumonia due to penicillin-resistant and erythromycin-resistant *Streptococcus pneumoniae*. J Antimicrob Chemother 2001; 47(1):33–42.

67. Sanchez C, Armengol R, Lite J, Mir I, Garau J. Penicillin-resistant pneumococci and community-acquired pneumonia. Lancet 1992; 339(8799):988.

68. Lonks JR, Medeiros AA. Emergence of erythromycin-resistant *Streptococcus pneumoniae*. Infect Med 1994; 11:415–424.

69. Lonks JR, Garau J, Gomez L, et al. Failure of macrolide antibiotic treatment in patients with bacteremia due to erythromycin-resistant *Streptococcus pneumoniae*. Clin Infect Dis 2002; 35(5):556–564.

70. Fogarty C, Goldschmidt R, Bush K. Bacteremic pneumonia due to multidrug-resistant pneumococci in 3 patients treated unsuccessfully with azithromycin and successfully with levofloxacin. Clin Infect Dis 2000; 31(2):613–615.

71. Kelley MA, Weber DJ, Gilligan P, Cohen MS. Breakthrough pneumococcal bacteremia in patients being treated with azithromycin and clarithromycin. Clin Infect Dis 2000; 31(4):1008–1011.

72. Waterer GW, Wunderink RG, Jones CB. Fatal pneumococcal pneumonia attributed to macrolide resistance and azithromycin monotherapy. Chest 2000; 118(6):1839–1840.

73. Musher DM, Dowell ME, Shortridge VD, et al. Emergence of macrolide resistance during treatment of pneumococcal pneumonia. N Engl J Med 2002; 346(8):630–631.

74. Ewig S, Ruiz M, Torres A, et al. Pneumonia acquired in the community through drug-resistant *Streptococcus pneumoniae*. Am J Respir Crit Care Med 1999; 159(6):1835–1842.

75. Bishai W. The in vivo-in vitro paradox in pneumococcal respiratory tract infections. J Antimicrob Chemother 2002; 49(6):433–436.

76. Nuermberger E, Bishai WR. The clinical significance of macrolide-resistant *Streptococcus pneumoniae*: it's all relative. Clin Infect Dis 2004; 38(1):99–103.

77. Mandell LA, Bartlett JG, Dowell SF, File TM Jr, Musher DM, Whitney C; Infectious Diseases Society of America. Update of practice guidelines for the management of community-acquired pneumonia in immunocompetent adults. Clin Infect Dis 2003; 37(11):1405–1433.

78. Sorensen HT, Skriver MV, Pedersen L, Larsen H, Ebbesen F, Schonheyder HC. Risk of infantile hypertrophic pyloric stenosis after maternal postnatal use of macrolides. Scand J Infect Dis 2003; 35(2):104–106.

79. Ray WA, Murray KT, Meredith S, Narasimhulu SS, Hall K, Stein CM. Oral erythromycin and the risk of sudden death from cardiac causes. N Engl J Med 2004; 351(11):1089–1096.

80. Aubier M, Aldons PM, Leak A, et al. Telithromycin is as effective as amoxicillin/clavulanate in acute exacerbations of chronic bronchitis. Respir Med 2002; 96(11): 862–871.

81. Zervos MJ, Heyder AM, Leroy B. Oral telithromycin 800 mg once daily for 5 days versus cefuroxime axetil 500 mg twice daily for 10 days in adults with acute exacerbations of chronic bronchitis. J Int Med Res 2003; 31(3):157–169.

82. Fogarty C, de Wet R, Nusrat R. Five-day telithromycin is as effective as 10-day clarithromycin in the treatment of acute exacerbations of chronic bronchitis (#2230). Abstracts of the Thirteenth European Respiratory Society Annual Congress, Vienna, Austria, 2003.

83. Luterman M, Tellier G, Lasko B, Leroy B. Efficacy and tolerability of telithromycin for 5 or 10 days vs amoxicillin/clavulanic acid for 10 days in acute maxillary sinusitis. Ear Nose Throat J 2003; 82(8):576–580.

84. Buchanan PP, Stephens TA, Leroy B. A comparison of the efficacy of telithromycin versus cefuroxime axetil in the treatment of acute bacterial maxillary sinusitis. Am J Rhinol 2003; 17(6):369–377.

85. Carbon C, Moola S, Velancsics I, Leroy B, Rangaraju M, Decosta P. Telithromycin 800 mg once daily for seven to ten days is an effective and well-tolerated treatment for community-acquired pneumonia. Clin Microbiol Infect 2003; 9(7):691–703.

86. van Rensburg DJ, Matthews PA, Leroy B. Efficacy and safety of telithromycin in community-acquired pneumonia. Curr Med Res Opin 2002; 18(7):397–400.

87. Fogarty CM, Kohno S, Buchanan P, Aubier M, Baz M. Community-acquired respiratory tract infections caused by resistant pneumococci: clinical and bacteriological efficacy of the ketolide telithromycin. J Antimicrob Chemother 2003; 51(4):947–955.

88. Hagberg L, Torres A, van Rensburg D, Leroy B, Rangaraju M, Ruuth E. Efficacy and tolerability of once-daily telithromycin compared with high-dose amoxicillin for treatment of community-acquired pneumonia. Infection 2002; 30(6):378–386.

89. Tellier G, Niederman MS, Nusrat R, Patel M, Lavin B. Clinical and bacteriological efficacy and safety of 5 and 7 day regimens of telithromycin once daily compared with a 10 day regimen of clarithromycin twice daily in patients with mild to moderate community-acquired pneumonia. J Antimicrob Chemother 2004; 54(2):515–523.

90. Pullman J, Champlin J, Vrooman PS Jr. Efficacy and tolerability of once-daily oral therapy with telithromycin compared with trovafloxacin for the treatment of community-acquired pneumonia in adults. Int J Clin Pract 2003; 57(5):377–384.

91. Aventis. Data on file. Ketek (Telithromycin). Briefing Document for the FDA Anti-Infective Drug Products Advisory Committee Meeting (January 2003). Executive Summary. Aventis, Bridgewater, NJ, 2003.

92. Dunbar LM, Carbon C, Nusrat, R. Telithromycin displays high clinical efficacy in outpatients with mild to moderate community-acquired pneumonia caused by atypical/intracellular pathogens (#984). Abstracts of the Thirteenth European Respiratory Society Annual Congress, Vienna, Austria, 2003.

93. Mandell LA, Wunderink RG, Anzueto A, et al. Infectious Diseases Society of America/American Thoracic Society consensus guidelines on the management of community-acquired pneumonia in adults. Clin Infect Dis 2007; 44(suppl 2):S27–72.

Metronidazole and Clindamycin for Anaerobic Infections

Elizabeth D. Hermsen
Department of Pharmaceutical and Nutrition Care, The Nebraska Medical Center, Omaha, Nebraska, U.S.A.

John C. Rotschafer
Department of Experimental and Clinical Pharmacology, University of Minnesota College of Pharmacy, Minneapolis, Minnesota, U.S.A.

METRONIDAZOLE

Pharmacology

Metronidazole [1-(2-hydroxyethyl)-2-methyl-5-nitroimidazole], a nitroimidazole antimicrobial, was introduced in 1960 and rapidly became the treatment of choice for *Trichomonas vaginalis* (1). Metronidazole has been used clinically for over 40 years and was initially regarded as an antiprotozoal agent, used for such infections as trichomoniasis, amebiasis, and giardiasis. Several years later, in the 1970s, the antibacterial activity of metronidazole versus obligate anaerobes was realized (2–5). Since that time, metronidazole has been used extensively for a multitude of anaerobic infections, such as those involving *Clostridium difficile* and Bacteroides spp. As metronidazole has been in use for over four decades, a superfluity of references concerning basic knowledge about metronidazole, including the mechanism of action, spectrum of activity, pharmacokinetics, adverse drug effects, clinical uses, and resistance exist, but pharmacodynamic data are lacking.

Although metronidazole is known to exhibit bactericidal activity against obligate anaerobes, the mechanism of action has not been clearly explicated. Metronidazole, a prodrug, must undergo intracellular nitroreduction in order to become active. Therefore, in the unchanged form, metronidazole is not pharmacologically active (1). The cytotoxic intermediates that are formed during reduction are thought to be accountable for killing the bacterial cells. Metronidazole possesses activity versus both dividing and nondividing cells because the reduction process depends on ongoing energy metabolism but not on cell multiplication (1).

Antimicrobial Spectrum

Anaerobic bacteria of the *Bacteroides fragilis* group are typically regarded as the most clinically significant anaerobic pathogens due to the presence of multidrug-resistance and the frequency with which they are involved in infectious diseases, many of which are polymicrobial in nature. Another clinically significant anaerobe is *C. difficile*, which is a frequent cause of nosocomial diarrhea and/or pseudomembranous colitis, termed *C. difficile*–associated diarrhea (CDAD) (6). Metronidazole is highly effective versus both of these medically relevant anaerobes. Furthermore, metronidazole possesses antibacterial activity against *Peptostreptococcus,* Prevotella spp., Fusobacterium spp., and Porphyromonas spp. among others (6,7).

While metronidazole exhibits activity against multiple obligate anaerobes, the drug is not considered to be active against aerobic bacteria. Conversely, some researchers argue that metronidazole exhibits activity against *Escherichia coli* when present in a mixed infection with *B. fragilis* (8,9). Interestingly, Chrystal et al. found increased effectiveness of metronidazole against *B. fragilis* in the presence of *E. coli* (10). The authors hypothesize that this effect may have been due to the ability of cytotoxic intermediates to be formed in the *E. coli* and to diffuse into the medium to kill the Bacteroides spp. These differences in activity could not be confirmed in a more recent mixed infection study (11). However, the relationships between the various microorganisms and the different antibiotics involved in polymicrobial infections are difficult to separate.

Pharmacokinetics

Metronidazole is almost completely absorbed, with a bioavailability of over 90%, when given via the oral route (12). Metronidazole is a fairly small molecular entity with a molecular weight of 171.16 (13). The protein binding of metronidazole is low (<20%), and the steady state volume of distribution in adults is 0.51 to 1.1 L/kg (12,13). The elimination half-life of metronidazole is approximately 6 to 12 hours (13,14). Metronidazole undergoes metabolism in the liver to form five known metabolites. The two major metabolites of metronidazole are the hydroxy metabolite [1-(2-hydroxyethyl)-2-hydroxymethyl-5-nitroimidazole] and the acid metabolite (2-methyl-nitroimidazole-1-acetic acid). The hydroxy metabolite exhibits 30% to 65% of the anaerobic activity of the parent compound (12).

Resistance

Worldwide metronidazole resistance among anaerobes is estimated to be <5% (15,16). In a recent multicenter study in the United States, all of the 2673 isolates of *B. fragilis* group species studied were susceptible to metronidazole [minimum inhibitory concentration (MIC) <8 mg/L] (15). In another study of 542 blood isolates of the *B. fragilis* group, metronidazole was the only agent that was active against all of the tested isolates (17). Of note, one study found that exposure to low levels of metronidazole (4 mg/L) increased both the virulence and the viability of the exposed *B. fragilis* group species (18).

Four genes (chromosomally-borne *nimB* and plasmid-borne *nimA*, *nimC*, and *nimD*) of Bacteroides spp. are commonly associated with metronidazole resistance (19–21) The conversion of the nitro group of metronidazole to an amino group, foregoing the formation of the toxic nitroradicals, is the proposed mechanism of resistance mediated by these genes (19,20). Transfer of these genes has been reported among different Bacteroides spp. and between *Bacteroides* and *Prevotella* (21).

A study conducted by Aldridge et al. found that 6% of *Peptostreptococcus* isolates were not susceptible to metronidazole (7). Peláez et al. documented that 6.3% of the studied 415 *C. difficile* isolates were resistant to metronidazole when using a resistance breakpoint of ≥32 mg/L (22). Conversely, no resistance to metronidazole was found among 186 *C. difficile* isolates from a geriatric population in a more recent study conducted by Drummond et al. (6).

Pharmacodynamics

The notion of pharmacodynamics materialized much after the standard dosing regimen for metronidazole (500–1000 mg q6–8h) was established. The bactericidal activity of metronidazole is concentration dependent, and metronidazole exhibits a prolonged postantibiotic effect (PAE) (>3 hours) after single doses (12,23–25). A combination of these factors, along with a long half-life and a favorable safety profile, provides for much manipulation of metronidazole dose and dosage interval. Because of these pharmacokinetic and pharmacodynamic characteristics, more convenient regimens of larger doses (e.g., ≤1500 mg) given every 12 hours or once daily are feasible and appear to result in similar bactericidal activity (23,24,26).

Four metronidazole dosing regimens, including a once-daily regimen and a standard thrice-daily regimen, were evaluated in an in vitro pharmacodynamic model to compare their activity against Bacteroides spp (24). All of the studied regimens achieved ≥99.9% reduction in bacterial load, which equates to bactericidal activity, by 12 hours. No differences were reported for the rate or the extent of bacterial killing for the four regimens, leading to the conclusion that the once-daily regimen and the standard thrice-daily regimen were equally effective.

In a more recent study, three regimens of metronidazole, 500 mg q8h, 1000 mg once daily, and 1500 mg once daily, were studied in combination with levofloxacin in 18 healthy volunteers (26). Bactericidal activity was evaluated against various *B. fragilis* group spp., *Peptostreptococcus asaccharalyticus*, and *E. coli*. The investigators found that the regimens resulting in the same total daily dose of metronidazole (500 mg q8h and 1500 mg once daily) produced similar activity against the tested anaerobes.

Although resistance to metronidazole among anaerobes remains low (16), it should always be a concern. Interestingly, a recent study analyzed the kill-kinetics of various antimicrobials, including metronidazole, against strains of *B. fragilis* with varying resistance patterns, two of which were metronidazole resistant (MIC >256 mg/L) (27). Surprisingly, the metronidazole resistant strains were killed by metronidazole when concentrations of 16 mg/L or more were utilized. Perhaps a more clear understanding and application of pharmacodynamics to metronidazole dosing may further elucidate this phenomenon.

Clinical Application
Clinical Uses of Metronidazole for Anaerobic Infections
Metronidazole's high oral bioavailability, low potential for selecting for vancomycin-resistant *Enterococcus*, and low cost make it the drug of choice for CDAD (6,22). Nonetheless, recent data suggest that metronidazole should not be empirically recommended for CDAD because of the unnecessary antibiotic exposure and selection pressure for those patients in whom *C. difficile* is not the causative agent (28). Metronidazole has been used successfully to treat a variety of anaerobic infections, including bacteremia, endocarditis, meningitis, brain abscesses, and mixed aerobic–anaerobic infections, although the addition of an antibiotic effective against aerobic bacteria is necessary for the latter (2,3,5,13,29,30).

Toxicity and Adverse Reactions
Gastrointestinal disturbances, including mild nausea, a bad/metallic taste in the mouth, or furring of the tongue, are the most common adverse events occurring with standard dosing of metronidazole (13). Vaginal and/or urethral burning, dark/discolored urine, and central nervous system symptoms, such as headache, ataxia, vertigo, somnolence, and depression, occur very rarely with metronidazole use (31). Metronidazole is known for causing a disulfiram-like reaction with the concurrent ingestion of alcohol (31). However, Visapää et al. reported a lack of

disulfiram-like properties of metronidazole when given concomitantly with etha-nol (32), and this reaction has also been disputed by others (33).

Summary

After more than 40 years of clinical use, metronidazole remains a mainstay in the treatment of anaerobic infections. Metronidazole has an excellent pharmacokinetic profile, with low protein binding, good oral bioavailability, and a long half-life. Although metronidazole has been in use for many years, resistance of anaerobes to metronidazole remains under 5%, which further cements metronidazole's role in the treatment of infections due to anaerobes. The dosing regimens of metroni-dazole were formulated before pharmacodynamics emerged as a science, but metronidazole's concentration-dependent bactericidal activity, prolonged PAE, and favorable pharmacokinetic and safety profiles allow for dosing manipulation toward more convenient regimens (e.g., $\leq$1500 mg every 24 hours).

CLINDAMYCIN
Pharmacology

Clindamycin (7-chloro-7-deoxylincomycin), a lincosamide antibiotic, has been used clinically for over 30 years. Clindamycin exerts an antibacterial effect through the inhibition of protein synthesis by binding to the 50S ribosomal subunit. Clindamy-cin is used for various gram-positive and anaerobic infections, and is an option for patients who are allergic to β-lactams. Clindamycin's use in anaerobic infections is the focus of this chapter. As in the case of metronidazole, a plethora of information regarding basic knowledge about clindamycin exist, although pharmacodynamic data are sparse.

Antimicrobial Spectrum

Clindamycin exhibits antimicrobial activity versus various aerobic gram-positive organisms as well as gram-positive and gram-negative anaerobic bacteria (34). Similar to metronidazole, clindamycin is effective against the clinically important anaerobe, Bacteroides spp. Additionally, clindamycin possesses antibacterial activ-ity against *Peptostreptococcus*, Fusobacterium spp., Propionibacterium spp., Eubac-terium spp., Actinomyces spp., and most strains of C. *perfringens* among others.

Pharmacokinetics

When given orally, clindamycin is well absorbed, with a bioavailability of approxi-mately 90%. Clindamycin is highly protein bound (>90%), and the steady state volume of distribution in adults is approximately 0.79 L/kg (35). The elimination half-life of clindamycin is approximately 2 to 2.4 hours (34,35). Moreover, clinda-mycin undergoes metabolism in the liver to form active metabolites (34). Although the use of clindamycin in AIDS patients is not discussed in this chapter, one should recognize that the pharmacokinetic parameters are significantly different in AIDS patients, for both intravenous and oral administration (35).

Resistance

Resistance is the primary concern with the use of clindamycin. A study evaluating the susceptibility of 186 C. *difficile* isolates to clindamycin reported that 66.7% of the isolates were resistant and 24.7% were intermediately resistant (6). Moreover, approximately 22% of B. *fragilis* group spp. are considered to be resistant to clindamycin (15,17). Of note, Aldridge et al. found that clindamycin-intermediate or -resistant isolates are more likely to have decreased susceptibility to other agents (17). However, this effect does not seem to hold true with metronidazole. In other words, clindamycin-intermediate or -resistant isolates do not show decreased susceptibility to metronidazole.

Pharmacodynamics

Although clindamycin has been in clinical use for several years, relatively little was known about its pharmacodynamic properties against anaerobic bacteria until recently. Klepser et al. evaluated the activity of clindamycin against *B. fragilis* in an in vitro model and found that clindamycin exhibited concentration-independent activity (36). This differs from the findings of an earlier study conducted by Aldridge and Stratton in which the investigators reported concentration-dependent activity for clindamycin against *B. fragilis* (37). However, the latter studied a much smaller range of antibiotic concentrations and also reported concentration-dependent activity of two cephalosporins, which are well-documented concentration-independent agents.

The findings of Klepser et al. (36) suggest that doses of 300 mg q8-12h may be more appropriate than the standard regimens used for clindamycin, ranging from 600 mg q6-8h to 900 mg q8h to 1200 mg q12h. A follow-up study conducted by Klepser et al. confirmed the effectiveness of a 300 mg q8-12h dosing regimen against *B. fragilis* by obtaining serum inhibitory and bactericidal titers from 12 healthy volunteers (38). Although aerobic infections are not discussed in this chapter, similar findings have been found with clindamycin against *Staphylococcus aureus* and *Streptococcus pneumoniae* (39). Less drug exposure, decreased likelihood of adverse events, and lower cost are three of the many advantages of using a lower total dose of clindamycin.

Summary

Like metronidazole, clindamycin has been in use for a number of years. Clindamycin is used for anaerobic bacterial infections of the abdomen, bone, lung, skin and skin structures, and pelvis. Moreover, clindamycin is often used for the treatment of gram-positive infections and in patients who cannot tolerate β-lactams. Resistance is of utmost importance when considering clindamycin use. Many *C. difficile* and *B. fragilis* group isolates are resistant to clindamycin, and such isolates have a higher likelihood of having decreased susceptibility to other agents as well. Recent pharmacodynamic studies have demonstrated concentration-independent bactericidal activity with clindamycin, suggesting a dosing regimen of 300 mg every 8 to 12 hours may be more appropriate than the currently recommended regimens (36,38). Utilizing this pharmacodynamic data to adjust clindamycin dosing regimens would translate into less antibiotic pressure, which may help to prevent the development of further resistance.

REFERENCES

1. Muller M. Metronidazole: its action on anaerobes. Infect Dis 1979.
2. Galgiani JN, Busch DF, Brass C, Rumans LW, Mangels JI, Stevens DA. *Bacteroides fragilis* endocarditis, bacteremia and other infections treated with oral or intravenous metronidazole. Am J Med 1978; 65:284–289.

3. George WL, Kirby BD, Sutter VL, Wheeler LA, Mulligan ME, Finegold SM. Intravenous metronidazole for treatment of infections involving anaerobic bacteria. Antimicrob Agents Chemother 1982; 21:441–449.

4. Ralph ED, Kirby WMM. Unique bactericidal action of metronidazole against *Bacteroides fragilis* and *Clostridium perfringens*. Antimicrob Agents Chemother 1975; 8:409–414.

5. Tally FP, Sutter VL, Finegold SM. Treatment of anaerobic infections with metronidazole. Antimicrob Agents Chemother 1975; 7:672–675.

6. Drummond LJ, McCoubrey J, Smith DG, Starr JM, Poxton IR. Changes in sensitivity patterns to selected antibiotics in *Clostridium difficile* in geriatric in-patients over an 18-month period. J Med Microbiol 2003; 52:259–263.

7. Aldridge KE, Ashcraft D, Cambre K, Pierson CL, Jenkins SG, Rosenblatt JE. Multicenter survey of the changing in vitro antimicrobial susceptibilities of clinical isolates of *Bacteroides fragilis* group, prevotella, fusobacterium, porphyromonas, and peptostreptococcus species. Antimicrob Agents Chemother 2001; 45:1238–1243.

8. Ingham HR, Hall CJ, Sisson PR, Tharagonnet D, Selkon JB. The activity of metronidazole against facultatively anaerobic bacteria. J Antimicrob Chemother 1980; 6:343–347.

9. Onderonk AB, Louie TJ, Tally FP, Bartlett JG. Activity of metronidazole against *Escherichia coli* in experimental intra-abdominal sepsis. J Antimicrob Chemother 1979; 5:201–210.

10. Chrystal EJT, Koch RL, McLafferty MA, Goldman P. Relationship between metronidazole metablolism and bactericidal activity. Antimicrob Agents Chemother 1980; 18:566–573.

11. Pendland SL, Jung R, Messick CR, Schriever CA, Patka J. In vitro bactericidal activity of piperacillin, gentamicin, and metronidazole in a mixed model containing *Escherichia coli*, *Enterococcus faecalis*, and *Bacteroides fragilis*. Diagn Microbiol Infect Dis 2002; 43:149–156.

12. Lamp KC, Freeman CD, Klutman NE, Lacy MK. Pharmacokinetics and pharmacodynamics of the nitroimidazole antimicrobials. Clinical Pharmacokinetics 1999; 36:353–373.

13. Robbie MO, Sweet RL. Metronidazole use in obstetrics and gynecology: a review. Am J Obstet Gynecol 1983; 145:865–881.

14. Ralph ED, Clarke JT, Libke RD, Luthy RP, Kirby WM. Pharmacokinetics of metronidazole as determined by bioassay. Antimicrob Agents Chemother 1974; 6:691–696.

15. Snydman DR, Jacobus NV, McDermott LA, et al. National survey on the susceptibility of *Bacteroides fragilis* group: report and analysis of trends for 1997–2000. Clin Infect Dis 2002; 35:S126–S134.

16. Finegold SM, Wexler HM. Present status of therapy for anaerobic infections. Clin Infect Dis 1996; 23:S9–S14.

17. Aldridge KE, Ashcraft D, O'Brien M, Sanders CV. Bacteremia due to *Bacteroides fragilis* group: distribution of species, β-lactamase production, and antimicrobial susceptibility patterns. Antimicrob Agents Chemother 2003; 47:148–153.

18. Diniz CG, Arantes RM, Cara DC, et al. Enhanced pathogenicity of susceptible strains of the *Bacteroides fragilis* group subjected to low doses of metronidazole. Microbes Infect 2003; 5:19–26.

19. Carlier JP, Sellier N, Rager MN, Reysset G. Metabolism of a 5-nitroimidazole in susceptible and resistant isogenic strains of *Bacteroides fragilis*. Antimicrob Agents Chemother 1997; 41:1495–1499.

20. Reysset G. Genetics of 5-nitroimidazole resistance in *bacteroides* species. Anaerobe 1996; 2:59–69.

21. Trinh S, Reysset G. Detection by PCR of the *nim* genes encoding 5-nitroimidazole resistance in Bacteroides spp. J Clin Microbiol 1996; 34:2078–2084.

22. Peláez T, Alcala L, Alonso R, Rodriguez-Creixems M, Garcia-Lechuz JM, Bouza E. Reassessment of *Clostridium difficile* susceptibility to metronidazole and vancomycin. Antimicrob Agents Chemother 2002; 46:1647–1650.

23. Craig WA. Pharmacokinetic/pharmacodynamic parameters: rationale for antibacterial dosing of mice and men. Clin Infect Dis 1998; 26:1–12.

24. Lewis RE, Klepser ME, Ernst EJ, Snabes MA, Jones RN. Comparison of oral immediate-release (ir) and extended-release (er) metronidazole bactericidal activity against Bacteroides spp. Using an in vitro model of infection. Diagn Microbiol Infect Dis 2000; 37:51–55.

25. Valdimarsdottir M, Erlendsdottir H, Gudmundsson S. Postantibiotic effects with *Bacteroides fragilis* determined by viable counts and CO2 generation. Clin Microbiol Infect 1997; 3:82–88.

26. Sprandel KA, Schriever CA, Pendland SL, et al. Pharmacokinetics and pharmacodynamics of intravenous levofloxacin at 750 milligrams and various doses of metronidazide in healthy adult subjects. Antimicrob Agents Chemother 2004; 48:4597.

27. Rodloff AC, Seidel C, Ackermann G, Schaumann R. Presented at the 42nd Interscience Conference on Antimicrobial Agents and Chemotherapy, San Diego, CA, 2002.

28. Vasa CV, Glatt AE. Effectiveness and appropriateness of empiric metronidazole for *Clostridium difficile*-associated diarrhea. Am J Gastroenterol 2003; 98:354–358.

29. Brook I. Meningitis and shunt infection caused by anaerobic bacteria in children. Pediatr Neurol 2002; 26:99–105.

30. Warner JF, Perkins RL, Cordero L. Metronidazole therapy of anaerobic bacteremia, meningitis, and brain abscess. Arch Intern Med 1979; 139:167–169.

31. Finegold SM. Metronidazole. Ann Intern Med 1980; 93:585.

32. Visapää JP, Tillonen JS, Kaihovaara PS, Salaspuro MP. Lack of disulfiram-like reaction with metronidazole and ethanol. Ann Pharmacother 2002; 36:971–974.

33. Williams CS, Woodcock KR. Do ethanol and metronidazole interact to produce a disulfiram-like reaction? Ann Pharmacother 2000; 34:255–257.

34. Falagas ME, Gorbach SL. Clindamycin and metronidazole. Med Clin North Am 1995; 79:845–867.

35. Gatti G, Flaherty J, Bubp J, White J, Borin M, Gambertoglio J. Comparative study of bioavailabilities and pharmacokinetics of clindamycin in healthy volunteers and patients with aids. Antimicrob Agents Chemother 1993; 37:1137–1143.

36. Klepser ME, Banevicius MA, Quintiliani R, Nightingale CH. Characterization of bactericidal activity of clindamycin against *Bacteroides fragilis* via kill curve methods. Antimicrob Agents Chemother 1996; 40:1941–1944.

37. Aldridge KE, Stratton CW. Bactericidal activity of ceftizoxime, cefotetan, and clindamycin against cefoxitin-resistant strains of the *Bacteroides fragilis* group. J Antimicrob Chemother 1991; 28:701–705.

38. Klepser ME, Nicolau DP, Quintiliani R, Nightingale CH. Bactericidal activity of low-dose clindamycin administered at 8- and 12-hour intervals against *Staphylococcus aureus*, *Streptococcus pneumoniae*, and *Bacteroides fragilis*. Antimicrob Agents Chemother 1997; 41:630–635.

39. Lewis RE, Klepser ME, Ernst EJ, Lund BC, Biedenbach DJ, Jones RN. Evaluation of low-dose, extended-interval clindamycin regimens against *Staphylococcus aureus* and *Streptococcus pneumoniae* using a dynamic in vitro model of infection. Antimicrob Agents Chemother 1999; 43:2005–2009.

Darren Abbanat, Mark Macielag, and Karen Bush
Johnson & Johnson Pharmaceutical Research & Development, L.L.C., Raritan, New Jersey, U.S.A.

INTRODUCTION

With the rapid emergence of both vancomycin-resistant enterococci (VRE) and methicillin (oxacillin)-resistant *Staphylococcus aureus* (MRSA) beginning in the late 1980s, new agents with antibacterial activity against these pathogens were avidly sought (1). The first agents developed in the 1990s to address this medical need were the streptogramin combination quinupristin-dalfopristin, known commercially as Synercid® (2,3) and the oxazolidinone linezolid known as Zyvox® (4). Of these two agents, linezolid has a broader set of therapeutic indications and is used more extensively.

Streptogramins are antibiotics originating from natural products that have been used as both human and veterinary therapeutic agents. Because of their use in animals, resistance to these agents was already present in the environment at the time quinupristin–dalfopristin was introduced clinically. In contrast, with the introduction of linezolid, the oxazolidinones have been hailed as the first novel antibiotic class to be introduced in the past 30 years. In spite of this, enterococcal resistance to linezolid arose rather rapidly upon extended therapy, but resistance in the staphylococci has been slow to emerge. Although both agents have inherent toxicities associated with their use, they have both been used to treat a variety of infections caused by gram-positive pathogens. In this chapter, their microbiological attributes, together with their pharmacokinetic (PK) and pharmacodynamic (PD) properties, will be described.

STREPTOGRAMINS
Structures
The streptogramin antibiotics in clinical use or in development consist of a mixture of two structurally distinct compounds, one belonging to the Group A streptogramins and the other to the Group B streptogramins. The Group A streptogramins, which include the semisynthetic derivative dalfopristin are macrocyclic polyketide-amino acid hybrids. They are typified by the presence of a 2,4-disubstituted oxazole and an (E,E)-allylic dienamide functionality (Fig. 1).

Quinupristin is a member of the Group B streptogramin class of antibiotics, which are cyclic depsipeptides with a characteristic ester linkage between the C-terminal L-phenylglycine residue and L-threonine at the amino terminus. Additionally, the amino terminus is capped with a 3-hydroxypicolinic acid moiety that appears to be involved in binding of divalent cations (Fig. 1) (5).

Although certain naturally occurring streptogramins had been used as oral antibiotics since the 1960s, their physicochemical properties made them unsuitable for intravenous administration. Synthetic modification of the natural products

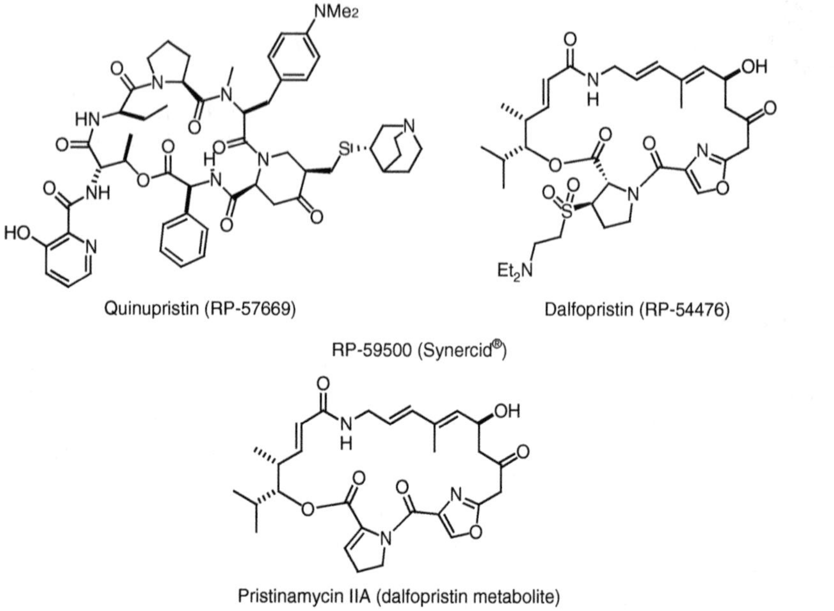

Quinupristin (RP-57669)   Dalfopristin (RP-54476)

RP-59500 (Synercid®)

Pristinamycin IIA (dalfopristin metabolite)

FIGURE 1 Molecular structures of the streptogramins quinupristin, dalfopristin, and pristinamycin IIA.

pristinamycin IIA and pristinamycin IA by chemists at Rhone-Poulenc afforded dalfopristin and quinupristin, respectively, the components of RP-59500, each of which contains a protonatable amine to enhance aqueous solubility (6). Interestingly, one of the primary metabolites of dalfopristin in humans is pristinamycin IIA (Fig. 1), the natural product from which dalfopristin was derived chemically (7).

Mechanism of Action
Both the Group A and the Group B streptogramins bind to the bacterial 50S ribosomal subunit and inhibit protein synthesis by interfering with the elongation cycle (8). Group A streptogramins block peptide bond formation by interfering with the proper positioning of aminoacyl-tRNA and peptidyl-tRNA at the A and P sites within the peptidyl transferase center of the ribosome (9). Antibiotic binding occurs only when both donor and acceptor sites are free of bound substrate. Thus, Group A streptogramins do not inhibit the function of ribosomes actively engaged in protein synthesis (10). The Group A streptogramins have been reported to inhibit the function of bacterial ribosomes even after dissociation from the peptidyl transferase center, thereby producing an extended postantibiotic effect (PAE) (11). Group B streptogramins cause the premature dissociation of peptidyl-tRNA from the ribosome in a manner similar to that of macrolide antibiotics such as erythromycin A (12). Biochemical experiments have shown that erythromycin A can readily displace virginiamycin S (a Group B streptogramin) from *Escherichia coli* ribosomes (13). Furthermore, methylation of adenine residue A2058 of 23S rRNA (*E. coli* numbering system) by the erythromycin ribosomal methylase renders ribosomes resistant to both macrolides and streptogramin B antibiotics (MLS$_B$ resistance), providing additional evidence that the binding sites of these antibiotics overlap (14).

The Group A and Group B streptogramins act synergistically to inhibit bacterial growth and are marketed as a combination product (e.g., Synercid, quinupristin–dalfopristin). The activity of the quinupristin–dalfopristin combination is generally about 10-fold higher than the activity of either drug alone, depending on the bacterial strain (15). Although each of the components is bacteriostatic on its own, the combination is bactericidal against staphylococci and streptococci (16). Early binding experiments had shown that preincubation of ribosomes with virginiamycin M (Group A streptogramin) led to a 5- to 10-fold increase in the association constant of the ribosomes for virginiamycin S (Group B streptogramin) (11). Affinity for virginiamycin S increased in the presence of virginiamycin M, not only for wild-type ribosomes but also for ribosomes resistant to Group B streptogramins due to methylation of A2058 (17). It is now generally accepted that the synergistic activity between Group A and Group B streptogramins is due, at least in part, to a conformational change upon binding Group A streptogramins, which increases the affinity of the ribosome for the Group B component.

Many of the mechanistic conclusions drawn from the results of early biochemical experiments have been corroborated by recent crystal structures of virginiamycin M bound to the large ribosomal subunit of *Haloarcula marismortui* and quinupristin–dalfopristin bound to the 50S ribosomal subunit of *Deinococcus radiodurans* (18,19). Both virginiamycin M and dalfopristin bind within the peptidyl transferase center, contacting nucleotides that comprise the A and P sites. Interestingly, dalfopristin was shown to induce a nonproductive, yet stable, conformational change of the universally conserved nucleotide U2585 within the peptidyl transferase center of *D. radiodurans*, possibly accounting for the prolonged effect of Group A streptogramins on ribosomal function (19). Quinupristin binds to the entrance of the peptide exit tunnel of the ribosome, in a similar fashion as the macrolide antibiotics, where it acts to block the passage of the nascent peptide chain during the translocation step (19). The synergistic activity of quinupristin and dalfopristin appears to be a consequence of hydrophobic interactions between the two streptogramin components as well as cooperative positioning of nucleotide A2062 to insure optimal antibiotic binding (19).

Clinical Use

Quinupristin–dalfopristin is currently the only streptogramin combination approved by the Food and Drug Administration (FDA) for therapeutic use in humans. It is indicated for the treatment of patients with serious or life-threatening infections associated with vancomycin-resistant *Enterococcus faecium* bacteremia. Quinupristin–dalfopristin has also been approved for treatment of complicated skin and skin-structure infections caused by *S. aureus* (methicillin-susceptible) or *Streptococcus pyogenes* (20).

In Vitro Susceptibility

Quinupristin–dalfopristin is active against most antibiotic-susceptible or multi-drug-resistant staphylococci, irrespective of the associated mechanisms of resistance. Synergy of the streptogramin combination is maintained in the presence of inducibly or constitutively expressed *erm* (21,22). In a comparison of MLS-susceptible or *erm*-containing *S. aureus* strains, the presence of *erm* increased quinupristin minimum inhibitory concentration (MIC) values up to 32-fold, while strains with *erm* retained susceptibility to quinupristin-dalfopristin, with MIC values increasing at most fourfold (Table 1).

TABLE 1 In Vitro Activities of Quinupristin–Dalfopristin with Comparative Agents

Organism	Compound	N	Range	MIC (μg/mL) MIC$_{50}$	MIC$_{90}$	Percentage susceptible	Refs.
Staphylococcus aureus							
MLS-S[a]	Quinupristin	3	2	NA[b]	NA	NB[c]	21
	Dalfopristin	3	2–4	NA	NA	NB	21
	Q-D[d]	3	0.25–0.5	NA	NA	S[a]	21
Erm(A, B, or C)	Quinupristin	13	2–64	$\geq$32	64	NB	21
	Dalfopristin	13	0.5–4	NR	4	NB	21
	Q-D	13	0.25–1	NR	1	S[a]	21
S. aureus[e]							
CA-MSSA	Vancomycin	1592	$\leq$0.12 to 2	1	1	100	23
	Ceftriaxone	1592	0.5 to >32	4	4	98.9	23
	Clindamycin	1592	$\leq$0.06 to >8	0.12	0.12	94.5	23
	Chloramphenicol	1592	$\leq$2 to >16	8	8	96.5	23
	Ciprofloxacin	1592	$\leq$0.25 to >4	0.25	1	91.3	23
	Q-D	1592	$\leq$0.25 to 2	0.25	0.5	99.9	23
CA-MRSA	Vancomycin	652	0.5 to 4	1	1	100	23
	Ceftriaxone	652	0.5 to >32	>32	>32	3.8	23
	Clindamycin	652	$\leq$0.06 to >8	>8	>8	33.7	23
	Chloramphenicol	652	$\leq$2 to >16	8	16	83.3	23
	Ciprofloxacin	652	$\leq$0.25 to >4	>4	>4	15.2	23
	Q-D	652	$\leq$0.25 to 1	0.5	1	100	23
N-MSSA	Vancomycin	706	0.5 to 2	1	1	100	23
	Ceftriaxone	706	$\leq$0.25 to >32	4	4	98.7	23
	Clindamycin	706	$\leq$0.06 to >8	0.12	0.25	91.6	23
	Chloramphenicol	706	$\leq$2 to >16	8	8	95.7	23
	Ciprofloxacin	706	$\leq$0.25 to >4	$\leq$0.25	2	89.7	23
	Q-D	706	$\leq$0.25 to 1	0.25	0.5	100	23
N-MRSA	Vancomycin	548	0.25 to 2	1	2	100	23
	Ceftriaxone	548	$\leq$0.25 to >32	>32	>32	3.8[f]	23
	Clindamycin	548	$\leq$0.06 to >8	>8	>8	23.5	23
	Chloramphenicol	548	4 to >16	8	16	79.6	23
	Ciprofloxacin	548	$\leq$0.25 to >4	>4	>4	6.0	23
	Q-D	548	$\leq$0.25 to 1	0.5	1	100	23
VISA	Vancomycin	19	4 to 8	4	8	NR[g]	24
	Daptomycin	19	1 to 8	2	8	NR	24
	Q-D	19	$\leq$0.5 to 2	$\leq$0.5	$\leq$0.5	NR	24
VRSA	Vancomycin	2	32, 1024	NR	NR	R[a]	25, 26
	Daptomycin	2	0.5, 1	NR	NR	NB	25, 26
	Minocycline	2	0.12, 0.25	NR	NR	S	25, 26
	Q-D	2	$\leq$1	NR	NR	S	25, 26
MS-CoNS	Vancomycin	109	NR	1	2	100	27
	Teicoplanin	109	NR	2	8	95.4	27
	Clindamycin	109	NR	0.06	0.12	92.7	27

(Continued)

TABLE 1 In Vitro Activities of Quinupristin–Dalfopristin with Comparative Agents *(Continued)*

Organism	Compound	N	Range	MIC (µg/mL) MIC_{50}	MIC_{90}	Percentage susceptible	Refs.
	Chloramphenicol	109	NR	4	8	95.4	27
	Tetracycline	109	NR	≤ 4	>8	78.0	27
	TMP—SMX[h]	109	NR	0.25	>1	82.4	27
	Q-D	109	NR	0.25	0.25	100	27
MR-CoNS	Vancomycin	389	NR	2	2	100	27
	Teicoplanin	389	NR	2	16	90.0	27
	Clindamycin	389	NR	>8	>8	43.2	27
	Chloramphenicol	389	NR	8	>16	59.9	27
	Tetracycline	389	NR	≤ 4	>8	75.8	27
	TMP-SMX	389	NR	>1	>1	38.6	27
	Q-D	389	NR	0.25	0.5	97.7	27
VS-*Enterococcus faecalis*	Vancomycin	121	2–4	2	2	100	28
	Teicoplanin	121	0.5	0.5	0.5	100	28
	Ampicillin	121	1–64	1	1	99.2	28
	Chloramphenicol	121	2–128	8	16	88.4	28
	Doxycycline	121	1–32	8	16	38	28
	Ciprofloxacin	121	1–64	2	64	47.9	28
	Q-D	121	1–32	8	16	1.7	28
VR-*E. faecalis*	Vancomycin	81	16–512	512	512	0.0	28
	Teicoplanin	81	0.5–128	64	128	43.2	28
	Ampicillin	81	1–128	1	2	97.5	28
	Chloramphenicol	81	4–64	8	64	77.8	28
	Doxycycline	81	1–32	8	16	35.8	28
	Ciprofloxacin	81	1–64	32	64	1.2	28
	Q-D	81	1–128	8	32	2.5	28
VS-*Enterococcus faecium*	Vancomycin	42	2	2	2	100	28
	Teicoplanin	42	0.5–1	0.5	1	100	28
	Ampicillin	42	1–128	64	128	21.4	28
	Chloramphenicol	42	2–32	4	8	90.5	28
	Doxycycline	42	1–32	1	8	73.8	28
	Ciprofloxacin	42	1–64	64	64	9.5	28
	Q-D	42	0.25–16	1	4	69.0	28
VR-*E. faecium*	Vancomycin	616	8–512	256	512	0.0	28
	Teicoplanin	616	0.5–128	32	32	22.2	28
	Ampicillin	616	1–256	64	128	3.2	28
	Chloramphenicol	616	1–64	4	8	99.1	28
	Doxycycline	616	1–32	4	16	60.7	28
	Ciprofloxacin	616	8–64	64	64	0.0	28
	Q-D	616	0.25–64	1	2	75.8	28
Streptococcus pyogenes	Penicillin	117	≤ 0.03	≤ 0.03	≤ 0.03	100	29
	Erythromycin	117	≤ 0.015 to >2	0.06	>2	82.1	29
	Clindamycin	117	≤ 0.25 to >1	≤ 0.25	≤ 0.25	94.9	29
	Daptomycin	117	≤ 0.015 to 0.12	0.06	0.06	100	29
	Q-D	117	≤ 0.12 to 0.5	≤ 0.12	≤ 0.12	100	29

(Continued)

TABLE 1 In Vitro Activities of Quinupristin–Dalfopristin with Comparative Agents *(Continued)*

Organism	Compound	N	Range	MIC (μg/mL) MIC$_{50}$	MIC$_{90}$	Percentage susceptible	Refs.
Streptococcus	Penicillin	56	≤0.03 to 2	≤0.03	0.06	96.4	29
agalactiae	Erythromycin	56	0.03 to >2	0.06	2	85.7	29
	Clindamycin	56	≤0.25 to >1	≤0.25	>1	87.5	29
	Daptomycin	56	0.06–1	0.25	0.5	NB	29
	Q-D	56	≤0.12 to 16	0.25	0.5	96.4	29
Streptococcus pneumoniae[i]	Penicillin	3304	NR	≤0.03	2	65.0	30
	Erythromycin	3304	NR	≤0.25	8	76.9	30
	Clindamycin	3304	NR	≤0.25	≤0.25	93.0	30
	Daptomycin	179	0.06–1	0.12	0.25	NB	29
	Q-D	3304	NR	0.5	0.5	99.9	30

[a]S, susceptible; R, resistant.
[b]Not applicable.
[c]No breakpoints have been assigned.
[d]Quinupristin-dalfopristin.
[e]CA, community-acquired; N, nosocomial.
[f]Dependent on breakpoint definition. Using the NCCLS definition, all MRSA are resistant.
[g]Not reported.
[h]Trimethoprim-sulfamethoxazole.
[i]Not listed in the Food and Drug Administration approved label.
Abbreviations: MIC, minimum inhibitory concentration; MLS, macrolides and streptogramin; MSSA, methicillin-susceptible *S. aureus*; MRSA, methicillin (oxacillin)-resistant *S. aureus*; VISA, *S. aureus* expressing a vancomycin-intermediate; VR, vancomycin-resistant; VRSA, vancomycin-resistant *S. aureus*; MLS, macrolide-lincosamide streptogramin; VISA, vancomycin intermediate *S. aureus*.

As expected, *erm* had no effect on dalfopristin MIC values. In one study (Table 1), community-derived or nosocomial methicillin-susceptible *S. aureus* (MSSA) strains were at least 99% susceptible to quinupristin-dalfopristin, with an MIC$_{90}$ value of 0.5 μg/mL. Susceptibility of these isolates to vancomycin, ceftriaxone, clindamycin, chloramphenicol, and ciprofloxacin was at least 90%. Community-acquired or nosocomial MRSA strains were pan-susceptible to quinupristin–dalfopristin and vancomycin, with an MIC$_{90}$ value of approximately 1 μg/mL. Against *S. aureus* expressing a vancomycin-intermediate *S. aureus* phenotype (Table 1), the MIC$_{90}$ (concentration inhibiting 90% of the strains tested) value of quinupristin–dalfopristin was ≤0.5 μg/mL. Against the two recent vancomycin-resistant *S. aureus* strains expressing the *van*(A) gene, quinupristin–dalfopristin MIC values remained similar to those of vancomycin-intermediate or -susceptible strains, suggesting that the streptogramin combination will remain efficacious against this resistance genotype as well. Potencies of quinupristin–dalfopristin against methicillin-susceptible and -resistant coagulase-negative staphylococci were similar to those observed for *S. aureus*.

Quinupristin–dalfopristin is active against most *E. faecium* isolates, but it lacks clinically useful activity against *Enterococcus faecalis* (Table 1). In one North American surveillance study (28), 69% and 76% of vancomycin-susceptible or -resistant *E. faecium* strains, respectively, were susceptible to quinupristin-dalfopristin. Only chloramphenicol and linezolid inhibited a higher proportion of susceptible strains among these isolates. As expected, in this study quinupristin–dalfopristin was ineffective against *E. faecalis* strains, with 97.5% of strains determined to be nonsusceptible.

With its predominantly gram-positive spectrum of activity, quinupristin–dalfopristin is also recommended for the treatment of skin and soft-tissue infections caused by *S. pyogenes*. In the study described by Fluit et al. (Table 1), all *S. pyogenes* strains tested were susceptible to quinupristin-dalfopristin, with an MIC_{90} value $\leqslant 0.12$ μg/mL, similar to penicillin and daptomycin. Against these isolates, erythromycin nonsusceptibility was 18%, while clindamycin nonsusceptibility was 5%, indicating that quinupristin–dalfopristin retains activity against macrolide- and lincosamide-resistant isolates. Susceptibility of *Streptococcus agalactiae* isolates was similar to that observed for *S. pyogenes*, although the MIC_{90} value of quinupristin–dalfopristin for *S. agalactiae* was at least fourfold higher than for *S. pyogenes*. Although not approved for treatment of *Streptococcus pneumoniae* infections, MIC values of quinupristin–dalfopristin for pneumococci are similar to those obtained with *S. agalactiae* (Table 1).

The potency of the streptogramin combination against *Moraxella catarrhalis* strains is comparable to that for staphylococci and *S. agalactiae*, with an MIC_{90} value of 0.5 μg/mL. Against *Haemophilus influenzae*, quinupristin–dalfopristin is less active, with an MIC_{50} (concentration inhibiting 50% of the strains tested) value $\geqslant 2$ μg/mL (22,30).

Resistance

Resistance to streptogramins is mediated by four distinct mechanisms: modification of the ribosome target site, efflux, enzymatic inactivation of streptogramin A or B components, or impermeability of the bacterial envelope. It has been proposed that acquired resistance to quinupristin–dalfopristin requires at a minimum the presence of the streptogramin A *vat* or *vga* resistance gene (31). Recent studies (see below) suggest that resistance to the streptogramin combination can also occur in pneumococci through selective mutations to 23S rRNA or to the ribosomal proteins L4 or L22 (32).

Resistance to streptogramins A or B can result from modifications to the 23S rRNA or specific ribosomal proteins through one of two mechanisms: (*i*) methylation of A2058 (*E. coli* numbering) by an Erm dimethylase, resulting in an MLS_B phenotype (resistance to macrolides, lincosamides, and streptogramin B antibiotics) and (*ii*) mutational changes to the 23S rRNA or 50S ribosomal subunit proteins L4 and L22. As previously discussed (Table 1), methylation of A2058 does not confer resistance to streptogramin A type antibiotics, and strains with *erm* remain susceptible to quinupristin–dalfopristin (33), although bactericidal activity of quinupristin–dalfopristin may be compromised (21). Mutational changes to 23S rRNA or to L4 or L22 ribosomal proteins have resulted in intermediate or resistant phenotypes. In one PROTEKT study of 7746 macrolide-resistant pneumococcal strains, 6 isolates of 77 with 23S rRNA, L4, or L22 mutations were found to be resistant to quinupristin-dalfopristin, with MIC values of 4 or 8 μg/mL. Five distinct combinations of ribosomal mutations correlated with this resistance, including (*i*) 23S rRNA C2611G (three of four RNA copies), (*ii*) 23S rRNA A2059G (four of four copies) with G95D in L22, (*iii*) an L22 tandem duplication $_{109}$RTA-HIT$_{114}$, (*iv*) in L4, a K68S with a deletion of the four amino acids $_{69}$GTGR$_{72}$, and (*v*) in L4, $_{69}$GT$_{70}$ changed to $_{69}$VP$_{70}$ (32). In another study, quinupristin–dalfopristin resistance in two clinical *S. aureus* isolates resulted from an L22 deletion of

$_{79}$GP$_{80}$, or a 21 bp duplication in L22 inserting the amino acids $_{100}$SAINKRT$_{101}$; in vitro selection of quinupristin–dalfopristin resistance in a susceptible *S. aureus* strain resulted in comparable deletions and insertions in a similar region of the L22 protein (34). Other studies describe mutations that selectively inhibit streptogramin B activity but are not known to render the isolates quinupristin–dalfopristin resistant, including the six amino acid L4 insertion $_{71}$GREKGTGR$_{72}$, the L4 substitution $_{69}$GTG$_{71}$ to $_{69}$TPS$_{71}$ (35), and C2611A, A2058G, A2062C, and A2142G substitutions in 23S rRNA (36,37).

Efflux of streptogramin A is mediated in staphylococci by *vga*(A) or *vga*(B) (38), and in *E. faecalis* by the *lsa* gene (39). The Lsa, Vga, and Msr efflux proteins are all members of the ATP transporter superfamily. Expression of *lsa* in *E. faecalis* renders most strains resistant to quinupristin-dalfopristin. In a study by Lina et al., *vga*-mediated resistance to quinupristin–dalfopristin in CoNS was frequently associated with an *erm* gene, although no resistance to the streptogramin combination in *S. aureus* from *vga* or *vga-erm* was identified (40). Efflux of streptogramin B antibiotics in staphylococci and *E. faecium* is mediated by *msr*(A) and *msr*(C) genes, respectively (38,41,42). In *S. aureus*, the presence of *msr* was not sufficient to cause resistance to quinupristin–dalfopristin (43).

Inactivation of streptogramin A occurs through the activity of the *vat* gene product [Vat(A–E)], which mediates the transfer of an acetyl group to the hydroxyl moiety of streptogramin A (38,44,45). The Vat(A–C) acetyltransferases are found in staphylococci, while Vat(D) [formerly Sat(A)] and Vat(E) [formerly Sat(G)] are found in enterococci (38). In one study, cloning of *vat*(D) into a *S. aureus* isolate or an *E. faecium* isolate increased dalfopristin MIC values eightfold and quinupristin–dalfopristin MIC values twofold (33). In another study, quinupristin–dalfopristin resistance among 13 clinical staphylococcal isolates resulted from the combination of *vat* and *erm*, and in some cases other resistance factors as well (40).

Inactivation of streptogramin B occurs in staphylococci and rarely in *E. faecium* through the action of the Vgb(A) and Vgb(B) lyases (33,38), which cleave the macrocyclic ring of the streptogramin B component at the lactone ester linkage (46). In experiments where *vgb* was cloned into an *E. faecium* or a *S. aureus* isolate, quinupristin MIC values increased eightfold, and quinupristin–dalfopristin MIC values increased fourfold (33). Resistance to quinupristin–dalfopristin (MIC = 16 µg/mL) was observed in one clinical *E. faecium* isolate containing *vgb*, *vat*, and *erm* genes (33).

Pharmacokinetics/Pharmacodynamics (PK/PD)
PDs In Vitro and In Vivo
Although the streptogramins have not been investigated pharmacodynamically to the same extent as other antibacterial classes such as the fluoroquinolones or β-lactams, a number of PD studies have been conducted both in vitro and in vivo. PAEs in vitro were demonstrated against four *S. aureus* clinical isolates when exposed to quinupristin–dalfopristin at 1x, 2x, and 4x the MIC (1). When two MRSA strains were exposed to 4x the quinupristin–dalfopristin MIC for 80 minutes, the measured PAE in vitro was five hours. In an in vitro model simulating human PKs, quinupristin–dalfopristin demonstrated potent bactericidal activity against both penicillin-susceptible and -resistant *S. pneumoniae*, with cidality observed within two hours following simulation of a single 7 mg/kg intravenous dose (1).

Craig et al. studied the drug combination both in vitro and in a murine thigh model (1,47). When MICs and MBCs were compared, the two values were identical for *S. pneumoniae* and exhibited only a one dilution difference (MBC twofold higher) for *S. aureus*. In the mouse thigh infection model, where neutropenic mice were first infected with selected pathogenic bacteria and then sacrificed to determine bacterial counts, the *in vivo* PAE for the streptogramin combination was 10 hours for *S. aureus* and 9.1 hours for *S. pneumoniae*. Efficacy was not dependent upon dosing schedule, for example, 50% maximal efficacy was achieved at the same total dose administered once, twice, or four times in 24 hours (48). The operative PD parameter correlating with therapeutic efficacy was proposed to be an area under the plasma concentration–time curve (AUC)/MIC, in spite of the lack of observed concentration-dependent killing (48,49). Efficacy was also shown to be related to the immune state of the host in studies, with lower PK/PD values in animals producing neutrophils (47).

Clinical PKs
Clinically, quinupristin–dalfopristin is formulated as a fixed 30:70 (w:w) combination that is infused for 60 minutes at 7.5 mg/kg either q8h for infections caused by vancomycin-resistant *E. faecium* or q12h for skin infections caused by grampositive pathogens (20). Both streptogramins are nonenzymatically converted to active metabolites with lower antimicrobial activity than their respective parent molecules (7). The active metabolite pristinamycin IIA is formed in nonacidified blood by the rapid metabolism of dalfopristin in all drug-treated subjects (7), but the in vivo formation of cysteine- or glutathione-quinupristin conjugates is variable from study to study (50). Because of the unpredictability of formation of active metabolites, the PD properties of both parent and daughter moieties must be considered, thereby making the analysis more complex than for other antibiotic combinations. For this reason, HPLC analyses were generally conducted to quantify the amounts of each of the three major components (quinupristin, dalfopristin. and pristinamycin IIA), but bioanalyses were performed to determine total antibacterial activity associated with the full complex of streptogramin entities. It was noted that blood samples needed to be acidified immediately in dilute hydrochloric acid upon collection to prevent further degradation of dalfopristin to pristinamycin IIA ex vivo (7).

Initial PK studies in healthy human volunteers were conducted using nine doses of the fixed 30:70 quinupristin–dalfopristin combination ranging from 1.4 to 29.4 mg/kg (total drug). HPLC analyses were used to analyze drug concentrations for each of the major components and the pristinamycin IIA metabolite, but bioassays were used to determine total antibacterial activity. In these studies, both AUC and C_{max} increased in a dose-dependent manner for total drug, quinupristin and the pristinamycin metabolite at all doses, and for dalfopristin at doses through 7 mg/kg (7). Peak blood levels for total drug ranged from 0.95 to 24.2 µg/mL. The synergistic effect of all the drugs and the biological activity of the metabolites were demonstrated by the observation that total antibacterial activity in these studies generally exceeded the sum of each of the measured components; in addition, the half-life for the total drug (1.27–1.53 hours) was higher than the $t_{1/2}$ observed for quinupristin (0.56–0.61 hours) or pristinamycin IIA (0.75–0.95 hours) (7).

Steady-state PK parameters in healthy male volunteers were obtained for quinupristin and dalfopristin, together with their metabolites, following nine doses of 7.5 mg/kg on a q12h schedule or 10 doses of 7.5 mg/kg on a q8h dosing schedule (Table 2).

The results showed proportionality between the q12h and q8h dosing regimens. Dalfopristin and its metabolites exhibited higher peak drug concentrations

TABLE 2 Steady-State Pharmacokinetic Properties of Quinupristin and Dalfopristin and Their Metabolites Following Dosing Two Groups of Healthy Young Adult Male Volunteers[a]

Parameter[b]	Quinupristin + metabolites	Dalfopristin + metabolites
C_{max} (mg/L)	3.20 ($\pm$0.67)[c]	7.96 ($\pm$1.30)
AUC (mg/L hr)	7.20 ($\pm$1.24)	10.57 ($\pm$2.24)
$t_{1/2}$ (hr)	3.07 ($\pm$0.51)	1.04 ($\pm$0.20)

[a]Each group received 7.5 mg/kg of the quinupristin–dalfopristin combination intravenously q12h or q8h for a total of 9 or 10 doses, respectively.
[b]C_{max}, maximum plasma drug concentration; AUC, area under the plasma concentration–time curve from zero to infinity; $t_{1/2}$, half-life.
[c]$\pm$Standard deviation.

and exposure, but a shorter half-life compared to quinupristin and its metabolites (2,20). At steady state, quinupristin had a larger mean volume of distribution than did dalfopristin, 0.45 and 0.24 L/kg, respectively (2). Both AUC and C_{max} values for the parent compounds increased approximately 20% when steady-state parameters are compared to parameters obtained after a single dose (51).

Special populations were studied to determine whether dosing regimens would need to be adjusted for these specific groups. No statistically significant differences in PK parameters were observed between male and female healthy volunteers aged 20 to 36 years. PK parameters were compared between healthy elderly volunteers (69–74 years) and a cohort of healthy young volunteers (19–34 years), both infused with a single 7.5 mg/kg dose of quinupristin-dalfopristin. No statistically significant differences were observed between the two groups, indicating that the combination can be used with no alteration in dosing regimen in healthy geriatric populations (51).

Although low renal clearance of the two drugs was observed in both laboratory animals and humans (52), a study was designed to compare PK properties between healthy volunteers and patients with severe renal insufficiency. In these studies (Table 3), AUC, half-lives, mean residence times, and plasma clearances for both quinupristin and dalfopristin were not significantly different in healthy volunteers compared to patients with renal failure, although there was a tendency for dalfopristin to have a higher AUC in the patient population (53).

Significant differences in C_{max} were observed between the two study populations for dalfopristin alone, and for total drug concentrations (Table 4), where 30% to 35% higher peak concentrations were observed in the renally impaired patients.

For quinupristin and its metabolites, all reported PK properties except t_{max} were significantly higher in the patients with renal insufficiency (Table 4). Based on the data for quinupristin, dalfopristin, and its major metabolite, the study concluded that no major effect was observed on quinupristin or dalfopristin kinetics in patients with renal impairment; thus, no alteration in dosing is required for patients with renal impairment (53). Disposition of the two parent streptogramins and their metabolites was also seen to be unaffected when quinupristin–dalfopristin was dosed in ambulatory peritoneal dialysis patients compared to healthy volunteers (54).

OXAZOLIDINONES
Structures
The oxazolidinone antibacterial agents are totally synthetic compounds of relatively low molecular weight (<500 Da) and structural complexity. Virtually all

TABLE 3 Pharmacokinetic Properties of Quinupristin and Dalfopristin and Its Active Metabolite Pristinamycin IIA in Healthy Volunteers (N = 13) or Patients With Chronic Renal Insufficiency (N = 13)[a]

Parameter[b]	Healthy volunteers			Patients with chronic renal insufficiency		
	Quinupristin	Dalfopristin	Pristinamycin IIa	Quinupristin	Dalfopristin	Pristinamycin IIa
C_{max} (mg/L)	2.72 ± 0.50	7.24 ± 1.91	1.47 ± 0.39	3.07 ± 0.63	9.14 ± 2.61^c	1.16 ± 0.56
Median t_{max} (hr)	0.92	0.50	0.92	0.92	0.92	1.05
AUC (mg/L hr)	3.31 ± 0.59	7.71 ± 2.60	2.22 ± 0.54	3.43 ± 0.72	10.1 ± 3.49	1.92 ± 0.94
$t_{1/2\lambda 1}$ (hr)	0.18 ± 0.05	0.17 ± 0.04	0.37 ± 0.08	0.19 ± 0.08	0.18 ± 0.06	0.50 ± 0.42
$t_{1/2\lambda z}$ (hr)	0.91 ± 0.18	0.98 ± 0.48	1.37 ± 0.85	0.83 ± 0.22	0.90 ± 0.59	1.14 ± 0.91
MRT	1.10 ± 0.09	0.86 ± 0.10	1.64 ± 0.33	1.09 ± 0.12	0.89 ± 0.07	1.75 ± 0.55
CL (L/hr/kg)	0.71 ± 0.14	0.74 ± 0.21	ND[d]	0.68 ± 0.14	0.58 ± 0.21	ND
V_{SS} (L/kg)	0.42 ± 0.08	0.27 ± 0.10	ND	0.40 ± 0.11	0.23 ± 0.10	ND
V_Z (L/kg)	0.92 ± 0.24	1.09 ± 0.74	ND	0.82 ± 0.28	0.79 ± 0.70	ND

[a]Drug concentrations were determined by high performance liquid chromatography (HPLC) after a 1 hr intravenous infusion of 7.5 mg/kg quinupristin–dalfopristin. Values are reported as mean ± standard deviations (53).
[b]C_{max}, maximum plasma drug concentration; t_{max}, time at which C_{max} occurred; AUC, area under the plasma concentration–time curve from zero to infinity; $t_{1/2\lambda 1}$, apparent distribution half-life; $t_{1/2\lambda z}$, apparent elimination half-life; MRT, mean residence time; CL, plasma clearance; V_{SS}, volume of distribution at steady state; V_Z, apparent volume of distribution during the terminal phase.
[c]$P < 0.05$ compared to healthy volunteers.
[d]Not determined.

TABLE 4 Pharmacokinetic Properties of Quinupristin, Dalfopristin and Their Active Metabolites in Healthy Volunteers ($N = 13$) or Patients with Chronic Renal Insufficiency ($N = 13$)[a]

Parameter[b]	Healthy volunteers		Patients with chronic renal insufficiency	
	Quinupristin + metabolites	Dalfopristin + metabolites	Quinupristin + metabolites	Dalfopristin + metabolites
C_{max} (mg/L)	2.20 ± 0.52	7.30 ± 1.74	3.00 ± 0.75^c	9.51 ± 2.39^c
Median t_{max} (hr)	0.92	0.92	0.92	0.92
AUC (mg/L hr)	3.86 ± 0.93	9.40 ± 2.56	5.32 ± 1.29^d	11.7 ± 3.8
$t_{1/2\lambda z}$ (hr)	1.32 ± 0.32	0.81 ± 0.21	2.38 ± 1.26^c	0.77 ± 0.15
MRT	1.81 ± 0.29	1.07 ± 0.10	2.58 ± 0.97^c	1.07 ± 0.10

[a]Drug concentrations were determined by bioassays after a 1 hr intravenous infusion of 7.5 mg/kg quinupristin–dalfopristin. Values are reported as means ± standard deviations (53).
[b]C_{max}, maximum plasma drug concentration; t_{max}, time at which C_{max} occurred; AUC, area under the plasma concentration–time curve from zero to infinity; $t_{1/2\lambda z}$, apparent elimination half-life; MRT, mean residence time.
[c]$P < 0.05$ compared to healthy volunteers.
[d]$P < 0.01$ compared to healthy volunteers.

members of this class possess a five-membered oxazolidin-2-one ring with an aromatic moiety and a functionalized methylene at the N3- and C5-positions of the heterocycle, respectively (55). The stereochemistry of the C5-substituent appears to be critically important, with the S-enantiomer having much greater antibacterial activity than the R-enantiomer (56). Linezolid is the first and only oxazolidinone approved for therapeutic use by the FDA. The oxazolidinone ring of linezolid is substituted with N3-3-fluoro-4-(4-morpholinyl)phenyl and C5-(S)-acetamidomethyl groups (Fig. 2).

DA-7867 is a more recent analog that currently appears to be in preclinical development. Although DA-7867 contains the same C5-substituent as linezolid, it has a more complex 3-fluoro-4-[6-(1-methyl-1H-tetrazol-5-yl)-3-pyridinyl]phenyl substituent at N3 (Fig. 2). Relatively few bioisosteres for the oxazolidinone ring have been described to date, and there are no reports of analogs derived from such isosteric replacement in clinical trials (55).

Mechanism of Action

Mechanistic studies of DuP-721 (the first oxazolidinone antibacterial agent to enter clinical trials) in whole cells revealed a specific inhibitory effect on protein synthesis, not DNA or RNA synthesis (57). Subsequently, the inhibition of protein synthesis by linezolid was demonstrated in a permeabilized *E. coli* strain and in *acrAB* or *tolC E. coli* mutants, indicating that the diminished activity in gram-negative bacteria was likely due to efflux and not due to the absence of the site of action of the drug (58,59). Linezolid also inhibited coupled transcription-translation in a cell-free system from *S. aureus* (59).

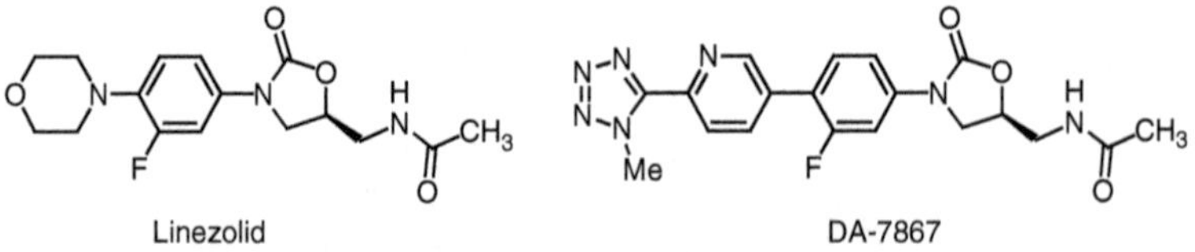

FIGURE 2 Molecular structures of the oxazolidinones linezolid and DA-7867.

Early binding studies suggested that the ribosome was the site of action of the oxazolidinones, as radio-labeled eperezolid, a close structural analog of linezolid, bound to the 50S subunit and could be displaced from the ribosome by chloramphenicol, lincomycin, and clindamycin (60). The mechanism of action was clearly distinct from the known ribosomal antibiotics, however, because the oxazolidinones were active against bacterial isolates resistant to other protein synthesis inhibitors (61). Mechanistic studies clearly showed that linezolid did not have an effect on the elongation or termination reactions of protein synthesis, nor did it inhibit the synthesis of N-formyl-methionyl-tRNA (59). In contrast, the drug prevented the formation of the N-formyl-methionyl-tRNA-ribosome-mRNA ternary complex, an essential step in the initiation of bacterial protein synthesis (62).

Initial attempts to define the exact site at which oxazolidinones bound within the bacterial ribosome relied on cross-linking and foot-printing studies conducted with isolated ribosomes in vitro (63). The binding site was mapped to nucleotides in 23S rRNA located near the ribosomal E-site as well as portions of the 16S rRNA of the small ribosomal subunit. More recent studies conducted in intact, actively growing *S. aureus* have shown that an oxazolidinone photoprobe was able to cross-link the universally conserved nucleotide A 2062 as well as large ribosomal subunit protein L27, both of which are associated with the peptidyl transferase center (64). The latter result is in accord with genetic studies of linezolid-resistant isolates, which place the mutations within the central loop of domain V of the ribosome, an essential component of the ribosomal peptidyl transferase (65–67). In fact, most of the mutations map to a small region of the ribosome adjacent to the nucleotides that comprise the P-site (68,69).

In light of the above evidence, it has been suggested that the oxazolidinones may interfere with the positioning of the initiator fMet-tRNA in the P-site of 70S ribosomes, thereby preventing the formation of the first peptide bond (64). The specific details of the binding will only become clear upon successful cocrystallization of linezolid or another oxazolidinone with the bacterial ribosome or its components.

Clinical Use

The in vitro antibacterial activity of linezolid against antibiotic-susceptible and -resistant gram-positive bacteria permits its use in infections caused by organisms such as MRSA, VRE, and multidrug-resistant *S. pneumoniae*. Linezolid is specifically indicated for the treatment of infections caused by linezolid-susceptible strains of vancomycin-resistant *E. faecium*, including bacteremia. It can be used in nosocomial pneumonia caused by *S. aureus* or *S. pneumoniae*, with combination therapy indicated if gram-negative organisms are present. It has been approved for the treatment of community-acquired pneumonia due to *S. pneumoniae* or MSSA. Linezolid has also been approved for use in skin infections, including uncomplicated skin and skin structure infections caused by MSSA or *S. pyogenes*, as well as complicated skin and skin structure infections, including diabetic foot infections, without concomitant osteomyelitis, due to *S. aureus*, *S. pyogenes*, or *S. agalactiae* (70).

In Vitro Susceptibility

The relevant clinical spectrum of linezolid is similar to that of quinupristin-dalfopristin, with activity primarily against gram-positive isolates, including antibiotic-susceptible and -resistant staphylococci, streptococci, E. faecium, and (in contrast to quinupristin–dalfopristin) E. faecalis species. Linezolid is also active in vitro against certain gram-negative strains, including M. catarrhalis and certain anaerobes and against mycobacteria and Nocardia species, although it is not clinically approved for these organisms. Linezolid is less potent against H. influenzae, limiting its utility for the empiric treatment of respiratory tract infections. DA-7867, a new oxazolidinone under development, demonstrates a similar spectrum of activity to linezolid, with MIC_{90} values that are at least 8- to 16-fold lower than linezolid against streptococci, staphylococci, enterococci, H. influenzae, and M. catarrhalis (Table 5); potency of DA-7867 against these latter two gram-negative species may be sufficient to support clinical efficacy.

Linezolid is similarly active against antibiotic-susceptible and -resistant E. faecium and E. faecalis isolates, including those expressing resistance to ciprofloxacin, doxycycline, ampicillin, and glycopeptides (Table 5). In a study by Zhanel et al. (28), the proportion of E. faecium strains nonsusceptible to linezolid was approximately 0.5% for vancomycin-resistant isolates and approximately 2.4% for vancomycin-susceptible isolates. All E. faecalis strains were susceptible to linezolid, regardless of their glycopeptide susceptibility.

In a recent evaluation of community-acquired and nosocomial-derived MSSA and MRSA isolates, linezolid demonstrated potent activity, regardless of the origin or antibiotic susceptibility of the strain (Table 5). Essentially all strains, including those nonsusceptible to penicillin, doxycycline, trimethoprim-sulfamethoxazole, and ciprofloxacin, were susceptible to linezolid, vancomycin, and quinupristin–dalfopristin. In another study with CoNS (Table 5), a similar pattern of activity was observed, with the activity of linezolid conserved regardless of the antibiotic resistance noted within the strain populations.

In recent studies (72,73), >99% of streptococci (including S. pneumoniae, S. pyogenes, and S. agalactiae) were susceptible to linezolid, including penicillin-, macrolide-, trimethoprim-sulfamethoxazole-, and levofloxacin-resistant strains (Table 5). Against these isolates, linezolid had a higher percentage of coverage than most of the other approved antibiotics evaluated.

Linezolid demonstrates potent activity in vitro against several clinically important facultatively anaerobic or anaerobic species, including Pasteurella multocida, Actinomyces spp., Peptostreptococcus spp., and Bacteroides fragilis (Table 5), although this compound is not approved for use against these organisms.

Resistance

To date, acquired resistance to linezolid is limited to mutational changes to the 23S rRNA, and mechanisms responsible for resistance to nonoxazolidinone protein synthesis inhibitors (including those for macrolides, lincosamides, streptogramins, tetracycline, chloramphenicol, and aminoglycosides) do not appear to confer cross-resistance to linezolid (61). Among gram-positive human pathogens, intrinsic resistance to linezolid is rare (77). Furthermore, data suggest that the emergence of resistance among relevant pathogens may be slow, presumably governed by the need to acquire simultaneous or stepwise mutations in the multiple copies of ribosomal RNA genes per genome. Two in vitro studies with staphylococcal strains determined that spontaneous resistance frequencies to linezolid were $<8 \times 10^{-11}$ and $<10^{-9}$, respectively (78,79). Additionally, one resistant E. faecium was obtained after five serial passages using the spiral gradient plate technique (78). In other linezolid-resistant S. aureus and enterococcal strains, associated mutations

TABLE 5 In Vitro Activities of the Oxazolidinones Linezolid and DA-7867 with Comparative Agents

Organism	Compound	N	MIC (μg/mL) Range	MIC$_{50}$	MIC$_{90}$	Percentage susceptible	Refs.
VS-*Enterococcus*	Vancomycin	121	2–4	2	2	100	28
faecalis	Teicoplanin	121	0.5	0.5	0.5	100	28
	Ampicillin	121	1–64	1	1	99.2	28
	Doxycycline	121	1–32	8	16	38	28
	Ciprofloxacin	121	1–64	2	64	47.9	28
	Linezolid	121	0.5–2	2	2	100	28
	DA-7867	49	≤0.06 to 0.12	0.12	0.12	NB[a]	71
VR-*E. faecalis*	Vancomycin	81	16–512	512	512	0.0	28
	Teicoplanin	81	0.5–128	64	128	43.2	28
	Ampicillin	81	1–128	1	2	97.5	28
	Doxycycline	81	1–32	8	16	35.8	28
	Ciprofloxacin	81	1–64	32	64	1.2	28
	Linezolid	81	1–2	1	2	100	28
	DA-7867	10	≤0.06 to 0.12	≤0.06	0.12	NB	71
VS-*Enterococcus*	Vancomycin	42	2	2	2	100	28
faecium	Teicoplanin	42	0.5–1	0.5	1	100	28
	Ampicillin	42	1–128	64	128	21.4	28
	Doxycycline	42	1–32	1	8	73.8	28
	Ciprofloxacin	42	1–64	64	64	9.5	28
	Linezolid	42	0.25–4	2	2	97.6	28
	DA-7867	29	≤0.06 to 0.12	0.12	0.12	NB	71
VR-*E. faecium*	Vancomycin	616	8–512	256	512	0.0	28
	Teicoplanin	616	0.5–128	32	32	22.2	28
	Ampicillin	616	1–256	64	128	3.2	28
	Doxycycline	616	1–32	4	16	60.7	28
	Ciprofloxacin	616	8–64	64	64	0.0	28
	Linezolid	616	0.25–16	1	2	99.5	28
	DA-7867	30	≤0.06 to 0.12	0.12	0.12	NB	71
Staphylococcus aureus[b]							
CA-MSSA	Vancomycin	1592	≤0.12 to 2	1	1	100	23
	Penicillin	1592	≤0.016 to >32	4	32	17.7	23
	Doxycycline	1592	≤0.5 to >4	≤0.5	≤0.5	99.0	23
	Ciprofloxacin	1592	≤0.25 to >4	0.25	1	91.3	23
	TMP-SFX[c]	1592	≤0.5 to >2	≤0.5	≤0.5	96.8	23
	Linezolid	1592	0.12–4	2	2	100	23
CA-MRSA	Vancomycin	652	0.5–4	1	1	100	23
	Penicillin	652	≤0.016 to >32	32	>32	0.5	23
	Doxycycline	652	≤0.5 to >4	≤0.5	2	96.1	23
	Ciprofloxacin	652	≤0.25 to >4	>4	>4	15.2	23
	TMP-SFX	652	≤0.5 to >2	≤0.5	≤0.5	92.2	23
	Linezolid	652	0.5–16	2	2	99.8	23
N-MSSA	Vancomycin	706	0.5–2	1	1	100	23
	Penicillin	706	≤0.016 to >32	4	32	17.4	23

(Continued)

TABLE 5 In Vitro Activities of the Oxazolidinones Linezolid and DA-7867 with Comparative Agents *(Continued)*

Organism	Compound	N	MIC (µg/mL) Range	MIC_{50}	MIC_{90}	Percentage susceptible	Refs.
	Doxycycline	706	≤0.5 to >4	≤0.5	≤0.5	99.0	23
	Ciprofloxacin	706	≤0.25 to >4	≤0.25	2	89.7	23
	TMP-SFX	706	≤0.5 to >2	≤0.5	≤0.5	96.6	23
	Linezolid	706	0.12–4	2	2	100	23
MSSA	DA-7867	33	0.12–0.25	0.25	0.25	NB	71
N-MRSA	Vancomycin	548	0.25–2	1	2	100	23
	Penicillin	548	0.12 to >32	32	>32	0.2	23
	Doxycycline	548	≤0.5 to >4	≤0.5	1	96.8	23
	Ciprofloxacin	548	≤0.25 to >4	>4	>4	6.0	23
	TMP-SFX	548	≤0.5 to >2	≤0.5	≤0.5	92.7	23
	Linezolid	548	0.5–4	2	2	100	23
MRSA	DA-7867	30	0.06–0.25	0.12	0.25	NB	71
VISA	Vancomycin	19	4–8	4	8	NR^d	24
	Daptomycin	19	1–8	2	8	NR	24
	Linezolid	19	1–8	2	4	NR	24
VRSA	Vancomycin	2	32, 1024	NA^e	NA	R^f	25, 26
	Daptomycin	2	0.5, 1	NA	NA	NB	25, 26
	Minocycline	2	0.12, 0.25	NA	NA	S^f	25, 26
	Linezolid	2	2, 1	NA	NA	S	25, 26
MS-CoNS	Vancomycin	109	NR	1	2	100	27
	Tetracycline	109	NR	≤4	>8	78.0	27
	TMP-SMX	109	NR	0.25	>1	82.4	27
	Linezolid	109	NR	2	2	100	27
	DA-7867	22	0.06 to 0.25	0.12	0.12	NB	71
MR-CoNS	Vancomycin	389	NR	2	2	100	27
	Tetracycline	389	NR	≤4	>8	75.8	27
	TMP-SMX	389	NR	>1	>1	38.6	27
	Linezolid	389	NR	1	2	100	27
	DA-7867	29	≤0.03 to 0.25	0.12	0.25	NB	71
Streptococcus	Penicillin	10,012	≤0.06 to 16	NR	2	64.6	72
pneumoniae	Erythromycin	10,012	≤0.06 to >256	NR	16	71.9	72
	Clindamycin	10,012	≤0.25 to >64	NR	≤0.25	91.5	72
	TMP-SMX	10,012	≤0.25 to >4	NR	>4	65.2	72
	Levofloxacin	10,012	≤0.012 to 256	NR	1	98.8	72
	Linezolid	10,012	≤0.06 to 4	NR	2	99.9	72
	DA-7867	22	≤0.008 to 0.12	0.03	0.03	NB	71
Streptococcus	Penicillin	4508	≤0.06 to 0.12	NR	≤0.06	100	72
pyogenes	Erythromycin	4508	≤0.06 to >256	NR	0.12	94.0	72
	Clindamycin	4508	≤0.25 to >64	NR	≤0.25	99.4	72
	TMP-SMX	4508	≤0.25 to >4	NR	0.5	NB	72
	Levofloxacin	4508	0.25–8	NR	1	99.9	72
	Linezolid	4508	≤0.06 to 2	NR	2	100	72
	DA-7867	15	≤0.008 to 0.12	0.06	0.12	NB	71

(Continued)

TABLE 5 In Vitro Activities of the Oxazolidinones Linezolid and DA-7867 with Comparative Agents *(Continued)*

Organism	Compound	N	MIC (μg/mL) Range	MIC$_{50}$	MIC$_{90}$	Percentage susceptible	Refs.
Streptococcus agalactiae	Penicillin	318	ND	0.03	0.06	100	73
	Erythromycin	318	ND	≤0.06	8	69.8	73
	Clindamycin	318	ND	≤0.06	>8	88.1	73
	Levofloxacin	318	ND	0.5	1	100	73
	Linezolid	318	ND	1	1	100	73
	DA-7867	15	≤0.008 to 0.12	0.06	0.12	NB	71
Haemophilus influenzae[g]	Ampicillin	2948[h]	≤0.12 to >16	0.25	>16	82.9	74
	Amox.-Clav.[i]	2948[h]	≤0.12 to 8	0.5	1	99.9	74
	Azithromycin	2948[h]	≤0.06 to >16	1	2	99.8	74
	Levofloxacin	2948[h]	≤0.008 to 2	0.015	0.015	100	74
	TMP-SMX	2948[h]	≤0.03 to >16	0.06	4	81.3	74
	Linezolid	2948[h]	0.06 to >8	8	>8	NB	74
	DA-7867	24	0.25–4	2	2	NB	71
Moraxella catarrhalis[g]	Ampicillin	1131	≤0.12 to >16	8	16	13	74
	Amox.-Clav.	1131	≤0.12 to 1	0.12	0.25	100	74
	Azithromycin	1131	≤0.06 to 0.25	≤0.06	≤0.06	100	74
	Levofloxacin	1131	0.015–1	0.03	0.03	100	74
	TMP-SMX	1131	≤0.03 to 4	0.25	0.5	97.8	74
	Linezolid	1131	0.25–8	4	4	NB	74
	DA-7867	24	0.25–8	0.5	0.5	NB	71
Pasteurella multocida subsp. *multocida*	Azithromycin	30	0.125–1	0.5	1	NB	75
	Amox.-Clav.	30	0.06–0.25	0.25	0.25	NB	75
	Linezolid	30	1–2	2	2	NB	75
Anaerobes							
Actinomyces spp.[g]	Vancomycin	22	0.5–1	0.5	1	NB	76
	Cefoxitin	22	≤0.03 to 1	0.125	0.5	NB	76
	Clindamycin	22	≤0.03 to 0.5	0.06	0.25	NB	76
	Metronidazole	22	≤0.03 to >128	32	>128	NB	76
	Linezolid	22	0.5–8	0.5	0.5	NB	76
Clostridium difficile[g]	Vancomycin	18	0.5–4	1	2	NB	76
	Cefoxitin	18	128 to >128	128	>128	NB	76
	Clindamycin	18	2 to >128	4	>128	NB	76
	Metronidazole	18	0.25–1	0.5	1	NB	76
	Linezolid	18	2–16	2	16	NB	76
Lactobacillus spp.[g]	Vancomycin	37	0.25 to >32	4	>32	NB	76
	Cefoxitin	37	≤0.06 to >128	64	>128	NB	76
	Clindamycin	37	≤0.03 to >128	0.06	2	NB	76
	Metronidazole	37	0.5 to >128	>128	>128	NB	76
	Linezolid	37	0.5–16	4	8	NB	76
Peptostreptococcus spp.[g]	Vancomycin	13	0.125–1	0.5	NR	NB	76
	Cefoxitin	13	≤0.03 to 8	0.5	4	NB	76
	Clindamycin	13	≤0.03 to 0.25	≤0.03	0.25	NB	76

(Continued)

TABLE 5   In Vitro Activities of the Oxazolidinones Linezolid and DA-7867 with Comparative Agents *(Continued)*

Organism	Compound	N	MIC (μg/mL)			Percentage susceptible	Refs.
			Range	MIC$_{50}$	MIC$_{90}$		
	Metronidazole	13	0.06 to >128	0.25	2	NB	76
	Linezolid	13	0.5–16	1	2	NB	76
Bacteroides	Vancomycin	17	16–128	64	128	NB	76
fragilis group[g]	Cefoxitin	17	4–128	32	64	NB	76
	Clindamycin	17	≤0.03 to >128	2	>128	NB	76
	Metronidazole	17	0.5–4	1	2	NB	76
	Linezolid	17	2–4	4	4	NB	76

[a]No breakpoints have been assigned.
[b]CA, community-acquired; N, nosocomial.
[c]Trimethroprim-sulfamethoxazole.
[d]Not reported.
[e]Not applicable.
[f]R, resistant; S, susceptible.
[g]Not listed in the Food and Drug Administration approved label.
[h]Includes 489 β-lactamase-positive isolates.
[i]Amoxicillin-clavulanic acid.
Abbreviations: MIC, minimum inhibitory concentration; MSSA, methicillin-susceptible *S. aureus*; MRSA, methicillin (oxacillin)-resistant *S. aureus*; VISA, vancomycin intermediate *S. aureus*; VR, vancomycin-resistant; VRSA, vancomycin-resistant *S. aureus*.

that mapped to domain V of the 23S rRNA included G2447U, G2505A, C2512U, G2513U, G2576U, and C2610G (*E. coli* numbering) (67,80).

Clinical resistance to linezolid, although uncommon, has been observed most frequently in VRE (81–83). In surveillance studies conducted in 2002, eight linezolid-resistant clinical isolates were identified from a population of 9833 grampositive strains (0.08%) (84), four were *E. faecium* and two were *E. faecalis*. Linezolid resistance was associated with a G2576U mutation in the 23S rRNA in those isolates where the mechanism was defined. Frequently, patients who develop resistance during linezolid treatment have indwelling devices and receive multiple weeks of therapy, thus allowing selection of resistant populations over time (85).

In *S. aureus*, linezolid-resistant MRSA strains were first reported from Boston and London (86–89). Consistent with results described for the enterococci, three of the linezolid-resistant MRSA strains had a G2576U 23S rRNA mutation (86,87,89). In a study describing several related linezolid-resistant MRSA isolates (88), resistance was associated with a T2500A mutation in two or three of the 23S rRNA genes, and in two isolates, with the concomitant loss of one of the six 23S rRNA genes. Interestingly, in this latter report, isolates exhibiting the linezolid-resistance genotype could not be detected in the patient following a seven-month evaluation period in the absence of linezolid, although the patient appeared to remain colonized by the original linezolid-susceptible MRSA strain. In another study, Meka et al. demonstrated that the G2576U mutation observed in four of five 23S rRNA copies was found only in two of five copies after 30 passages in antibiotic free medium, and in one of five 23S rRNA copies after 60 passages in antibiotic-free medium; linezolid MIC values decreased from 16 to 8 to 2 μg/mL, respectively (89). These results indicate that in some cases, mutations conferring clinical resistance to linezolid may be lost in the absence of linezolid exposure. Furthermore, this and other studies suggest that the MIC values of linezolid-resistant enterococcal or staphylococcal isolates increase proportionately with the number of 23S rRNA copies retaining the G2576U mutation (83,89,90).

Pharmacokinetics/Pharmacodynamics
PD Models
Linezolid PD studies have been conducted in neutropenic mice whose thighs were infected with one of eight *S. pneumoniae* strains of varying penicillin susceptibility, or each of four staphylococcal strains, two MSSA and two MRSA (91). When doses of 20 or 80 mg/kg were administered, linezolid produced only a bacteriostatic effect against *S. aureus* ($\leq$0.5 $\log_{10}$ reduction in bacterial load), with <2.0 $\log_{10}$ reduction in colony-forming units per thigh for the pneumococcal infections. Higher rates of killing were observed with escalating doses. No PAE was observed for *S. pneumoniae*, but PAEs of 3.2 to 3.4 hours were observed for the two doses in the staphylococcal infections. Little effect was seen on PAE regardless of whether free drug levels or total drug levels were considered. In the staphylococcal infection model, there was no clear discrimination among the three PD parameters related to efficacy: percentage time the plasma drug concentration exceeds the MIC, AUC/MIC, or peak drug concentration/MIC. Consideration of total drug levels compared to free drug levels did not change the conclusions. For the pneumococcal infection model, AUC/MIC gave the strongest relationship with efficacy. The 24-hour static doses against the pneumococcal strains were 18 to 72 mg/kg/day with 24-hour AUC/MIC ratios ranging from 22 to 97. Higher doses, 95 to 119 mg/kg/day, were required for a 24-hour static dose against the four *S. aureus* strains, with the AUC/MIC ratio ranging from 59 to 167. Based on compilations of the data, a PD goal was set for a 24-hour AUC/MIC (static effect) of 82.9 $\pm$ 57 against *S. aureus* and 48.3 $\pm$ 29 for *S. pneumoniae*.

PD efficacy of linezolid was also determined in an immunocompetent rat pneumococcal pneumonia model (92). In this study, only two doses, 25 and 50 mg/kg dosed twice daily, were evaluated, following infection with a standard *S. pneumoniae* strain. Survival was monitored, as was the bacterial count in bronchoalveolar lavage fluid postinfection. The results from these determinations were reasonably consistent with the Andes et al. study, with linezolid free-fraction PD parameters reported as 147 for AUC/MIC, 24.3 for C_{max}/MIC, and 39% time above the MIC. None of these parameters were clearly preferable as the defining PD parameter predictive of outcome.

Clinical PKs
PK parameters determined from Phase I clinical studies with single- and multiple-dose oral or intravenous linezolid are summarized in Tables 6 and 7 (93).

In these studies men and nonpregnant women from 18 to 55 years old were randomized between placebo and oral (375, 500 or 625 mg) or intravenous (500 or 625 mg) linezolid. Active compound was dosed once for single-dose data and then twice a day beginning on day 2 to obtain steady-state data. The oral doses were selected because of the availability of drug in 125 mg tablets, and approximated the approved clinical doses of 400 and 600 mg, dosed twice a day. Single-dose data indicated rapid absorption with peak linezolid concentrations at one to two hours after oral dosing and immediately after the end of infusion for intravenous dosing. Both C_{max} and AUC increased proportionally with the oral and intravenous doses. An absolute bioavailability of 103% was determined. Mean steady-state parameters

TABLE 6 Pharmacokinetic Properties of Linezolid Following a Single Oral or Intravenous Dose in Healthy Adult Volunteers[a]

	Oral dosing			Intravenous dosing	
Parameter[b]	375 mg	500 mg	625 mg	500 mg	625 mg
C_{max} (mg/L)	8.21 ± 2.07	10.4 ± 253	12.7 ± 3.36	11.7 ± 2.3	13.4 ± 1.73
t_{max} (hr)	1.67 ± 0.88	1.38 ± 0.92	1.33 ± 0.61	0.5 ± 0	0.5 ± 0.10
AUC_{0-12} (mg/L hr)	–	–	–	63.2 ± 14.3	79.2 ± 227.8
$AUC_{0-\infty}$ (mg/L hr)	65.5 ± 24.9	74.3 ± 27.9	102 ± 29.7	65.8 ± 17	83.6 ± 34.7
$t_{1/2}$ (hr)	4.98 ± 1.18	4.59 ± 1.83	4.87 ± 1.44	4.68 ± 1.66	4.42 ± 2.36
λz (hr^{-1})	0.15 ± 0.051	0.17 ± 0.069	0.15 ± 0.049	0.16 ± 0.052	0.194 ± 0.083
CL (mL/min)	108 ± 42.1	125 ± 43.3	112 ± 45.2	133 ± 32	138 ± 38.7
V_d (L)	44.3 ± 14.4	45 ± 10.8	45 ± 13.9	47.3 ± 9.74	46 ± 11.2

[a]Mean values with standard deviations are reported (93).
[b]C_{max}, maximum plasma drug concentration; t_{max}, time at which C_{max} occurred; AUC_{0-12}, area under the plasma concentration curve from zero to 12 hr; $AUC_{0-\infty}$, area under the plasma concentration–time curve from zero to infinity; $t_{1/2}$, half-life; λz, terminal elimination rate constant; CL, apparent oral clearance; V_d, volume of distribution at steady state (V_{ss}).

were obtained after two days of dosing, with trough concentrations of linezolid ranging from 3.5 to 8.0 µg/mL. A target MIC of 4 µg/mL would cover most staphylococcal and enterococcal strains; this concentration was exceeded at steady state for 10 to 16 hours following oral dosing, and for 9 to 10 hours after intravenous administration. Steady-state elimination half-lives were independent of dose and were similar regardless of the route of administration (4.8–5.7 hours). When these data were combined with the in vitro PAE of 1.8 to 3.0 hours (94), clinical doses of 400 and 600 mg b.i.d. were fully supported. In the linezolid Phase I studies, considerable variability was observed in clearance values, after both oral and intravenous dosing. It was determined that this was due to a high variability in nonrenal clearance, proposed to be due to nonenzymatic metabolism of linezolid (93).

In healthy pediatric populations, the disposition of linezolid was found to be age-dependent, with total body clearance and volume of distribution significantly

TABLE 7 Steady-State Pharmacokinetic Properties of Linezolid Following Multiple Oral and Intravenous Doses in Healthy Adult Volunteers[a]

	Oral dosing			Intravenous dosing	
Parameter[b]	375 mg b.i.d.	500 mg b.i.d.	625 mg b.i.d.	500 mg b.i.d.	625 mg b.i.d.
C_{max} (mg/L)	13.1 ± 2.9	15.3 ± 3.72	18.8 ± 6.24	14.4 ± 3.1	15.7 ± 2.6
C_{min} (mg/L)	3.9 ± 1.85	5.04 ± 2.38	8.02 ± 3.63	3.51 ± 1.36	3.84 ± 2.46
t_{max} (hr)	1.0 ± 0.32	1.25 ± 0.99	2.12 ± 1.12	0.51 ± 0.03	0.51 ± 0.03
AUC_{0-12} (mg/L hr)	82.8 ± 22.6	99.2 ± 36.7	147 ± 58	81.2 ± 19.6	93.4 ± 32.3
$t_{1/2}$ (hr)	5.4 ± 1.0	5.7 ± 2.0	5.4 ± 0.90	5.6 ± 1.3	4.8 ± 1.7
λz (hr^{-1})	0.132 ± 0.023	0.135 ± 0.048	0.13 ± 0.022	0.128 ± 0.028	0.158 ± 0.047
CL (L/hr/kg)	80.6 ± 23.0	91.5 ± 25.3	78.2 ± 23.3	108 ± 28.7	123 ± 40.3
V_{ss} (L)	37.7 ± 7.91	43.3 ± 15.4	36.1 ± 10.5	45 ± 7.7	45.5 ± 4.87

[a]Mean values with standard deviations are reported (93). Fasted subjects were dosed orally once on day 1 followed by twice-a-day beginning on day 2 for 14.5 days. Intravenous doses were infused for 30 min on day 1 and then every 12 hr for 7.5 days beginning on day 2.
[b]C_{max}, maximum plasma drug concentration; C_{min}, minimum (trough) plasma drug concentration; t_{max}, time at which C_{max} occurred; AUC_{0-12}, area under the plasma concentration curve from zero to 12 hr; $t_{1/2}$, half-life; λz, terminal elimination rate constant; CL, apparent oral clearance; V_{ss}, volume of distribution at steady state.

greater than that for adults (95). Twelve hours after a single 10 mg/kg dose, plasma concentrations were below the MIC_{90} value for selected pathogens that would be of importance. These data support multiple doses per 24-hour period. Linezolid half-lives were shorter in pediatric patients up to the age of 11 years ($t_{1/2}$ of 2.9 hours) compared to adults or adolescents ($t_{1/2}$ of 4.1 hours). More rapid clearance was also observed in the younger population, 3.8 mL/min/kg for the 0 to 11-year-age group compared to a clearance of 2.1 mL/min/kg for adolescents (96).

Drug–drug interaction studies were conducted with various agents that might be coadministered with linezolid. Because linezolid is a weak monoamine oxidase (MAO) inhibitor, its PKs were studied in the presence of common over-the-counter medications that also interact with MAO (97). When linezolid was coadministered with sympathomimetics (pseudoephedrine and phenylpropanolamine), the plasma concentrations of these latter agents were increased minimally, but at a level of statistical significance. Plasma concentrations of dextrorphan, the primary metabolite of the serotonin reuptake inhibitor dextromethorphan, were decreased minimally, but also statistically significantly, after coadministration of linezolid and dextromethorphan (97). None of the drugs affected the PKs of linezolid. Concomitant dosing of linezolid (600 mg) with the antioxidant vitamin C (1000 mg) or vitamin E (800 IU) showed that none of the linezolid PK parameters was significantly affected by the vitamins (98). This study demonstrated that the antioxidant vitamins did not affect the overall reactive oxygen species balance, nor did they affect the in vivo clearance of linezolid.

In clinical trials, empiric therapy included the combination of aztreonam, an antibacterial agent targeted against gram-negative pathogens only, and linezolid with a purely gram-positive spectrum. In a single-dose randomized study, healthy subjects received a single 30 minute intravenous infusion of either linezolid alone (375 mg) or aztreonam alone (1000 mg), or the combination of 375 mg linezolid with 1000 mg of aztreonam (99). After analysis of the PK parameters, the only statistically significant differences were an 18% increase in linezolid peak concentrations and a 7% decrease in the apparent elimination rate of aztreonam. These differences were not judged to be clinically significant, and no dosing adjustments were made in clinical studies.

Special patient populations have been studied to determine the PK and PD parameters of linezolid. Examples include hospitalized children with community-acquired pneumonia (100), a VRE-infected patient undergoing dialysis (101), cystic fibrosis patients (102), patients with serious illness who were enrolled in compassionate use programs (103), and patients with hepatic and renal dysfunction (104). When 66 children aged 1 to 12 years were dosed with 10 mg/kg linezolid every 12 hours to treat community-acquired pneumonia, mean peak and trough concentrations were 9.5 ± 4.8 and $0.8 \pm 1.2\,\mu g/mL$, respectively (100). In this study, a 92% clinical cure was reported, indicating that the drug was effective in treating serious pediatric infections. In the dialysis patient with *E. faecium* peritonitis, linezolid concentrations in the peritoneal dialysis fluid tended to increase with treatment, with a mean concentration of $7.6\,\mu g/mL$, well above the susceptibility breakpoint of $4\,\mu g/mL$ (101). A second study of subjects on hemodialysis concluded that no adjustment of linezolid dosing was required (105). However, in a study of 12 adult cystic fibrosis patients (102), the PD target of AUC/MIC of 83 was not achieved in any of the patients after a single 600 mg dose of linezolid (range 16.3–52.2, mean of 28.1). The recommendation was made that this patient population should be dosed more frequently than twice-a-day. Studies of patients enrolled under the compassionate use program showed that higher success rates were observed when AUC/MIC values ranged from 80 to 120 for bacteremia, lower respiratory tract infection, and skin infections, and when the linezolid concentration remained above the MIC during the full dosing interval (103). In patients with hepatic dysfunction (104) or renal failure (105), the overall linezolid PK parameters were not significantly changed to require dosage adjustments.

Newer Oxazolidinones

DA-7867 is an investigational oxazolidinone that was shown to have 70.8% oral bioavailability in rats (106). PK parameters were dose-independent, with approximately 22% of the drug eliminated via intestinal first-pass effect. Following two-week oral dosing at 2 mg/kg/day, DA-7867 accumulated in rats with the AUC increasing from 1430 µg/min/mL following a single 2 mg/kg dose to 1880 µg/min/mL after multiple dosing. Tissue to plasma levels were less than 1.0, indicating low accumulation of drug in the tissues (107).

REFERENCES

1. Low DE. Quinupristin/dalfopristin: spectrum of activity, pharmacokinetics, and initial clinical experience. Microb Drug Resist 1995; 1:223–234.
2. Allington DR, Rivey MP. Quinupristin/dalfopristin: a therapeutic review. Clin Ther 2001; 23:24–44.
3. Bonfiglio G, Furneri PM. Novel streptogramin antibiotics. Expert Opin Invest Drugs 2001; 10:185–198.
4. Stevens DL, Dotter B, Madaras-Kelly K. A review of linezolid: the first oxazolidinone antibiotic. Expert Rev Antiinfect Ther 2004; 2:51–59.
5. Di Giambattista M, Cocito C. Action of ions and pH on the binding of virginiamycin S to ribosomes. Biochim Biophys Acta 1983; 757:92–100.
6. Barriere JC, Paris JM. From the Michael reaction to the clinic. R Soc Chem Spec Publ 1997; 198:27–41.
7. Etienne SD, Montay G, Le Liboux A, Frydman A, Garaud JJ. A Phase I, double-blind, placebo-controlled study of the tolerance and pharmacokinetic behaviour of RP 59500. J Antimicrob Chemother 1992; 30:123–131.
8. Beyer D, Pepper K. The streptogramin antibiotics: update on their mechanism of action. Expert Opin Invest Drugs 1998; 7:591–599.
9. Chinali G, Moureau P, Cocito CG. The action of virginiamycin M on the acceptor, donor, and catalytic sites of peptidyltransferase. J Biol Chem 1984; 259:9563–9568.
10. Chinali G, Di Giambattista M, Cocito C. Ribosome protection by tRNA derivatives against inactivation by virginiamycin M: evidence for two types of interaction of tRNA with the donor site of peptidyl transferase. Biochemistry 1987; 26:1592–1597.
11. Parfait R, Cocito C. Lasting damage to bacterial ribosomes by reversibly bound virginiamycin M. Proc Natl Acad Sci U S A 1980; 77:5492–5496.
12. Chinali G, Nyssen E, Di Giambattista M, Cocito C. Inhibition of polypeptide synthesis in cell-free systems by virginiamycin S and erythromycin. Evidence for a common mode of action of type B synergimycins and 14-membered macrolides. Biochim Biophys Acta 1988; 949:71–78.
13. Moureau P, Engelborghs Y, Di Giambattista M, Cocito C. Fluorescence stopped flow analysis of the interaction of virginiamycin components and erythromycin with bacterial ribosomes. J Biol Chem 1983; 258:14233–14238.
14. Weisblum B. Inducible resistance to macrolides, lincosamides, and streptogramin type B antibiotics: the resistance phenotype, its biological diversity, and structural elements that regulate expression—a review. J Antimicrob Chemother 1985; 16:63–90.
15. Neu HC, Chin NX, Gu JW. The in-vitro activity of new streptogramins, RP 59500, RP 57669 and RP 54476, alone and in combination. J Antimicrob Chemother 1992; 30:83–94.

16. Delgado G Jr, Neuhauser MM, Bearden DT, Danziger LH. Quinupristin-dalfopristin: an overview. Pharmacotherapy 2000; 20:1469–1485.

17. Canu A, Leclercq R. Overcoming bacterial resistance by dual target inhibition: the case of streptogramins. Curr Drug Targets—Infect Disord 2001; 1:215–225.

18. Hansen JL, Moore PB, Steitz TA. Structures of five antibiotics bound at the peptidyl transferase center of the large ribosomal subunit. J Mol Biol 2003; 330:1061–1075.

19. Harms JM, Schlunzen F, Fucini P, Bartels H, Yonath A. Alterations at the peptidyl transferase centre of the ribosome induced by the synergistic action of the streptogramins dalfopristin and quinupristin. BMC Biol 2004; 2:4. (http://www.biomedcentral.com/1741–7007/2/4)

20. Synercid (quinupristin-dalfopristin). King Pharmaceuticals. PDR Electronic Library, Thomson Micromedex. Greenwood Village, Colorado (2005).

21. Clarebout G, Nativelle E, Bozdogan B, Villers C, Leclercq R. Bactericidal activity of quinupristin-dalfopristin against strains of *Staphylococcus aureus* with the MLS(B) phenotype of resistance according to the *erm* gene type. Int J Antimicrob Agents 2004; 24:444–449.

22. Pankuch GA, Kelly LM, Lin G, et al. Activities of a new oral streptogramin, XRP 2868, compared to those of other agents against *Streptococcus pneumoniae* and *Haemophilus* species. Antimicrob Agents Chemother 2003; 47:3270–3274.

23. Fritsche TR, Jones RN. Antimicrobial activity of tigecycline (GAR-936) tested against 3498 recent isolates of *Staphylococcus aureus* recovered from nosocomial and community-acquired infections. Int J Antimicrob Agents 2004; 24:567–571.

24. Jevitt LA, Smith AJ, Williams PP, et al. In vitro activities of daptomycin, linezolid, and quinupristin-dalfopristin against a challenge panel of staphylococci and enterococci, including vancomycin-intermediate *Staphylococcus aureus* and vancomycin-resistant *Enterococcus faecium*. Microb Drug Resist 2003; 9:389–393.

25. Chang S, Sievert DM, Hageman JC, et al. Infection with vancomycin-resistant *Staphylococcus aureus* containing the vanA resistance gene. N Engl J Med 2003; 348:1342–1347.

26. Tenover FC, Weigel LM, Appelbaum PC, et al. Vancomycin-resistant *Staphylococcus aureus* isolate from a patient in Pennsylvania. Antimicrob Agents Chemother 2004; 48:275–280.

27. Gales AC, Andrade SS, Sader HS, Jones RN. Activity of mupirocin and 14 additional antibiotics against staphylococci isolated from Latin American hospitals: report from the SENTRY antimicrobial surveillance program. J Chemother 2004; 16:323–328.

28. Zhanel GG, Laing NM, Nichol KA, et al. Antibiotic activity against urinary tract infection (UTI) isolates of vancomycin-resistant enterococci: results from the 2002 North American Vancomycin Resistant Enterococci Susceptibility study (NAVRESS). J Antimicrob Chemother 2003; 52:382–388.

29. Fluit AC, Schmitz FJ, Verhoef J, Milatovic D. Daptomycin in vitro susceptibility in European gram-positive clinical isolates. Int J Antimicrob Agents 2004; 24:59–66.

30. Jones RN, Biedenbach DJ. Comparative activity of garenoxacin (BMS 284756), a novel desfluoroquinolone, tested against 8,331 isolates from community-acquired respiratory tract infections: North American results from the SENTRY Antimicrobial Surveillance Program (1999–2001). Diagn Microbiol Infect Dis 2003; 45:273–278.

31. Hershberger E, Donabedian S, Konstantinou K, Zervos MJ. Quinupristin-dalfopristin resistance in gram-positive bacteria: mechanism of resistance and epidemiology. Clin Infect Dis 2004; 38:92–98.

32. Farrell DJ, Morrissey I, Bakker S, Buckridge S, Felmingham D. In vitro activities of telithromycin, linezolid, and quinupristin-dalfopristin against *Streptococcus pneumoniae* with macrolide resistance due to ribosomal mutations. Antimicrob Agents Chemother 2004; 48:3169–3171.

33. Bozdogan B, Leclercq R. Effects of genes encoding resistance to streptogramins A and B on the activity of quinupristin-dalfopristin against *Enterococcus faecium*. Antimicrob Agents Chemother 1999; 43:2720–2725.

34. Malbruny B, Canu A, Bozdogan B, et al. Resistance to quinupristin-dalfopristin due to mutation of L22 ribosomal protein in *Staphylococcus aureus*. Antimicrob Agents Chemother 2002; 46:2200–2207.

35. Tait-Kamradt A, Davies T, Appelbaum PC, et al. Two new mechanisms of macrolide resistance in clinical strains of *Streptococcus pneumoniae* from Eastern Europe and North America. Antimicrob Agents Chemother 2000; 44:3395–3401.
36. Johnston NJ, Mukhtar TA, Wright GD. Streptogramin antibiotics: mode of action and resistance. Curr Drug Targets 2002; 3:335–344.
37. Wang G, Taylor DE. Site-specific mutations in the 23S rRNA gene of *Helicobacter pylori* confer two types of resistance to macrolide-lincosamide-streptogramin B antibiotics. Antimicrob Agents Chemother 1998; 42:1952–1958.
38. Roberts MC, Sutcliffe J, Courvalin P, Jensen LB, Rood J, Seppala H. Nomenclature for macrolide and macrolide-lincosamide-streptogramin B resistance determinants. Antimicrob Agents Chemother 1999; 43:2823–2830.
39. Singh KV, Weinstock GM, Murray BE. An *Enterococcus faecalis* ABC homologue (Lsa) is required for the resistance of this species to clindamycin and quinupristin-dalfopristin. Antimicrob Agents Chemother 2002; 46:1845–1850.
40. Lina G, Quaglia A, Reverdy M-E, Leclercq R, Vandenesch F, Etienne J. Distribution of genes encoding resistance to macrolides, lincosamides, and streptogramins among staphylococci. Antimicrob Agents Chemother 1999; 43:1062–1066.
41. Werner G, Hildebrandt B, Witte W. The newly described msrC gene is not equally distributed among all isolates of *Enterococcus faecium*. Antimicrob Agents Chemother 2001; 45:3672–3673.
42. Portillo A, Ruiz-Larrea F, Zarazaga M, Alonso A, Martinez JL, Torres C. Macrolide resistance genes in *Enterococcus* spp. Antimicrob Agents Chemother 2000; 44:967–971.
43. Kim HB, Lee B, Jang H-C, et al. A high frequency of macrolide-lincosamide-streptogramin resistance determinants in *Staphylococcus aureus* isolated in South Korea. Microb Drug Resist 2004; 10:248–254.
44. Rende-Fournier R, Leclercq R, Galimand M, Duval J, Courvalin P. Identification of the satA gene encoding a streptogramin A acetyltransferase in *Enterococcus faecium* BM4145. Antimicrob Agents Chemother 1993; 37:2119–2125.
45. Sugantino M, Roderick SL. Crystal structure of Vat(D): an acetyltransferase that inactivates streptogramin group A antibiotics. Biochemistry 2002; 41:2209–2216.
46. Mukhtar TA, Koteva KP, Hughes DW, Wright GD. Vgb from *Staphylococcus aureus* inactivates streptogramin B antibiotics by an elimination mechanism not hydrolysis. Biochemistry 2001; 40:8877–8886.
47. Andes D, Craig WA. Animal model pharmacokinetics and pharmacodynamics: a critical review. Int J Antimicrob Agents 2002; 19:261–268.
48. Carbon C. Quinupristin/dalfopristin: a review of its activity in experimental animal models of infection. J Antimicrob Chemother 1997; 39:115–119.
49. Craig WA. Pharmacokinetic/pharmacodynamic parameters: rationale for antibacterial dosing of mice and men. Clin Infect Dis 1998; 26:1–10.
50. Bearden DT. Clinical pharmacokinetics of quinupristin/dalfopristin. Clin Pharmacokinet 2004; 43:239–252.
51. Lomaestro BM, Briceland LL. Streptogramins and their potential role in geriatric medicine. Drug Aging 1998; 13:443–465.
52. Bergeron M, Montay G. The pharmacokinetics of quinupristin/dalfopristin in laboratory animals and in humans. J Antimicrob Chemother 1997; 39:129–138.
53. Chevalier P, Rey J, Pasquier O, et al. Pharmacokinetics of quinupristin/dalfopristin in patients with severe chronic renal insufficiency. Clin Pharmacokinet 2000; 39:77–84.
54. Johnson CA, Taylor CA 3rd, Zimmerman SW, et al. Pharmacokinetics of quinupristin-dalfopristin in continuous ambulatory peritoneal dialysis patients. Antimicrob Agents Chemother 1999; 43:152–156.
55. Hutchinson DK. Recent advances in oxazolidinone antibacterial agent research. Expert Opin Ther Pat 2004; 14:1309–1328.
56. Gregory WA, Brittelli DR, Wang CLJ, et al. Antibacterials. Synthesis and structure-activity studies of 3-aryl-2-oxooxazolidines. 1. The B group. J Med Chem 1989; 32:1673–1681.

57. Eustice DC, Feldman PA, Slee AM. The mechanism of action of DuP 721, a new antibacterial agent: effects on macromolecular synthesis. Biochem Biophys Res Commun 1988; 150:965–971.
58. Buysse JM, Demyan WF, Dunyak DS, Stapert D, Hamel JC, Ford CW. Mutation of the AcrAB antibiotic efflux pump in *Escherichia coli* confers susceptibility to oxazolidinone antibiotics. 36th Interscience Conference on Antimicrobial Agents and Chemotherapy, New Orleans, LA, 1996.
59. Shinabarger DL, Marotti KR, Murray RW, et al. Mechanism of action of oxazolidinones: effects of linezolid and eperezolid on translation reactions. Antimicrob Agents Chemother 1997; 41:2132–2136.
60. Lin AH, Murray RW, Vidmar TJ, Marotti KR. The oxazolidinone eperezolid binds to the 50S ribosomal subunit and competes with binding of chloramphenicol and lincomycin. Antimicrob Agents Chemother 1997; 41:2127–2131.
61. Fines M, Leclercq R. Activity of linezolid against Gram-positive cocci possessing genes conferring resistance to protein synthesis inhibitors. J Antimicrob Chemother 2000; 45:797–802.
62. Swaney SM, Aoki H, Ganoza MC, Shinabarger DL. The oxazolidinone linezolid inhibits initiation of protein synthesis in bacteria. Antimicrob Agents Chemother 1998; 42:3251–3255.
63. Matassova NB, Rodnina MV, Endermann R, et al. Ribosomal RNA is the target for oxazolidinones, a novel class of translational inhibitors. RNA 1999; 5:939–946.
64. Colca JR, McDonald WG, Waldon DJ, et al. Cross-linking in the living cell locates the site of action of oxazolidinone antibiotics. J Biol Chem 2003; 278:21972–21979.
65. Kloss P, Xiong L, Shinabarger DL, Mankin AS. Resistance mutations in 23 S rRNA identify the site of action of the protein synthesis inhibitor linezolid in the ribosomal peptidyl transferase center. J Mol Biol 1999; 294:93–101.
66. Xiong L, Kloss P, Douthwaite S, et al. Oxazolidinone resistance mutations in 23S rRNA of *Escherichia coli* reveal the central region of domain V as the primary site of drug action. J Bacteriol 2000; 182:5325–5331.
67. Prystowsky J, Siddiqui F, Chosay J, et al. Resistance to linezolid: characterization of mutations in rRNA and comparison of their occurrences in vancomycin-resistant enterococci. Antimicrob Agents Chemother 2001; 45:2154–2156.
68. Aoki H, Ke L, Poppe SM, et al. Oxazolidinone antibiotics target the P site on *Escherichia coli* ribosomes. Antimicrob Agents Chemother 2002; 46:1080–1085.
69. Bobkova EV, Yan YP, Jordan DB, Kurilla MG, Pompliano DL. Catalytic properties of mutant 23 S ribosomes resistant to oxazolidinones. J Biol Chem 2003; 278:9802–9807.
70. Zyvox (linezolid). Pfizer. PDR Electronic Library, Thomson Micromedex. Greenwood Village, Colorado (2005).
71. Yong D, Yum JH, Lee K, Chong Y, Choi SH, Rhee JK. In vitro activities of DA-7867, a novel oxazolidinone, against recent clinical isolates of aerobic and anaerobic bacteria. Antimicrob Agents Chemother 2004; 48:352–357.
72. Brown SD, Rybak MJ. Antimicrobial susceptibility of *Streptococcus pneumoniae*, *Streptococcus pyogenes*, and *Haemophilus influenzae* collected from patients across the USA, in 2001–2002, as part of the PROTEKT US study. J Antimicrob Chemother 2004; 54:i7–i15.
73. Biedenbach DJ, Stephen JM, Jones RN. Antimicrobial susceptibility profile among β-hemolytic *Streptococcus* spp. collected in the SENTRY antimicrobial surveillance program-North America, 2001. Diagn Microbiol Infect Dis 2003; 46:291–294.
74. Hoban D, Felmingham D. The PROTEKT surveillance study: antimicrobial susceptibility of *Haemophilus influenzae* and *Moraxella catarrhalis* from community-acquired respiratory tract infections. J Antimicrob Chemother 2002; 50:49–59.
75. Goldstein EJC, Citron DM, Merriam CV. Linezolid activity compared to those of selected macrolides and other agents against aerobic and anaerobic pathogens isolated from soft tissue bite infections in humans. Antimicrob Agents Chemother 1999; 43:1469–1474.
76. Citron DM, Merriam CV, Tyrrell KL, Warren YA, Fernandez H, Goldstein EJ. In vitro activities of ramoplanin, teicoplanin, vancomycin, linezolid, bacitracin, and four other

antimicrobials against intestinal anaerobic bacteria. Antimicrob Agents Chemother 2003; 47:2334–2338.

77. Livermore DM. Linezolid in vitro: mechanism and antibacterial spectrum. J Antimicrob Chemother 2003; 51(suppl 2):ii9–ii16.

78. Zurenko GE, Yagi BH, Schaadt RD, et al. In vitro activities of U-100592 and U-100766, novel oxazolidinone antibacterial agents. Antimicrob Agents Chemother 1996; 40: 839–845.

79. Kaatz GW, Seo SM. In vitro activities of oxazolidinone compounds U100592 and U100766 against *Staphylococcus aureus* and *Staphylococcus epidermidis*. Antimicrob Agents Chemother 1996; 40:799–801.

80. Swaney SM, Shinabarger DL, Schaadt RD, Bock JH, Slightom JL, Zurenko GE. Oxazolidinone resistance is associated with a mutation in the peptidyl transferase region of 23S rRNA (abstr). 38th Interscience Conference on Antimicrobial Agents and Chemotherapy, San Diego, CA, 1998.

81. Herrero IA, Issa NC, Patel R. Nosocomial spread of linezolid-resistant, vancomycin-resistant *Enterococcus faecium*. N Engl J Med 2002; 346:867–869.

82. Jones RN, Della-Latta P, Lee LV, Biedenbach DJ. Linezolid-resistant *Enterococcus faecium* isolated from a patient without prior exposure to an oxazolidinone: report from the SENTRY Antimicrobial Surveillance Program. Diagn Microbiol Infect Dis 2002; 42:137–139.

83. Ruggero KA, Schroeder LK, Schreckenberger PC, Mankin AS, Quinn JP. Nosocomial superinfections due to linezolid-resistant *Enterococcus faecalis*: evidence for a gene dosage effect on linezolid MICs. Diagn Microbiol Infect Dis 2003; 47:511–513.

84. Mutnick AH, Enne V, Jones RN. Linezolid resistance since 2001: SENTRY Antimicrobial Surveillance Program. Ann Pharmacother 2003; 37:769–774.

85. Moellering RC. Linezolid: the first oxazolidinone antimicrobial. Ann Intern Med 2003; 138:135–142.

86. Tsiodras S, Gold HS, Sakoulas G, et al. Linezolid resistance in a clinical isolate of *Staphylococcus aureus*. Lancet 2001; 358:207–208.

87. Wilson P, Andrews JA, Charlesworth R, et al.. Linezolid resistance in clinical isolates of *Staphylococcus aureus*. J Antimicrob Chemother 2003; 51:186–188.

88. Meka VG, Pillai SK, Sakoulas G, et al. Linezolid resistance in sequential *Staphylococcus aureus* isolates associated with a T2500A mutation in the 23S rRNA gene and loss of a single copy of rRNA. J Infect Dis 2004; 190:311–317.

89. Meka VG, Gold HS, Cooke A, et al. Reversion to susceptibility in a linezolid-resistant clinical isolate of *Staphylococcus aureus*. J Antimicrob Chemother 2004; 54:818–820.

90. Marshall SH, Donskey CJ, Hutton-Thomas R, Salata RA, Rice LB. Gene dosage and linezolid resistance in *Enterococcus faecium* and *Enterococcus faecalis*. Antimicrob Agents Chemother 2002; 46:3334–3336.

91. Andes D, van Ogtrop ML, Peng J, Craig WA. In vivo pharmacodynamics of a new oxazolidinone (linezolid). Antimicrob Agents Chemother 2002; 46:3484–3489.

92. Gentry-Nielsen MJ, Olsen KM, Preheim LC. Pharmacodynamic activity and efficacy of linezolid in a rat model of pneumococcal pneumonia. Antimicrob Agents Chemother 2002; 46:1345–1351.

93. Stalker DJ, Jungbluth GL, Hopkins NK, Batts DH. Pharmacokinetics and tolerance of single- and multiple-dose oral or intravenous linezolid, an oxazolidinone antibiotic, in healthy volunteers. J Antimicrob Chemother 2003; 51:1239–1246.

94. Munckhof WJ, Giles C, Turnidge JD. Post-antibiotic growth suppression of linezolid against Gram-positive bacteria. J Antimicrob Chemother 2001; 47:879–883.

95. Kearns GL, Abdel-Rahman SM, Blumer JL, et al; Pediatric Pharmacology Research Unit N. Single dose pharmacokinetics of linezolid in infants and children. Pediatr Infect Dis J 1178; 19:1178–1184.

96. Lyseng-Williamson KA, Goa KL. Linezolid: in infants and children with severe Gram-positive infections. Paediatr Drugs 2003; 5:419–429.

97. Hendershot PE, Antal EJ, Welshman IR, Batts DH, Hopkins NK. Linezolid: pharmacokinetic and pharmacodynamic evaluation of coadministration with pseudoephedrine

HCl, phenylpropanolamine HCl, and dextromethorphan HBr. J Clin Pharmacol 2001; 41:563–572.

98. Gordi T, Tan LH, Hong C, et al. The pharmacokinetics of linezolid are not affected by concomitant intake of the antioxidant vitamins C and E. J Clin Pharmacol 2003; 43:1161–1167.

99. Sisson TL, Jungbluth GL, Hopkins NK. A pharmacokinetic evaluation of concomitant administration of linezolid and aztreonam. J Clin Pharmacol 1999; 39:1277–1282.

100. Kaplan SL, Patterson L, Edwards KM, et al; Linezolid Pediatric Pheumonia Study Group P, Pharmacia and Upjohn. Linezolid for the treatment of community-acquired pneumonia in hospitalized children. Pediatr Infect Dis J 2001; 20:488–494.

101. DePestel DD, Peloquin CA, Carver PL. Peritoneal dialysis fluid concentrations of linezolid in the treatment of vancomycin-resistant *Enterococcus faecium* peritonitis. Pharmacotherapy 2003; 23:1322–1326.

102. Bosso JA, Flume PA, Gray SL. Linezolid pharmacokinetics in adult patients with cystic fibrosis. Antimicrob Agents Chemother 2004; 48:281–284.

103. Rayner CR, Forrest A, Meagher AK, Birmingham MC, Schentag JJ. Clinical pharmaco-dynamics of linezolid in seriously ill patients treated in a compassionate use pro-gramme. Clin Pharmacokinet 2003; 42:1411–1423.

104. Stalker DJ, Jungbluth GL. Clinical pharmacokinetics of linezolid, a novel oxazolidi-none antibacterial. Clin Pharmacokinet 2003; 42:1129–1140.

105. Brier ME, Stalker DJ, Aronoff GR, et al. Pharmacokinetics of linezolid in subjects with renal dysfunction. Antimicrob Agents Chemother 2003; 47:2775–2780.

106. Bae SK, Chung WS, Kim EJ, et al. Pharmacokinetics of DA-7867, a new oxazolidinone, after intravenous or oral administration to rats: intestinal first-pass effect. Antimicrob Agents Chemother 2004; 48:659–662.

107. Bae SK, Chung WS, Kim EJ, et al. Pharmacokinetics, blood partition and protein binding of DA-7867, a new oxazolidinone. Biopharm Drug Dispos 2004; 25:127–135.

13 Pharmacokinetics and Pharmacodynamics of Tetracyclines

David Andes and William A. Craig
University of Wisconsin and William S. Middleton Memorial Veterans Hospital, Madison, Wisconsin, U.S.A.

INTRODUCTION

Tetracyclines were discovered in the 1940s and have been in clinical use since the 1950s. Tetracyclines have been widely used in clinical practice because of their broad spectrum of activity and relatively low toxicity. The microbial spectrum includes gram-positive and gram-negative bacteria, intracellular chlamydiae, mycoplasma, rickettsiae, and several parasites such as malaria (1–5). However, due to the emergence of drug resistance, the early generation compounds remain the preferred choice for a relatively small number of disease states (3,6–9). The predominant clinical use of these agents remains as an alternative treatment of community-acquired respiratory tract infection, skin and skin structure, and sexually transmitted diseases (4,10–12). Tetracyclines, however, remain the drug-of-choice for several less-common infections including Lyme disease and brucellosis (4,13,14). Ten tetracycline compounds have been marketed in the last five decades and new derivatives (glycylcyclines) are in clinical development (15). Glycylcyclines are new tetracycline analogs derived from minocycline. These compounds exhibit the same spectrum of activity as tetracyclines and remain active against many pathogens resistant to tetracyclines (5,16–18). These newer generation compounds are in clinical development for treatment of respiratory, skin, and intra-abdominal infections. The interest and common use of pharmacodynamics began after development of the early generation tetracyclines. Thus, it was not until the recent development of the new glycylcycline derivatives that most of the work detailing the pharmacodynamic characteristics of this class has been undertaken.

MECHANISM OF ACTION

Tetracyclines reversibly inhibit bacterial protein synthesis by binding to the ribosomal complex (15,19). The 30s ribosomal subunit is the binding target for these compounds. Drug binding to the ribosome prevents the entry of aminoacyl-transfer RNA to the A site of the ribosome. Inhibition of this interaction prohibits incorporation of amino acids onto elongating peptide chains.

SPECTRUM OF ACTIVITY

The tetracyclines have broad-spectrum activity, which includes gram-positive and gram-negative bacteria, atypical respiratory pathogens, rickettsiae, spirochetes, and some parasites (3–5,15). In general, among the commonly available tetracycline compounds, the relative order of potency against gram-positive bacteria is minocycline > doxycycline > tetracycline. The gram-positive spectrum includes most *Staphylococcus aureus,* many *S.* epidermidis, and most streptococci. Activity against methicillin-resistant *S. aureus,* β-lactam–resistant pneumococci, and enterococcal species is limited. Activity against gram-negative bacteria is more variable. The class demonstrates intrinsic activity against many common nosocomial gram-negative organisms such as *Escherichia coli* and *Klebsiella* spp. However, the emergence of resistance has rendered the class ineffective as an empiric option in therapy targeted toward these pathogens. Still, tetracyclines remain potent against a wide spectrum of gram-negative species encountered in the community including *Yersinia pestis, Vibrio cholera, Francisella tularensis, Pasteurella multocida,* and *Haemophilus ducreyi.* The common intracellular bacteria *Mycoplasma pneumoniae* and all *Chlamydia* spp. remain exquisitely susceptible to the tetracyclines.

The new generation glycylcyclines exhibit enhanced spectrum activity against a variety of tetracycline-susceptible and -resistant bacteria (5,16). The activity of this group does not appear to be appreciably effected by either drug efflux or ribosomal resistance mechanisms. The glycylcyclines are also active against bacteria resistant to other drug classes including β-lactam–resistant pneumococci, staphylococci, and glycopeptide-resistant enterococci.

MECHANISMS OF RESISTANCE

Tetracycline drug resistance emerged in gram-negative enteric bacteria soon after the onset of clinical use (2,8,15). Resistance in both gram-negative and gram-positive bacterial pathogens has limited the empiric use of tetracyclines. Two resistance mechanisms are responsible for the majority of clinically significant resistance (2). Numerous drug efflux pumps and alterations in binding to the ribosome have been the predominant mechanisms described. Reduced cell permeability due to changes in outer membrane porins and chemical modification of the tetracycline molecular has also been reported. With few exceptions, resistance to tetracycline in bacteria is associated with acquisition of resistance genes from other organisms via transposons, plasmids, or integrons. The nomenclature for tetracycline resistance determinants includes an alphabetic designation following the *tet* gene symbol. The reported efflux genes include *tet* A, B, C, D, E, G, H, I, J, K, L, Y, and Z and the ribosomal protection genes include *tet* O, S, M, Q, T, P, and W. Glycylcyclines maintain potent activity against organisms with both efflux and ribosomal protection-resistance mechanisms (5,16).

PHARMACOKINETICS

Tetracycline is only available as an oral formulation. The most common tetracycline regimens include 250 and 500 mg every six hours. Both doxycycline and minocycline can be given either orally or intravenously. The most common maintenance regimen of both compounds is 100 to 200 mg by mouth once daily. The glycylcyclines are only available for parenteral use. The recommended maintenance regimen of tigecycline is 100 mg followed by 50 mg i.v. twice daily.

Oral absorption of tetracyclines occurs in the stomach and proximal small intestine. Oral bioavailability with these compounds is relatively high, ranging from 75% to 100%. Food can reduce the absorption of both tetracycline and doxycycline considerably (up to 50%) but has less of an effect on minocycline. Binding to divalent cations such as Ca^{2+}, Fe^{2+}, and Mg^{2+} can reduce absorption of each of the compounds from this class.

Serum pharmacokinetic indices among the commonly used tetracyclines are relatively similar (Table 1) (20–23). Following an oral 500 mg dose of tetracycline or a 200 mg dose of either doxycycline or minocycline, peak levels in the range of 3 to 4 mg/L are observed in two to four hours. The elimination half-lives for all of the tetracyclines are relatively long. The elimination half-life of tetracycline in various reports ranges from 8 to 11 hours and is the shortest among the class. For doxycycline and minocycline, elimination half-lives are more prolonged, ranging from 12 to 25 hours. The new glycylcyclines exhibit the longest half-lives ranging from 36 to 67 hours. The tetracyclines are eliminated by both renal and hepatic mechanisms to varying degrees. Both tetracycline and doxycycline have substantial renal elimination, with concentrations of active drug in urine ranging from 40% to 60%. Conversely, both minocycline and the glycylcycline derivatives achieve minimal active urine concentrations (5–15%). They are renally filtered at the glomerulus but primarily reabsorbed because of their high lipid solubility. The tetracycline class is not effectively dialyzed. Yet, dose adjustment in patients with renal insufficiency is not necessary for tigecycline, doxycycline, or minocycline. Tetracycline, however, is contraindicated in patients with a creatinine clearance of less than 30 mL/min. Dose adjustments are not necessary in patients with hepatic dysfunction.

The reported tissue distribution of tetracycline derivatives varies depending primarily upon the lipophilicity of the individual drugs (22–27). Penetration in most tissues has been predicted fairly well by serum concentration. For some tissue sites, distribution measurements in tissue have been lower than in serum. For example, concentrations of tetracycline and doxycycline in respiratory secretions have been reported to be lower than in serum. Reports of doxycycline concentrations in sputum, sinus secretions, and middle ear fluid have ranged from 20% to 60% of concomitant serum values (25,27). Measurement of doxycycline concentration in the epithelial lining fluid has not been reported. However, epithelial lining fluid (ELF) measurement of minocycline has been undertaken and concentrations three- to fivefold in excess of those in serum have been observed (28). One would anticipate a similar relationship between serum and ELF concentrations for doxycycline. Recent studies with the glycylcycline derivative, tigecycline, reported ELF exposures, which are 30% higher than those in serum.

Tissue distribution studies in other tissue sites have demonstrated higher concentrations of the more lipophilic minocycline and doxycycline than tetracycline (5,29,30). Studies examining penetration into the eye and central nervous system have reported variable, but therapeutic, concentrations with each of the tetracyclines, particularly in association with inflamed tissues.

TABLE 1 Pharmacokinetics and Pharmacodynamics of Tetracyclines in Humans

Drug	Regimen	C_{max} (mg/L)	Half-life (hr)	AUC (mg hr/L)	Protein binding	PK/PD breakpoint[a]
Tetracycline	500 mg q6hr	4.0	8–11	40	20–65%	0.5 (1)
Doxycycline	200 mg qday	4	12–20	84	80–93%	0.5 (1)
Minocycline	200 mg qday	3.1	16–25	50	75%	0.5 (1)
Tigecycline	50 mg qday	0.63	42	5.5	71–89%	0.12 (0.25)

[a]Value in parentheses represents the MIC based upon the PD target in the presence of neutrophils.
Abbreviations: AUC, area under concentration curve; PK, pharmacokinetics; PD, pharmacodynamics.

Protein-binding assays have demonstrated modest binding and relatively similar values (5,31,32). Protein binding reports with tetracycline have varied mostly depending upon the specific study from as low as 20% to as high as 65%. The range of binding values for the other common derivatives includes doxycycline at 80% to 95%, minocycline at 75%, and tigecycline at 71% to 89%.

PHARMACODYNAMICS—IN VITRO AND ANIMAL MODELS
Time Course Activity
Antimicrobial pharmacodynamics examines the relationship between a measure of drug exposure and antimicrobial effect (33). There are two major pharmacodynamic characteristics, which determine the relationship between drug concentration and efficacy over time. The first is whether organism killing is enhanced by increasing the concentrations of drug. The second is the presence and duration of antimicrobial effects, which can persist following drug exposure, termed postantibiotic effects. Three patterns of activity have been described based upon these two characteristics. One pattern is characterized by concentration-dependent killing and prolonged persistent effects. Optimal activity with drugs exhibiting this pattern is enhanced by maximizing drug concentrations. Drug classes distinguished by this pattern of activity include the quinolones and aminoglycosides. A second pharmacodynamic pattern is characterized by time-dependent killing and brief postantibiotic effects. Efficacy with drugs with this pattern of activity, such as the β-lactams, is increased by prolonging the duration of drug exposure. The final pattern of activity is similarly marked by time-dependent killing; however, these drugs also produce prolonged persistent effects. The long postantibiotic effects allow more widely spaced dosing intervals, and efficacy is improved by maximizing the drug exposure or the area under concentration curve (AUC). This pharmacodynamic pattern drives efficacy for drugs from the macrolides, (azalides) oxazolidinones, and streptogramin classes.

Both in vitro and in vivo investigations have been recently undertaken to characterize the time course activity of several tetracycline compounds. The in vitro time course activity of doxycycline was examined over a wide concentration range. Concentrations near and several fold higher than the minimum inhibitory concentration (MIC) had minimal impact on the extent of killing (34). The pharmacodynamic activity of doxycycline against *Streptococcus pneumoniae* has been also been examined in a neutropenic murine thigh infection model (35). Single-dose time kill studies similarly demonstrated minimal enhancement in the extent or rate of killing with escalating doxycycline dose levels. Despite the lack of concentration-dependent killing, however, regrowth of organisms following in vivo exposures was suppressed for prolonged periods of time, with in vivo postantibiotic durations ranging from 3.3 to 13 hours.

Pharmacodynamic study with the tetracycline derivative, minocycline, has been undertaken using an in vitro *S. aureus* time kill model (36). Minocycline exposures two- and fourfold in excess of the MIC resulted in postantibiotic suppression of regrowth of 1.4 and 3.8 hours, respectively. In vivo study with minocycline using a murine thigh infection model, reported extremely prolonged postantibiotic effects, ranging from 16 to 20 hours following doses of 5 and 10 mg/kg, respectively (37).

Similar pharmacodynamic investigations have been undertaken with the glycylcycline derivatives. An in vitro time kill study with the glycylcycline derivative, DMG-DMDOT, against *S. aureus* reported postantibiotic effects ranging from 1.7 to 3.9 hours following exposures 2× and 4× the MIC (36). In vivo time kill investigation with this derivative demonstrated even longer postantibiotic durations ranging from 4.4 to 9.0 hours following doses of 5 and 10 mg/kg, respectively (36). Similar, in vitro models have been used to examine the impact of dose escalation of the glycylcycline and tigecycline against enterococcal species (38). These studies reported maximal tigecycline killing at a concentration only two times the MIC of *Enterococcus faecalis* and *Enterococcus faecium*. No enhancement in killing was observed following exposures up to 20-fold in excess of the MIC. Modest in vitro postantibiotic effects were observed over this range of exposures ranging from 1 to 4.5 hours. The mouse thigh model has also been used to characterize the pharmacodynamics of two glycylcyclines (GAR 936 = tigecycline and WAY 152,288) (37). In vivo studies were undertaken against numerous *S. pneumoniae*, *S. aureus*, and gram-negative bacilli. Escalating dose levels had minimal impact on the extent of organism killing. A similar relationship between dose level and in vivo effect was observed in an enterococcal endocarditis model in rabbits (38). However, similar to the other tetracycline derivatives, in vivo time kill studies in mice demonstrated prolonged postantibiotic effects with both compounds against *S. pneumoniae* and *E. coli*, ranging from five to nine hours.

Results from both in vitro and in vivo studies with several tetracycline derivatives have demonstrated congruent pharmacodynamic characteristics. Tetracycline pharmacodynamic activity is distinguished by time-dependent killing and prolonged persistent effects. Previous investigations with compounds exhibiting time-dependent killing and prolonged post-antibiotic effect (PAE) have found that the pharmacodynamic parameter predictive of efficacy is the 24-hour area under the concentration curve in relation to the MIC (33).

Dose Fractionation

Dose fractionation studies in which the same total dose levels are divided into smaller doses and administered at different dosing frequencies have been useful for defining or confirming which pharmacodynamic parameter (% time above MIC, AUC/MIC, and C_{max}/MIC) is important for antimicrobial activity (33). The dose fractionation design allows the investigator to better discern the impact of these parameters by reducing the interrelationships observed with simple dose escalation. For example, increasing the dose level of drug A will linearly increase both the C_{max}/MIC and AUC/MIC and also increase the T > MIC. If efficacy in these fractionation studies is optimal with large infrequent doses, the C_{max}/MIC parameter is the best predictor of activity. Conversely, if smaller and more frequently administered regimens prove most effective, the T > MIC parameter is likely the important parameter. Lastly, if efficacy is dependent upon the dose of drug, but independent of the dosing frequency, the 24-hour AUC/MIC parameter best describes the dose–response relationship. A number of animal model dose fractionation studies have been performed with drugs from the tetracycline class of antimicrobials.

Minocycline study in the murine thigh infection model included therapy over a 100-fold dose range, fractionated using four dosing intervals against either *S. pneumoniae* or *E. coli* (39). The results demonstrated that outcome was dependent on the dose level and independent of dosing interval. Nonlinear regression analysis of the pharmacodynamic relationships (T > MIC, AUC/MIC, and C_{max}/MIC vs. outcome) showed the importance of the 24-hour AUC/MIC parameter for best describing the minocycline dose–response relationship.

Doxycycline studies in the neutropenic thigh model using dose fractionation similarly demonstrated that outcome was dependent upon the total drug exposure, and conversely the dosing interval did not appreciably impact in vivo efficacy (40). As the dosing interval was lengthened from every six hours to every 24 hours, efficacy remained similar for each of the total dose levels (Fig. 1). The dose response data regressed most strongly with the 24-hour AUC/MIC parameter (Fig. 2).

Similar in vivo study with the glycylcycline derivative, tigecycline, demonstrated the importance of both concentration and time of exposure in dose fractionation studies. Tigecycline efficacy was similar for 6, 12, and 24 hourly regimens (37). However, the 48 hourly regimen was less effective. The ineffectiveness of the prolonged 48 hourly regimen is not surprising given the relatively short elimination half-life in small rodents. The impact of dose fractionation was similar in a study with the glycylcycline, WAY 152,288. However, for this derivative, the shift in efficacy was not observed until the dosing interval was lengthened to every 48 hours. Regression of the data sets with each of the pharmacodynamic parameters similarly suggested the importance of both the 24-hour AUC/MIC and time above MIC parameters. However, when the 48 hourly data are eliminated, the 24-hour AUC/MIC is more clearly the predictive pharmacodynamic parameter. Lung infection models with both *S. pneumoniae* and *Pseudomonas aeruginosa* using dose escalation and dose fractionation reported a similar reduction in lung burden when similar total dose levels were fractionated once or twice daily (41–43).

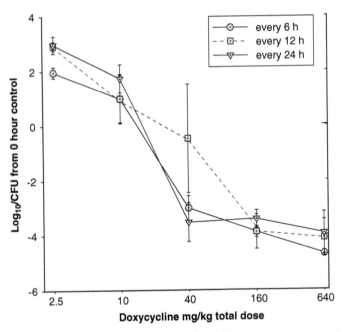

FIGURE 1 Impact of dose and dosing interval on efficacy of doxycycline against *Streptococcus pneumoniae* in a murine infection model. Each symbol represents the mean change in cfu/thigh from two mice compared to burden of organisms at the start of therapy. The error bars represent the standard deviation. The horizontal dashed line represents the burden of organisms in thighs of mice at the start of therapy.

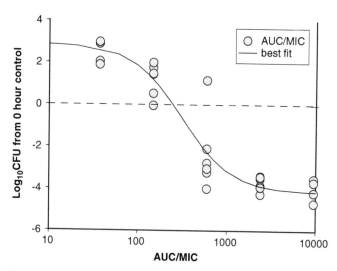

FIGURE 2 Relationships between doxycycline 24-hour AUC/MIC and efficacy against *Strepto-coccus pneumoniae* in the murine thigh infection model. Each symbol represents the mean change in cfu/thigh from two mice compared to burden of organisms at the start of therapy. The horizontal dashed line represents the burden of organisms in thighs of mice at the start of therapy. The solid sigmoid curve represents a best-fit line. *Abbreviations*: AUC, area under concentration curve; MIC, minimum inhibitory concentration.

Pharmacodynamic Target Determination

The magnitude of the drug exposure in the context of the important pharmacody-namic parameter can also be elucidated in preclinical studies. To put it simply, what AUC/MIC, C_{max}/MIC, or T>MIC value (how much drug) is needed to achieve the desired treatment end point. Results from these in vitro and in vivo investigations have proven to be similar among the preclinical models and most importantly have been similar in clinical investigation (33).

Fewer pharmacodynamic target studies have been undertaken with tetracy-cline derivatives. Animal models studies with doxycycline examined the pharma-codynamic target (24-hour AUC/MIC value) associated with in vivo efficacy (40). Study with 11 pneumococci with widely varying doxycycline susceptibility (ranging more than 30-fold) found that the free-drug 24-hour AUC/MIC value associated with a net bacteriostatic effect (static dose) was near 25. The pharmaco-dynamic value associated with a 2 $\log_{10}$ cfu/thigh reduction in organism burden was near 50. The doxycycline pharmacodynamic exposures (24-hour AUC/MIC) necessary for efficacy were not impacted by drug resistance to β-lactams, macro-lides, or tetracycline.

Similar studies have been undertaken to identify the impact of the host neutrophil on treatment efficacy (44). Study with doxycycline in a murine model found than the presence of neutrophils enhanced the in vivo activity of doxycy-cline compared to study in neutropenic animals. The 24-hour AUC/MIC value associated with a static effect was two- to threefold lower in non-neutropenic mice infected with pneumococci. In addition, the duration of the postantibiotic effect was two to four hours longer in non-neutropenic mice.

Pharmacodynamic target investigation has also been undertaken with gly-cylcycline derivatives against *S. pneumoniae* and *S. aureus* in the neutropenic murine thigh model (37). The 24-hour AUC/MIC exposure necessary to produce a net static effect was similar for two glycylcycline compounds. The 24-hour AUC/MIC target for *S. pneumoniae* was 13.5 ± 7.8. The value against *S. aureus* was somewhat (but not statistically significant) higher (24-hour AUC/MIC 38 ± 16). A substudy of therapeutic end points also demonstrated a strong relationship between the microbiologic burden of organisms after 24 hours of therapy and animal survival after five days. The glycylcycline exposure associated with a net bacteriostatic effect was predictive of 50% survival in neutropenic mice. Maximal survival was observed in association with a near 2 $\log_{10}$ reduction in organism burden in the thighs of mice.

PHARMACODYNAMICS—CLINICAL

There are no clinical investigations that allow analysis of the pharmacodynamics of tetracycline, doxycycline, or minocycline. However, one can consider the pharmacodynamic target identified in animals (24-hour AUC/MIC 25) in relation to the pharmacokinetics of these compounds in humans (Table 1). For example, a dosing regimen of 200 mg once daily regimen of minocycline or doxycycline would provide a free-drug exposure in serum equivalent to this AUC/MIC target for organisms with MICs up to 0.5 mg/L (Table 1). The presence of white cells would reduce the required target further and allow coverage of organisms up to an MIC of 1 mg/L. The current susceptibility breakpoint of 2 mg/L is a dilution higher than what pharmacodynamic analysis would predict.

Development of the new glycylcycline derivative, tigecycline, has included population pharmacokinetics and pharmacodynamic analysis of clinical trials. For example, recent investigation examined the relationship between tigecycline pharmacokinetics in patients and treatment efficacy in streptococcal and staphylococcal skin and skin-structure infections (45). Among the 35 patients observed in the treatment trial, clinical and microbiologic outcomes were strongly linked to the glycylcycline 24-hour AUC/MIC. Patients with 24-hour AUC/MIC exposures in excess of 12.5 were 13 times more likely to experience a favorable treatment outcome. This value fits nicely with the 13.5 to 36 target identified in neutropenic animals, which should be at least twofold lower in normal animals.

THERAPEUTIC USES

The tetracyclines are the drugs of choice or effective alternative therapy for a wide variety of bacterial, chlamydial, mycoplasmal, and rickettsial infections (4). The major infectious syndromes for which the tetracyclines remain a viable first line option include Lyme disease, brucellosis, ehrlichia infections, chlamydia infections, rickettsial infections, *Helicobacter pylori*, and vibrio infections. The most common use of tetracyclines as an alternative regimen is in the treatment of upper and lower respiratory tract infections, in particular infections in otherwise healthy young adults in which *M. pneumoniae* are more common etiologic agents. Doxycycline in particular remains an alternative treatment option for outpatient management of community-acquired pneumonia in guidelines from several profession societies including the Infectious Diseases Society of America and the American Thoracic Society (11). Relatively large clinical trials have suggested that doxycycline is equivalent to therapy with both macrolides and quinolone antibiotics (46–48). However, empiric use of these agents for community-acquired pneumonia has become less common due to the increased prevalence of resistant *S. pneumoniae*. There are few, if any, clinical data describing the outcome of tetracycline therapy in patients infected with resistant strains.

The glycylcyclines are being developed for a number of indications including respiratory, skin, and intra-abdominal infections. The single trial published thus far is an open label study in skin and skin-structure infections in 160 patients (49). Two dose levels (25 and 50 mg i.v. every 12 hours) were examined and clinical success was observed in 67% and 74%, respectively.

SUMMARY

Both in vitro and animal model pharmacodynamic studies have demonstrated time-dependent antimicrobial activity. However, time course studies have also identified prolonged suppression of regrowth or postantibiotic effects. The pharmacodynamic parameter shown to correlate with efficacy is the 24-hour AUC/MIC ratio. The AUC/MIC ratio of free drug that predicts efficacy in animal infection models is approximately 15 to 25. Since the AUC is the product of concentration over time, this value is essentially like averaging a drug concentration at the MIC over a 24-hour period or 1×24.

REFERENCES

1. Bouchillon SK, Hoban DJ, Johnson BM, et al. In vitro evaluation of tigecycline and comparative agents in 3049 clinical isolates: 2001 to 2002. Diagn Microbiol Infect Dis 2005; 51:291–295.
2. Chopra I, Roberts MC. Tetracycline antibiotics: mode of action, applications, molecular biology, and epidemiology of bacterial resistance. Microbiol Mol Bio Rev 2001; 65: 232–260.
3. Poulsen RK, Knudsen JC, Petersen MB. In vitro activity of 6 macrolides, clindamycin, and tetracycline on *Streptococcus pneumoniae* with different penicillin susceptibilities. APMIS 1996; 104:227–233.
4. Standiford HC. Tetracyclines and chloramphenicol. In: Mandell ed. Principles and Practice of Infectious Disease. 5th ed. Churchill Livingston, Inc, 2000:336–347.
5. Zhanel GG, Homenuik K, Nichol K, et al. The glycylcyclines, a comparative review with tetracyclines. Drugs 2004; 64:63–88.
6. Doern GV, Heilmann KP, Huynh HK, Rhomberg PR, Coffman SL, Brueggemann AB. Antimicrobial resistance among clinical isolates of *Streptococcus pneumoniae* in the United States during the 1999–2000, including a comparison of resistance rates since 1994–1995. Antimicrob Agents Chemother 2001; 45:1721–1729.
7. Doern GV. Trends in antimicrobial susceptibility of bacterial pathogens of the respiratory tract. Am J Med 1995; 99:3S–7S.
8. Finland M. Changing patterns of susceptibility of common bacterial pathogens to antimicrobial agents. Ann Intern Med 1972; 76:1009.
9. Murray BE. Resistance of Shigella, Salmonella, and other selected enteric pathogens to antimicrobial agents. Rev Infect Dis 1986; 8(suppl 2):S172–S181.
10. Norrby SR, the Nordic Atypical Pneumonia Study Group. Atypical pneumonia in the Nordic countries: aetiology and clinical results of a trial comparing fleroxacin and doxycycline. J Antimicrob Chemother 1997; 39:499–508.
11. Mandell LA, Bartlett JG, Dowell SF, File TM, Musher DM, Whitney C. Update of practice guidelines for the management of community-acquired pneumonia in immunocompetent adults. Clin Infect Dis 2003; 37:1405–1433.
12. Centers for Disease Control and Prevention. Guidelines for treatment of sexually transmitted diseases. MMWR 1998; 47:1–116.

13. Weber K, Pfister HW. Clinical management of Lyme borreliosis. Lancet 1994; 343:1017–1020.
14. Pappas G, Akritidis N, Bosilkovski M, Tsianos E. Brucellosis. New Engl J Med 2005; 352:2325–2336.
15. Roberts MC. Tetracycline therapy: update. Clin Infect Dis 2003; 36:462–467.
16. Tally FT, Ellestad GA, Testa RT. Glycylcyclines: a new generation of tetracyclines. J Antimicrob Chemother 1995; 35:449–452.
17. Eliopoulos GM, Wennersten CB, Cole G, Moellering RC. In vitro activities of two glycylcyclines against gram–positive bacteria. Antimicrob Agents Chemother 1994; 38:534–541.
18. Milatovic D, Schmitz FJ, Verhoef J, Fluit AC. Activities of the glycylcycline tigecycline (gar-936) against 1,924 recent european clinical bacterial isolates. Antimicrob Agents Chemother 2003; 47:400–404.
19. Chopra I, Hawkey PM, Hinton M. Tetracyclines, molecular and clinical aspects. J Antimicrob Chemother 1992; 29:245–277.
20. Fabre J, Milek E, Kalopoulos P, et al. The kinetics of tetracyclines in man II. Excretion, penetration in normal and inflammatory tissues, behavior in renal insufficiency and hemodialysis. In: Doxycycline: A compendium of clinical evaluations. New York: Pfizer Laboratories, 1972:19–28.
21. MacDonald H, Kelley RG, Allen ES, et al. Pharmacokinetic studies on minocycline in man. Clin Pharmacol Ther 1973; 14:852.
22. Saivin S, Houin G. Clinical pharmacokinetics of doxycycline and minocycline. Clin Pharm 1988; 15:355–366.
23. Muralidharan G, Micalizzi M, Speth J, Raible D, Troy S. Pharmacokinetics of tigecycline after single and multiple doses in healthy subjects. Antimicrob Agents Chemother 2005; 49:220–229.
24. Saux MC, Mosser J, Pontagnier H, Leng B. Pharmacokinetics of doxycycline polyphosphate after oral multiple dosing in humans. Eur J Drug Metab Pharm 1982; 7:123–130.
25. Hitt JA, Gerding DN. Sputum antimicrobial levels and clinical outcome in bronchitis. Semin Respir Infect 1991; 6:122–128.
26. Gerding DN, Hughes CE, Bamberger DM, Foxworth J, Larson TA. Extravascular antimicrobial distribution and the respective blood concentrations in humans. In: Lorian VL, ed. Antibiotics in Laboratory Medicine. Williams & Wilkins, 1996:835–899.
27. MacArthur CG, Johnson AJ, Allen ES, et al. The absorption and sputum penetration of doxycycline. J Antimicrob Chemother 1978; 4:509–514.
28. Cunha BA. Antibiotic pharmacokinetic considerations in pulmonary infections. Semin Respir Infect 1991; 6:168–182.
29. Yim CW, Flynn NM, Fitzgerald FT. Penetration of oral doxycycline into the cerebrospinal fluid of patients with latent or neurosyphilis. xycAntimicrob Agents Chemother 1985; 28:347.
30. Wood WS, Kipnis GP. The concentrations of tetracycline, chlortetracycline and oxytetracycline in the cerebrospinal fluid after intravenous administration. In: Welch H, Marti-Ibanex F, eds. Antibiotics Annual, 1953–1954. New York: Medical Encyclopedia, 1953:98–101.
31. Bennett JV, Mickewait JS, Barrett JE, et al. Comparative serum binding of four tetracyclines under simulated in vivo conditions. Antimicrob Agents Chemother 1965; 5:180–182.
32. Kunin CM. Comparative serum binding distribution and excretion of tetracycline and a new analogue methacycline. Proc Soc Exp Biol med 1962; 110:311.
33. Craig WA. Pharmacokinetic/pharmacodynamic parameters: rationale for antibacterial dosing of mice and men. Clin Infect Dis 1998; 26:1–12.
34. Cunha BA, Domenico P, Cunha CB. Pharmacodynamics of doxycycline. Clin Microbiol Infect 2000; 6:270–273.
35. Christianson J, Andes D, Craig WA. Characterization of the pharmacodynamics of doxycycline against *Streptococcus pneumoniae* in a murine thigh-infection model. 41th Interscience Conference on Antimicrobial Agents and Chemotherapy, American Society for Microbiology 2001.

36. Totsuka K, Shimizu K. In vitro and in vivo postantibiotic effects of CL 331,928 (DMG-DMDOT), a new glycylcycline against *Staphylococcus aureus* [abstr F114]. 34th Interscience Conference on Antimicrobial Agents and Chemotherapy, Orlando, FL, 1994.

37. Van Ogtrop ML, Andes D, Stamstad TJ, et al. In vivo pharmacodynamic activities of two glycylcyclines (GAR-936 and WAY 152,288) against various gram-positive and gram-negative bacteria. Antimicrob Agents Chemother 2000; 44:943–949.

38. Lefort A, Lafaurie M, Massias L, et al. Activity and diffusion of tigecycline (GAR-936) in experimental enterococcal endocarditis. Antimicrob Agents Chemother 2003; 47: 216–222.

39. Walker R, Andes D, Ebert S, Conklin R, Craig W. Pharmacodynamic comparison of 6-demethyl 6-deoxytetracycline and minocycline in an animal infection model [abstr F116]. 34th Interscience Conference on Antimicrobial Agents and Chemotherapy, American Society for Microbiology, Orlando, FL, 1994.

40. Christianson J, Andes D, Craig W. Magnitude of the 24-h AUC/MIC required for efficacy of doxycycline against *Streptococcus pneumoniae* in a murine thigh infection model [abstr 475, p 120]. 39th Infectious Diseases Society of America.

41. Mikels SM, Brown AS, Breden L, et al. In vivo activities of GAR-936 (GAR), gentamicin, piperacillin, alone and in combination in a murine model of *Pseudomonas aeruginosa* pneumonia. 39th Interscience Conference on Antimicrobial Agents and Chemotherapy, Chicago, IL, 2000.

42. Mikels SM, Brown AS, Breden L, et al. Therapeutic efficacy of GAR-963, a novel glycylcycline, in murine infections. 38th Interscience Conference on Antimicrobial Agents and Chemotherapy, San Diego, CA, 1999.

43. Murakami K, Tateda K, Matsumoto T, et al. Therapeutic efficacy of a novel tetracycline derivative, glycylcycline, against penicillin-resistant *Streptococcus pneumoniae* in a mouse model of pneumonia. J Antimicrob Chemother 2000; 46:629–631.

44. Christianson J, Andes D, Craig WA. Impact of neutrophils on pharmacodynamic activity of clindamycin and doxycycline against *Streptococcus pneumoniae*. 42nd Interscience Conference on Antimicrobial Agents Chemotherapy, 2002.

45. Meagher AK, Passarell JA, Cirincione BB, et al. Exposure-response analysis of the efficacy of tigecycline in patients with complicated skin and skin-structure infections [abstr P1184].

46. Ailani RK, Agastya G, Mukunda BA, Shekar R. Doxycycline is a cost effective therapy for hospitalized patients with community acquired pneumonia. Arch Intern Med 1999; 159:266–270.

47. Biermann C, Loken A, Risse R. Comparison of spiramycin and doxycycline in the treatment of lower respiratory infections in general practice. J Antimicrob Chemother 1988; 22(suppl B):155–158.

48. Harazim H, Wimmer J, Mittermayer HP. An open randomized comparison of ofloxacin and doxycycline in lower respiratory tract infections. Drugs 1987; 34(suppl 1):71–73.

49. Postier RG, Green SL, Klein SR, Ellis-Grosse EJ, Loh E. Results of a multicenter, randomized, open-label efficacy and safety study of two doses of tigecycline for complicated skin and skin-structure infections in hospitalized patients. Clin Therapeutics 2004; 26:704–714.

14 The Clinical Pharmacology of Nucleoside Reverse Transcriptase Inhibitors

Jennifer J. Kiser and Courtney V. Fletcher
University of Colorado Health Sciences Center, Denver, Colorado, U.S.A.

MECHANISM OF ACTION

Nucleoside reverse transcriptase inhibitors (NRTIs) include thymidine analogs such as stavudine (d4T) and zidovudine (AZT or ZDV); cytosine analogs such as emtricitabine (FTC), lamivudine (3TC), and zalcitabine (ddC); the inosine derivative didanosine (ddI); and the guanosine analog abacavir sulfate (ABC). Tenofovir disoproxil fumarate (TDF or PMPA) is an adenosine-derived nucleotide reverse transcriptase inhibitor. As a class, NRTIs require stepwise phosphorylation to the 5′-triphosphate, which is their pharmacologically active moiety. This intracellular phosphorylation occurs by cytoplasmic or mitochondrial kinases and phosphotransferases. The active triphosphate then inhibits viral replication through competitive binding to the viral enzyme reverse transcriptase and chain reaction termination after incorporation into the proviral DNA due to the modified 3′-hydroxyl group (1,2). The triphosphate anabolite is also a potential source of toxicity through inhibition of mitochondrial DNA polymerase (3).

SPECTRUM OF ANTIVIRAL ACTIVITY

The activity spectrum of the NRTIs includes HIV types one and two (4). The NRTIs lamivudine (3TC) and emtricitabine (FTC) also have activity against hepatitis B virus (HBV). The antiviral spectrum of activity for the nucleotide, tenofovir, includes HIV types one and two, various other retroviruses, and HBV (4). Examples of the range of in vitro susceptibility values (concentration required to inhibit viral replication by 50%, IC_{50}, in μM) for select agents against wild-type HIV are abacavir, 0.07 to 5.8; lamivudine, 0.002 to 15; tenofovir, 0.04 to 8.5; and zidovudine, 0.01 to 0.048. For comparison purposes, the IC_{50} range for the protease inhibitor lopinavir is 0.004 to 0.027 μM and the IC_{90} range for the non-NRTI efavirenz is 0.0017 to 0.025 μM.

MECHANISM OF RESISTANCE

HIV exhibits a very high turnover rate; 10 million new viruses are produced daily. The fidelity of reverse transcriptase is poor, and many mistakes are made during the replication process. These errors in the final DNA product contribute to rapid mutation of the virus and allow drug resistance to evolve. It is estimated that approximately 70% of treatment-experienced patients have some degree of resistance to antiretroviral drugs (5). Transmission of resistant virus is also an emerging concern. Recent data suggest the presence of resistance in 14.5% of newly diagnosed, treatment naïve patients including NRTI resistance in 7.1% (6).

The resistance patterns of older NRTI combinations are well characterized. When zidovudine or stavudine is combined with lamivudine, the M184V mutation emerges rapidly. Despite the presence of this mutation, maintaining lamivudine (and probably emtricitabine) in a regimen has been associated with continued virologic benefit, presumably due to a lowered probability of developing thymidine analogue mutations (TAMs). There are six TAMs, which seem to occur in two distinct pathways. The M41L, L210W, and T215Y/F pathway is the most prevalent and results in the greatest loss of susceptibility compared with the D67N, K70R, and K219Q/E/N/R pathway. The most commonly observed mutations with zidovudine plus didanosine or didanosine plus stavudine are TAMs. Also, the multinucleoside resistance mutations, Q151M and the T69 insertion, are more common with didanosine than lamivudine-containing regimens.

The K65R and L74V mutations appear to be associated with combinations of the more recently approved NRTIs. When abacavir, tenofovir, or didanosine is used with lamivudine (or emtricitabine) in the absence of thymidine analogues, the mutations likely to develop include M184V, K65R, and L74V. The signature mutation of tenofovir is K65R. K65R can also emerge with abacavir, though L74V is more common. Zidovudine appears to protect against the emergence of K65R and L74V mutations.

In recent years, there have been some alterations in the prevalence of antiretroviral drug resistance mutations. These changes are likely a reflection of how our treatment of HIV with the NRTIs has evolved. Consistent with the decline in the use of the thymidine analogues, a continued reliance on lamivudine, and an increase in the use of tenofovir, the prevalence of M184V and Q151M mutations has remained relatively constant, while the L74V mutation, the 69 insertion, and TAMs have decreased. The K65R and Y115F mutations have increased in prevalence, though they are still relatively uncommon (7).

The International AIDS Society and Department of Health and Human Services guidelines recommend resistance testing in cases of treatment failure, primary infection, and those with chronic infection of less than two years duration (8,9). Antiretroviral drug resistance can be assessed using genotypic or phenotypic tests, both of which have advantages as well as limitations. Genotypic tests identify mutations in a patient's viral reverse transcriptase and protease genes, which are known to be associated with the development of antiretroviral drug resistance. Phenotypic testing determines the degree to which a drug inhibits replication of the patient's virus, with the results provided as the fold change in IC_{50} compared with wild-type reference. Phenotypic tests are more expensive, but provide quantitative information on the degree of resistance and an assessment of the impact of the combination of mutations on overall susceptibility. Phenotypic and genotypic tests only detect the predominant HIV population in the patient, so minority populations cannot be accurately assessed, and the tests usually require that patients have viral loads of at least 500 copies/mL.

SUMMARY OF PHARMACOKINETIC PROPERTIES
NRTI as a Class

Rational use of antiretroviral drugs requires knowledge of their pharmacokinetic properties. A number of the pharmacokinetic properties of the NRTIs are listed in Table 1. The NRTIs are prodrugs, which require intracellular phosphorylation to exert their antiviral effects. Therefore, the intracellular NRTI triphosphate moiety is

TABLE 1 Plasma and Intracellular Pharmacokinetic Properties of the Nucleos(t)ide Reverse Transcriptase Inhibitors

	Route of metabolism/elimination	Food effects	Dose	Plasma PK			Intracellular PK		Pertinent ARV drug interactions
				Mean AUC[a] (mcg*hr/mL)	Median C_{ss} (ng/mL)	$t_{1/2}$ (hrs)	Average ICTP concentrations (fmol/10^6 cells)	$t_{1/2}$ (hrs)	
ABC (abacavir)	Metabolized by alcohol dehydrogenase and glucuronyl transferase	With or without food	300 mg b.i.d. 600 mg q.d.	6.02 11.95	500 500	1.5 ND	88–220 ND	20 ND	TPV decreases plasma ABC AUC by 44%
DDI (didanosine)	Hydrolyzed to hypoxanthine by purine nucleoside phosphorylase	Without food	400 mg q.d.	2.4	101	1.4	8	24	Should be administered 2 hrs apart from TPV Avoid use with D4T (additive toxicity) and TDF (antagonism)
FTC (emtricitabine)	Eliminated via a combination of glomerular filtration and active tubular secretion	With or without food	200 mg q.d.	10	460	10	1410–1550	39	Avoid use with 3TC (antagonism)
3TC (lamivudine)	Eliminated unchanged in urine by active organic cationic secretion	With or without food	150 mg b.i.d. 300 mg q.d.	4.53 8.44	440 380	5 ND	3550 3610	22 ND	Avoid use with FTC (antagonism)
D4T (stavudine)	Eliminated via a combination of active tubular secretion and glomerular filtration	With or without food	40 mg b.i.d.	2.6	110	1.4	31	7	Avoid use with DDI (additive toxicity) and ZDV (antagonism)
TDF	Eliminated unchanged via a combination of glomerular filtration and active tubular secretion	With food	300 mg q.d.	3.3	140	17	85–110	150	LPV/r, ATV, and ATV/r all increase TDF AUC by ~30% TDF decreases ATV AUC by 25% Avoid use with DDI (antagonism)
ZDV (zidovudine)	Metabolized by glucuronyl transferases and 3' reduction	With food	600 mg q.d. in divided doses	2	190	1.1	51	7	RTV, NFV, TPV decrease plasma ZDV AUC by 25–40% Avoid use with D4T (antagonism)

[a]AUC for the dosing interval, fmol/10^6 cells = femtomoles per million cells, LPV/r = lopinavir/ritonavir, ATV = atazanavir, ATV/r = atazanavir/ritonavir, RTV = ritonavir, NFV = nelfinavir, TPV = tipranavir, ND = no data.

Abbreviations: AUC, area under the concentration–time curve; PK, pharmacokinetics; TDF, tenofovir disoproxil fumarate; q.d., once daily; b.i.d., twice daily; ICTP, intracellular triphosphate.

the clinically relevant compound. However, measuring intracellular concentrations of the NRTIs is both difficult and expensive, and intracellular data for these compounds have only recently become available. The NRTIs are all dosed either once or twice daily, excluding zalcitabine, which is given thrice daily, is poorly tolerated, rarely used, and therefore, will not be discussed further in this chapter. The NRTIs have minimal binding to plasma and serum proteins. Most have short plasma half-lives, with the exception of emtricitabine and tenofovir, which have plasma half-lives of 10 and 17 hours, respectively. Intracellular half-lives range from seven hours for zidovudine and stavudine to perhaps 150 hours for tenofovir.

In general, the NRTIs have a low propensity for drug interactions, at least compared with the non-NRTIs and protease inhibitors, since the NRTIs are not substrates, inhibitors, or inducers of CYP enzymes. However, they are not devoid of interactions. NRTIs that are analogues for the same nucleic acid often demonstrate antagonism. Zidovudine and stavudine are thymidine analogues, which are phosphorylated by the same kinases; therefore antagonism of phosphorylation and thereby anti-HIV activity occurs between these drugs both in vitro and in vivo, and they should not be given together (10). Didanosine and tenofovir are both adenosine analogues, and recently these drugs have demonstrated increased virologic failure rates and CD4 depletion in combination (11). A didanosine and tenofovir-containing regimen should be used cautiously, if at all. Tenofovir also exhibits some unexplained drug interactions with the protease inhibitors. Lopinavir/ritonavir, atazanavir, atazanavir/ritonavir, and the investigational protease inhibitor, brecanavir plus ritonavir, have all been shown to increase the area under the concentration–time curve (AUC) of tenofovir by approximately 30% (12–14). These interactions appear to occur at the renal level, via protease inhibitor inhibition of renal transporters responsible for tenofovir efflux from the kidneys into the urine (14,15). Didanosine tablets, which contain antacid buffers to improve drug absorption, predispose this drug to absorption interactions. Tipranavir, a recently approved protease inhibitor, decreases plasma concentrations of zidovudine and abacavir by approximately 40%. The mechanism for this interaction is unclear. It is also unknown if these decreases in plasma concentrations translate into reduced intracellular concentrations. Additional drug interaction data for the NRTIs can be found in the following section. Also, the Department of Health and Human Services Guidelines for the Use of Antiretroviral Agents in HIV-1 Infected Adults and Adolescents contain continually updated antiretroviral drug interaction tables (9).

In addition to drug–drug interactions, there are a few drug–food interactions to consider with this class of agents. Didanosine must be taken on an empty stomach, whereas zidovudine tolerability and tenofovir exposures are improved when these drugs are taken with food. When possible, antiretroviral drugs with symmetrical pharmacokinetic properties (i.e., similar half-lives and food requirements) should be used in combination to simplify dosing regimens and potentially improve adherence.

Hallmark toxicities for the NRTI class include peripheral neuropathy, lipoatrophy, and hepatomegaly/steatosis with lactic acidemia. All the NRTIs have black box warnings in their product labeling for lactic acidosis and severe hepatomegaly, but the drugs have differing propensities for these adverse effects based on their affinities for mammalian mitochondrial DNA polymerase γ. The NRTIs have been ranked based on their rates of incorporation into mammalian mitochondrial DNA polymerase γ and rates of exonuclease removal of each analog after incorporation. Zalcitabine, didanosine, and stavudine have 13- to 36-fold tighter binding and ineffective exonuclease removal relative to the other NRTIs, and thus these drugs are more likely to be associated with mitochondrial toxicites (16). Due to their overlapping toxicities, the combination of didanosine and stavudine is not recommended. Additional drug-specific adverse effects are discussed in the following section.

Individual Agents

Lamivudine is a dideoxy-cytidine analogue. It has a molecular weight of 229.3 g/mol. Lamivudine can be given once or twice daily. Once daily dosing results in a similar AUC, but 66% higher maximum concentrations (C_{max}) and 53% lower troughs in the plasma. Intracellular lamivudine triphosphate AUC and C_{max} are similar with once daily and twice daily dosing, but the intracellular trough is lower in the once versus twice daily regimen. Despite the differences in pharmacokinetics, in combination with zidovudine and efavirenz, the proportion of antiretroviral naïve patients with undetectable viral loads at 48 weeks was equivalent with once versus twice daily lamivudine (17). Absolute bioavailability of the drug is 86 ± 16% (mean ± SD). The apparent volume of distribution is 1.3 ± 0.4 L/kg, which is independent of dose and does not correlate with body weight. Less than 36% of the drug is protein bound. The majority of lamivudine is eliminated unchanged in the urine by active organic cationic secretion. Total clearance and renal clearance (mean ± SD) are 398.5 ± 69.1 and 199.7 ± 56.9 mL/min, respectively. The mean plasma elimination half-life ranges from five to seven hours. Lamivudine exposure, C_{max}, and half-life increase with diminishing renal function; thus it is recommended that the dosage of lamivudine be reduced in patients with renal impairment. Hemodialysis, continuous ambulatory peritoneal dialysis, and automated peritoneal dialysis have negligible effects on lamivudine clearance; therefore, beyond creatinine clearance guided adjustments, no additional modifications appear necessary. The effects of continuous hemodialysis are unknown. Lamivudine pharmacokinetics are not altered with impaired hepatic function; however, the safety and efficacy have not been established in those with decompensated liver disease. Lamivudine is an extremely well-tolerated medication, but it can cause exacerbations of hepatitis B following drug discontinuation. Lamivudine has also been associated, though the risk is quite low, with the development of pancreatitis in children.

Abacavir is a dideoxy-guanosine analogue with a molecular weight of 670.76 g/mol. Abacavir is converted intracellularly to its active metabolite, carbovir triphosphate. The mean absolute bioavailability of abacavir is 83%. The apparent volume of distribution is 0.86 ± 0.15 L/kg. Binding of abacavir to human plasma proteins is approximately 50%. Abacavir is metabolized by alcohol dehydrogenase (to form the 5′ carboxylic acid) and glucuronyl transferase (to form the 5′ glucuronide). The observed plasma elimination half-life (following a single dose) is estimated to be 1.5 ± 0.63 hours. Total clearance is 0.8 ± 0.24 L/hr/kg (mean ± SD). The pharmacokinetics of abacavir have not been assessed in patients with impaired renal function, but renal excretion of unchanged abacavir is only a minor route of elimination. Abacavir AUC and half-life are increased 89% and 58%, respectively, in patients with mild hepatic impairment, thus the dosage should be reduced to 200 mg twice daily in these patients. Abacavir can cause a hypersensitivity reaction characterized by at least two of the following: fever, rash, gastrointestinal symptoms, constitutional symptoms (generalized malaise, fatigue, and achiness), and/or respiratory symptoms. The symptoms usually appear within the first six weeks of abacavir initiation and the median time to onset is nine days. The hypersensitivity reaction occurred in 8% of 2670 patients in nine clinical trials, and there is a genetic predisposition for the development of this adverse effect (18). This reaction can be fatal if not recognized early, thus subjects who develop the abacavir hypersensitivity reaction should discontinue abacavir immediately and never be rechallenged with the drug.

Zidovudine is a synthetic analogue of the naturally occurring nucleoside, thymidine. The molecular weight of zidovudine is 267.24 g/mol. The mean absolute bioavailability of this drug is 64%. Zidovudine should be taken with food to decrease nausea. The apparent volume of distribution is 1.6 ± 0.6 L/kg. Binding of zidovudine to human plasma proteins is less than 25%. The major metabolite of zidovudine is the 5' glucuronidated metabolite (GZDV) formed by uridine diphosphate-glucuronosyl transferase. The AUC of GZDV is threefold greater than the AUC of zidovudine, but GZDV is not an active compound. Urinary recovery of ZDV and GZDV accounts for 14 and 74%, respectively, of a dose. A second metabolite, 3'-amino-3'-deoxythymidine, is formed via hepatic 3'-reduction of the azide moiety. The observed plasma elimination half-life of zidovudine is between 0.5 and 3 hours. Systemic and renal clearance of zidovudine are 1.6 ± 0.6 and 0.34 ± 0.05 L/hr/kg (mean $\pm$ SD), respectively. No dosage adjustments are necessary for patients with creatinine clearances greater than or equal to 15 mL/min. In patients undergoing hemodialysis or peritoneal dialysis, the recommended dose is 100 mg every six to eight hours. There are insufficient data to recommend a dose in patients with mild to moderate hepatic impairment or cirrhosis. However, since zidovudine is primarily eliminated via hepatic metabolism, a dose reduction may be necessary. Tipranavir, ritonavir, nelfinavir, and rifampin all decrease zidovudine plasma concentrations. Atovaquone, fluconazole, methadone, probenecid, and valproic acid all increase zidovudine plasma concentrations. Zidovudine causes headache, malaise, anorexia, nausea, and vomiting in a greater proportion of patients than placebo. Cytopenias are a major dose-limiting adverse effect of zidovudine therapy because zidovudine is toxic to the myeloid and erythroid precursors in bone marrow. Myopathy may also occur with this drug.

Stavudine is also a thymidine analogue. The molecular weight of stavudine is 224.2 g/mol. Binding of stavudine to serum proteins was negligible over the concentration range of 0.01 to 11.4 µg/mL. Stavudine oral bioavailability is estimated at $86.4 \pm 18.2\%$. The volume of distribution is 46 ± 21 L. Total body and renal clearance of stavudine are 594 ± 164 and 237 ± 98 mL/min, respectively. The elimination half-life of stavudine in plasma following an oral dose is 1.6 ± 0.23 hours. About $42 \pm 14\%$ of a dose is recovered in the urine. Approximately 40% of a stavudine dose is eliminated via a combination of active tubular secretion and glomerular filtration. Stavudine concentrations are increased in patients with renal impairment; thus dosage adjustments are necessary for those with creatinine clearances of <50 mL/min. Stavudine pharmacokinetics were not altered in five non–HIV-infected subjects with hepatic impairment secondary to cirrhosis following a single 40 mg dose.

Didanosine is a synthetic nucleoside analogue of the naturally occurring nucleoside deoxyadenosine. Its molecular weight is 236.2 g/mol. Binding of didanosine to plasma proteins is less than 5%. The oral bioavailability of didanosine is $42 \pm 12\%$. The apparent volume of distribution is 1.08 ± 0.22 L/kg. The systemic and renal clearance of the drug are 13 ± 1.6 and 5.5 ± 2.1 mL/min/kg, respectively. The plasma elimination half-life of didanosine is 1.5 hours. About $18 \pm 8\%$ of the drug is recovered in the urine in a dosing interval. In the presence of food, the didanosine C_{max} and AUC are reduced 46% and 19%, thus didanosine should be taken on an empty stomach. Purine nucleoside phosphorylase (PNP) catalyzes the reversible phosphorolysis of didanosine to hypoxanthine. Tenofovir monophosphate is a potent inhibitor of PNP in the presence of didanosine. Thus didanosine concentrations are increased in the presence of tenofovir. It is theorized that the combination of tenofovir and didanosine may also cause purine deoxyribonucleotide triphosphates to accumulate in T-cells (19). Large amounts of dATP and dGTP can impair T-cell maturation and differentiation, and hence cause the CD4 cell depletion seen with this combination. Ribavirin increases didanosine triphosphate concentrations in vitro, and toxicities have been noted in patients receiving this combination (20). The dose of didanosine should be reduced in patients with impaired renal function and/or receiving hemodialysis. It is unknown if hepatic impairment significantly alters didanosine pharmacokinetics. Toxicities associated with didanosine include pancreatitis, peripheral neuropathy, and hepatitis.

TDF is an acyclic nucleotide diester analog of adenosine monophosphate. The molecular weight of the TDF prodrug tablet is 635.52 g/mol. The molecular weight of the actual tenofovir compound is 288 g/mol. A high fat meal increases tenofovir AUC by 40%. The prodrug, TDF, is a substrate for P-glycoprotein (21). In vitro binding of tenofovir to human plasma or serum proteins is less than 0.7% and 7.2%, respectively. The apparent volume of distribution of tenofovir is 1.3 ± 0.6 and 1.2 ± 0.4 L/kg following intravenous administration of 1 and 3 mg/kg of tenofovir, respectively. The plasma and intracellular half-lives of tenofovir and tenofovir diphosphate are 17 and 150 hours, respectively. About $32 \pm 10\%$ of a dose is recovered in the urine over 24 hours. The systemic and renal clearance of tenofovir are (mean $\pm$ SD) 37.4 ± 14.5 L/hr and 12 ± 3.5 L/hr, respectively (15). Tenofovir is eliminated via a combination of glomerular filtration and active tubular secretion. Tenofovir enters renal proximal tubule cells via human organic anion transporters 1 and 3 and exits the cells into the urine via multidrug resistance protein 4 (22). Tenofovir concentrations are significantly increased in patients with renal impairment, thus tenofovir should be used cautiously (if at all) in patients with creatinine clearance values <50 mL/min and the dosage interval increased. The pharmacokinetics of tenofovir are not significantly altered in patients with hepatic impairment, and therefore dose adjustments are not necessary. Tenofovir is a well-tolerated medication. However, some patients do develop renal toxicities including acute renal failure and Fanconi syndrome with this drug. The patients at risk for renal toxicities and the mechanism by which these toxicities occur remain to be elucidated.

Emtricitabine is a dideoxy-cytidine analogue. The molecular weight of emtricitabine is 247.24 g/mol. Structurally, emtricitabine is very similar to lamivudine, the only difference between these compounds being the addition of fluorine to emtricitabine. The mean absolute bioavailability of emtricitabine is 93%. Emtricitabine may be taken without regard to meals. In vitro binding of emtricitabine to plasma proteins is less than 4%, 86% of a dose is eliminated in the urine and 14% in the feces. Emtricitabine is eliminated via a combination of active tubular secretion and glomerular filtration. The plasma and intracellular emtricitabine and emtricitabine triphosphate half-lives are 10 and 39 hours, respectively. The effect of hepatic impairment on emtricitabine pharmacokinetics has not been assessed. The concentrations of emtricitabine are increased in renal impairment and thus dose adjustments are necessary in patients with creatinine clearance values <50 mL/min. Emtricitabine can cause skin discoloration.

CLINICAL PHARMACOKINETICS AND CLINICAL PHARMACODYNAMICS
NRTI as a Class

The era of antiretroviral therapy began in 1987 following the demonstration that the NRTI, zidovudine, when given to persons with AIDS demonstrated a survival benefit. In this study, individuals with AIDS or AIDS-related complex were randomized to receive either zidovudine 1500 mg/day or placebo. Zidovudine reduced the probability of developing an opportunistic infection from 43% to 23% ($P < 0.001$) and significantly improved survival. Furthermore, after 12 weeks of treatment, zidovudine recipients had a higher CD4 count (68 cells/µL) compared with those who received placebo (33 cells/µL) (23). NRTIs remain today, fundamental components of almost all therapeutic strategies for the treatment of HIV infection. Indeed, preferred therapy for the antiretroviral-naïve HIV-infected person involves the use of two NRTIs with either a non-NRTI or a protease inhibitor (9).

Monotherapy with zidovudine has not only been shown to improve survival in patients who have AIDS, but also been demonstrated to effect a dramatic reduction in the transmission of HIV from the pregnant woman to her newborn baby. AIDS Clinical Trials Group (ACTG) protocol 076 randomized 477 HIV-infected pregnant women (14 to 34 weeks' gestation) to either zidovudine or placebo. The zidovudine regimen consisted of antepartum zidovudine (100 mg five times daily) plus a continuous infusion of zidovudine during labor (2 mg/kg intravenously over one hour followed by 1 mg/kg/hr), and zidovudine for the newborn (2 mg/kg orally every six hours for six weeks). The HIV transmission rate was 25.5% among those that received placebo, but it was 8.3% when the mothers and their babies received zidovudine. This difference corresponds to a two-third reduction in the risk of maternal-to-infant HIV transmission. Adverse reactions associated with zidovudine therapy in the study were minimal: hemoglobin concentrations were significantly lower at birth in infants whose mothers received zidovudine, but this difference disappeared by 12 weeks of age; there was no difference in minor or major structural abnormalities in the two groups (24). An abbreviated course of zidovudine (i.e., given during labor or in the first 48 hours of life) can also substantially reduce transmission and may be easier for the patient to take (25).

Since 1987, the pharmacotherapy of HIV has changed rapidly as new agents, including those with different mechanisms of action, became available and treatment paradigms evolved. NRTIs have played a key role in this evolution, and certain milestones in the development of antiretroviral therapeutics have provided insight into clinical pharmacodynamic characteristics of the NRTIs. The introduction of didanosine and zalcitabine in 1991 and 1992, respectively, allowed experimentation with various dual NRTI combinations. One of the most significant milestones in the evolution of antiretroviral therapy was the finding that combining two NRTIs had a synergistic effect and provided better immunologic and virologic improvements compared with a single NRTI. The ACTG protocol 175 was a randomized, double-blind, placebo-controlled study comparing four treatment arms: zidovudine (200 mg t.i.d.), didanosine (200 mg b.i.d.), zidovudine plus didanosine, and zidovudine plus zalcitabine (0.75 mg t.i.d.) (26). This study enrolled 2467 HIV-infected adults (1067 antiretroviral-naive patients and 1400 with previous therapy) with CD4 counts between 200 and 500 cells/µL. The median duration of treatment was 118 weeks. Primary end points for this study were greater than 50% decline in CD4 count, development of AIDS, or death. Of patients receiving zidovudine only, 32% progressed to the primary end point compared with 22% on didanosine monotherapy versus 18% and 20% for zidovudine plus didanosine, and zidovudine plus zalcitabine, respectively. When zidovudine was used alone, the incidence of AIDS-defining events was 16% compared with 11% to 12% in the other three arms. The mortality rate in the zidovudine-only group was 9% compared with 7% in the zidovudine plus zalcitabine group and 5% in the two didanosine-containing arms. ACTG protocol 175 demonstrated that the combined regimen of zidovudine and didanosine or zalcitabine was superior to zidovudine monotherapy in immunologic and virologic parameters, particularly in patients with no previous antiretroviral therapy.

The demonstrated synergy of a dual NRTI combination led to investigations of triple NRTI combinations. The triple combination of zidovudine, lamivudine, and abacavir (now commercially available in one combination tablet) has been extensively evaluated. An open-label study in 195 antiretroviral-naïve HIV-infected individuals evaluated this regimen versus the combination of zidovudine and lamivudine plus the protease inhibitor nelfinavir (27). At week 48, based on an intent to treat analysis, plasma HIV RNA was less than 50 copies/mL in 54/95 (57%) of the triple NRTI recipients compared with 53/91 (58%) of the dual NRTI plus nelfinavir recipients. These data provided evidence for synergy among these three NRTIs and that this combination had antiviral activity comparable to that of nelfinavir plus two NRTIs. This study and others provided support for a large randomized, controlled trial of a triple NRTI regimen.

ACTG Study 5095 evaluated the triple NRTI regimen of zidovudine, lamivudine, and abacavir versus two NRTI regimens plus the non-NRTI, efavirenz; zidovudine and lamivudine plus efavirenz, and zidovudine, lamivudine, and abacavir plus efavirenz (28). 1147 antiretroviral-naïve individuals were enrolled into this randomized, double-blind study. A scheduled review by a data and safety monitoring board found that after a median of 32 weeks of therapy, 82/382 (21%) of recipients of the triple NRTI regimen experienced virologic failure compared with 85/765 (11%) of those who received one of the efavirenz-containing NRTI regimens. The time to virologic failure was significantly shorter in the triple-NRTI recipients. The proportion of subjects who achieved levels of HIV RNA < 50 copies/mL at week 48 was 61% among the triple NRTI recipients and was 83% among the efavirenz plus NRTI recipients. These results led to the early discontinuation of the triple NRTI regimen in this study; both efavirenz plus NRTI arms were allowed to continue to planned completion. While this study found a high rate of virologic efficacy for this triple NRTI regimen, it was inferior to a regimen containing efavirenz and either two or three NRTIs.

The ACTG 5095 study reconfirmed the synergy of the triple NRTI regimen of zidovudine, lamivudine, and abacavir; however, other studies have shown that not all dual and triple NRTI regimens are synergistic. For example, the combination of tenofovir and didanosine has shown high rates of virologic failure when combined with efavirenz. In a small study by Maitland and colleagues in antiretroviral-naïve persons, 5 (12%) of 41 patients assigned to the tenofovir/didanosine/efavirenz arm exhibited virologic failure by week 12, compared with none of 36 patients in the lamivudine/didanosine/efavirenz arm (11). Other investigations of the tenofovir and didanosine NRTI combination (plus a third agent) have reported even higher rates of virologic failure, from 28% to 47%. These data strongly suggest an antagonism of the anti-HIV effect when tenofovir and didanosine are combined. The mechanism of this antagonism is not understood, but tenofovir monophosphate, as previously discussed, is an inhibitor of PNP and it has been hypothesized that it may induce a state of high endogenous nucleotides, which would alter the ratio of exogenous anti-HIV nucleotide to endogenous natural nucleotide and thereby reduce the antiviral effect (19). Whatever the mechanism, the combination of tenofovir and didanosine should be used cautiously, if at all.

Properties of Individual Agents

Optimal use of antiretroviral drugs involves determination of their therapeutic indices (i.e., the concentration range, which maximizes efficacy while minimizing toxicity). To date, the majority of concentration-effect studies in HIV medicine have been conducted with protease inhibitors and non-NRTIs. There are few data with the NRTIs given that measuring the intracellular moieties of these drugs is exceptionally challenging. There are however dose-effect relationships with the drugs. Stavudine, didanosine, and zidovudine all underwent dose de-escalation during clinical development due to unacceptably high rates of toxicity. Didanosine doses above 9.6 mg/kg/day were frequently associated with peripheral neuropathy, pancreatitis, or hepatitis (29). In phase I and II trials of stavudine, doses in excess of 2 mg/kg/day demonstrated rates of peripheral neuropathy between 41 and 66 per 100 patient years (30). The first studies of zidovudine used 1500 mg daily versus the 600 mg daily dosage currently used in clinical practice (23). Despite dramatic increases in survival, these higher doses were associated with substantial toxicity, mainly peripheral blood cytopenias (31). Later studies showed a reduced dosage of zidovudine was equally effective and less toxic (32,33). The only study to date that has examined a possible relationship between intracellular zidovudine concentrations and toxicity found that higher zidovudine mono-, di-, and triphosphate concentrations in peripheral blood mononuclear cells were associated with reduced hemoglobin levels during zidovudine monotherapy (34). There are also data relating zidovudine plasma and intracellular pharmacokinetics with immunological response (35). In a comparison of concentration-controlled versus standard dose zidovudine therapy, concentration-controlled therapy (with a target plasma level of 0.7 µM) provided higher plasma concentrations of zidovudine [0.76 µM (12% CV) vs. 0.62 µM (32% CV)] and higher intracellular levels of zidovudine triphosphate (160 fmol/million cells vs. 92 fmol/million cells) versus the standard 500 mg per day dose. Subjects in the concentration-controlled group had better CD4 cell count responses (22% increase vs. 7% decrease) versus those on the standard of care. There were also no differences in tolerability between groups. A follow-up study was performed by the same group evaluating concentration-controlled combination antiretroviral therapy. In 33 antiretroviral naïve subjects receiving lamivudine, zidovudine, and indinavir, zidovudine triphosphate concentrations of >30 fmol/million cells and lamivudine triphosphate concentrations >7017 fmol/million cells were independently predictive of a more durable virologic response (1). Lamivudine and zidovudine triphosphate concentrations were also found to be 1.6- and 2.3-fold higher in this study in women compared to men. This observation translated clinically, in that the time to reach undetectable viral load was twice as fast in women (56 days vs. 112 days). Subjects with CD4 cell counts of less than 100 cells/mm^3 at baseline also had higher triphosphate concentrations in this study. These findings provide a pharmacologic basis for previous observations that women and those with more compromised immunologic status are more likely to develop NRTI toxicities (36,37).

Tenofovir is the most recently approved NRTI. It offers significant clinical advantages over older drugs in this class in that it is highly efficacious and well

TABLE 2 FDA-Approved Dosing Regimens for Nucleos(t)ide Reverse Transcriptase Inhibitors

Abbreviation (generic name)	Brand names	Formulation(s)	Adult dosage	Pediatric dosage	Dose adjustments	Adverse effects
ABC (abacavir)	Ziagen, Trizivir, Epzicom	Pediatric oral solution, tablets, combination tablets	300 mg b.i.d. (Ziagen, Trizivir), 600 mg q.d. (Ziagen, Epzicom)	>3 mo–16 yr 8 mg/kg b.i.d. (max 300 mg b.i.d.)	Mild hepatic impairment 200 mg b.i.d., dosing not established in moderate to severe hepatic impairment Dosing not established in patients with renal impairment	Hypersensitivity reaction
DDI (didanosine)	Videx, Videx EC	Pediatric powder for oral solution, buffered tablets, and enteric-coated capsules	Buffered tablets: ≥60 kg 200 mg b.i.d. <60 kg 125 mg b.i.d. Capsules: ≥60 kg 400 mg q.d. <60 kg 250 mg q.d. Capsules with TDF: ≥60 kg 250 mg q.d. <60 kg 200 mg q.d.	2 wk–8 mo 100 mg/m² b.i.d. >8 mo 120 mg/m² b.i.d. Videx EC has not been studied in children	Dosing not established in patients with hepatic impairment Dose adjustments are necessary for patients with CrCl < 60 mL/min	Pancreatitis Lactic acidosis Peripheral neuropathy
FTC (emtricitabine)	Emtriva, Truvada, Atripla	Capsule (Emtriva), combination tablet (Truvada, Atripla)	200 mg q.d.	Studies ongoing, not FDA-approved	Dosing not established in patients with hepatic impairment. Dose adjustments are necessary when CrCl < 50 mL/min	Exacerbations of HBV following drug discontinuation
3TC (lamivudine)	Epivir, Combivir, Epzicom, Trizivir	Oral solution, tablets, combination tablets	150 mg b.i.d. (Epivir, Combivir, Trizivir), 300 mg q.d. (Epivir, Epzicom)	<30 days 2 mg/kg b.i.d., pediatric (<16 yr) 4 mg/kg (max 150 mg) b.i.d.	No dose adjustments are necessary in patients with hepatic impairment. Dose adjustments are necessary when CrCl < 50 mL/min	Exacerbations of HBV following drug discontinuation

(Continued)

Table 2 FDA-Approved Dosing Regimens for Nucleostide Reverse Transcriptase Inhibitors (*Continued*)

Abbreviation (generic name)	Brand names	Formulation(s)	Adult dosage	Pediatric dosage	Dose adjustments	Adverse effects
D4T (stavudine)	Zerit	Capsules and oral solution	≥60 kg 40 mg b.i.d. <60 kg 30 mg b.i.d.	Birth–13 days 0.5 mg/kg b.i.d. 14 days to 30 kg 1 mg/kg b.i.d.	Limited data, but PK does not appear to be altered in hepatic impairment Doses should be adjusted for CrCl < 50 mL/min	Lipoatrophy Lactic acidosis Peripheral neuropathy
TDF (tenofovir disoproxil fumarate)	Viread, Truvada, Atripla	Tablet and combination tablet (Truvada, Atripla)	300 mg q.d.	Studies ongoing, not FDA-approved	No adjustments necessary for hepatic impairment Doses should be adjusted for CrCl <50 mL/min	Renal toxicities Exacerbations of HBV following drug discontinuation
ZDV (zidovudine)	Retrovir, Combivir, Trizivir	Capsules, tablets, syrup, and concentrate for i.v. injection/infusion	600 mg daily in divided doses (usually 300 mg b.i.d.)	Birth–6 wks 2 mg/kg po q6h, may be given i.v. at 1.5 mg/kg infused over 30 min, q6h 6 wks–12 yrs 160 mg/m² q8h (max 200 q8h)	Limited data with hepatic impairment but concentrations likely increased No adjustments necessary for CrCl ≥ 15 mL/min, ESRD 100 mg q6–8h	Hematologic toxicities Myopathy

Abbreviations: q.d., once daily; b.i.d., twice daily; i.v., intravenous; CrCl, creatinine clearance; PK, pharmacokinetics; ESRD, end stage renal disease; po, orally; q6h, every 6 hours; q6–8h, every 6 to 8 hours; q8h, every 8 hours; HBV, hepatitis B virus.

tolerated and has many desirable pharmacokinetic properties including a sufficiently long plasma and intracellular half-life to allow for once daily dosing. Tenofovir is an acyclic nucleotide. Structurally, it is similar to adefovir and cidofovir. Nephrotoxicity, which manifests as changes in laboratory markers of renal tubular function, is the main clinical toxicity of these drugs. Tenofovir, however, does not inhibit renal cell growth and epithelium integrity to the same extent as adefovir and cidofovir in vitro, and clinical trials have shown a very low incidence of renal toxicities with tenofovir (38–40). Nevertheless, there are case reports of tenofovir-induced renal dysfunction (41) and retrospective cohorts have shown greater declines in renal function with NRTI regimens including tenofovir compared to other NRTI-based regimens (42). Use of other nephrotoxic drugs and underlying renal dysfunction appear to predispose patients to tenofovir-induced renal dysfunction, but other risk factors have not been definitively linked to the toxicity. In many of the case reports, tenofovir plasma concentrations were elevated, so it is likely that this is a concentration-dependent toxicity, but further studies are needed to determine the exact mechanisms for this toxicity and the patients at risk. There are no studies to date evaluating associations between plasma and intracellular tenofovir concentrations and the development of renal toxicity.

Optimal use of antiretroviral agents means finding an acceptable balance between maximizing therapeutic benefits and reducing adverse effects. Further studies of the associations between NRTI intracellular concentrations and clinical outcomes are desperately needed.

THERAPEUTIC USES, FDA-APPROVED INDICATIONS, AND DOSING REGIMENS

There are eight commercially available, Food and Drug Administration (FDA)-approved NRTIs, though zalcitabine is rarely used. Table 2 highlights the FDA-approved dosing regimens for these drugs. Five NRTIs are approved for use in children including abacavir, didanosine, lamivudine, stavudine, and zidovudine, though the approved age groups and dosing strategies (weight vs. body surface area) vary widely. Dose finding, efficacy, and safety studies of tenofovir and emtricitabine in children are ongoing. With the exception of abacavir and zidovudine, the NRTIs are predominately renally eliminated and thus require dose adjustments for impaired renal function. Abacavir and zidovudine are metabolized in the liver, and therefore may require dose adjustments in patients with hepatic impairment. Drug-specific adverse effects for the NRTIs are also listed in Table 2. The Department of Health and Human Services Guidelines for the Use of Antiretroviral Agents in HIV-1 Infected Adults and Adolescents is a frequently updated resource for information on the treatment of HIV (9).

REFERENCES

1. Anderson PL, Kakuda TN, Kawle S, Fletcher CV. Antiviral dynamics and sex differences of zidovudine and lamivudine triphosphate concentrations in HIV-infected individuals. Aids 2003; 17(15):2159–2168.
2. Kakuda TN. Pharmacology of nucleoside and nucleotide reverse transcriptase inhibitor-induced mitochondrial toxicity. Clin Ther 2000; 22(6):685–708.
3. Lewis W, Simpson JF, Meyer RR. Cardiac mitochondrial DNA polymerase-g is inhibited competitively and noncompetitively by phosphorylated zidovudine. Circ Res Feb 1994; 74(2):344–348.

4. De Clercq E. Antiviral drugs in current clinical use. J Clin Virol 2004; 30(2):115–133.
5. Pillay D, Green H, Matthias R, et al. Estimating HIV-1 drug resistance in antiretroviral-treated individuals in the United Kingdom. J Infect Dis 2005; 192(6):967–973.
6. Bennet D, McCormick L, Kline R. US surveillance of HIV drug resistance at diagnosis using HIV diagnostic sera. 12th Conference on Retroviruses and Opportustic Infections, Boston, MA, Feb 22–25, 2005, Abstract 674.
7. Wainberg MA, Turner D. Resistance issues with new nucleoside/nucleotide backbone options. J Acquir Immune Defic Syndr 2004; 37(suppl 1):S36–S43.
8. Aberg JA, Gallant JE, Anderson J, et al. Primary care guidelines for the management of persons infected with human immunodeficiency virus: recommendations of the HIV Medicine Association of the Infectious Diseases Society of America. Clin Infect Dis 2004; 39(5):609–629.
9. DHHS guidelines for the use of antiretroviral agents in HIV-infected adults and adolescents. Available at: http://www.aidsinfo.hiv.gov. Accessed June 20, 2006.
10. Havlir DV, Tierney C, Friedland GH, et al. In vivo antagonism with zidovudine plus stavudine combination therapy. J Infect Dis 2000; 182(1):321–325.
11. Maitland D, Moyle G, Hand J, et al. Early virologic failure in HIV-1 infected subjects on didanosine/tenofovir/efavirenz: 12-week results from a randomized trial. Aids 2005; 19(11):1183–1188.
12. Kearney BP, Mittan A, Sayre J, et al. Pharmacokinetic Drug Interaction and Long Term Safety Profile of Combination Tenofovir DF and Lopinavir/ritonavir. 43rd ICAAC, Chicago, IL, Sep 14–17, 2003, Abstract A–1617.
13. Taburet AM, Piketty C, Chazallon C, et al. Interactions between atazanavir-ritonavir and tenofovir in heavily pretreated human immunodeficiency virus-infected patients. Antimicrob Agents Chemother 2004; 48(6):2091–2096.
14. Ford SL, Murray SC, Anderson MT, Ng-Cashin J, Johnson MA, Shelton MJ. Tenofovir renal clearance decreased following co-administration with brecanavir/ritonavir. 7th International Workshop on Clinical Pharmacology of HIV Therapy, Lisbon, Portugal, April 20–22, 2006, Abstract 38.
15. Kiser J, Carten M, Wolfe P, King T, Delahunty T, Bushman L. Effect of lopinavir/ritonavir on the renal clearance of tenofovir in HIV-infected patients. 13th Conference on Retroviruses and Opportunistic Infections, Denver, CO, Feb 5–8, 2006, Abstract 570.
16. Johnson AA, Ray AS, Hanes J, et al. Toxicity of antiviral nucleoside analogs and the human mitochondrial DNA polymerase. J Biol Chem 2001; 276(44):40847–40857.
17. DeJesus E, McCarty D, Farthing CF, et al. Once-daily versus twice-daily lamivudine, in combination with zidovudine and efavirenz, for the treatment of antiretroviral-naive adults with HIV infection: a randomized equivalence trial. Clin Infect Dis 2004; 39(3):411–418.
18. Mallal S, Nolan D, Witt C, et al. Association between presence of HLA-B*5701, HLA-DR7, and HLA-DQ3 and hypersensitivity to HIV-1 reverse-transcriptase inhibitor abacavir. Lancet 2002; 359(9308):727–732.
19. Kakuda TN, Anderson PL, Becker SL. CD4 cell decline with didanosine and tenofovir and failure of triple nucleoside/nucleotide regimens may be related. Aids 2004; 18(18):2442–2444.
20. Moreno A, Quereda C, Moreno L, et al. High rate of didanosine-related mitochondrial toxicity in HIV/HCV-coinfected patients receiving ribavirin. Antivir Ther 2004; 9(1):133–138.
21. Ray AS, Tong L, Robinson KL, Kearney BP, Rhodes GR. Role of intestinal absorption in increased tenofovir exposure when tenofovir disoproxil fumarate is co-administered with atazanavir or lopinavir/ritonavir. 7th International Workshop on Clinical Pharmacology of HIV Therapy, Lisbon, Portugal, April 20–22, 2006, Abstract 49.
22. Ray AS, Cihlar T, Robinson KL, et al. Mechanism of active tubular secretion of tenofovir and a potential for a renal drug-drug interaction with HIV protease inhibitors. 7th International Workshop on Clinical Pharmacology of HIV Therapy, Lisbon, Portugal, April 20–22, 2006, Abstract 39.

23. Fischl MA, Richman DD, Grieco MH, et al. The efficacy of azidothymidine (AZT) in the treatment of patients with AIDS and AIDS-related complex. A double-blind, placebo-controlled trial. N Engl J Med 1987; 317(4):185–191.

24. Connor EM, Sperling RS, Gelber R, et al. Reduction of maternal-infant transmission of human immunodeficiency virus type 1 with zidovudine treatment. Pediatric AIDS Clinical Trials Group Protocol 076 Study Group. N Engl J Med 1994; 331(18): 1173–1180.

25. Wade NA, Birkhead GS, Warren BL, et al. Abbreviated regimens of zidovudine prophylaxis and perinatal transmission of the human immunodeficiency virus. N Engl J Med 1998; 339(20):1409–1414.

26. Hammer SM, Katzenstein DA, Hughes MD, et al. A trial comparing nucleoside monotherapy with combination therapy in HIV-infected adults with CD4 cell counts from 200 to 500 per cubic millimeter. AIDS Clinical Trials Group Study 175 Study Team. N Engl J Med 1996; 335(15):1081–1090.

27. Matheron S, Descamps D, Boue F, et al. Triple nucleoside combination zidovudine/lamivudine/abacavir versus zidovudine/lamivudine/nelfinavir as first-line therapy in HIV-1-infected adults: a randomized trial. Antivir Ther 2003; 8(2):163–171.

28. Gulick RM, Ribaudo HJ, Shikuma CM, et al. Triple-nucleoside regimens versus efavirenz-containing regimens for the initial treatment of HIV-1 infection. N Engl J Med 2004; 350(18):1850–1861.

29. Yarchoan R, Pluda JM, Thomas RV, et al. Long-term toxicity/activity profile of 2′,3′-dideoxyinosine in AIDS or AIDS-related complex. Lancet 1990; 336(8714):526–529.

30. Skowron G. Biologic effects and safety of stavudine: overview of phase I and II clinical trials. J Infect Dis 1995; 171(suppl 2):S113–S117.

31. Richman DD, Fischl MA, Grieco MH, et al. The toxicity of azidothymidine (AZT) in the treatment of patients with AIDS and AIDS-related complex. A double-blind, placebo-controlled trial. N Engl J Med 1987; 317(4):192–197.

32. Fischl MA, Parker CB, Pettinelli C, et al. A randomized controlled trial of a reduced daily dose of zidovudine in patients with the acquired immunodeficiency syndrome. The AIDS Clinical Trials Group. N Engl J Med 1990; 323(15):1009–1014.

33. Collier AC, Bozzette S, Coombs RW, et al. A pilot study of low-dose zidovudine in human immunodeficiency virus infection. N Engl J Med 1990; 323(15):1015–1021.

34. Stretcher BN, Pesce AJ, Frame PT, Greenberg KA, Stein DS. Correlates of zidovudine phosphorylation with markers of HIV disease progression and drug toxicity. Aids 1994; 8(6):763–769.

35. Fletcher CV, Acosta EP, Henry K, et al. Concentration-controlled zidovudine therapy. Clin Pharmacol Ther 1998; 64(3):331–338.

36. Currier JS, Spino C, Grimes J, et al. Differences between women and men in adverse events and CD4+ responses to nucleoside analogue therapy for HIV infection. The AIDS Clinical Trials Group 175 Team. J Acquir Immune Defic Syndr 2000; 24(4):316–324.

37. Lichtenstein KA, Delaney KM, Armon C, et al. Incidence of and risk factors for lipoatrophy (abnormal fat loss) in ambulatory HIV-1-infected patients. J Acquir Immune Defic Syndr 2003; 32(1):48–56.

38. Squires K, Pozniak AL, Pierone G Jr, et al. Tenofovir disoproxil fumarate in nucleoside-resistant HIV-1 infection: a randomized trial. Ann Intern Med 2003; 139(5 Pt 1):313–320.

39. Gallant JE, DeJesus E, Arribas JR, et al. Tenofovir DF, emtricitabine, and efavirenz vs. zidovudine, lamivudine, and efavirenz for HIV. N Engl J Med 2006; 354(3):251–260.

40. Izzedine H, Hulot JS, Vittecoq D, et al. Long-term renal safety of tenofovir disoproxil fumarate in antiretroviral-naive HIV-1-infected patients. Data from a double-blind randomized active-controlled multicentre study. Nephrol Dial Transplant 2005; 20(4):743–746.

41. Zimmermann AE, Pizzoferrato T, Bedford J, Morris A, Hoffman R, Braden G. Tenofovir-associated acute and chronic kidney disease: a case of multiple drug interactions. Clin Infect Dis 2006; 42(2):283–290.

42. Gallant JE, Parish MA, Keruly JC, Moore RD. Changes in renal function associated with tenofovir disoproxil fumarate treatment, compared with nucleoside reverse-transcriptase inhibitor treatment. Clin Infect Dis 2005; 40(8):1194–1198.

15 Pharmacodynamics of Antivirals

George L. Drusano
Ordway Research Institute, Albany, New York, U.S.A.

INTRODUCTION

Although pharmacodynamic relationships for antibacterial agents have been sought and identified for a long period of time, it has been a general feeling that the treatment of viral and fungal infections was different in kind. A corollary to this is that delineation of pharmacodynamic relationships would be much more difficult.

In reality, the exact principles that govern the delineation of pharmacokinetic/pharmacodynamic (PK/PD) relationships for antibacterials also govern antifungals and, as will be shown here, antivirals.

The first issue, as always in the development of dynamics relationships, is to decide upon an end point. Most of the data for this chapter have been derived from studies of the human immunodeficiency virus (HIV), type 1. Consequently, this chapter addresses the in vitro and in vivo data for this virus. Other viruses should behave in a similar fashion (as is seen with cytomegalovirus). An exception to this rule is hepatitis C virus (among others), because of our inability to obtain more than one round of viral replication in vitro.

With HIV-1, there are a number of end points that may be chosen. Some are:

- Survivorship
- Change in CD_4 count consequent to therapy
- Change in viral load consequent to therapy
- Change in the hazard of emergency of resistance consequent to therapy
- Durability of maintaining the viral load below detectable levels.

Survivorship is, as always, the ultimate test of an intervention. With the advent of potent HIV chemotherapy, this disease process has been converted from a relatively rapid killer to one that will likely take its toll over a number of decades. Consequently, survivorship may not be the best end point to examine for clinical studies, because of the long lead times. In this chapter, we concentrate on end points two to five for clinical studies.

In anti-HIV chemotherapy, as in antibacterial chemotherapy, one can generate PD relationships from in vitro approaches as well as from clinical studies. One difference between these areas is the ease of employing animal systems for the generation of dynamic relationships. For antibacterials, there are many examples of the generation of dynamic relationships for different drug classes (1–3). For HIV, the cost and ethical implications of using simian models has prevented much work in this area. The McCune model and its variants are attractive, but again, because of the cost of maintaining the system, little has been done to develop dynamic relationships for HIV in animal systems. Other retroviruses have been employed (4,5), but it is difficult to draw inferences for human chemotherapy of HIV from a different pathogen with different pathogenic properties.

Consequently, in this chapter, in vitro but not animal model systems are also reviewed to add insight gleaned from these systems.

IN VITRO SYSTEMS

Although one might consider determination of viral 50% effectiveness concentration (EC-50) or EC-90 to be a PD system measurement, this chapter takes the view that this is simply a static measure of drug potency. It will play an important role in determining the final shape of the PD relationship when added to other measures (see below). Alone, it is merely a measure of compound potency.

For in vitro systems to be true PD systems, there must be the possibility of changing drug concentrations over time to examine the effect for a pathogen of known susceptibility of the drug employed. The only system with peer-reviewed publication for HIV-1 is shown in Figure 1. In addition it is important to factor in the effect of protein binding, as only the free drug is virologically active.

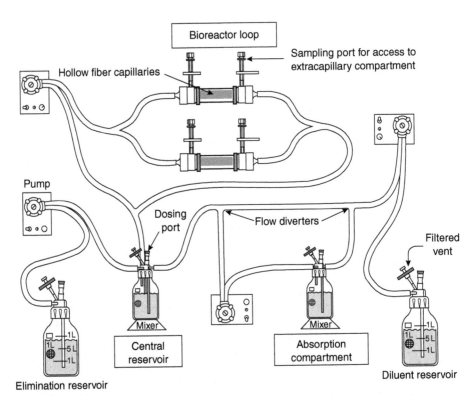

FIGURE 1 Schematic diagram of the in vitro pharmacodynamic model system. One of the two systems enclosed within a single incubator housed in a biological safety cabinet is shown. Cells are grown, and samples are removed from the extracapillary compartment of HF bioreactors. Constant infusion and oral or intravenous bolus doses are introduced through the dosing ports in the diluent reservoir, absorption compartment, or central reservoir, respectively. Exposure of cells to fluctuating concentrations of drug is affected by programmed dilution of drug within the central reservoir, while the volume of the central compartment is maintained constant by elimination. The mean pore diameter of the HF capillaries (10 kDa) would prevent HIV or HIV-infected cells from exiting the bioreactor and from circulating through the tubing. *Abbreviation*: HF, hollow fiber.

PROTEIN BINDING

The effect of protein binding has been examined in greatest detail for its effect on the anti-HIV potency of a drug by this laboratory. Bilello et al. (6) examined an HIV-1 protease inhibitor (A-80987) and determined the effect of protein binding on free-drug concentration, on drug uptake into infected cells, and, finally, on the virological effect of the drug. This set of experiments has a "transitive logic" organization.

In the first experiment (Fig. 2), increasing concentrations of the binding protein, α-1 acid glycoprotein, were introduced into the test system and the concentration of unbound drug was determined. As can be seen, increasing binding protein concentration leads to monotonically decreasing free-drug concentrations.

In Figure 3, the amount of drug that penetrates infected CEM cells is shown as a function of the free fraction of the drug. Increased external free-drug concentration results in an increased amount of intracellular drug (stop oil experiments) in a

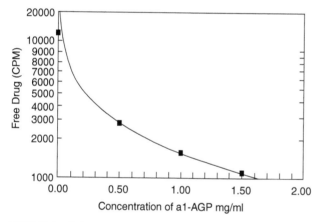

FIGURE 2 Relationship between increasing α-1 acid glycoprotein concentration and A80987 free-drug concentration.

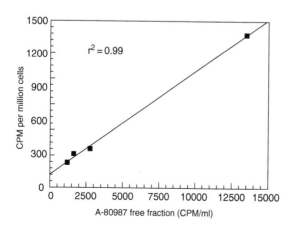

FIGURE 3 Relationship between increasing A-80987 free fraction and intracellular A80987.

linear function. Finally (Fig. 4), the amount of intracellular drug is related to the decrement of p24 output, as indexed through an inhibitory sigmoid-E_{max} model. It is clear through this series of experiments that only the free-drug concentration can induce a decrease in viral production.

These data make it clear that it is important to interpret drug concentrations in the hollow fiber system as representing the external free-drug concentrations necessary to induce the desired antiviral effect.

The most obvious place to initiate evaluation of anti-HIV agents is with the nucleoside analogs and with the HIV-1 protease inhibitors. The first published hollow fiber evaluation of an antiviral was the nucleoside analog stavudine (d4T). Bilello et al. (7) examined issues of dose finding and schedule dependency for this drug.

It is important to examine the development of stavudine in order to place its evaluation in the hollow fiber system in the proper perspective. The initial clinical evaluation of stavudine was initiated at a total daily dose of 2 mg/kg/day. The initial schedule chosen was every eight hours. Whereas the initial dose had an effect, dose escalation to 12 mg/kg/day produced no change in effect but much higher rates of drug-related neuropathy (8).

After this, the dose was deescalated to 4 mg/kg/day and the schedule was lengthened to every 12 hours. Further dose de-escalation was taken from this point on a 12-hour schedule, and nine dosing cohorts were eventually examined. This process took approximately two years.

The hollow fiber evaluation took approximately three months. The identified dose was ultimately the dose chosen by the phase I/II trial and has stood the test of time and usage. It is, at least to our knowledge, the first prospective identification of a drug dose and schedule from an in vitro test system with clinical validation. The outcome of the experiments is illustrated in Figure 5. This evaluation makes clear that for nucleoside analogs, the pharmacodynamically linked variable is area under the plasma concentration–time curve (AUC)/EC-90. With matching AUCs (Fig. 5), there was no difference in outcome between the exposure being given in a continuous infusion mode and half the exposure every 12 hours (data not shown). The reason is likely the phosphorylation of the parent compound into the virologically active form of the molecule (the triphosphate for

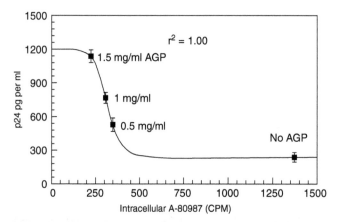

FIGURE 4 Relationship (inhibitory sigmoid E_{max}) between intracellular A-80987 and HIV viral output from infected peripheral blood mononuclear cells (PBMCs).

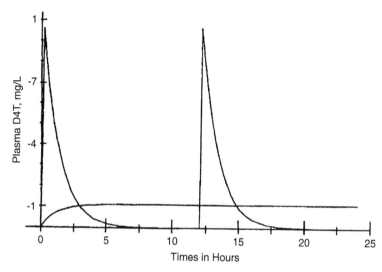

FIGURE 5 The concentration–time profile for stavudine in the hollow fiber unit is displayed for a 1 mg/(kg day) dose administered either as a continuous infusion or as a 0.5 mg/kg administered every 12 hours. The AUCs developed are identical. The ability to suppress viral replication was identical for the regimens.

stavudine, or diphosphate for prephosphorylated compounds such as tenofovir). It is clear from Figures 6 and 7 that there is full effect at 0.5 mg/kg every 12 hours, irrespective of starting challenge. At half the exposure (0.25 mg/kg q12hr), there is a hint of loss of control late in the experiment. Finally, at 0.125 mg/kg q12hr, there is no discernible antiretroviral effect.

HIV-1 protease inhibitors have also been examined in this system. The first to be examined was the early Abbott inhibitor A-77003. This drug was a proof-of-principle agent and was administered intravenously as a continuous infusion. Consequently, the hollow fiber evaluation was performed as a continuous infusion. When protein binding was taken into account, the concentrations required for effect were above those tolerable clinically. Consequently, the drug was predicted to fail. Indeed, an extensive phase I/II evaluation came to this conclusion (9,10).

The next protease inhibitor evaluated was amprenavir (141W94). Here (11), time > EC-90 was demonstrated to be the pharmacodynamically linked variable. Dosing intervals of q12hr and q8hr with matching AUCs were compared to continuous infusion of the same AUC. Continuous infusion provided the most robust viral control, followed closely by q8hr dosing. Every 12-hour dosing lost a significant fraction of the control of viral turnover. This outcome was to be expected, as HIV-1 protease inhibitors are reversible inhibitors, freely moving across cell membranes without the requirement for energy or a transporter, and do not require activation (phosphorylation) for effect as the nucleoside analogs do.

A third protease inhibitor was examined, the potent once-daily PI atazanavir (BMS 232632) (12). In addition to the hollow fiber unit evaluation, the technique of Monte Carlo simulation was also brought to bear on the outcome of these experiments.

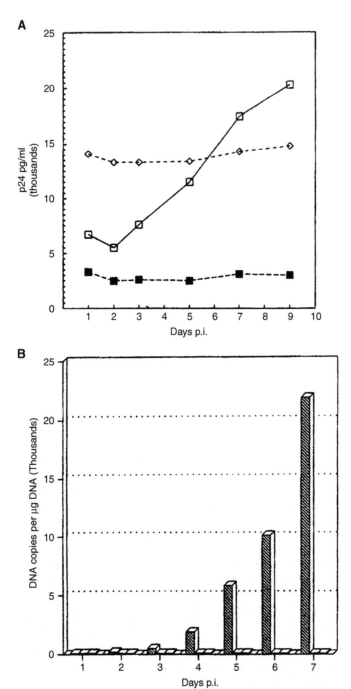

FIGURE 6 Panel **a** displays the increase in p24 output over time in a hollow fiber tube that was an untreated control [□], in a treated (0.5 mg/kg q12hr) tube where 1/1000 cells were chronically HIV-infected at start [■] or where 1/100 cells were HIV-infected at start [◇]. Panel **b** shows numbers of HIV deoxyribonucleic acid copies in treated and untreated hollow fiber units.

A

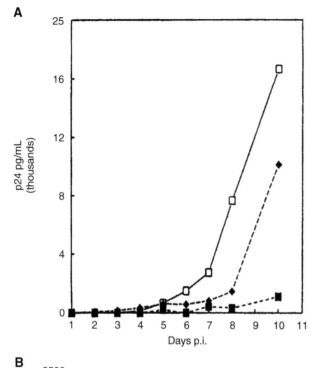

B

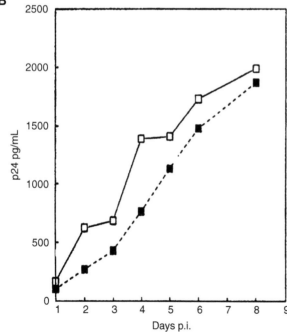

FIGURE 7 Panel **a** displays the increase in p24 output over time in a hollow fiber tube that was untreated control [□], in a treated 90.25 mg/kg q12hr) tube where 1/1000 cells were chronically HIV-infected at start [■] or where 1/100 cells were HIV-infected at start [◇]. Panel **b** shows p24 output in a treated hollow fiber unit 90.125 mg/kg q12hr [■]) and untreated control [□].

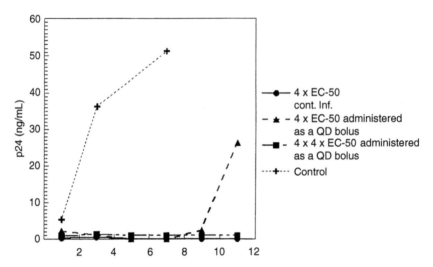

FIGURE 8 Effect of BMS 232632 (atazanavir) on HIV replication. Three infected hollow fiber units were treated with BMS 232632. One tube was treated with a concentration of four times the EC-50 as a continuous infusion. This produced a 24-hour AUC of 4×24×EC-50. The second tube received the same 24-hour AUC but was given in a peak-and-valley mode once daily. The third tube received an exposure calculated a priori to provide a time > EC-90 that would give essentially the same suppression as the continuous infusion of 4×EC-50. *Abbreviations*: AUC, area under the plasma concentration–time curve; EC, effectiveness concentration.

In Figure 8, it is clear that time > EC-90 is the pharmacodynamically linked variable. In the one instance, complete control of viral replication in vitro is achieved with a continuous infusion regiment at 4×EC-50 (about EC-90–95). This same daily AUC administered as a bolus allows breakthrough growth. Four times this daily AUC administered as a bolus regains control of the viral replication. Indeed, this latter part of the experiment demonstrates that free-drug concentrations need to exceed the EC-90 for approximately 80% to 85% of a dosing interval in order to maintain control of the HIV turnover. This served as the therapeutic target for further evaluation.

The sponsor provided PK data for atazanavir administered to normal volunteers at doses of 400 and 600 mg orally, once daily at steady state. Population PK analysis was performed, and the mean parameter vector and covariance matrix were employed to perform Monte Carlo simulation. The ability to attain the therapeutic target was assessed, accounting for the population variability in the handling of the drug. This is demonstrated in Figure 9. The viral isolate susceptibilities to atazanavir are displayed as EC-50 values. However, an internal calculation corrects for the difference between EC-50 and EC-90 as well as for the protein binding of the drug. The fractional target attainment is for free drug being greater that the nominal EC-90 for 85% of the dosing interval (the therapeutic target determined in Fig. 8). As can be seen, as the viral isolates become less and less susceptible to atazanavir, the greater the difference between the 400 and 600 mg doses. The sponsor also determined the viral susceptibility to 43 clinical isolates from a phase I/II clinical trial of the drug in patients who were HIV-treatment naïve. As can be seen, all had EC-50 values below 2 nM. Consequently, this allows us to take an expectation over the distribution of measured EC-50 values to determine the fraction of patients who will

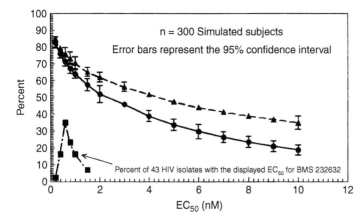

FIGURE 9 A Monte Carlo simulation of 1000 subjects performed three times was employed to estimate the fraction of these subjects whose concentration–time curve would produce maximal viral suppression on the basis of the data presented in Figure 2. The evaluation was performed for doses of (●) 400 mg and (▲) 600 mg. Forty-three isolates from a clinical trail of atazanavir were tested by the Virologics Phenosense assay.

attain a maximal response to the drug, under the assumptions that the viral susceptibility distribution is correct for naïve patients and that the kinetics in normal volunteers and its distribution are a fair representation of the kinetics of the drug in infected but treatment-naïve patients. When this calculation is performed, approximately 69% of subjects taking the 400 mg dose will attain the exposure target, whereas slightly greater than 74% of patients taking 600 mg will attain the exposure target.

A clear lesson learned regarding the chemotherapy of HIV is that combination chemotherapy is more effective than monotherapy. However, little has been done to determine optimal combination dosing regimens. An initial effort to determine the interaction of antivirals in a fully parametric system was published by Drusano et al. (13). The Greco model was fit to the data from an in vitro examination of 141W94 (amprenavir, an HIV-1 protease inhibitor) plus 1592U89 (abacavir, a nucleoside analog). This interaction is displayed in Figures 10–12. Figure 10 displays the full effect surface from these two agents. The effects of protein binding are taken into account as the effects are developed in the presence of physiological amounts of the binding proteins human albumin and human α-1 acid glycoprotein. The drug interaction can be seen by subtracting the theoretical additive surface, and the synergy surface is seen in Figure 11. It is important to note that there is synergy across all concentrations of the agents. As a model was fit to the data, the weighted residuals are displayed in Figure 12 to demonstrate that the actual regression process was unbiased. The actual degree of interaction is given by estimation of the interaction parameter, α, which is 1.144. The 95% confidence bound about the interaction parameter is from 0.534 to 1.754. As this boundary does not cross zero, the synergy is significant at the 0.05 level.

This fully parametric analysis was employed in combination with Monte Carlo simulation to examine whether the drugs at the doses used clinically would be synergistic and whether the dosing interval affected the outcome (14). The answer to both these questions was found to be yes. The data are displayed in Chapter 14.

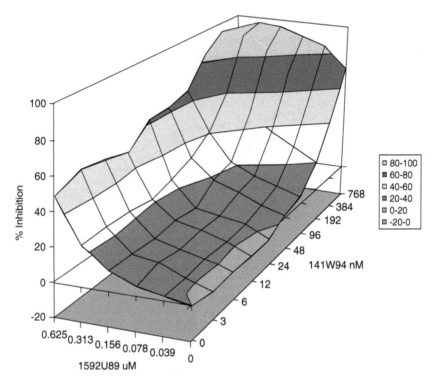

FIGURE 10   1592U89 and 141W94 combination study with albumin (40 mg/mL) and α-1 acid glycoprotein (1 mg/mL). A 3-D response surface representing the interaction from the in vitro matrix is displayed. Percent inhibition data from a 3-(4.5-dimethylthiazol-2-yl)-2,5-diphenyltetrazolium bromide assay with HIV-1$_{IIIB}$ and MT-2 cells is displayed.

CLINICAL STUDIES OF ANTIVIRAL PHARMACODYNAMICS

As noted previously, the first decision required for determining a PD relationship in the clinic is the choice of an end point. In this section, the change in CD_4 counts, the change from baseline viral load, prevention of resistance, and durability of maintenance of viral loads below delectability are the end points examined. Where possible, they are examined for both nucleoside analogs and HIV-1 protease inhibitors, both alone and in combination.

Use of nucleoside analogs as monotherapy occurred mostly at a time when viral copy number determinations were not freely available. Consequently, virtually all the available studies employed change in DC_4 cell count or p24 as the dynamic end point.

One of the first studies in this regard examined dideoxyinosine (ddI) use in a naïve patient population (15). No concentration-effect response was found for CD_4 cells, but a clear relationship was discerned between the number of CD_4 cells present at baseline and the number of cells that returned with the initiation of ddI therapy (Fig. 13). In addition, this study found a relationship between ddI exposure (as indexed to the AUC) and the fall in p24, both within patients and across the population (Fig. 14). The finding reported in the study just cited was also found for zidovudine (16).

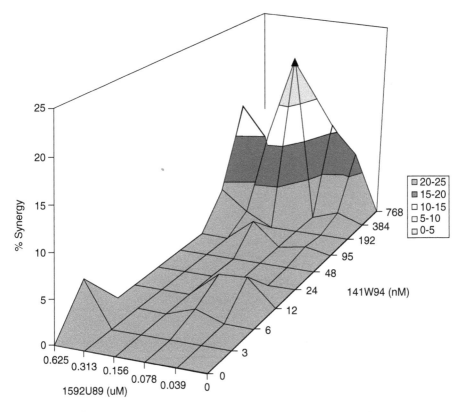

FIGURE 11 Synergy plot drawn from the data displayed in Figure 10. Plotted at the 95% confidence level.

More recently, Fletcher et al. (17) examined the relationship between ddI AUC and fall in viral load in children (Fig. 15). This study is flawed by the fact that the relationships were developed in patients receiving combination chemotherapy without any effort being made to account for the interaction between drugs.

With regard to emergence of resistance, there was one important study, largely ignored, that demonstrated the importance of viral susceptibility for effect. Kozal et al. (18) examined patients switching from zidovudine to didanosine (ddI). Over half of such patients developed a mutation at codon 74 by week 24 of ddI therapy that is known to confer ddI resistance. The effect on the number of CD_4 cells is demonstrated in Figure 16. At the time of appearance of the mutation, the CD_4 count dips below the baseline number of CD_4 cells, indicating that the increase in EC-90 attendant to the mutation drives a loss of virological effect.

There is considerably more information relating exposure to effect for the HIV-1 protease inhibitors. Because of the time at which they were studied, virtually all of the data link some measure of exposure to the change from baseline in the viral load. Early publications examining the relationship between indinavir exposure and the CD_4 count were published by Stein and Drusano (19,20). In the

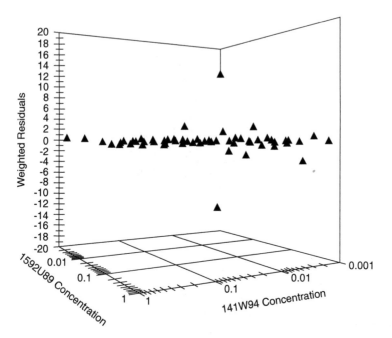

FIGURE 12   Weighted residual plot from the fully parametric analysis. The residuals are scattered about the zero line without bias.

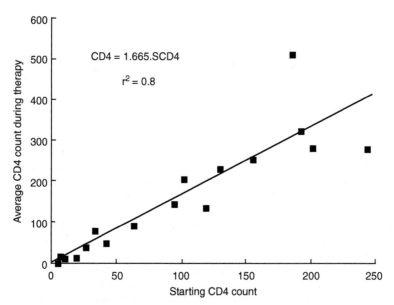

FIGURE 13   Linear regression between the number of CD$_4$-positive T-lymphocytes during therapy with dideoxyinosine and the baseline CD$_4$ count.

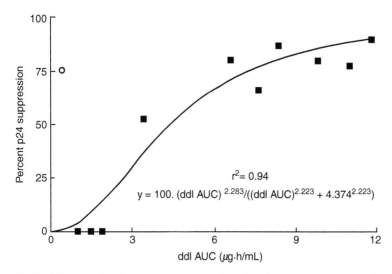

FIGURE 14 Relation between the suppression of p24 antigen and the steady-state area under the plasma concentration–time curve of dideoxyinosine.

first of these publications, it was demonstrated that the return of CD_4 count was related to the baseline CD_4 count, as with nucleoside analogs. In the second, it was demonstrated that the return of CD_4 cells had another component attached to it. The model was expanded to include the decline in viral load. The return of CD_4

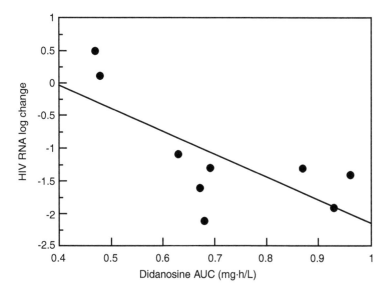

FIGURE 15 Didanosine AUC versus baseline to week 24 changes in plasma HIV ribonucleic acid levels. The solid line represents the line of best fit as determined with linear regression; the equation for this line is $y = 1.337 - (3.47 \times AUC)$; r^2 0.51; $p = 0.03$. *Abbreviation*: AUC, area under the plasma concentration–time curve.

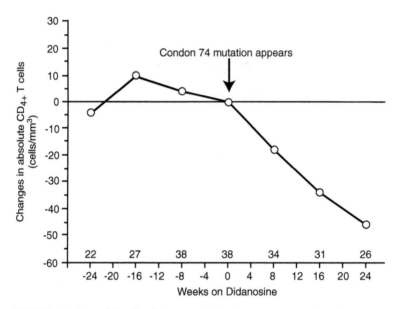

FIGURE 16 Mean CD_{4+} T-cell changes before the appearance of the HIV-1 reverse transcriptase mutation at codon 74 and CD_{4+} T-cell changes after the appearance of the mutation in 38 patients switched from zidovudine to didanosine.

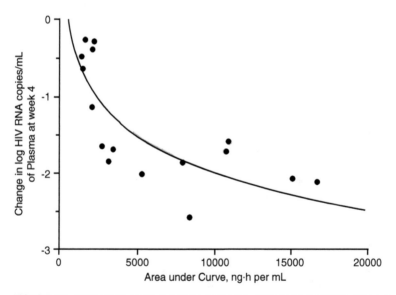

FIGURE 17 Drug levels at week 4 (area under the plasma concentration–time curve to 24 hours) plotted against the decrease in plasma HIV ribonucleic acid levels at week 4 for each patient in whom pharmacokinetics was studied. The line represents best fit and was determined using the least squares algorithm, $r = 0.801$, expressed as the Pearson correlation coefficient. (o) Patients receiving 3600 mg of saquinavir per day; (•) patients receiving 7200 mg of saquinavir per day.

cell count was better explained by the larger model of baseline CD_4 count plus viral load change as the independent variables than either alone. Later data demonstrated that a subset of patients usually with lower baseline CD_4 counts would not respond with higher CD_4 counts, even with suppression of viral replication to below assay sensitivity.

The first published paper examining the relationship between drug exposure and viral load decline studied high-dose saquinavir. Shapiro et al. (21) demonstrated a relationship between the saquinavir AUC and the change in viral load (Fig. 17). A problem with this analysis is that the form of the function is not specified. Consequently, interpretation is difficult.

Shortly thereafter, Stein et al. published a small phase I/II study of indinavir (22). This was the first "high-dose" indinavir study (2400 mg/day) and was the first to demonstrate robust viral suppression with this drug. In addition, the authors examined the relationship between indinavir exposure and both CD_4 cell return and viral load decline. These relationships are shown in Figure 18. One should not draw the conclusion that HIV suppression is linked to AUC (as this has a slightly better r^2). In this study, the drug was administered on a fixed dose and schedule, maximizing the colinearity (i.e., one could not make the AUC rise without also increasing the C_{min}). The in vitro studies simulated different doses and schedules, minimizing the colinearity, and definitively show that time > EC-90 is the dynamically linked variable. This is correlated with C_{min}. The comparison of the in vitro and in vivo results also raises the issue of the importance of the EC-90.

Drusano et al. demonstrated for both indinavir and amprenavir (23) that normalizing the measure of drug exposure to the EC-50 (or EC-90) of a particular patient's isolate decrease the variance and increased the r^2. This makes sense, as the amount of exposure needed to suppress a sensitive isolate will, on first principles, be less than that needed to suppress a resistant isolate (see above for the case of ddI and CD_4 cells).

An issue that is not addressed by these studies is the duration of HIV suppression and the closely linked issue of suppression of emergence of resistance. The first study to examine this issue was that of Kempf et al. (24), who showed that obtaining a viral load below the detectability limit of the assay was important to the duration of control of the infection. Shortly, thereafter, Drusano et al. (25) examined this issue both for protease inhibitor monotherapy with indinavir and, for the first time, for combination therapy that included indinavir. The results for monotherapy are displayed in Figure 19. It is clear that patients who attain viral loads that are below the detectability of the assay have the lowest hazard of losing control of the infection or emergence of resistance. The time to loss of control of infection is clearly related to the nadir viral load attained.

Of greater interest is the situation with combination chemotherapy, as this is the clinical norm. This study also examined the influence of different combination regimens on the hazard of loss-of-control of the viral infection. Combination regimens of zidovudine-indinavir, zidovudine-didanosine-indinavir, and zidovudine-lamivudine-indinavir were examined and compared to their monotherapy arms in a stratified analysis. The results are displayed in Table 1. Only the regimen of zidovudine-lamivudine-indinavir remained significantly different from monotherapy after adjustment for the fall in viral copy number.

This result caused the in vitro investigation of this regimen (26). A fully parametric analysis demonstrated that the interaction among all three drugs was key to the effect obtained in the clinical studies with this regimen. This is shown in

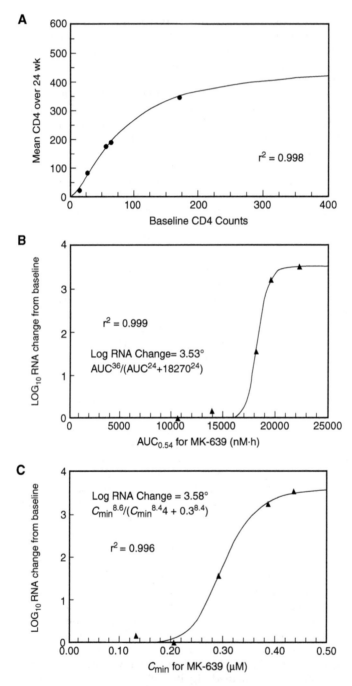

FIGURE 18 Modeling of data using a sigmoid E_{max} relationship and inhomogeneous differential equations. (**A**) The relationship of baseline CD_4 lymphocyte count to the average CD_4 lymphocyte count obtained over 24 weeks of therapy. (**B**) The relationship of the inhibition of HIV generation to total drug exposure (area under the plasma concentration–time curve). (**C**) The relationship of the inhibition of HIV generation to C_{min} serum concentration.

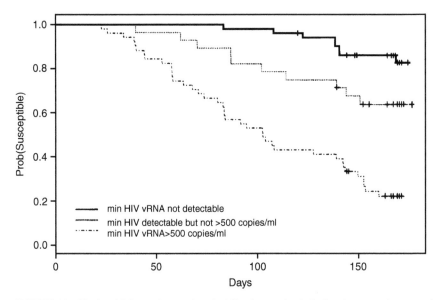

FIGURE 19 Kaplan–Meier estimate of probability that patient's isolate is not resistant to therapy (lack of sustained increase of ≥ 0.75 $\log_{10}$ copies/mL of HIV-1 RNA from patient's minimum level) at a given study day, stratified by minimum HIV-1 RNA achieved. Plot shows that probability of remaining susceptible is largest for patients who achieve undetectable level of HIV-1 RNA. Vertical slashes on plot indicate censoring events. *Abbreviation*: RNA, ribonucleic acid.

Table 2, where the as represent the interaction parameters. There are three two-drug interaction parameters and one for the interaction of all three compounds. Indinavir plus zidovudine interact in an additive manner as the value is close to zero and the 95% confidence interval overlaps zero. The same is true of indinavir plus lamivudine. Zidovudine plus lamivudine has a value that is positive and a 95% confidence interval that does not overlap zero, indicating a degree of synergistic interaction that is statistically significant. Finally, the magnitude of the synergistic interaction for the three-drug term is very large (and significant). It may be this exceptionally strong synergistic interaction that explains the superb results seen with this particular three-drug combination.

It is of interest that lamivudine not only plays a key role in the regimen but is also its Achilles' heel. Holder et al. (27) demonstrated that when the triple regimen fails, in about 70% of cases, the failure is due to an M184V mutation that produces high-level resistance to lamivudine. So that although lamivudine appears to be a

TABLE 1   Effect of Combination Therapy Vs. Indinavir Monotherapy Before and After Adjusting for the Effect of the Minimum Level of HIV-1 RNA

Combination therapy	P before adjustment	Coefficient[a]	Hazard ratio[b] (95% CI)	P
IDV/AZT	0.264	-0.35 ± 0.52	0.705 (0.254, 1.957)	0.497
IDV/AZT/ddl	0.002	-1.36 ± 0.090	0.258 (0.044, 1.514)	0.105
IDV/AZT/3TC	<0.001	-1.68 ± 0.80	0.186 (0.039, 0.893)	0.016

[a]Estimate + S. E.
[b]Hazard ratio is versus IDV monotherapy group in the same study.
Abbreviations: ddi, dideoxyinosine; IDV, indinavir; AZT, zedovodine; 3TC, lamivodine.

TABLE 2 In Vitro Assessment of Drug Interaction of AZT-3TC-Indinavir

Parameter[a]	Estimate	95% Confidence interval
E_{con}	98.99	97.8 to 100.2
$IC_{50, IND}$	146.9	128.30 to 165.60
m_{IND}	1.711	1.393 to 2.030
$IC_{50, AZT}$	118.4	108.2 to 128.60
m_{AZT}	12.89	5.576 to 20.20
$IC_{50, 3TC}$	1029.0	1018 to 1041
M_{3TC}	68.75	36.9 to 100.6
$\alpha INC, AZT$	0.0001301	−0.6191 to 0.6194
$\alpha IND, 3TC$	0.6881	−0.05189 to 1.428
$\alpha AZT, 3TC$	0.9692	0.9417 to 0.9966
$\alpha IND, AZT, 3TC$	8.94	3.434 to 14.45

[a]E_{con}, effect seen in the absence of drug (percent); IC_{50} concentration of drug necessary to reduce HIV-1 turnover by half when used alone (nM); m, slope parameter, corresponding to the rate of rise of effect with increasing drug concentration; α, interaction parameter.

key part of this therapeutic regimen and its synergy, it also has the lowest genetic barrier to resistance. If there is some nonadherence to the regimen, enough rounds of viral replication may occur to allow the point mutant (M184V) to be amplified in the total population. When this clone becomes dominant in the population, lamivudine will lose most of its contribution to the regimen. Most of the synergy is also lost, and the result is viral rebound. As Holder et al. demonstrated, this occurs most of the time with mutation solely affecting the low genetic barrier drug.

It is obvious, then, that optimal chemotherapeutic regimens for HIV (and other viral pathogens) are likely to require explicit modeling of the interaction of the drugs in the regimens.

Burger et al. (28) also examined this combination. In a multivariate logistic regression with attaining a viral load below the limit of assay detectability as the end point, they demonstrated that baseline viral load, indinavir trough concentrations, and prior HIV-1 protease inhibitor use influenced the probability of attaining this end point. Later, the concept of inhibitory quotient was introduced as an aid to optimizing therapy for protease inhibitors (29).

All of the above has been related to HIV. Other viruses can also have a PD evaluation elucidated. For CMV, there is a clear-cut dynamic relationship (30) that has been set forth in Chapter 14.

Finally, a breakthrough in modeling the relationship between drug exposure and the amount of triphosphate produced was published by Zhou et al. (31). These investigators combined a superb assay for intracellular mono-di- and triphosphate of lamivudine with sophisticated population PK analysis to identify model parameters for the production of triphosphate. They identified a clear Michaelis–Menten step in the pathway, which was different from that previously reported. The model fit the clinical data quite acceptably. This exercise has importance, not so much for itself, but as a method to follow for dose and schedule optimization for future nucleoside analogues. By identifying a clear Michaelis–Menten step, it is possible to stop dose escalation at a point where the maximum amount of the true anti-HIV moiety (triphosphate) is produced and to avoid unnecessary dose escalations, which only result in encumbering the patient with excess nucleoside analog-related toxicities.

In summary, viruses follow the same laws of physics as bacterial pathogens (fungal pathogens). It is important to delineate relationships both in vitro and in vivo between different measures of drug exposure and the end point that is deemed important. The in vitro investigations allow delineation of the true dynamically linked variable in a more straightforward manner. The in vivo investigations are important for validation. The future is in generating exposure-response relationships for combinations of agents. In this way, optimal therapy regimens can be generated to provide the greatest benefit for patients infected with viral pathogens.

REFERENCES

1. Vogelman B, Gudmundsson S, Leggett J, Turnidge J, Ebert S, Craig WA. Correlation of antimicrobial parameters with therapeutic efficacy in an animal model. J Infect Dis 1988; 158:831–847.
2. Drusano GL, Johnson DE, Rosen M, Standiford HC. Pharmacodynamics of a fluoroquinolone antimicrobial in a neutropenic rat model of Pseudomonas sepsis. Antimicrob Agents Chemother 1993; 37:483–490.
3. Louie A, Kaw P, Lui W, Jumbe N, Miller MH, Drusano GL. Pharmacodynamics of daptomycin in a murine thigh model of Staphylococcus aureus infection. Antimicrob Agents Chemother 2001; 45:845–851.
4. Bilello JA, Eiseman JL, Standiford HC, Drusano GL. Impact of dosing schedule upon suppression of a retrovirus in a murine model of AIDS encephalopathy. Antimicrob Agents Chemother 1994; 38:628–631.
5. Biello JA, Kort JJ, MacAuley C, Fredrickson TN, Yetter RA, Eiseman JL. ZDV delays but does not prevent the transmission of MAIDS by LP-BM% MuLV-infected macrophage-monocytes. J Acquir Immune Defic Syndr 1992; 5:571–576.
6. Bilello JA, Bilello PA, Stellrecht K, et al. The uptake and anti-HIV activity of A 80987, an inhibitor of the HIV-1 protease, is reduced by human serum 1-acid glycoprotein. Antimicrob Agents Chemother 1996; 40:1491–1497.
7. Bilello JA, Bauer G, Dudley MN, Cole GA, Drusano GL. The effect of 2,3-di-deoxy-2,3-didhydrothymidine (D4T) in an in vitro hollow fiver pharmacodynamic model system correlates with results of dose ranging clinical studies. Antimicrob Agnes Chemother 1994; 38:1386–1391.
8. Browne MJ, Mayer KH, Chafee SB, et al. 2',3' didehydro-3'-deoxythymidine (d4T') in patients with AIDS or AIDS-related complex: a phase I trial. J Infect Dis 1993; 167:21–29.
9. Bilello JA, Bilello PA, Kort JJ, Dudley MN, Leonard J, Drusano GL. Efficacy of constant infusion of A 77003, an inhibitor of the HIV protease in limiting acute HIV-1 infection in vitro. Antimicrob Agents Chemother 1995; 39:2523–2527.
10. Reedijk M, Boucher CA, van Brommel T, et al. Safety, pharmacokinetics and antiviral activity of A77002, a C2 symmetry-based human immunodeficiency virus protease inhibitor. Antimicrob Agents Chemother 1995; 39:1559–1564.
11. Preston SL, Piliero PJ, Bilello JA, Stein DS, Symonds WT, Drusano GL. In vitro-in vivo model for evaluating the antiviral activity of amprenavir in combination with ritonavir administered at 600 and 100 milligrams, respectively, every 12 hours. Antimicrob Agents Chemother 2003; 47:3393–3399.
12. Drusano GL, Bilello JA, Preston SL, et al. Hollow fiber unit evaluation of new human immunodeficiency virus (HIV)-1 protease inhibitor, BMS232632, for determination of the linked pharmacodynamic variable. J Infect Dis 2000; 183:1126–1129.
13. Drusano GL, D'Argenio DZ, Symonds W, et al. Nucleoside analog 1592U89 and human immunodeficiency virus protease inhibitor 141W94 are synergistic in vitro. Antimicrob Agents Chemother 1998; 42:2153–2159.
14. Drusano GL, D'Argenio DZ, Preston SL, et al. Use of drug effect interaction modeling with Monte Carlo simulation to examine the impact of dosing interval on the projected antiviral activity of the combination of abacavir and amprenavir. Antimicrob Agents Chemother 2000; 44:1655–1659.
15. Drusano GL, Yuen GJ, Lambert JS, Seidlin M, Dolin R, Valentine FT. Quantitative relationships between dideoxyinosine exposure and surrogate markers of response in a phase I trial. Ann Intern Med 1992; 116:562–566.

16. Drusano GL, Balis FM, Gitterman SR, Pizzo PA. Quantitative relationships between zidovudine exposure and efficacy and toxicity. Antimicrob Agents Chemother 1994; 8:1726–1731.

17. Fletcher CV, Brundage RC, Remmel RP, et al. Pharmacologic characteristics of indinavir, didanosine and stavudine in human immunodeficiency virus-infected children receiving combination chemotherapy. Antimicrob Agents Chemother 2000; 44:1029–1034.

18. Kozal MJ, Kroodsma K, Winters MA, et al. Didanosine resistance in HIV-infected patients switching from zidovudine to didanosine monotherapy. Ann Intern Med 1994; 121:263–268.

19. Stein DS, Drusano GL. Modeling of the change in CD_4 lymphocyte cell counts in patients before and after administration of the human immunodeficiency virus protease inhibitor indinavir. Antimicrob Agents Chemother 1997; 41:449–453.

20. Drusano GL, Stein DS. Mathematical modeling of the interrelationship of CD_4 lymphocyte count and viral load changes introduced by the protease inhibitor indinavir. Antimicrob Agents Chemother 1998; 42:359–361.

21. Shapiro JM, Winters MA, Stewart F, et al. The effect of high-dose saquinavir on viral load and CD_{4+} T-cell counts in HIV-infected patients. Ann Intern Med 1996; 124:1039–1050.

22. Stein DS, Fish DG, Bilello JA, et al. A 24 week open label phase I evaluation of the HIV protease inhibitor MK-639. AIDS 1996; 10:485–492.

23. Drusano GL, Sadler BM, Millars J, et al. and the 141W94 International Product Development Team. Pharmacodynamics of 141W94 as determined by short term change in HIV RNA: Influence of viral isolate baseline EC-50 [abstr A-16]. 37th Interscience Conference on Antimicrobial Agents and Chemotherapy, Toronto, ON, Canada, Sept 28–Oct 1, 1997.

24. Kempf DJ, Rode RA, Xu Y, et al. The duration of viral suppression during protease inhibitor therapy for HIV-1 infection is predicted by plasma HIV-1 RNA at the nadir. AIDS 1998; 12: F9–F14.

25. Drusano GL, Bilello JA, Stein DS, et al. Factors influencing the emergence of resistance to indinavir: role of virologic, immunologic, and pharmacologic variables. J Infect Dis 1998; 178:360–367.

26. Snyder S, D'Argenio DZ, Weislow O, Bilello JA, Drusano GL. The triple combination indinavir, zidovudine-lamivudine is highly synergistic. Antimicrob Agents Chemother 2000; 44:1051–1058.

27. Holder DJ, Condra JH, Schleif WA, Chodakewitz J, Emini EA. Virologic failure during combination therapy with crixivan and RT inhibitors is often associated with expression of resistance-associated mutations in RT only. Conf Retroviruses Opportun Infect 1999; 6:160 [abstr 492].

28. Burger DM, Hoetelmans RMW, Hugen PWH, et al. Low Plasma concentrations of indinavir are related to virological treatment failure in HIV-1-infected patients on indinavir containing triple therapy. Antiviral Ther 1998; 3:215–220.

29. Hoefnagel JG, Koopmans PP, Burger DM, Schurman R, Galama JM. Role of inhibitory quotient in HIV Therapy. Antivir Ther 2005; 10:879–892.

30. Drusano GL, Aweeka F, Gambertoglio J, et al. Relationship between foscarnet exposure, baseline cytomegalovirus retinitis in HIV-positive patients. AIDS 1996; 10:1113–1119.

31. Zhou Z, Rodman JH, Flynn PM, Robbins BL, Wilcox CK, D'Argenio DZ. Model for intracellular lamivudine metabolism in peripheral blood mononuclear cells ex vivo and in human immunodeficiency virus type 1-infected adolescents. Antimicrob Agents Chemother 2006; 2686–2694.

16 Antifungal Agents Pharmacokinetics and Pharmacodynamics of Amphotericin B

David Andes

University of Wisconsin and William S. Middleton Memorial Veterans Hospital, Madison, Wisconsin, U.S.A.

INTRODUCTION

Amphotericin B (AmB) was the first available systemic antifungal drug. The deoxycholate formation has been in use since the mid-1950s and remained the primary systemic antifungal until the development of the triazoles in the late 1980s. AmB was the first-line treatment for most life-threatening systemic antifungal infections both because it was the only choice and also because of its potency and broad spectrum of activity. The predominant factor, which has made use of this compound difficult, is the associated dose-limiting nephrotoxicity. More recent development of less-toxic lipid formulations of AmB has significantly improved the therapeutic drug window for this drug class (1). The fungal disease for which AmB remains the best first-line option is cryptococcal meningitis (2). However, the drug continues to be recommended in most treatment guidelines as a first-line alternative for invasive Candida, Aspergillus, and severe endemic fungal infections (3–5). Pharmacokinetics and pharmacodynamics have not traditionally been considered in the development of antifungal dosing regimens. Recent in vitro and in vivo investigations have examined the pharmacodynamic characteristics of the polyene antifungals. Results from these studies should be useful for design of rationale dosing strategies to both optimize efficacy and limit toxicity.

MECHANISM OF ACTION

AmB is a polyene macrolide antifungal natural product of the actinomycete bacterium *Streptomyces nodosus*. The compound consists of seven conjugated bonds, an internal ester, a free carboxyl group, and a glycoside chain with a primary amino group. AmB acts primarily by binding to ergosterol, the principal sterol in the plasma membrane of fungi, leading to formation of membrane pores, leakage of monovalent ions, other intracellular contents, and subsequent fungal cell death (6). A second mechanism of action may involve oxidative damage to the cell through a cascade of oxidative reactions linked to its own oxidation resulting in formation of free radicals or an increase in membrane permeability (7).

Three lipid formulations were developed and approved for use in the 1990s. While each of the formulations involves complexing to a lipid entity, the specific molecules vary markedly. The active compound in AmB colloidal dispersion (ABCD) is complexed to cholesterol sulfate resulting in the formation of 48 nm disc-like structures. AmB lipid complex (ABLC) forms ribbon-like particles. The lipid components of ABLC include dimyristoyl phosphatidylcholine and dimyristoyl phosphatidylglycerol in a 7:3 ratio. The lipid moiety in liposomal AmB is a mixture of phosphatidylcholine:distearoyl:phosphatidylglycerol:cholesterol in a 2:0.8:1:0.4 ratio, forming a unilamellar liposome.

SPECTRUM OF ACTIVITY

AmB exhibits broad-spectrum antifungal activity, which includes most fungi pathogenic to humans (8–10). The microbial spectrum includes *Candida* spp., *Cryptococcus neoformans*, *Aspergillus* spp., and the endemic fungi. AmB potency against *Candida* spp. is species dependent. The relative order of activity against these pathogens is *Candida albicans* > *Candida tropicalis* = *Candida parapsilosis* > *Candida krusei* > *Candida glabrata* > *Candida lusitaniae*. The major gaps in coverage are limited to less common, yet emerging pathogens including *Trichosporon* spp., certain *Fusarium* spp., *Aspergillus terreus*, and *Scedosporium prolificans* (11,12).

MECHANISMS OF RESISTANCE

Drug resistance to AmB is uncommon and secondary polyene resistance has not been a significant clinical problem to date (11,12). However, drug-resistant isolates have been reported and characterized. Intrinsic resistance to AmB has been reported in *Trichosporon beigelii*, *C. lusitaniae*, *S. prolificans*, and certain dematiaceous fungi. Laboratory strains of Candida, Cryptococcus, and Aspergillus have been evolved following drug exposure in vitro. Resistance has been associated with qualitative or quantitative variations in membrane sterols. Most polyene-resistant clinical isolates have reduced ergosterol membrane content and subsequent increases in other sterols. AmB has lower affinity for certain of these sterols such as fecosterol and episterol than for ergosterol. Based on an analysis of sterol composition, several clinical isolates of *C. albicans* may be defective in ERG2 or ERG3. There is evidence to suggest that alterations in the membrane structure or in the sterol-to-phospholipid ratio in the membrane may also be associated with resistance to AmB.

PHARMACOKINETICS

AmB formulations are only available for intravenous administration. Absorption of AmB from the gastrointestinal tract and following intramuscular administration is negligible. Conventional AmB is solubilized with deoxycholate as a micellar suspension. Following intravenous infusion, AmB is released from the carrier molecule, distributed with lipoproteins, and is subsequently taken up by organs of the mononuclear phagocytic system (13–20). Following a dose of 0.6 mg/kg i.v., peak concentrations of AmB are near 1 mg/L and the area under the curve (AUC) is around 17 mg*hr/L. The drug follows biphasic elimination from the bloodstream with a beta elimination half-life of around 24, followed by a terminal half-life of up to 15 days. Tissue accumulation and redistribution appear to account for most AmB disposition. Concentrations in most tissues are in the therapeutic range, including the brain parenchyma (21). However, concentrations in body fluids other than plasma are low including the cerebrospinal fluid (CSF) and vitreous. The drug is only very slowly excreted in the urine and bile. AmB metabolites have not been identified (13).

Depending on the composition of the lipid moiety, electrical charge, particle size, and configuration, each of the four AmB formulations possesses unique pharmacokinetic characteristics (17,22–25). However, all three distribute preferentially to organs of the mononuclear phagocytic system and functionally spare the kidney. Whether these distinct kinetic features translate into different pharmacodynamic properties and clinical effectiveness is largely unknown. Compared with AmB deoxycholate, both ABCD and ABLC have lower C_{max}, shorter circulating half-life, smaller AUC, and a larger volume of distribution consistent with uptake into tissues. Following administration of doses of 5 mg/kg, peak levels of ABLC and ABCD in serum are 1.7 and 3.1 mg/L, respectively. In contrast, the small unilamellar formulation, liposomal AmB, is more slowly cleared from the bloodstream, achieves much higher C_{max} and AUC values, but has a smaller volume of distribution (18). The peak serum concentration of L-AmB following a 5 mg/kg dose exceeds 80 mg/L.

Independent of the formulation and based upon the prolonged elimination half-life, all AmB preparations are usually administered once daily. AmB deoxycholate is most commonly dosed at 0.5 to 1.0 mg/kg i.v. every 24 hours (13,16–18). Because of their reduced nephrotoxicity, lipid formulations allow for delivery of higher doses than the deoxycholate preparation. However, animal models have also demonstrated that higher dosages are usually required for equivalent antifungal efficacy (26–29). Each of the lipid formulations is administered at dose levels ranging from 3 to 5 mg/kg. Few case series and pharmacokinetic studies have examined the utility, safety, and kinetics of lipid formulations following dose levels up to 15 mg/kg (24,30). While these higher doses were fairly well tolerated, rigorous efficacy data in support of these regimens are lacking. In addition, pharmacokinetic studies have demonstrated extreme nonlinearity at these higher levels such that the added AmB exposure at the high end of this dose range is negligible (23).

Serum pharmacokinetic indices for the four AmB formulations are shown in Table 1. Due to the absence of appreciable renal and hepatic elimination, dose adjustment with each of the preparations in patients with organ dysfunction is not indicated. AmB is not effectively dialyzed.

Protein-binding assays have demonstrated a high degree of binding to albumin (95%). Binding to albumin does not appear to be different among the formulations. However, binding to lipoproteins and the lipid carrier molecules is variable (22). The impact of the differences in lipid binding on treatment efficacy remains unclear.

PHARMACODYNAMICS—IN VITRO AND ANIMAL MODELS

Characterization of the pharmacodynamics of AmB occurred long after the development of the drug and dosing conventions were already established (28,31,32). The only consideration of pharmacokinetic properties in dosing regimen selection involved an apparent attempt to achieve "adequate concentrations" at the site of infection while minimizing treatment-related toxicities. Complex pharmacodynamic interactions between the host, the antifungal, and the pathogen had not been characterized and thus were not considered when constructing therapeutic regimens.

The first major hurdle for robust pharmacodynamic study of antifungals was the development of reproducible in vitro susceptibility testing methodology (33). The current Clinical Laboratory Standards Institute (formerly NCCLS) method was only developed and approved in the 1990s. Since the creation of the M-27

TABLE 1 Amphotericin B Pharmacokinetics

Formulation	Carrier	%AmB	Dose	C_{max}	AUC	t1/2	Spleen	Lung	Kidney
AmB	Deoxycholate	34	0.6	1.1	17	24	80	15	82
ABLC	DMPC	35	5	1.7	14	173	3008	77	44
–	DMPG	–	–	–	–	–	–	–	–
ABCD	Cholesterol	50	5	3.1	43	30	850	27	87
–	Sulfate	–	–	–	–	–	–	–	–
L-AmB	HSPC	10	5	83	555	24	765	31	74
–	DSPG	–	–	–	–	–	–	–	–

Dose mg/kg, AUC mg*hr/L, t1/2 hours, organ concentration mg/L.
Abbreviations: DMPC, dimyristoyl phospitidylcholine; DMPG, dimyristoyl phospitidylcglycerol; HSPC, hydrogenated soy phosphatidylcholine; DSPG, distearoyl phosphitidylcholine; AmB, amphotericin B; ABLC, AmB lipid complex; ABCD, AmB colloidal dispersion; AUC, area under the curve.

method, both experimental models in vitro and in vivo and clinical trial data have been able to demonstrate a correlation between drug dose, the minimum inhibitory concentration (MIC), and outcome (34–37). These investigations have been important for describing the relative potencies of these antifungal drugs against a number of relevant pathogens. Further characterization of the pharmacodynamics of the polyenes soon followed.

Time Course Activity
The field of antimicrobial pharmacodynamics examines the relationship between drug exposure and antimicrobial activity or host toxicity (38). There are two major factors, which are used to define the relationship between drug concentration and microbiologic activity over time. The first is whether organism killing is enhanced by increasing drug concentration. The second is the presence and duration of antimicrobial effects that persist following brief drug exposure (postantibiotic or postantifungal effects) (39).

In Vitro
Quantitation of the impact of AmB concentrations on antifungal activity has been examined in a number of time kill in vitro studies (40–44). These studies have demonstrated that as the concentration of AmB is increased, both the rate and the extent of antifungal killing are enhanced. For example, Di Bonaventura et al. examined the activity of several antifungals including AmB deoxycholate in an in vitro time kill model against nonalbicans Candida species (45). Following exposure to AmB concentrations 0.125 to 8 times the MIC, viable counts were determined every 2 to 16 hours over a 24-hour period. More than a 4 log reduction in organism burden was observed within two hours and maximal killing was noted with AmB concentration only twofold in excess of the MIC. Similarly, Klepser et al. used an in vitro time kill model to characterize the impact of dose escalation of AmB against *C. albicans* and *C. neoformans* (41,42). These investigators also observed maximal killing of organisms with AmB concentrations two to four times the MIC. Klepser et al. also noted that the rate of organism killing increased as the drug concentration was escalated from 0.125 times the MIC to 4 times the MIC. The concentration-dependent effects of AmB have also been demonstrated against filamentous fungal pathogens. Lewis et al. observed enhanced activity against Aspergillus, Fusarium, and Scedosporium species with escalating AmB concentrations using a hyphal viability assay (43).

While supra-MIC concentrations of AmB have been shown to enhance killing, the polyenes have been shown to exhibit effects on fungal organisms at concentration below the MIC (41,42,46). Detection of growth inhibition in these models has been reported at concentrations one-half to one-quarter lower than the MIC. The impact of sub-MIC concentrations has also been examined at the cellular level. Nosanchuk et al. observed alteration in the capsule of *C. neoformans* and enhanced organism phagocytosis by macrophages following sub-MIC AmB exposure (47). Nugent and Couchot found that sub-MIC concentrations of AmB inhibited germ tube formation in *C. albicans* (48).

The growth dynamics after removal of antifungal drug in vitro was first reported by Turnidge et al. (46). After exposure of *C. albicans* and *C. neoformans* to AmB concentrations ranging from 0.5 to 32 times the MIC for only 0.5 to 2 hours, regrowth of the organisms was inhibited for an additional 2.8 to 10.4 hours. The duration of these postantibiotic effects was longer with higher AmB concentrations than with lower concentrations in these studies (Fig. 1). Similar investigations have confirmed the presence of prolonged post-antibiotic effect (PAE) following AmB exposure and confirmed the impact of AmB concentration on the duration of these phenomena (40). Concentration-dependent growth suppression has also been observed against filamentous fungi, including *Aspergillus* spp. and zygomycetes (49,50). In vitro time course study with the lipid formulations of AmB has not been reported.

In Vivo

Numerous in vivo fungal models have been used to examine the impact of polyene dose escalation on treatment efficacy (28,29,32,51). Van't Wout observed a marked increase in killing of *C. albicans* in the kidneys of neutropenic mice as the dose of AmB was increased over a 10-fold dose range (52). Organism burden in kidneys was reduced by 2.5 $\log_{10}$ with the lower doses of AmB, but the extent of killing increased to more the 4 $\log_{10}$ cfu/kidneys at the higher dose levels. Clemons et al. have similarly examined the effect of dose escalation with conventional AmB as well as each of the three available lipid formulations in systemic murine models of Aspergillosis and cryptococcosis (26,27). Over a 10-fold dose range (1–10 mg/kg/day), burden of organisms in the kidneys, brains, spleen, liver, and lung were reduced in a dose-dependent fashion. A similar relationship between dose level and effect has been reported from studies using rabbit models of pulmonary Aspergillosis and coccidioidal meningitis (51,53).

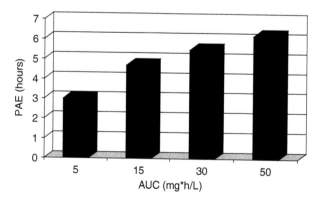

FIGURE 1 Impact of amphotericin B exposure (AUC) on the duration of the postantibiotic effect against *Candida albicans* in vitro. *Abbreviations*: PAE, post-antibiotic effect; AUC, area under the concentration curve.

In vivo time kill studies have also been undertaken with AmB (32). As the dose of AmB was increased 20-fold, the burden in kidneys of neutropenic mice was determined over 48 hours. Both the rate and the extent of antifungal activity increased with dose escalation. Measurement of in vivo PAE demonstrated extremely prolonged growth suppression ranging from 20 to more than 30 hours. For the lowest dose level studied, serum levels never rose to the MIC, yet prolonged growth suppression was observed. Thus the PAE for this lower dose level was due entirely to sub-MIC effects. The PAEs demonstrated in this in vivo study were much longer than those described in vitro. It is possible that the longer PAEs observed in vivo as compared to in vitro may be explained by potent sub-MIC effects previously described in vitro.

Results from both in vitro and in vivo studies with each of the AmB preparations have demonstrated congruent pharmacodynamic characteristics. AmB pharmacodynamic activity is distinguished by concentration-dependent killing and prolonged persistent effects. Previous investigations with compounds exhibiting this pattern of activity have found that the pharmacodynamic parameter predictive of efficacy is either the peak level in relation to the MIC or the AUC/MIC ratio (31,38). Dose fractionation studies and data regression with each of the pharmacodynamic parameters, which followed were utilized to confirm the importance of these parameters.

Dose Fractionation

Dose fractionation studies in which the same total dose levels are divided into smaller doses and administered at varying dosing frequencies have been useful for determining which pharmacodynamic parameter (% time above MIC, AUC/MIC, C_{max}/MIC) is important for antimicrobial activity (31). Dose fractionation studies provide the opportunity to reduce the interrelationships among the three pharmacodynamic parameters that are observed with simple dose escalation. For example, increasing the dose level of a drug will linearly increase both the C_{max}/MIC and AUC/MIC and also increase the T > MIC. If efficacy in these fractionation studies is optimal with large infrequent doses, the C_{max}/MIC parameter is the best predictor of activity. Conversely, if smaller and more frequently administered regimens prove most effective, the T > MIC parameter is likely the important parameter. Lastly, if efficacy is dependent upon the dose of drug, but independent of the dosing frequency, the 24-hour AUC/MIC parameter best describes the dose–response relationship. Several in vitro and animal models studies have examined the impact of AmB dose fractionation on the extent of antifungal killing.

Sokol-Anderson et al. studied the impact of AmB fractionation of three dose levels on outcome in an in vitro *C. albicans* model (54). For each assay, including viable counts and two surrogates of antifungal activity, K^+ leakage, and intracellular protein leakage, the single large concentration resulted in enhanced antifungal effect compared to administration of the total concentration in smaller fractions. Dose fractionation has also been investigated in an in vivo systemic candidiasis model (32). A 250-fold AmB dose range was fractionated into 1, 3, or 6 doses over 72 hours in neutropenic mice. Increasing doses produced a concentration-dependent reduction in organism burden. The most widely spaced dosing interval was more efficacious at each of the dose levels examined. The total amount of AmB necessary to produce a net static effect or 1 log reduction in viable organisms was 4.8- to 7.6-fold smaller when administered as single large

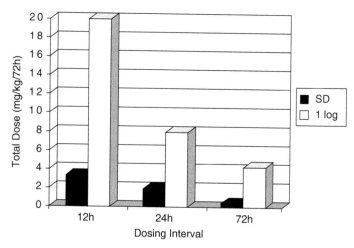

FIGURE 2 Impact of the amphotericin B dosing interval on efficacy against *Candida albicans* in a disseminated candidiasis model. The solid vertical bars represent the dose needed to produce no net change compared to the started of therapy or the static dose. The hollow vertical bars represent the dose needed to result in a 1 log reduction in organism burden in the kidneys of mice compared to the start of therapy.

doses as opposed to smaller dose levels given two to six times during the treatment period (Fig. 2). These principles have been validated in vivo using the liposomal preparation of AmB in murine candidiasis models of systemic infection, soft-tissue infection, and pneumonia (29,55,56). For example, in a disseminated candidiasis model, over a dose range of 4 to 20 mg/kg, larger doses administered once every 72 hours were similarly or more effective as smaller dose levels of L-AmB given daily in a systemic neutropenic murine candidiasis model (56).

Nonlinear regression analysis of dose fractionation data from a model of disseminated candidiasis was undertaken to further examine the relationship between each of the pharmacodynamic parameters and outcome. Regression of the dose fractionation data sets from a murine disseminated candidiasis model with each of the three pharmacokinetic/pharmacodynamic parameters demonstrated the peak/MIC parameter best described the dose–response relationship (Fig. 3). The dosing regimens maximizing this parameter deliver large AmB doses infrequently.

Pharmacodynamic Target Determination

The amount of drug in the context of the predictive pharmacodynamic parameter can also be determined in preclinical studies. Simply put, what AUC/MIC, C_{max}/MIC, or T > MIC value is needed to achieve the desired treatment end point. Results from these investigations have proven to be similar among the preclinical models and most importantly have been similar in clinical investigation (31,38).

Numerous in vitro time kill studies with both yeast and filamentous fungi have observed maximal AmB activity when concentrations exceeded the MIC by a factor of two to four (41–43). In vivo study in a neutropenic murine model was undertaken to determine the AmB deoxycholate exposure associated with various

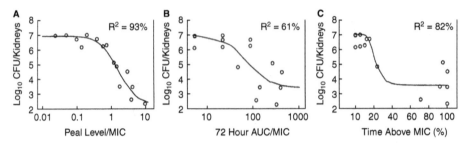

FIGURE 3 Relationship between the amphotericin B exposure expressed as the peak/MIC ratio, AUC/MIC ratio, or the percent time above MIC and efficacy against *Candida albicans* in a neutropenic murine disseminated candidiasis model. Each hollow symbol represents the mean $\log_{10}$ cfu/kidneys from two mice. The solid line represents the best-fit line. The R2 is the coefficient of determination. *Abbreviations*: AUC, area under the concentration curve; MIC, minimum inhibitory concentration.

killing end points (32). When peak serum AmB concentrations exceeded the MIC for the five Candida isolates by two- to fourfold, killing of greater than 1 $\log_{10}$ cfu/organ was observed in a neutropenic disseminated candidiasis mouse model. The highest AmB dose levels examined produced a peak level in serum in relation to the MIC (peak/MIC) of 10, which resulted in nearly a 2 $\log_{10}$ cfu/organ reduction in viable burden.

Similar studies intended to specifically identify a pharmacodynamic target for the lipid associated AmB preparations have not been undertaken. However, a number of animal model investigations have demonstrated that higher doses of a liposomal preparation of AmB on a milligram per kilogram basis is necessary to obtain the same antifungal effect as AmB deoxycholate (26,27,51). Some experts have suggested these preparations are from three- to fivefold less potent. Among the three lipid associated preparations, there have not been clear differences in efficacy based on the magnitude of exposure in the majority of infection models. Comparative study of polyene efficacy in relation to drug exposure using a *Candida meningitis* model is the single published example in which differential activity has been quantified. Groll et al. assay concentrations of conventional AmB and each of the available lipid formulations in the serum, CSF, and brain parenchyma. The concentrations of AmB deoxycholate and liposomal AmB were significantly higher in the brain tissue, and treatment resulted in reduction of organism burden below the limit of detection. Conversely, treatment with ABLC and ABCD regimens, which achieved lower concentrations in the brain tissue, resulted in residual *C. meningitis* in the model (21).

PHARMACODYNAMICS—CLINICAL

There are no clinical investigations that allow direct analysis of the pharmacodynamics of AmB preparations. However, one can consider the peak/MIC pharmacodynamic target ratio of two to four identified in vitro and in animal models relative to human pharmacokinetics and MIC distribution in the community. Doses of 0.7 to 1 mg/kg of conventional AmB would be expected to achieve this ratio for organisms with MICs in the range of 0.25 to 1 mg/L. Two retrospective clinical analyses have demonstrated a relationship between AmB in vitro susceptibility and outcome. Lass-Florl et al. reported AmB treatment outcome of

29 patients with invasive Aspergillosis (36). Ninety-five percent of patients with AmB MICs higher than 2 mg/L died. In a similar analysis of candidemia outcome, AmB failures were associated with an MIC greater than 1 mg/L (34). Pharmacodynamic data from in vitro and in vivo studies would have predicted these treatment failures based on estimated AmB kinetics in humans (peak/MIC ratios from only 0.25 to 1.0).

Early analysis of AmB dosing, pharmacokinetics, and safety demonstrated the ability to safely maximize peak serum concentrations by administering double the usual daily dose every other day (16). In fact, this dosing schedule was commonly recommended in earlier antifungal dosing guidelines, but has been replaced more recently by daily dosing. Whether this dosing strategy would be similarly or more effective or less toxic than the most commonly used regimens is unclear.

THERAPEUTIC USES

AmB has traditionally been the drug of choice for most all life-threatening fungal infections. Dose-limiting toxicities have made use of the polyenes difficult. The development of less toxic and similarly efficacious lipid AmB preparations, triazoles, and echinocandins has reduced the use of conventional AmB deoxycholate. Despite these antifungal advances, AmB remains the treatment of choice for cryptococcal meningitis. In addition, most treatment guidelines continue to recommend AmB as a first-line choice for systemic candidiasis, Aspergillosis, and life-threatening endemic fungal infections.

SUMMARY

Both in vitro and animal model pharmacodynamic studies have demonstrated concentration-dependent killing and prolonged postantibiotic effects associated with AmB exposure. The pharmacodynamic parameter shown to correlate with efficacy is the peak/MIC ratio. A dosing strategy in which large doses are administered infrequently would optimize this pharmacodynamic parameter. The peak/MIC ratio associated with efficacy from in vitro and in animal infection models is in the range of two to four.

REFERENCES

1. Wong-Beringer A, Jacobs RA, Guglielmo BJ. Lipid formulations of amphotericin B: clinical efficacy and toxicities. Clin Infect Dis 1998; 27:603–618.
2. Saag MS, Graybill RJ, Larsen RA, et al. Practice guidelines for the management of cryptococcal disease. Clin Infect Dis 2000; 30:710–718.
3. Pappas PG, Rex JH, Sobel JD, et al. Guidelines for treatment of candidiasis. Clin Infect Dis 2004; 38:161–189.
4. Stevens DA, Kan VL, Judson MA, et al. Practice guidelines for diseases caused by *Aspergillus*. Clin Infect Dis 2000; 30:696–709.
5. Wheat J, Sarosi G, McKinsey D, et al. Practice guidelines for the management of patients with histoplasmosis. Clin Infect Dis 2000; 30:688–695.
6. Brajtburg J, Powderly WG, Kobayashi GS, Medoff G. Amphotericin B: current understanding of mechanisms of action. Antimicrob Agents Chemother 1990; 34:183–188.
7. Brajtburg J, Elberg S, Schwartz DR, et al. Involvement of oxidative damage in erythrocyte lysis induced by amphotericin B. Antimicrob Agents Chemother 1985; 27:172–176.

8. Diekema DJ, Messer SA, Hollis RJ, Jones RN, Pfaller MA. Activities of caspofungin, itraconazole, posaconazole, ravuconazole, voriconazole, and amphotericin B against 448 recent clinical isolates of filamentous fungi. J Clin Microbiol 2003; 41:3623–3626.
9. Pfaller MA, Messer SA, Boyken L, et al. Global trends in the antifungal susceptibility of *Cryptococcus neoformans* (1990 to 2004). J Clin Microbiol 2005; 43(5):2163–2167.
10. Pfaller MA, Diekema DJ, Messer SA, Boyken L, Hollis RJ, Jones RN. International Fungal Surveillance Participant Group. In vitro activities of voriconazole, posaconazole, and four licensed systemic antifungal agents against Candida species infrequently isolated from blood. J Clin Microbiol 2003; 41:78–83.
11. Vanden Bossche H, Dromer F, Improvisi I, Lozano-Chiu M, Rex JH, Sanglard D. Antifungal drug resistance in pathogenic fungi. Med Mycol 1998; 36(suppl 1):119–128.
12. White TC, Marr KA, Bowden RA. Clinical, cellular, and molecular factors that contribute to antifungal drug resistance. Clin Microbiol Rev 1998; 11:382–402.
13. Atkinson AJ, Bennett JE. Amphotericin B pharmacokinetics in humans. Antimicrob Agents Chemother 1978; 13:271–276.
14. Bekersky I, Fielding RM, Dressler DE, Lee JW, Buell DN, Walsh TJ. Pharmacokinetics, excretion, and mass balance of liposomal amphotericin B and amphotericin B deoxycholate in humans. Antimicrob Agents Chemother 2002; 46:828–833.
15. Bekersky I, Boswell GW, Hiles R, Fielding RM, Buell D, Walsh TJ. Safety, toxicokinetics, and tissue distribution of long-term intravenous liposomal amphotericin B: a 91 day study in rats. Pharmaceutical Res 2000; 17:1494–1502.
16. Bindschadler DD, Bennett JE. A pharmacologic guide to the clinical use of amphotericin B. J Infect Dis 1969; 120:427–436.
17. Groll AH, Gea-Banacloche JC, Glasmacher A, Just-Nuebing G, Mashmeyer G, Walsh TJ. Clinical pharmacology of antifungal compounds. Infect Dis Clin N Am 2003; 17:159–191.
18. Janknegt R, de Marie S, Bakker Woudenberg IA, Crommelin DJ. Liposomal and lipid formulations of amphotericin B. Clin Pharmacokinet 1992; 23:279–291.
19. Polak A. Pharmacokinetics of amphotericin B and flucytosine. Postgrad Med J 1979; 55:667–670.
20. Proffitt RT, Satorius A, Chiang SM, Sullivan L, Adler-Moore JP. Pharmacology and toxicology of a liposomal formulation of amphotericin B in rodents. J Antimicrob Chemother 1991; 28(suppl B):49–61.
21. Groll AH, Giri N, Petraitis V, et al. Comparative efficacy and distribution of lipid formulations of amphotericin B in experimental *Candida albicans* infection of the central nervous system. J Infect Dis 2000; 182:274–282.
22. Bekersky I, Fielding RM, Dressler DE, Lee JW, Buell DN, Walsh TJ. Plasma protein binding of amphotericin B and pharmacokinetics of bound versus unbound amphotericin B after administration of intravenous liposomal amphotericin B and amphotericin B deoxycholate. Antimicrob Agents Chemother 2002; 46:834–840.
23. Walsh TJ, Jackson AJ, Lee JW, et al. Dose-dependent pharmacokinetics of amphotericin B lipid complex in rabbits. Antimicrob Agents Chemother 2000; 44:2068–2076.
24. Walsh TJ, Goodman JL, Pappas P, et al. Safety, tolerance, and pharmacokinetics of high-dose liposomal amphotericin B in patients infected with *Aspergillus* species and other filamentous fungi: maximum tolerated dose study. Antimicrob Agents Chemother 2001; 45:3487–3496.
25. Nagata MP, Gentry CA, Hampton EM. Is there a therapeutic or pharmacokinetic rationale for amphotericin B dosing in systemic *Candida* infections? Ann Pharmacother 1996; 30:811–818.
26. Clemons KV, Stevens DA. Comparison of fungizone, amphotec, ambisome, and abelcet for treatment of systemic murine cryptococcosis. Antimicrob Agents Chemother 1998; 42:899–902.
27. Clemons KV, Stevens DA. Comparative efficacies of four amphotericin B formulations—fungizone, amphotec, ambisome, abelcet—against systemic murine aspergillosis. Antimicrob Agents Chemother 2004; 48:1047–1050.
28. Groll AH, Piscitelli SC, Walsh TJ. Antifungal pharmacodynamics: concentration-effect relationships in vitro and in vivo. Pharmacotherapy 2001; 21:133–148.
29. Hoffman JL, Lewis RE, Ernst EJ, et al. In vivo pharmacodynamics of liposomal formulations of amphotericin B in a neutropenic murine lung infection model. Pharmacotherapy 2000; 20:357–358.

30. Walsh TJ, Yeldandi V, McEvoy M, et al. Safety, tolerance, and pharmacokinetics of a small unilamellar liposomal formulation of amphotericin B in neutropenic patients. Antimicrob Agents Chemother 1998; 42:2191–2198.
31. Andes D. Pharmacodynamics of antifungals In vivo pharmacodynamics of antifungal drugs in treatment of candidiasis. Antimicrob Agents Chemother 2003; 47:1179–1186.
32. Andes D, Stamstad T, Conklin R. Pharmacodynamics of amphotericin B in a neutropenic mouse disseminated candidiasis model. Antimicrob Agents Chemother 2001; 45:922–926
33. National Committee for Clinical Laboratory Standards. 1997. Reference method for broth dilution antifungal susceptibility testing of yeasts. Approved standard. National Committee for Clinical Laboratory Standards, Wayne, PA.
34. Nguyen MH, Clancy CJ, Ye VL, et al. Do in vitro susceptibility data predict the microbiologic response to amphotericin B? Results of a prospective study of patients with *Candida* fungemia. J Infect Dis 1997; 177:425–430.
35. Rex JH, Pfaller MA, Rinaldi MG, Polak A, Galgiani JN. Antifungal susceptibility testing. Clin Microbiol Rev 1993; 6:367–381.
36. Lass-Florl C, Kofler G, Kropshofer G, et al. In vitro testing of susceptibility to amphotericin B is a reliable predictor of clinical outcome in invasive aspergillosis. J Antimicrob Chemother 1998; 42:497–502.
37. Anaissie EJ, Karyotakis NC, Hachem R, Dignani MC, Rex JH, Paetznick V. Correlation between in vitro and in vivo activity of antifungal agents against *Candida* species. J Infect Dis 1994; 170:384–389.
38. Craig WA. Pharmacokinetic/pharmacodynamic parameters: rationale for antibacterial dosing of mice and men. Clin Infect Dis 1998; 26:1–12.
39. Craig WA, Gudmundsson S. Postantibiotic effect. In: Lorian V, ed. Antibiotics in Laboratory Medicine. 5th ed. Baltimore, MD: The Williams and Wilkins Co., 2005; 296–329.
40. Ernst E, Klepser ME, Pfaller MA. Postantifungal effects of echinocandin, azole, and polyene antifungal agents against *Candida albicans* and *Cryptococcus neoformans*. Antimicrob Agents Chemother 2000; 44:1108–1111.
41. Klepser ME, Wolfe EJ, Pfaller MA. Antifungal pharmacodynamic characteristics of fluconazole and amphotericin B against *Cryptococcus neoformans*. J Antimicrob Chemother 1998; 41:397–401.
42. Klepser ME, Wolfe EJ, Jones RN, Nightingale CH, Pfaller MA. Antifungal pharmacodynamic characteristics of fluconazole and amphotericin B tested against *Candida albicans*. Antimicrob Agents Chemother 1997; 41:1392–1395.
43. Lewis RE, Wiederhold NP, Klepser ME. In vitro pharmacodynamics of amphotericin B, itraconazole, and voriconazole against *Aspergillus, Fusarium,* and *Scedosporium* spp. Antimicrob Agents Chemother 2005; 49:945–951.
44. Lewis RE, Lund BC, Klepser ME, Ernst EJ, Pfaller MA. Assessment of antifungal activities of fluconazole and amphotericin B administered alone and in combination against *Candida albicans* by using a dynamic in vitro mycotic infection model. Antimicrob Agents Chemother 1998; 42:1382–1386.
45. Di Bonaventura G, Spedicato I, Picciani C, D'Antonio D, Piccolomni R. In vitro pharmacodynamic characteristics of amphotericin B, caspofungin, fluconazole, and voriconazole against bloodstream isolates of infrequent *Candida* species from patients with hematologic malignancies. Antimicrob Agents Chemother 2004; 48:4453–4456.
46. Turnidge JD, Gudmundsson S, Vogelman B, Craig WA. The postantibiotic effect of antifungal agents against common pathogenic yeasts. J Antimicrob Chemother 1994; 34:83–92.
47. Nosanchuk JD, Cleare W, Franzot SP, Casadevall A. Amphotericin B and fluconazole affect cellular charge, macrophage phagocytosis, and cellular morphology of *Cryptococcus neoformans* at subinhibitory concentrations. Antimicrob Agents Chemother 1999; 43:233–239.
48. Nugent KM, Couchot KR. Effects of sublethal concentrations of amphotericin B on *Candida albicans*. J Infect Dis 1986; 154:665–669.
49. Vitale RG, Mouton JW, Afeltra J, Meis JFGM, Verweij PE. Method for measuring postantifungal effect in *Aspergillus* species. Antimicrob Agents Chemother 2002; 46:1960–1965.
50. Vitale RG, Meis JF, Mouton JW, Verweij PE. Evaluation of the post-antifungal effect (PAFE) of amphotericin B and nystatin against 30 zygomycetes using two different media. J Antimicrob Chemother 2003; 52(1):65–70.

51. Clemons KV, Sobel RA, Williams PL, Pappagianis D, Stevens DA. Efficacy of intravenous liposomal amphotericin B against coccidioidal meningitis in rabbits. Antimicrob Agents Chemother 2002; 46:2420–2426.

52. Van't Wout J, Mattie H, Van Furth R. Comparison of the efficacies of amphotericin B, fluconazole, and itraconazole against systemic *Candida albicans* infection in normal and neutropenic mice. Antimicrob Agents Chemother 1989; 33:147–151.

53. Allende MC, Lee JW, Francis P, et al. Dose dependent antifungal activity and nephrotoxicity of amphotericin B colloidal dispersion in experimental pulmonary aspergillosis. Antimicrob Agents Chemother 1994; 38:518–522.

54. Sokol-Anderson ML, Brajtburg J, Medoff G. Sensitivity of *Candida albicans* to amphotericin B administered as single or fractionated doses. Antimicrob Agents Chemother 1986; 29: 701–702.

55. Lewis RE, Klepser ME, Piscitelli SC, et al. In vivo activity of high-dose liposomal amphotericin B in a neutropenic murine candidal thigh infection model [abstr]. In: Abstracts of the 39th Interscience Conference on Antimicrobial Agents and Chemotherapy, San Francisco, CA, 1999.

56. Adler-Moore JP, Olson JA, Proffitt RT. Alternative dosing regimens of liposomal amphotericin B effective in treating murine systemic candidiasis. J Antimicrob Chemother 2004; 54:1096–1102.

Pharmacokinetics and Pharmacodynamics of Azoles

Johan W. Mouton

Department of Medical Microbiology and Infectious Diseases, Canisius Wilhelmina Hospital, Nijmegen, The Netherlands

INTRODUCTION

The azoles comprise a wide spectrum of antifungal drugs with varying activity and pharmacological properties. Some of the agents, in particular the older ones, can only be used topically and were mainly used against nail and skin infections, while others can be used as systemic drugs. Some can be used intravenously, but oral formulations exist as well. For therapeutical purposes, a distinction needs to be made between those that are primarily active against nonfilamentous fungi, e.g., Candida spp., and Cryptococcus spp. only, such as fluconazole, and azoles that are active against filamentous fungi. Within the group of filamentous fungi however, large differences are readily apparent. In addition, because of their growth characteristics, laboratory techniques differ between the two groups to determine activity of the drugs. In that respect, nonfilamentous fungi behave more like bacteria as will be explained below.

Candida infections now rank among the top five of bloodstream infections with considerable mortality (1–3). It is therefore not surprising that most of our understanding of pharmacokinetic–PD (PK–PD) relationships stems from the activity of azoles against yeasts, in particular Candida species. In this chapter, the guideline will therefore be the exposure–response relationship of azoles against Candida species. Where applicable, differences for filamentous fungi will be highlighted.

The primary goal of this chapter is to give an overview of exposure–response relationships of azoles in vitro and in vivo. After a brief description and overview of the class, the pharmacokinetics of the major azoles is portrayed in order to discuss the exposure of the drug in human. Next, concentration–effect relationships are described, and finally the exposure–response relationship in vivo, both in animals and in men.

BRIEF HISTORY OF THE CLASS

During the Second World War, in the wake of the discovery of penicillin in the previous decade and the development to its clinical use, a quest for other drugs with activity against infectious agents led to the discovery in 1944 that benzimidazole possessed, apart from antibacterial, antimycotic activity (4). The activity was attributed initially to competitive inhibition between this drug and the purines adenine and guanine, but the mechanism of action later described proved to be inhibition of ergosterol synthesis through the cytochrome P-450 system (see below). In 1952, Jerchel (5) described several analogs that had profoundly increased antimicrobial activity. During the next decades, a large number of antifungals from this class were studied extensively to determine possible clinical use. The first azole to be used clinically was chlormidazole, used as a topical agent. Other compounds, still in use today, include clotrimazole and econazole but can be used as topical agents only. In general, imidazoles could not be used systemically, mainly because of hepatotoxicity and major drug interaction effects. A major area of application is hair, skin, and nail infections, because topical use is a good alternative in most cases. The only noticeable exceptions were miconazole (6) and, later, ketoconazole (7). Although the latter drug was widely acclaimed as a major improvement to existing therapeutic options because of its activity as well as availability as an oral formulation, there were still too many side effects for wide-scale systemic use.

A major breakthrough was the discovery of a second group of azoles, the triazoles. These differ from the imidazoles by a third nitrogen in the azole ring. Terconazole was the first agent undergoing clinical trials (8), followed by itraconazole and fluconazole. Of these, fluconazole is the only agent that is readily solvable in water. Since then, the number of azoles available has been increasing steadily: voriconazole, ravuconazole, posaconazole, and isavuconazole. Of these, voriconazole and posaconazole are currently available, ravuconazole is no longer pursued, and isavuconazole is still in the development phase. However, all of these suffer from the same drawbacks as itraconazole in that they are difficult to solve in water-based solutions and that oral formulations were the ones that were available first. Most of them use cyclodextrin to increase solvability in order to provide an intravenous formulation. A new approach was taken for isavuconazole. This compound is linked to an ester to form a prodrug. The ester is almost immediately cleaved by plasma enzymes resulting in the availability of active drug once taken up systemically (9). A summary of the PK characteristics of the azoles available for systemic use is shown in Table 1. The PK characteristics of each of these compounds are briefly discussed.

Fluconazole
Fluconazole was the first nontoxic antifungal agent available both as an oral and as an intravenous formulation and was the result of a development program aimed at the development of a broad-spectrum antifungal agent active by both the oral and the intravenous routes for the treatment of superficial and systemic infections (26). The volume of distribution is around 0.7 L/kg, indicating distribution over total body water. This is also reflected by a cerebrospinal fluid/plasma ratio of 0.86 (27) and microdialysis studies (28). Bioavailability is good, close to 100%. Even in intensive-care patients receiving fluconazole by a feeding tube (29) and critically ill surgical patients with invasive mycoses and compromised gastrointestinal function, there was no significant difference between exposures of intravenous and oral dosing (11). Elimination is primarily by renal excretion. Protein binding of fluconazole is relatively low, 12% in healthy volunteers as determined by ultrafiltration (20). Peculiarly, protein binding was increased to 23% in patients with chronic renal failure and cancer (30) and might be related to higher concentrations of α 1-acid glycoprotein in these patients (20).

Pharmacokinetics was extensively studied in children. The half-life was increased in neonates reflecting the renal function in these patients and was

TABLE 1 Overview of Pharmacokinetic Properties of Azoles

Drug	Formulation/dose	AUCs1101s/24 (mg.hr/L)	V_{dss}[a] (L/kg)	Half-life (hr)	Bioavailability	Fu	Major route elimination	References
Fluconazole	i.v. 400 mg	400	0.65	30	–	0.88	Renal	10–13
	Oral 400 mg	350	0.75	30	90–100	–	–	10–12
Itraconazole	Oral capsules 200 mg	13–15	–	36	40–100	–	Hepatic	14–17
	Oral solution 200 mg	18–22	–	–	60–100	–	–	18,19
	i.v. cyclodextrin 200 mg	32		26–65	–	0.03	–	14,20–22
	i.v. nanocrystals 200 mg	40	550 L	26–61	–	–	–	21
Voriconazole	i.v. 6 mg/kg	22	4.6	6	–	0.42	Hepatic	23
	Oral 400 mg	18–22	–	–	90–100	–	–	23
Posaconazole	Oral 400 mg	68	486 L	31	–	–	–	24
Isavuconazole	i.v. 100 mg	236	542 L	117	–	0.02	–	9,25
	Oral 100 mg	255	308	84.5	90–100	–	–	9,25

[a]V_{dss}/F for oral administration.
Note: The half-life after a single dose was 26 hours.
Abbreviation: i.v., intravenous.

329

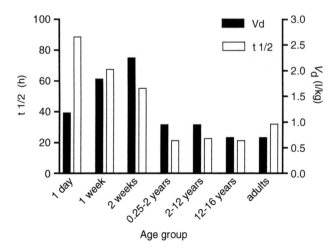

FIGURE 1 Half-life and volume of distribution of fluconazole in various age groups. Based on data in Refs. 31 and 32 presented at Trends in Medical Mycology, Berlin 2005. *Abbreviation*: V_d, volume of distribution.

slightly lower in other age groups (31,32). However, an age dependency exists and this does result in different exposures in various age groups when the same dose is administered (Fig. 1). In addition, the variation in plasma levels is much higher in some age groups, and this may lead to underdosing in some patients.

Itraconazole

The formulation used initially was an encapsulated form, but the absorption of itraconazole in a subset of immunocompromised patients was not optimal, and its pharmacokinetics varied considerably between patients (14,33–35). In addition, absorption was highly influenced by gastric pH and the use of antacids (36,37) as well as by concomitant food intake (15,16). An oral solution based on hydroxypropyl-beta-cyclodextrin became available in the late 1990s, which showed a more favorable PK profile (17,18,38–43) and is superior to capsules in preventing invasive Aspergillosis infections (44). However, the variability in pharmacokinetics was still high in some patient groups and oral treatment was not suitable for those high-risk patients who were unable to tolerate food and drink due to chemotherapy-induced mucosal barrier injury. The development of formulations for intravenous use was hampered by the poor solubility of itraconazole in water, but eventually a solution based on dextrin was made available in the United States and elsewhere. However, the presence of high amounts of dextrin can limit the use of higher doses even though the dextrin compound is readily eliminated (19) and can be dialyzed (45). This led to the exploration of other solubles. Lately, a new formulation using nanocrystals as a solvent has been administered successfully (21). However, this formulation is not (yet) available for commercial use. An interesting feature of itraconazole is its increase in terminal half-life after multiple dosing (14,21) reflecting extensive tissue distribution and possibly, a rate limiting step in its metabolization and/or excretion. However, concentrations of the OH metabolite did not support this. Protein binding of itraconazole in serum is relatively high and was 97% in both volunteers and patients with end-stage renal disease (20), while it was slightly lower in patients with diabetes mellitus (46). However the drug is also bound to erythrocytes, further limiting its availability (47). Pharmacokinetics in children was studied in a number of age groups (42). In general, the pharmacokinetics was well comparable between age groups, although the half-life in children between 0 and 2 years was slightly increased.

Voriconazole

Voriconazole was developed to enhance activity against moulds (48). It is available both as an oral and as an intravenous formulation. Its oral availability is estimated to be more than 90% and unaffected by gastric pH (49,50,52). Concomitant intake with food reduces the bioavailability around 20% as determined from a comparison of AUCtau (50). The half-life is six to seven hours and thus has by far the highest elimination rate of the triazoles. The drug is administered twice a day. Like itraconazole, it takes a number of days to reach steady state, even using loading doses on the first day, and shows nonlinear pharmacokinetics in that the C_{max} and AUC show a disproportional increase in dose (51), and accumulates more than can be predicted from single dose data (52). The AUCtau increases almost fivefold during this period (49). Voriconazole is primarily eliminated via metabolization. Around 80% of the metabolite is excreted by the kidneys, and the remainder can be retrieved in feces (23,53). Less than 2% is excreted unchanged into urine. Serum concentrations can vary widely, probably through individual differences in metabolization. A number of authors have therefore suggested that drug monitoring would be necessary for this agent, to ensure adequate serum concentrations as well as to prevent toxicity (54,55).

Posaconazole

Posaconazole is a new triazole that has recently become available as an oral formulation. The structure of posaconazole is not very different from that of itraconazole and the PK properties do not markedly differ. It is extensively distributed with a volume of distribution almost 10 times that of body water and has a long half-life of over 30 hours both after single and multiple doses. The increase in AUC is dose proportional up to 800 mg/day (56). Absorption is almost fourfold increased by taking the drug together with high-fat food and the oral suspension had a significantly increased bioavailability compared to a table formulation (57). Almost 80% of the drug can be retrieved from feces, indicating that biliary excretion is the major route of elimination. Most of the drug is excreted by the fecal route and most of the remainder as metabolites via the kidneys. Like itraconazole and voriconazole, there is a large variation in serum concentrations in individual patients (58), which may partly be due to differences in absorption, but also due to differences in metabolization. However, as yet there are no recommendations for therapeutic drug monitoring.

Isavuconazole

Isavuconazole (BAL4815) is another new triazole, but is not yet available. To enhance its solubility and improve its mode of administration, isavuconazole is administered as a prodrug (BAL8557), both for oral and for intravenous administration. In serum, the drug is rapidly converted by plasma esterases. The drug accumulates extensively in tissues with a volume of distribution similar to posaconazole and itraconazole (9,25). The half-life is by far the longest of the triazoles with values up to five days in healthy volunteers. Exposures after intravenous and oral administration are similar, indicating a bioavailability nearing 100% (25). The protein binding of isavuconazole is high with reported values of up to 98%.

Mode of Action

Triazole antifungal agents exert their action by inhibiting the conversion of lanosterol to ergosterol. It is thought that the nitrogen of the azole ring binds to the hem moiety of the fungal cytochrome P450 enzyme lanosterol 14-α-demethylase. Ergosterol is an essential component of the fungal membrane, comparable to that of cholesterol in the mammalian cell, and depletion of this molecule finally ensues in disruption of the membrane (59). Alternatively, severe depletion of ergosterol may interfere with other important functions in the cell.

CONCENTRATION–EFFECT RELATIONSHIPS IN VITRO

The most common concentration–effect relationship measured is susceptibility testing, i.e., determining the activity of the drug under standard conditions.

Susceptibility Testing of Yeasts

Susceptibility testing of antifungals differs fundamentally from susceptibility testing of antibacterials and this is perhaps one of the reasons that good correlations between in vitro activity and in vivo efficacy have been difficult to show until recently. If the activity or potency of a drug is measured in an in vitro system, there are basically two important issues to consider. The first is the method as such, including reproducibility, ease of use, etc. The second issue is the interpretation of the test result itself.

Two susceptibility methods are currently used. One is the method as published by Clinical Laboratory Standards Institute (CLSI) (60) in the United States and the other by the European Committee on Antimicrobial Susceptibility Testing (EUCAST) (61), in Europe. The major difference between the two methods is that the medium used by the EUCAST method includes the addition of more glucose to RPMI cell culture, thereby increasing the growth rate of the yeasts, and this permits reading after 24 hours instead of the 48 hours as prescribed by the CLSI method (Table 2). The results of the two methods have been shown to be reasonably comparable in that the resulting minimal inhibitory concentration (MIC) distributions show similar shapes and have a good correlation (62). Because the reading following the EUCAST method is after 24 hours, the resulting MICs are slightly lower than those of the CLSI (63). However, to determine the exact differences between the two methods, the differences in MICs of the individual strains should be compared (64). The difference in $\log_2$ dilutions with their confidence intervals will give a good idea of the difference in test result and, if there is a difference, whether this is important quantitatively. Unfortunately, this is very seldom performed. An example is shown in Table 3 for *Pichia anomala*.

The interpretation of the test result is a matter of long debate. As can be observed from Figure 2, the difference between values exerting a minimum and a maximum drug effect extends to four twofold dilutions and the MIC value thus varies with the definition of inhibition. While full inhibition of growth—as determined by eye or read by a machine—is used for antibacterials, and the difference between growth and no growth is one twofold dilution at most (in the same test),

TABLE 2 Comparison Between the EUCAST AFST Method and the CLSI AFST Method for Azoles

Method	Glucose supplementation (%)	Shape of well	Inoculum (CFU/mL)	Incubation time (hr)	Reading	End point for azole drugs
EUCAST AFST	2	Flat bottom	0.5×10^5–2.5×10^5	24	Spectrophotometric	Lowest concentration of drug that inhibits growth by 50% of that of the control
CLSI M27-A2	0.2	Round bottom	0.5×10^3–2.5×10^3	48	Visual	Lowest concentration of drug that inhibits growth substantially compared with that of the control

Abbreviations: EUCAST, European Committee on Antimicrobial Susceptibility Testing; CLSI, Clinical and Laboratory Standards Institute; AFST, antifungal susceptibility test.
Source: Adapted from Ref. 62.

TABLE 3 Distribution of 58 *Pichia anomala* Isolates According to Differences in MIC Results Obtained by EUCAST Method Compared with CLSI Method

Antifungal Drug	No of isolates for which EUCAST-MICs differed from CLSI-MICs							Agreement within no of dilution (%)		Mean difference (log$_2$ values)	95% Confidence Interval
	−3	−2	−1	0	1	2	3	±1	±2		
Fluconazole		6	8	33	9	2		86	100	−0.12	−0.36 to 0.12
Itraconazole	1	2	14	30	10		1	93	96	−0.14	−0.36 to 0.11
Voriconazole		3	23	26	4	2		95	100	−0.43	−0.62 to 0.25

Abbreviations: EUCAST, European Committee on Antimicrobial Susceptibility Testing; MIC, minimal inhibitory concentration; CLSI, Clinical and Laboratory Standards Institute.
Source: From Ref. 65.

the MIC for azoles as currently defined is 50% inhibition ("significant inhibition of growth"), either read by eye (introducing a subjective interpretation) or by an alternative technique. The use of 50% inhibition was arbitrary initially, and other % inhibition is being used for other drugs. For instance, full inhibition (no turbidity) is used for amphotericin B, even if the so-called trailing effect—an effect sometimes observed that represents a relatively slower decrease in turbidity at the higher concentrations as compared to lower concentrations and is fairly typical for azoles—is taken into account. However in terms of effect measurements and the concentration range over which the drugs exert their action, 50% inhibition is a very logical choice, since this equals to the EC50 of the drug. As can be observed from Figure 2, the concentration–effect relationship of, in this example, fluconazole follows a classical sigmoid concentration–effect relationship and is characterized by the EC50, the Hill coefficient, and the minimum and the maximum effect. This offers the possibility to determine the EC50 exactly, if a more precise way of measurement is used than reading by eye. For yeasts, the simplest and easiest alternative is to use optical density (OD) as a parameter. This process can even be automated and the EC50 determined automatically. In general, automation and standardization of the whole susceptibility procedure was shown to be more optimal than currently applied in most laboratories (66). However, the use of the EC50 as described here has not been applied systematically.

How then, should the MIC of an azole be interpreted? For antibacterials, the concentration–effect curve is relatively steep, as follows from a clear distinction between growth and no growth in the two adjoining wells in a two fold dilution tray, and the concentration where the MIC is read is inhibitory (at least), although the number of microorganisms may be anywhere between 0 and 10^7 (67). The interpretation of a value that is read at 50% inhibition is more difficult and how this value is to be translated to a static effect remains uncertain. Thus, even when the MIC results are reproducible, and the term inhibitory is used, the absolute value needs to be interpreted with far more caution, much more so than for antibacterials.

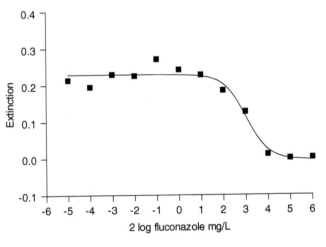

FIGURE 2 Relationship between optical density of *Candida albicans* and the effect of fluconazole demonstrating the four to five twofold dilutions between minimum and maximum effect. *Source:* Unpublished data.

Susceptibility Testing of Filamentous Fungi

Susceptibility testing of filamentous fungi bears some resemblance to yeasts, but there are a number of additional problems in interpretation of the test result that one needs to be aware of. The MIC of azoles is read at absence of growth (68), although there is an exception for some species. The concentration–effect relationships show a similar pattern as for yeasts, that is the difference between minimum effect and a maximum effect is around four twofold dilutions. However, while OD measurements reasonably well describe the biomass for yeasts and bear a reasonable correlation with the number of colony forming units (CFU), this is not true for filamentous fungi. In broth, the fungi grow as hyphens in an erratic and inhomogeneous fashion. Reading by eye is the current standard to interpret growth in each well but has a subjective interpretation and adds to the variability in testing. Alternatively, a number of other methods have been developed, which reflect the biomass of filamentous fungi, including spectrophotometric, colorimetric, flowcytometric, and radiometric assays (69,70). The most promising of these are spectrophotometric methods that have been developed based on the reduction of tetrazolium salts. An important characteristic of these drugs is that they are reduced to highly colored formazans that are produced when the tetrazolium salts receive electrons that are, for instance, produced within the mitochondria of cells with dehydrogenase activity (71). Thus, the formation of formazan reflects cell activity and thereby can be used as a measure of biomass. There are a number of dyes that have been introduced over the last couple of years, of which MTT and XTT are the most promising. A major advantage of these methods is that they permit reading after 24 hours of incubation instead of 48 or 72 hours. Also, these methods permit a much more precise measurement of activity of a drug. This is also an advantage in testing combinations of drugs (72). A number of studies have shown excellent agreement between these colorimetric methods and the CLSI method (73–75). An example is shown in Figure 3, demonstrating the relationship

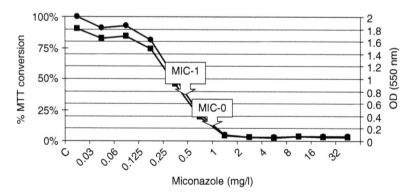

FIGURE 3 Example of the use of MTT in testing of susceptibility of *S. apiospermum* strain to miconazole. The curve with the circles shows the percentage of optical density reduction compared with the drug-free well and the curve with the squares presents the absorption of formed formazan at 550 nm. The MIC-0 and MIC-1 determined by CLSI method and were 1 and 0.5 mg/L, respectively. *Abbreviations*: MIC, minimal inhibitory concentration; CLSI, Clinical and Laboratory Standards Institute. *Source*: From Ref. 76.

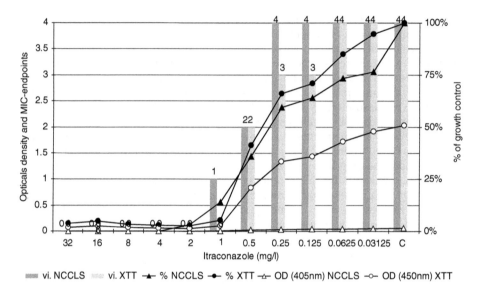

FIGURE 4 Results of susceptibility testing of an *Aspergillus ustus* strain against itraconazole by visual and spectrophotometric reading of CLSI and XTT method after 24 hours of incubation. The bars represent the MIC end points obtained by visual reading of the CLSI methods (*dark bars*) and the XTT method (*light bars*) in the scale from 0 (absence of growth or color) to 4 (no reduction of growth or color compared with drug-free control). The curves with the circles represent the OD at 450 nm (*open symbols*) and the relative OD (percentage) (*closed symbols*) obtained by the XTT method. The curves with the triangles represent the OD at 405 nm (*open symbols*) and the relative OD (percentage) (*closed symbols*) obtained by the CLSI method. *Abbreviations*: CLSI, Clinical and Laboratory Standards Institute; MIC, minimal inhibitory concentration; OD, optical density; NCCLS. *Source*: From Ref. 74.

between MTT conversion and the concentration of miconazole. Figure 4 shows the relationship between OD and itraconazole concentration using various methods of reading, both by eye and using a spectrophotometer.

The Postantifungal Effect
The determination of the postantifungal effect (PAFE) in vitro of yeast-like organisms is relatively straightforward and is similar to the method as described for bacteria. As described for bacteria, an inoculum of microorganisms is exposed to a certain concentration of drug (usually four or five times the MIC) for a limited time (usually one hour). The antimicrobial is then washed out by dilution or degraded by enzymes such as beta-lactamases or aminoglycosidases. The post-antibiotic effect (PAE) is then defined as the time it takes the curve to increase one 10log in comparison with the control (77). The original assumption behind the value of the PAE was that those drugs that do exhibit a PAE might be administered less frequently by making the incorrect inference that the drug would still be active once the concentrations declined below the MIC for a time period similar to the PAE (78,79). The term incorrect is used here, because the conditions under which the PAE is determined in vitro—delay in growth after one hour of exposure and immediate removal of the drug—are vastly different from the supposed effect, a decline below the MIC after several hours of exposure with a slow decline in

concentration dependent on the rate of elimination. To determine a "true" PAE, the concentration–time curve should be simulated in vitro, or the effect should be determined in vivo. Still, an in vitro PAE may have meaning in some situations, in particular when the PAE is relatively long.

The PAFE of azoles has been determined for Candida spp. and cryptococci following classic methodology (80,81). In general, all azoles display a moderate PAFE except for fluconazole. However, the PAFE is dependent on a number of critical factors, including stage of growth, temperature, and medium (82–86). The importance of the PAFE is therefore uncertain.

The PAFE of filamentous fungi is less straightforward for a number of reasons, but two of these are particularly important. The first is that it is almost impossible to perform reliable, reproducible cfu determinations of filamentous fungi. This is inherent to their mode of growth, i.e., the formation of filaments. Even the procedure of simple dilution will result in considerable interassay variation because of the disruption of the hyphae and the presence of multiple cells within one hyphae. The number of cfu therefore does not represent total biomass and is an unreliable method to determine relative biomass as well. Thus, to determine a PAFE, the use of cfu as a measure to compare a control curve to exposed fungi is not adequate. Three methods have been suggested to solve this. The first is the use of MTT and determine the delay in growth by comparing the growth control curve with that of the exposed curve (87). The second method uses (11) C-labeled amino acid accumulation in antifungal drug-pretreated mycelia (88). Azoles showed a PAFE of less than 0.5 hours, much less than comparators such as amphotericin B and caspofungin. Finally, a method has been proposed based on the production of CO_2 (89).

COMPARATIVE ACTIVITY OF AZOLES IN VITRO

The activity of azoles has been compared in a number of studies. Table 4 displays the comparative activity of fluconazole, itraconazole, voriconazole, and isavuconazole. In general, all azoles, except for fluconazole, display activity against both yeasts and filamentous fungi. The differences between itraconazole, voriconazole, and isavuconazole are only marginal. Itraconazole is slightly less active against the dermatophytes compared to the other two azoles.

Emergence of Resistance

Although emergence of resistance is not a topic in this chapter, a brief overview is given here. In general, resistance development is much less of a problem than in bacteria, which may be linked to the increased complexity of the organisms. For *Candida albicans*, a number of reports have indicated that emergence of resistance does occur during treatment and follows sequential steps that lead to increasing MICs (94). However, perhaps more importantly, a shift in species that are inherently less susceptible to azoles causing infections is observed in particular in non-HIV patients (95–97). This was shown to be dose or dosing regimen dependent for fluconazole (98). Thus, although the mechanism and path to more resistant strains may be different, the end result, patients being infected by less susceptible strains, is the same. For infections caused by *C. albicans*, decreased susceptibility may in some instances be treated with increased doses (see below). In *Candida glabrata*,

TABLE 4 Comparative Susceptibilities (mg/L) of Azoles for a Selected Number of Species

Species (n)	Fluconazole	Itraconazole	Voriconazole	Posaconazole	Isavuconazole
Trichophyton mentagrophytes (19)	32 (8–64)	1 (0.25–16)	0.5 (0.5–1)	–	1 (0.5–2)
Trichophyton rubrum (9)	4 (1–16)	0.5 (0.25–0.5)	0.25 (0.25–0.5)	–	0.25 (0.125–0.25)
Microsporum canis (12)	8 (2–32)	0.5 (0.063–1)	0.5 (0.25–0.5)	–	0.25 (0.125–0.5)
Candida albicans FLC susc (19)	0.25 (0.125–4)	0.012 (0.008–0.025)	0.008 (0.008–0.25)	–	0.008 (0.008–1)
Candida albicans FLC res (18)	64 (4–128)	2 (0.5–32)	2 (0.5–32)	–	2 (0.125–8)
Candida glabrata (19)	16 (4–128)	1 (0.5–32)	1 (0.5–8)	–	0.5 (0.032–8)
Candida krusei (5)	64 (5–64)	0.5 (0.063–2)	0.5 (0.063–4)	–	0.25 (0.25–2)
Aspergillus fumigatus (62)	–	0.25 (0.125–8.0)	0.25 (0.125–8.0)	–	0.5 (0.125–2)
Aspergillus terreus (18)	–	0.125 (0.06–0.5)	0.25 (0.25–0.5)	–	0.5 (0.25–0.5)
Aspergillus flavus (20)	–	0.25 (0.25–0.5)	0.5 (0.25–1)	–	0.5 (0.5–2.0)
Aspergillus niger (18)	–	0.5 (0.25–0.4)	0.25 (0.25–1.0)	–	0.5 (0.25–2.0)
Absidia spp. (n = 8)	–	4 (0.5 to > 16)	>16 (8→16)	0.5 (0.25 to > 16)	–
Mucor spp. (n = 19)	–	4 (1 to > 16)	>16 (8→16)	0.5 (0.25 to > 16)	–
Rhizopus (n = 11)	–	4 (0.05–16)	16 (8→16)	0.5 (0.25 to > 16)	–

upregulation of CDR1 and CDR2 genes results in resistance to other azoles as well and subtherapeutic doses of fluconazole may thus lead to emergence of resistance for all azoles (94). In filamentous fungi, emergence of resistance is very rare. The most prevalent mechanism of azole resistance in Aspergillus spp. appears to be due to specific mutations in *cyp51A* gene, leading to different modes of resistance and cross-resistance (99).

CONCENTRATION AND DOSE EFFECT RELATIONSHIPS IN VIVO

Since most of the early azoles were applied topically, concentration and dose–effect relationships were not studied in detail for these drugs. Even with the advent of miconazole and ketoconazole, the direction of research was more directed toward toxicity rather than efficacy. However, translational research from animals to humans focusing on PD relationships is increasing and indicates that the principles that apply for antibacterials (100) are equally valid for antifungals, in that one has to determine exposure–response relationships to determine efficacy, and from those relationship draw conclusions with respect to the application of adequate dosing regimens. These relationships are easiest determined in animal model systems. These can then be translated taking human pharmacokinetics into consideration. Alternatively, these relationships can be determined from analysis of studies in humans. Both approaches are discussed below. The relationships studied all apply to yeasts; proper studies involving filamentous fungi are still lacking for azoles.

PD Relationships in Animals

Exposure–response relationships of azoles have primarily been determined for yeasts. By administration of varying doses and dosing intervals, the relationship between dose or exposure and effect was determined following the same methods as described elsewhere in this volume for antibacterials. An example is shown in Figure 5 for fluconazole. Female CBA/J mice were infected intravenously with *C. albicans* two hours before starting therapy. Groups of two mice were treated for 72 hours with dosing regimens of fluconazole using twofold increasing total doses

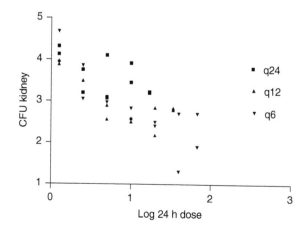

FIGURE 5 Relationship between the 24-hour total dose and effect of fluconazole in a non-neutropenic murine model of disseminated candidiasis for three dosing intervals. Difference in CFU at the start of therapy and after three days of therapy. *Abbreviation*: CFU, colony forming units. *Source*: Unpublished data.

administered i.p. at dosing intervals from 6, 12, and 24 hours. Total doses ranged from 1.25 to 160 mg/kg/24 hr. After 72 hours of therapy, kidneys were removed for CFU determination. Untreated control mice were sacrificed just before treatment and at $t = 72$ hr. Figure 5 shows the relationship between the 24-hour dose and number of cfu after 72 hours of treatment. From this relationship, the conclusion could be drawn that there is a direct relationship between 24-hour dose and response. However, a closer look at the relationship seems to indicate that the effect of the dosing regimen with the 24-hour interval is less pronounced than the dosing regimens with a 6- or 24-hour interval. There are several explanations for this, but the most obvious are differences in exposure due to PK properties of the drug. Importantly, the half-life of fluconazole in this experiment was 3.6 hours, and accumulation occurs therefore less during the 24-hour dosing regimens compared to shorter dosing intervals. To further characterize the exposure–response relationships, the fAUC/MIC, fC$_{max}$/MIC, and %fT$_{>MIC}$ were determined for each dosing regimen. Figure 6 shows the relationships between these PD indices (PDI) and the difference in colony forming units between start of treatment end of treatment. Each of the three indices shows some correlation with effect, although this is most pronounced for the fAUC/MIC ratio. However, there is also a reasonably good relationship with the %fT$_{>MIC}$.

From the exposure–response relationships, the dose and PDI can be calculated that results in no net change in cfu during treatment, the so-called static effect. This measure of antimicrobial activity has shown to be a good parameter to describe the overall activity of the drug. Here, the fAUC/MIC ratio that was needed to result in a net static effect was 24.3.

The exposure–response relationship is not only dependent on exposure but also on the susceptibility of the microorganism as signified by the MIC component in the PDI. Figure 7 (left panel) shows the relationship between dose and exposure for 10 Candida strains with different MICs for fluconazole in a neutropenic mouse model of infection. It is apparent that a clear relationship between dose and response does exist, and increasing doses result in increased killing of the Candida strains. However, the extent of killing differs for each strain. The relationship between AUC and effect would show similar curves. Figure 7 (right panel) shows the same relationship, but normalized to the MIC by determining the AUC/MIC ratio. From these experiments, it is obvious that the relationship between exposure and effect follows a typical pattern that can be conceived as a general

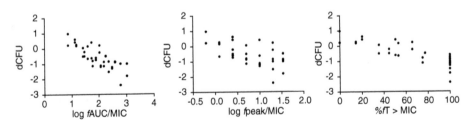

FIGURE 6 Relationship between three PDIs and effect of fluconazole in the same model. The 0 represents no net growth; lower values represent net killing. Each symbol represents data for one mouse (mean CFU of two kidneys). *Abbreviations*: AUC, area under the curve; dCFU, difference in CFU. *Source*: Unpublished data.

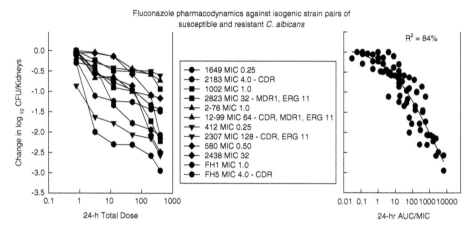

FIGURE 7 Relationship between total daily dose (*left panel*) and AUC/MIC (*right panel*) and dcfu for 10 isogenic Candida strains with varying MICs in a neutropenic mouse model of infection. The response curves are identical for all 10 strains when normalized to the MIC of each strain. *Abbreviations*: AUC, area under the curve; MIC, minimal inhibitory concentration; CFU, colony forming units. *Source*: From Ref. 101.

characteristic for the effect of fluconazole on Candida. Similar findings have been reported by other authors (102).

The relationship between exposure and effect has also been determined for other azoles. For all azoles, a relationship between AUC and effect has been demonstrated. However, the magnitude of the AUC that correlates with a static effect differs for each azole. The AUC needed for a static effect for posaconazole for instance is much larger than for fluconazole. The single most important reason of this difference is the difference in protein binding. When the magnitude of the PDIs resulting in a static effect is recalculated for the free fraction of the azole only, a similar pattern appears for each azole (Fig. 8) (101). For all four azoles, a

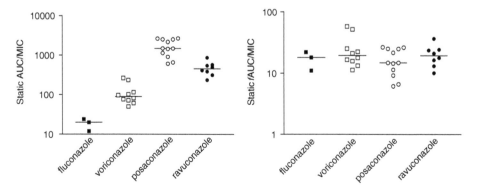

FIGURE 8 Exposures expressed as AUC/MIC ratio resulting in a static effect for total drug (*left panel*) and free drug (*right panel*) in a mouse model of infection. The static effect for unbound drug is comparable for all four agents. *Abbreviations*: AUC, area under the curve; MIC, minimal inhibitory concentration. *Source*: From Refs. 103–105.

*f*AUC/MIC ratio of around 25 is needed for a static effect. Thus, within the class of azoles, the effect of the various members of that class is very comparable. This is similar to the observations made for antibacterials. For instance, the *f*AUC/MIC ratio resulting in a static effect for fluoroquinolones and pneumococci is around 35 and the %*f*T$_{>MIC}$ for cephalosporins and enterobacteriaceae 35% to 39%. It must again be emphasized here that the value of 25 is the ratio of two factors, i.e., the *f*AUC and the MIC. The values established here were obtained using the CLSI method; the EUCAST method would result in higher values because the MICs are slightly lower.

PD Relationships in Humans

The relationship between exposure and effect has lately been determined in humans as well. An example is shown in Figure 9 showing the relationship between dose/MIC and efficacy of fluconazole treatment for esopharyngeal candidiasis (OPC) (107). In this study, 132 patients were treated with varying doses of fluconazole, and the effect of therapy was determined by culturing before and after therapy. The MIC of each Candida strain was also determined. As can be observed, a clear relationship between dose/MIC and response exists. The EI-50 (value of PDI that results in 50% effect) in this study was 43.7, while for a maximum response a value of close to 100 was needed. It must be borne in mind however that the MICs were determined following the EUCAST method, and the results thus do not translate directly to the results in a quantitative sense as obtained in other studies that used the CLSI method.

Recently, another study was published where the authors looked at the relationship between dose, exposure, and effect in patients with a candidaemia (Fig. 10). As can be observed, a good relationship was found between dose/MIC

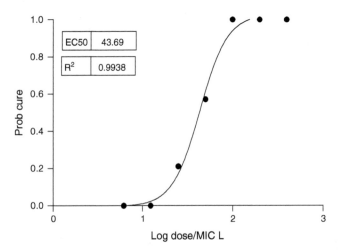

FIGURE 9 Relationship between dose and response in patients with oropharyngeal candidiasis. *Abbreviation*: EC-50, 50% effectiveness concentration. *Source*: From Ref. Rodriguez Tudela JL, Almirante B, Rodríguez-Pardo D, et al. Correlation of the MIC and Dose/MIC ratio of fluconazole to the therapeutic response in of patients with candidemia and mucosal candidosis. Submitted for publication.

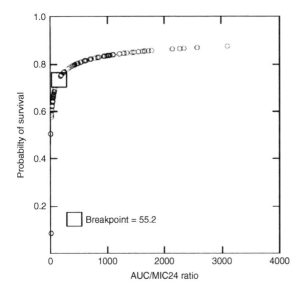

FIGURE 10 Relationship between AUC/MIC of fluconazole and survival. Using CART® analysis, a value of 55.2 was found to be discriminative between patients that had a high and a low probability of survival, respectively. *Abbreviations*: AUC, area under the curve; MIC, minimal inhibitory concentration. *Source*: From Ref. 106.

and mortality. Classification and regression tree analysis was performed in this study, and value of 55.2 was found to be discriminative between patients with a poor and a better outcome.

An analysis of compiled data from four studies that looked at the correlation between dose/MIC ratios and efficacy also seems to point in the same direction. Figure 11 shows an analysis of the data provided in a recent review by Pfaller

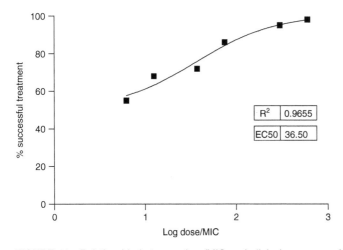

FIGURE 11 Relationship between dose/MIC and clinical response after treatment with fluconazole. The values for the dose/MIC were derived from the upper class limits of the dose/MIC ratios as provided in reference (108). The Hill equation was fit using Graphpad Prism 3.0 (Graphpad Inc., San Diego, California, U.S.). *Abbreviations*: MIC, minimal inhibitory concentration; EC-50, 50% effectiveness concentration.

et al. (107). The efficacy of fluconazole therapy was plotted as a function of the dose/MIC ratios. Subsequently, the Hill equation was used to fit to the data to determine the EI-50. From this analysis, it appears that the EI-50 is 35.5 and thus follows the same pattern as found for the other studies and further demonstrates this relationship between dose/MIC ratio and outcome.

These two studies show that the exposure–response relationships that have been observed in animal models of infection are comparable to those in men. In itself this would seem obvious, because antifungals exert their effect on the fungus and not on the host. Yet, host factors play major role in determining the outcome of fungal disease and are at least as important as antifungal therapy.

There are only few reports on the exposure–response relationships for other azoles.

For itraconazole, it was shown that a trough level of more than 0.5 mg/L during prophylaxis resulted in significantly less breakthrough infections in neutropenic patients (108). Since the dosing regimens in these patients were standardized, it is highly probable that the trough level correlate well with the AUC of itraconazole and that it was exposure itself that determined this relationship. It would be worthwhile to determine that relationship to resolve the correlation between AUC/MIC and efficacy. In another recent study, a difference in the efficacy of prophylaxis of Aspergillus infections was found between the oral solution and the capsule formulation of itraconazole (44). This was attributed to the difference in bioavailability, and thus exposure of the drug. For voriconazole, an analysis was performed on the results of the phase 2 and phase 3 trials (109). Although there were some differences in dosing regimens, a relationship between MIC and clinical response could be determined. In neither of these three studies a full quantitative analysis between a PDI and response was performed.

PREDICTION OF RESPONSE

The ultimate purpose of understanding PK–PD relationships is to be able to provide the choice of therapy and dosing regimen to optimally treat patients. As discussed in the previous paragraphs, this is dependent of three factors: the host, the exposure, and the susceptibility of the microorganism. While conditions of the host (e.g., neutropenia and age) can be regarded—from the exposure–response point of view—as external factors, the exposure–response relationship taking exposure and the susceptibility of the microorganism into account is the topic of this discussion.

Optimizing exposure–response relationships of azoles includes several steps and follows the same line of reasoning as discussed elsewhere in this volume (Chapter 2). The first is defining the optimal target exposures of the drug. Using the PK–PD relationships for azoles, these can be established as the percentage of the population that has reacted favorably to therapy to a certain drug exposure. For instance, a PK–PD target that results in 90% or 100% cure could be a reasonably target value. From Figure 9 one would conclude that a drug/AUC ratio of 100 would be the PK–PD target value for treatment of OPC (Rodriguez Tudela JL, Almirante B, Rodríguez-Pardo D, et al. Correlation of the MIC and Dose/MIC ratio of fluconazole to the therapeutic response in of patients with candidemia and mucosal candidosis. Submitted for publication). Alternatively, the results from

CART® analysis (Fig. 10) show that a value of 55.2 discriminates between a high and a low probability of successful treatment, and as a treating physician one would like to be reasonably certain that the patient would reach that PK–PD target. Since the PK–PD target depends on two parameters—the MIC and the exposure—both of these contribute to optimization of the dosing regimen. If the MIC of the target microorganism is known, or the distribution of MICs of the target microorganisms is known, the dosing regimen needed resulting in a high probability of cure can be derived from the PK properties of the drug. Alternatively, if a standard dosing regimen is applied, the PK properties of the drug will indicate what the highest MICs of the microorganisms may be in order to guarantee a high probability of cure. Since the dosing regimen is reasonably well established for most drugs, and the MIC of a microorganism causing the infection is not immediately known in most cases, the PK properties of the drug effectively determine whether there will be high probability of cure, given the MIC.

However, as explained elsewhere in this volume (Chapter 2), the pharmacokinetics in each patient does differ, both to differences in clearance and to differences in volume of distribution. To take these variance into the equations Monte Carlo simulations (MCS) can be used to determine the probability of attaining a specific PD target, and this technique has been used extensively over the last years to determine the probability of target attainment for antibacterials. For antifungals, MCS is still in its infancy. There are several reasons for that, the most important one that for a meaningful outcome of MCS, the underlying assumptions need to be reasonably well attested. As explained above, data involving the PK–PD target have become available only recently.

The construction of a PK model to reflect the dispersion of exposure within the target population, i.e., the population that is to be treated with the drug, is not different as described for antibacterials. Ideally, a PK population model is built to that purpose and the parameter estimates and measures of dispersion used to

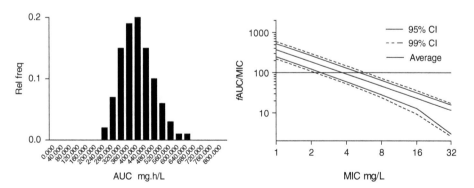

FIGURE 12 Results of Monte Carlo Simulations of fluconazole. Left panel: distribution of AUCs after a 400 mg dose. Right panel: Probabilities of target attainment for fluconazole i.v. for 400 mg/day in steady state with 95 and 99% confidence intervals. The horizontal line indicates the PK–PD target *f*AUC/MIC of 100. The following pharmacokinetic parameters were used to perform the MCS (10,111): Vd of 45L, CV 12%; half-life of 32 hours, CV 15%; and fraction unbound 88%. Presented at Trends in Medical Mycology, Berlin 2005. *Abbreviations*: i.v., intravenous; MIC, minimal inhibitory concentration.

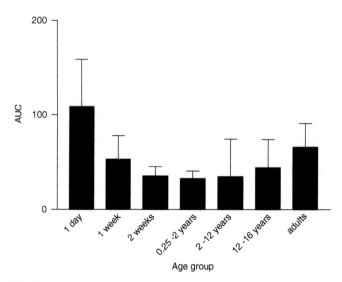

FIGURE 13 Exposure of fluconazole (total drug) in various age groups per unit dose (mg/kg). Administration of the same unit dose results in a lower exposure in some age groups. *Source*: Presented at Trends in Medical Mycology, Berlin 2005.

perform the MCS. Although there is ongoing discussion as to what the optimal approach is to obtain the simulations, this is—using adequate software—relatively straightforward, and the discussion here is mainly focused on whether the PK population model used to perform the MCS adequately reflects the population where the drug is used.

MCS has been performed for fluconazole and isavuconazole (110). Figure 12 shows the results of MCS of fluconazole. The left panel shows the distribution of AUCs in the population, while the right panel shows the relationship between MIC and ƒAUC/MIC. The confidence interval around the mean target attainment indicates that at a target value of 100, infections caused by Candida strains with MICs up to 2 mg/L have a high probability of cure. Thus, from this analysis, treatment of infections caused by Candida can be optimized, depending on the MIC of the microorganism and/or the species.

One of the downsides of the example above, and the conclusions derived, is the premises. The conclusions are based on a number of assumptions, namely, the estimates of the PK parameter values and the measures of dispersion, and the assumption that these apply to the population to be treated. However, the exposure may differ for a number of reasons. Well known are differences in renal function, but age also may play a significant role. An example is shown in Figure 13. From data in the literature, the AUC of fluconazole per unit dose was determined for a number of age groups. As can be observed, the values vary considerably per age group and are particularly low during adolescence. Thus, exposure in this age group may be suboptimal—and thus the probability of cure lower—when dosing regimens are inferred from parameter estimates in other age groups. Yet, few data or studies are generally available for most drugs. For fluconazole—this example—one would or should conclude that the dose should be increased in certain age groups.

CONCLUDING REMARKS

In this chapter, the PK, PD, and PK–PD properties of azoles were discussed. The pharmacokinetics of most azoles is characterized by a large volume of distribution, relatively high protein binding, and a large volume of distribution. Because of their physiochemical properties, they are poorly soluble in water, except for fluconazole. Susceptibility to azoles—and antifungals in general—by a relatively large concentration ranges between no visible effect and full inhibition. The definition of the MIC is therefore, although resulting in a reproducible assay, not comparable to that for antibacterials. PK–PD relationships are characterized by a relationship between AUC/MIC and the PDI value resulting in a static effect is comparable across the class for the unbound fraction of the drug. From the PK properties and the exposure of the azoles, and the relationship between exposure and effect as derived from animal models and human trials, the optimal dosing regimen, and/or the susceptibility of the microorganisms can be derived, although there is still a gap in our knowledge here with respect to filamentous fungi. The recently acquired understanding in PK–PD relationships of antifungals serves the benefit of patients.

REFERENCES

1. Pfaller MA, Jones RN, Messer SA, Edmond MB, Wenzel RP. National surveillance of nosocomial blood stream infection due to *Candida albicans*: frequency of occurrence and antifungal susceptibility in the SCOPE Program. Diagn Microbiol Infect Dis 1998; 31(1): 327–332.
2. Edmond MB, Wallace SE, McClish DK, Pfaller MA, Jones RN, Wenzel RP. Nosocomial bloodstream infections in United States hospitals: a three-year analysis. Clin Infect Dis 1999; 29(2):239–244.
3. Wisplinghoff H, Bischoff T, Tallent SM, Seifert H, Wenzel RP, Edmond MB. Nosocomial bloodstream infections in US hospitals: analysis of 24,179 cases from a prospective nationwide surveillance study. Clin Infect Dis 2004; 39(3):309–317.
4. Woolley DW. Some biological effects produced by benzimidazole and their reversal by purines. J Biological Chem 1944; 152:225–232.
5. Jerchel D, Fischer H, Kracht M. Zur Darstellung der Benzimidazole. Liebigs Annalen der Chemie 1952; 575:162–173.
6. Godefroi EF, Heeres J, Van Cutsem J, Janssen PA. The preparation and antimycotic properties of derivatives of 1-phenethylimidazole. J Med Chem 1969; 12(5):784–791.
7. Heeres J, Backx LJ, Mostmans JH, Van Cutsem J. Antimycotic imidazoles. part 4. Synthesis and antifungal activity of ketoconazole, a new potent orally active broad-spectrum antifungal agent. J Med Chem 1979; 22(8):1003–1005.
8. Heeres J, Hendrickx R, Van Cutsem J. Antimycotic azoles. 6. Synthesis and antifungal properties of terconazole, a novel triazole ketal. J Med Chem 1983; 26(4):611–613.
9. Schmitt-Hoffmann A, Roos B, Heep M, et al. Single-ascending-dose pharmacokinetics and safety of the novel broad-spectrum antifungal triazole BAL4815 after intravenous infusions (50, 100, and 200 milligrams) and oral administrations (100, 200, and 400 milligrams) of its prodrug, BAL8557, in healthy volunteers. Antimicrob Agents Chemother 2006; 50(1): 279–285.
10. Grant SM, Clissold SP. Fluconazole. A review of its pharmacodynamic and pharmacokinetic properties, and therapeutic potential in superficial and systemic mycoses. Drugs 1990; 39(6):877–916.
11. Buijk SL, Gyssens IC, Mouton JW, Verbrugh HA, Touw DJ, Bruining HA. Pharmacokinetics of sequential intravenous and enteral fluconazole in critically ill surgical patients with invasive mycoses and compromised gastro-intestinal function. Intensive Care Med 2001; 27(1):115–121.

12. Brammer KW, Farrow PR, Faulkner JK. Pharmacokinetics and tissue penetration of fluconazole in humans. Rev Infect Dis 1990; 12(suppl 3):S318–S326.
13. Brammer KW, Coakley AJ, Jezequel SG, Tarbit MH. The disposition and metabolism of [14C]fluconazole in humans. Drug Metab Dispos 1991; 19(4):764–767.
14. Hardin TC, Graybill JR, Fetchick R, Woestenborghs R, Rinaldi MG, Kuhn JG. Pharmacokinetics of itraconazole following oral administration to normal volunteers. Antimicrob Agents Chemother 1988; 32(9):1310–1313.
15. Zimmermann T, Yeates RA, Laufen H, Pfaff G, Wildfeuer A. Influence of concomitant food intake on the oral absorption of two triazole antifungal agents, itraconazole and fluconazole. Eur J Clin Pharmacol 1994; 46(2):147–150.
16. Zimmermann T, Yeates RA, Albrecht M, Laufen H, Wildfeuer A. Influence of concomitant food intake on the gastrointestinal absorption of fluconazole and itraconazole in Japanese subjects. Int J Clin Pharmacol Res 1994; 14(3):87–93.
17. Zhao Q, Zhou H, Pesco-Koplowitz L. Pharmacokinetics of intravenous itraconazole followed by itraconazole oral solution in patients with human immunodeficiency virus infection. J Clin Pharmacol 2001; 41(12):1319–1328.
18. Barone JA, Moskovitz BL, Guarnieri J, et al. Enhanced bioavailability of itraconazole in hydroxypropyl-beta-cyclodextrin solution versus capsules in healthy volunteers. Antimicrob Agents Chemother 1998; 42(7):1862–1865.
19. Zhou H, Goldman M, Wu J, et al. A pharmacokinetic study of intravenous itraconazole followed by oral administration of itraconazole capsules in patients with advanced human immunodeficiency virus infection. J Clin Pharmacol 1998; 38(7):593–602.
20. Arredondo G, Martinez-Jorda R, Calvo R, Aguirre C, Suarez E. Protein binding of itraconazole and fluconazole in patients with chronic renal failure. Int J Clin Pharmacol Ther 1994; 32(7):361–364.
21. Mouton JW, van Peer A, de Beule K, Van Vliet A, Donnelly JP, Soons PA. Pharmacokinetics of itraconazole and hydroxyitraconazole in healthy subjects after single and multiple doses of a novel formulation. Antimicrob Agents Chemother 2006; 50(12):4096–4102.
22. De Beule K, Van Gestel J. Pharmacology of itraconazole. Drugs 2001; 61(suppl 1):27–37.
23. Theuretzbacher U, Ihle F, Derendorf H. Pharmacokinetic/pharmacodynamic profile of voriconazole. Clin Pharmacokinet 2006; 45(7):649–663.
24. Courtney R, Pai S, Laughlin M, Lim J, Batra V. Pharmacokinetics, safety, and tolerability of oral posaconazole administered in single and multiple doses in healthy adults. Antimicrob Agents Chemother 2003; 47(9):2788–2795.
25. Schmitt-Hoffmann A, Roos B, Maares J, et al. Multiple-dose pharmacokinetics and safety of the new antifungal triazole BAL4815 after intravenous infusion and oral administration of its prodrug, BAL8557, in healthy volunteers. Antimicrob Agents Chemother 2006; 50(1):286–293.
26. Richardson K, Cooper K, Marriott MS, Tarbit MH, Troke PF, Whittle PJ. Discovery of fluconazole, a novel antifungal agent. Rev Infect Dis 1990; 12 (suppl 3):S267–S71.
27. Arndt CA, Walsh TJ, McCully CL, Balis FM, Pizzo PA, Poplack DG. Fluconazole penetration into cerebrospinal fluid: implications for treating fungal infections of the central nervous system. J Infect Dis 1988; 157(1):178–180.
28. Sasongko L, Williams KM, Day RO, McLachlan AJ. Human subcutaneous tissue distribution of fluconazole: comparison of microdialysis and suction blister techniques. Br J Clin Pharmacol 2003; 56(5):551–561.
29. Nicolau DP, Crowe H, Nightingale CH, Quintiliani R. Bioavailability of fluconazole administered via a feeding tube in intensive care unit patients. J Antimicrob Chemother 1995; 36(2):395–401.
30. Arredondo G, Calvo R, Marcos F, Martinez-Jorda R, Suarez E. Protein binding of itraconazole and fluconazole in patients with cancer. Int J Clin Pharmacol Ther 1995; 33(8):449–452.
31. Brammer KW, Coates PE. Pharmacokinetics of fluconazole in pediatric patients. Eur J Clin Microbiol Infect Dis 1994; 13(4):325–329.
32. Debruyne D. Clinical pharmacokinetics of fluconazole in superficial and systemic mycoses. Clin Pharmacokinet 1997; 33(1):52–77.

33. Barone JA, Koh JG, Bierman RH, et al. Food interaction and steady-state pharmacokinetics of itraconazole capsules in healthy male volunteers. Antimicrob Agents Chemother 1993; 37(4):778–784.

34. Poirier JM, Berlioz F, Isnard F, Cheymol G. Marked intra- and inter-patient variability of itraconazole steady state plasma concentrations. Therapie 1996; 51(2):163–167.

35. Poirier JM, Hardy S, Isnard F, Tilleul P, Weissenburger J, Cheymol G. Plasma itraconazole concentrations in patients with neutropenia: advantages of a divided daily dosage regimen. Ther Drug Monit 1997; 19(5):525–529.

36. Lange D, Pavao JH, Wu J, Klausner M. Effect of a cola beverage on the bioavailability of itraconazole in the presence of H2 blockers. J Clin Pharmacol 1997; 37(6):535–540.

37. Jaruratanasirikul S, Kleepkaew A. Influence of an acidic beverage (Coca-Cola) on the absorption of itraconazole. Eur J Clin Pharmacol 1997; 52(3):235–237.

38. Reynes J, Bazin C, Ajana F, et al. Pharmacokinetics of itraconazole (oral solution) in two groups of human immunodeficiency virus-infected adults with oral candidiasis. Antimicrob Agents Chemother 1997; 41(11):2554–2558.

39. Boogaerts MA, Maertens J, Van Der Geest R, et al. Pharmacokinetics and safety of a 7-day administration of intravenous itraconazole followed by a 14-day administration of itraconazole oral solution in patients with hematologic malignancy. Antimicrob Agents Chemother 2001; 45(3):981–985.

40. Barone JA, Moskovitz BL, Guarnieri J, et al. Food interaction and steady-state pharmacokinetics of itraconazole oral solution in healthy volunteers. Pharmacotherapy 1998; 18(2):295–301.

41. Schmitt HJ, Edwards F, Andrade J, Niki Y, Armstrong D. Comparison of azoles against aspergilli in vitro and in an experimental model of pulmonary aspergillosis. Chemotherapy 1992; 38(2):118–126.

42. de Repentigny L, Ratelle J, Leclerc JM, et al. Repeated-dose pharmacokinetics of an oral solution of itraconazole in infants and children. Antimicrob Agents Chemother 1998; 42(2):404–408.

43. Mattiuzzi GN, Kantarjian H, O'Brien S, et al. Intravenous itraconazole for prophylaxis of systemic fungal infections in patients with acute myelogenous leukemia and high-risk myelodysplastic syndrome undergoing induction chemotherapy. Cancer 2004; 100(3): 568–573.

44. Glasmacher A, Prentice A, Gorschluter M, et al. Itraconazole prevents invasive fungal infections in neutropenic patients treated for hematologic malignancies: evidence from a meta-analysis of 3,597 patients. J Clin Oncol 2003; 21(24):4615–4626.

45. Mohr JF, Finkel KW, Rex JH, Rodriguez JR, Leitz GJ, Ostrosky-Zeichner L. Pharmacokinetics of intravenous itraconazole in stable hemodialysis patients. Antimicrob Agents Chemother 2004; 48(8):3151–3153.

46. Arredondo G, Suarez E, Calvo R, Vazquez JA, Garcia-Sanchez J, Martinez-Jorda R. Serum protein binding of itraconazole and fluconazole in patients with diabetes mellitus. J Antimicrob Chemother 1999; 43(2):305–307.

47. Poirier JM, Cheymol G. Optimisation of itraconazole therapy using target drug concentrations. Clin Pharmacokinet 1998; 35(6):461–473.

48. Boucher HW, Groll AH, Chiou CC, Walsh TJ. Newer systemic antifungal agents: pharmacokinetics, safety and efficacy. Drugs 2004; 64(18):1997–2020.

49. Lazarus HM, Blumer JL, Yanovich S, Schlamm H, Romero A. Safety and pharmacokinetics of oral voriconazole in patients at risk of fungal infection: a dose escalation study. J Clin Pharmacol 2002; 42(4):395–402.

50. Purkins L, Wood N, Kleinermans D, Greenhalgh K, Nichols D. Effect of food on the pharmacokinetics of multiple-dose oral voriconazole. Br J Clin Pharmacol 2003; 56(suppl 1): 17–23.

51. Hoffman A, Danenberg HD, Katzhendler I, Shuval R, Gilhar D, Friedman M. Pharmacodynamic and pharmacokinetic rationales for the development of an oral controlled-release amoxicillin dosage form [In Process Citation]. J Controlled Release 1998; 54(1): 29–37.

52. Purkins L, Wood N, Greenhalgh K, Allen MJ, Oliver SD. Voriconazole, a novel wide-spectrum triazole: oral pharmacokinetics and safety. Br J Clin Pharmacol 2003; 56(suppl 1): 10–16.
53. Mar PIL. Label: voriconazole for injection, tablets, oral suspension. In: LAB-0271-12; 2005.
54. Saad AH, DePestel DD, Carver PL. Factors influencing the magnitude and clinical significance of drug interactions between azole antifungals and select immunosuppressants. Pharmacotherapy 2006; 26(12):1730–1744.
55. Smith J, Safdar N, Knasinski V, et al. Voriconazole therapeutic drug monitoring. Antimicrob Agents Chemother 2006; 50(4):1570–1572.
56. Courtney R, Sansone A, Smith W, et al. Posaconazole pharmacokinetics, safety, and tolerability in subjects with varying degrees of chronic renal disease. J Clin Pharmacol 2005; 45(2):185–192.
57. Courtney R, Wexler D, Radwanski E, Lim J, Laughlin M. Effect of food on the relative bioavailability of two oral formulations of posaconazole in healthy adults. Br J Clin Pharmacol 2004; 57(2):218–222.
58. Krishna G, Sansone-Parsons A, Martinho M, Kantesaria B, Pedicone L. Posaconazole plasma concentrations in juvenile patients with invasive fungal infection. Antimicrob Agents Chemother 2007. In press.
59. Georgopapadakou NH, Walsh TJ. Antifungal agents: chemotherapeutic targets and immunologic strategies. Antimicrob Agents Chemother 1996; 40(2):279–291.
60. National Committee for Clinical Laboratory Standards. Reference method for broth dilution antifungal susceptibility testing of yeasts. M27-A2. Wayne: National Committee for Clinical Laboratory Standards; 2002.
61. Rodriguez-Tudela JL, Barchiesi F, Bille J, et al. Method for the determination of minimum inhibitory concentration (MIC) by broth dilution of fermentative yeasts. Clin Microbiol Infect 2003; 9:I–VIII.
62. Rodriguez-Tudela JL, Donnelly JP, Pfaller MA, et al. Statistical analyses of correlation between fluconazole MICs for Candida spp. assessed by standard methods set forth by the European Committee on Antimicrobial Susceptibility Testing (E.Dis. 7.1) and CLSI (M27-A2). J Clin Microbiol 2007; 45(1):109–111.
63. Espinel-Ingroff A, Barchiesi F, Cuenca-Estrella M, et al. International and multicenter comparison of EUCAST and CLSI M27-A2 broth microdilution methods for testing susceptibilities of Candida spp. to fluconazole, itraconazole, posaconazole, and voriconazole. J Clin Microbiol 2005; 43(8):3884–3889.
64. Mouton JW, Voss A, van Elzakker EPM, et al. Linezolid susceptibility of glycopeptide-intermediately susceptible Staphylococcus aureus (GISA)—the Dutch experience. In: ECC-MID; 2003; Glasgow; 2003.
65. Ribeiro da Matta VL, de Souza Carvalho Melhem M, Colombo AL, et al. Susceptibility profile to antifungal drugs of *Pichia anomala* isolated from patients presenting nosocomial fungemia 10.1128/AAC.01038-06. Antimicrob Agents Chemother 2007;AAC.01038–06.
66. Odds FC, Vranckx L, Woestenborghs F. Antifungal susceptibility testing of yeasts: evaluation of technical variables for test automation. Antimicrob Agents Chemother 1995; 39(9):2051–2060.
67. Mouton JW, Vinks AA. PK-PD modelling of antibiotics in vitro and in vivo using bacterial growth and kill kinetics: the MIC versus stationary concentrations. Clinical Pharmacokinetics 2005. In press.
68. NCCLS. Development of In Vitro Susceptibility Testing Criteria and Quality Control Parameters; Approved Guideline-Second Edition. NCCLS document M23-A2. Wayne: NCCLS; 2001.
69. Cormican MG, Pfaller MA. Standardization of antifungal susceptibility testing. J Antimicrob Chemother 1996; 38(4):561–578.
70. Rex JH, Pfaller MA, Walsh TJ, et al. Antifungal susceptibility testing: practical aspects and current challenges. Clin Microbiol Rev 2001; 14(4):643–658, table of contents.
71. Altman FP. Tetrazolium salts and formazans. Prog Histochem Cytochem 1976; 9(3):1–56.
72. Meletiadis J, Verweij PE, TeDorsthorst DT, Meis JF, Mouton JW. Assessing in vitro combinations of antifungal drugs against yeasts and filamentous fungi: comparison of different drug interaction models. Med Mycol 2005; 43(2):133–152.

73. Meletiadis J, Mouton JW, Meis JF, Bouman BA, Donnelly JP, Verweij PE. Colorimetric assay for antifungal susceptibility testing of Aspergillus species. J Clin Microbiol 2001; 39(9): 3402–3408.
74. Meletiadis J, Mouton JW, Meis JF, Bouman BA, Donnelly PJ, Verweij PE. Comparison of spectrophotometric and visual readings of NCCLS method and evaluation of a colorimetric method based on reduction of a soluble tetrazolium salt, 2,3-bis {2-methoxy-4-nitro-5-[(sulfenylamino) carbonyl]-2h- tetrazolium-hydroxide}, for antifungal susceptibility testing of aspergillus species. J Clin Microbiol 20011; 39(12):4256–4263.
75. Meletiadis J, Mouton JW, Meis JF, Bouman BA, Verweij PE. Comparison of the Etest and the sensititre colorimetric methods with the NCCLS proposed standard for antifungal susceptibility testing of Aspergillus species. J Clin Microbiol 2002; 40(8):2876–2885.
76. Meletiadis J, Meis JF, Mouton JW, Donnelly JP, Verweij PE. Comparison of NCCLS and 3-(4,5-dimethyl-2-Thiazyl)-2, 5-diphenyl-2H-tetrazolium bromide (MTT) methods of in vitro susceptibility testing of filamentous fungi and development of a new simplified method. J Clin Microbiol 2000; 38(8):2949–2954.
77. Craig WA, Gudmundsson S. Post-antibiotic effect. In: Lorian V, ed. Antibiotics in Laboratory Medicine. 4th ed. Baltimore: Williams and Wilkins, 1996:296–329.
78. den Hollander JG, Fuursted K, Verbrugh HA, Mouton JW. Duration and clinical relevance of postantibiotic effect in relation to the dosing interval. Antimicrob Agents Chemother 1998; 42(4):749–754.
79. den Hollander JG, Mouton JW, van Goor MP, Vleggaar FP, Verbrugh HA. Alteration of postantibiotic effect during one dosing interval of tobramycin, simulated in an in vitro pharmacokinetic model. Antimicrob Agents Chemother 1996; 40(3):784–786.
80. Ernst EJ, Klepser ME, Pfaller MA. Postantifungal effects of echinocandin, azole, and polyene antifungal agents against *Candida albicans* and *Cryptococcus neoformans*. Antimicrob Agents Chemother 20000; 44(4):1108–1111.
81. Anil S, Ellepola AN, Samaranayake LP. Post-antifungal effect of polyene, azole and DNA-analogue agents against oral *Candida albicans* and *Candida tropicalis* isolates in HIV disease. J Oral Pathol Med 2001; 30(8):481–488.
82. Garcia MT, Llorente MT, Minguez F, Prieto J. Influence of pH and concentration on the postantifungal effect and on the effects of sub-MIC concentrations of 4 antifungal agents on previously treated Candida spp. Scand J Infect Dis 2000; 32(6):669–673.
83. Garcia MT, Llorente MT, Minguez F, Prieto J. Influence of temperature and concentration on the postantifungal effect and the effects of sub-MIC concentrations of four antifungal agents on previously treated Candida species. Chemotherapy 2000; 46(4):245–252.
84. Garcia MT, Llorente MT, Minguez F, Prieto J. Postantifungal effect and effects of sub-MIC concentrations on previously treated Candida sp. influence of growth phase. J Infect 2002; 45(4):263–267.
85. Garcia MT, Llorente MT, Minguez F, Prieto J. Post-antifungal effect and effects of sub-MIC concentrations on previously treated Candida spp.: influence of exposure time and concentration. Scand J Infect Dis 2002; 34(3):197–200.
86. Garcia MT, Llorente MT, Lima JE, Minguez F, Del Moral F, Prieto J. Activity of voriconazole: post-antifungal effect, effects of low concentrations and of pretreatment on the susceptibility of *Candida albicans* to leucocytes. Scand J Infect Dis 1999; 31(5):501–504.
87. Vitale RG, Mouton JW, Afeltra J, Meis JF, Verweij PE. Method for measuring postantifungal effect in Aspergillus species. Antimicrob Agents Chemother 2002; 46(6):1960–1965.
88. Manavathu EK, Ramesh MS, Baskaran I, Ganesan LT, Chandrasekar PH. A comparative study of the post-antifungal effect (PAFE) of amphotericin B, triazoles and echinocandins on *Aspergillus fumigatus* and *Candida albicans*. J Antimicrob Chemother 2004; 53(2):386–389.
89. Chryssanthou E, Sjolin J. Post-antifungal effect of amphotericin B and voriconazole against *Aspergillus fumigatus* analysed by an automated method based on fungal CO2 production: dependence on exposure time and drug concentration. J Antimicrob Chemother 2004; 54(5):940–943.
90. Warn PA, Sharp A, Denning DW. In vitro activity of a new triazole BAL4815, the active component of BAL8557 (the water-soluble prodrug), against Aspergillus spp. J Antimicrob Chemother 2006; 57(1):135–138.

91. Breuker I, Meis JF, Verweij PE, Mouton JW. In vitro activity of the new azole BAL4815 against clinical dermatophyte isolates. In: Microbiology ASf, ed. 45th ICAAC; 2005 October 30–November 2; Washington DC: Database Publishing Group. Inc; 2005:428.

92. Breuker I, Meis JF, Verweij PE, Mouton JW. In vitro activity of the new azole BAL4815 against clinical candida isolates comprising resistant *C. albicans* and less susceptible candida spp. In: Microbiology ASf, ed. 45th ICAAC; December 16–19; Washington DC: Database Publishing Group. Inc; 2005.

93. Torres-Narbona M, Guinea J, Martinez-Alarcon J, Pelaez T, Bouza E. In vitro activity of amphotericin B, caspofungin, itraconazole, posaconazole and voriconazole against 45 clinical isolates of zygomycetes: comparison of CLSI M-38 A, Sensititre YeastOne and the E-test 10.1128/AAC.01539-06. Antimicrob Agents Chemother 2006:AAC.01539–06.

94. Ghannoum MA, Rice LB. Antifungal agents: mode of action, mechanisms of resistance, and correlation of these mechanisms with bacterial resistance. Clin Microbiol Rev 1999; 12(4):501–517.

95. Alangaden G, Chandrasekar PH, Bailey E, Khaliq Y. Antifungal prophylaxis with low-dose fluconazole during bone marrow transplantation. The Bone Marrow Transplantation Team. Bone Marrow Transplant 1994; 14(6):919–924.

96. Nguyen MH, Peacock JE Jr, Morris AJ, et al. The changing face of candidemia: emergence of non-*Candida albicans* species and antifungal resistance. Am J Med 1996; 100(6):617–623.

97. Rex JH, Rinaldi MG, Pfaller MA. Resistance of Candida species to fluconazole. Antimicrob Agents Chemother 1995; 39(1):1–8.

98. Clancy CJ, Staley B, Nguyen MH. In vitro susceptibility of breakthrough Candida bloodstream isolates correlates with daily and cumulative doses of fluconazole. Antimicrob Agents Chemother 2006; 50(10):3496–3498.

99. Mellado E, Alcazar-Fuoli L, Garcia-Effron G, Alastruey-Izquierdo A, Cuenca-Estrella M, Rodriguez-Tudela JL. New resistance mechanisms to azole drugs in Aspergillus fumigatus and emergence of antifungal drugs-resistant A. fumigatus atypical strains. Med Mycol 2006; 44(suppl):367–371.

100. Ambrose PG, Bhavnani SM, Rubino CM, et al. Pharmacokinetics-pharmacodynamics of antimicrobial therapy: it's not just for mice anymore. Clin Infect Dis 2007; 44(1):79–86.

101. Andes D, Forrest A, Lepak A, Nett J, Marchillo K, Lincoln L. Impact of antimicrobial dosing regimen on evolution of drug resistance in vivo: fluconazole and *Candida albicans*. Antimicrob Agents Chemother 2006; 50(7):2374–2383.

102. Louie A, Drusano GL, Banerjee P, et al. Pharmacodynamics of fluconazole in a murine model of systemic candidiasis. Antimicrob Agents Chemother 1998; 42(5):1105–1109.

103. Andes D, Marchillo K, Stamstad T, Conklin R. In vivo pharmacodynamics of a new triazole, ravuconazole, in a murine candidiasis model. Antimicrob Agents Chemother 2003; 47(4):1193–1199.

104. Andes D, Marchillo K, Stamstad T, Conklin R. In vivo pharmacokinetics and pharmacodynamics of a new triazole, voriconazole, in a murine candidiasis model. Antimicrob Agents Chemother 2003; 47(10):3165–3169.

105. Andes D, Marchillo K, Conklin R, et al. Pharmacodynamics of a new triazole, posaconazole, in a murine model of disseminated candidiasis. Antimicrob Agents Chemother 2004; 48(1):137–142.

106. Pai MP, Turpin RS, Garey KW. Association of fluconazole area under the concentration-time curve/MIC and dose/MIC ratios with mortality in non-neutropenic patients with candidemia. Antimicrob Agents Chemother 2007; 51(1):35–39.

107. Pfaller MA, Diekema DJ, Sheehan DJ. Interpretive breakpoints for fluconazole and Candida revisited: a blueprint for the future of antifungal susceptibility testing. Clin Microbiol Rev 2006; 19(2):435–447.

108. Glasmacher A, Hahn C, Leutner C, et al. Breakthrough invasive fungal infections in neutropenic patients after prophylaxis with itraconazole. Mycoses 1999; 42(7–8):443–451.

109. Pfaller MA, Diekema DJ, Rex JH, et al. Correlation of MIC with outcome for Candida species tested against voriconazole: analysis and proposal for interpretive breakpoints. J Clin Microbiol 2006; 44(3):819–826.

110. Mouton JW, Schmitt-Hoffmann A, Punt NC. Monte Carlo Simulations of BAL8557: a New Water-soluble Azole with Antifungal Activity. In: Microbiology ASf, ed. 44th ICAAC; October 30–November 2; Washington DC; 2004:10.
111. Goa KL, Barradell LB. Fluconazole. An update of its pharmacodynamic and pharmacokinetic properties and therapeutic use in major superficial and systemic mycoses in immunocompromised patients. Drugs 1995; 50(4):658–690.

18 Glucan Synthase Inhibitors

Tawanda Gumbo
University of Texas Southwestern Medical Center, Division of Infectious Diseases, Dallas, Texas, U.S.A.

Fumiaki Ikeda
Infectious Diseases Department, Pharmacology Research Laboratories, Drug Discovery Research, Astellas Pharma, Inc., Osaka, Japan

Arnold Louie
Emerging Infections and Pharmacodynamics Laboratory, Ordway Research Institute, Albany Medical College, Albany, New York, U.S.A.

BRIEF HISTORY OF CLASS

Echinocandins are a novel class of antifungal agents that were first isolated from the broth of *Aspergillus* species, including *Aspergillus aculeatus* and *Aspergillus rugulovalvus*, in the mid-1970s (1,2). The echinocandins block cell wall formation by inhibiting the enzyme 1,3-β-D-glucan synthase, resulting in fungal death. These compounds are poorly absorbed through the gastrointestinal tract and, therefore, are administered intravenously. Since initial investigations centered on the therapeutic activity of these compounds against *Pneumocystis carinii* and *Candida* species, the compounds were named pneumocandins. However, the spectrum of activity includes some molds, including *Aspergillus* species (Table 1). The three echinocandins that have reached clinical development are semisynthetic cyclic hexapeptides with N-linked fatty acyl side chains attached to pneumocandin B$_0$ (Fig. 1) (7). Caspofungin was the first echinocandin to be licensed for clinical use (in many countries, including the United States). Micafungin is licensed for clinical use in Japan and in the United States, and anidulafungin is licensed in the United States.

The echinocandins are an important addition to the antifungal armamentarium for several reasons. First *Candida* species are now responsible for approximately 10% of nosocomial bloodstream infections (8,9). The number of cases of candidemia and deep-seated candidal infections continue to rise as the number of immunosuppressed patients increases. Second, although *Candida albicans* has traditionally been the most common *Candida* species encountered in clinical practice, the proportion of infections due to non–*C. albicans* species has now surpassed that of *C. albicans* (9). Non–*C. albicans* species such as *Candida glabrata*, *Candida guilliermondii*, *Candida lusitaniae*, and *Candida krusei* tend to be resistant to standard antifungal agents (10), including amphotericin B and fluconazole, and some patients do poorly even if treated (11). In addition, Gumbo et al. (12) and Clancy et al. (13) have demonstrated that even after apparent therapeutic success with fluconazole or amphotericin B, complications of candidemia may occur many months later. Third, there has been

TABLE 1 MIC_{90} (mg/L) of Echinocandins Compared with Other Antifungal Agents in Fungal Clinical Isolates

Species (no. of isolates)	Drug				
	Caspofungin	Micafungin	Anidulafungin	Fluconazole	Amphotericin B
Candida spp.	–	–	–	–	–
C. albicans (733)	0.5	0.03	0.03	2	0.25
C. glabrata (458)	1	0.06	0.13	32	0.5
C. parapsilosis (391)	2	2	2	2	0.5
C. tropicalis (307)	1	0.06	0.13	16	0.5
C. krusei (50)	2	0.25	0.13	>64	0.5
C. lusitaniae (20)	2	2	0.25	2	0.5
C. dubliniensis (18)	0.5	0.03	0.06	0.5	0.13
Cryptococcus neoformans (10)	>16	–	>16	16	1
(20)	–	>64	–	4	0.5
Aspergillus spp.	–	–	–	–	–
A. fumigatus (28)	0.25	–	<0.03	–	1
(39)	–	0.03	–	>64	2
(256)	0.06[a]	–	–	–	1
A. flavus (19)	0.06	–	<0.03	–	1
(11)	–	0.03	–	>64	2
(30)	0.06[a]	–	–	–	2
A. niger (9)	0.25	–	<0.03	–	0.5
(11)	–	0.03	–	>64	1
(29)	0.06[a]	–	–	–	1
A. glaucus (8)	0.12	–	<0.03	–	1
A. terreus (6)	–	0.015	–	>64	2
(16)	0.06[a]	–	–	–	2
A. versicolor (20)	0.12[a]	–	–	–	2
Penicillium spp. (35)	0.12[a]	–	–	–	2
Pseudaallescheria boydii (6)	1.3	–	2.5	–	2.6
Rhizopus arrhizus (5)	>16	–	>16	–	0.57

[a]Minimal effective concentration.
Source: From Refs. 3,4–6.

an increase in the number of patients at risk for infection with molds, such as *Aspergillus* species. Unfortunately, poor clinical outcome is common in patients with mold infections who are treated with amphotericin B. Fourth, because of the unique mechanism of action of echinocandins, cross-resistance with other antifungal drug classes is not anticipated. Finally, amphotericin B has infusion-related side effects and is nephrotoxic, which increase patient morbidity and health-care costs (14,15). In contrast, the echinocandins have excellent safety profiles. These and other problems of azole- and polyene-based therapies have led to the development of echinocandins.

During the same period that echinocandins were being developed, the science of pharmacodynamics started to be applied to anti-infective agents. Pharmacodynamic studies have led to improvement in dosing of many antibacterial and antiviral compounds (16). It is therefore important to understand the pharmacokinetic-pharmacodynamic properties of the echinocandins, so that we may optimize the therapeutic impact of this new class of antifungal agents.

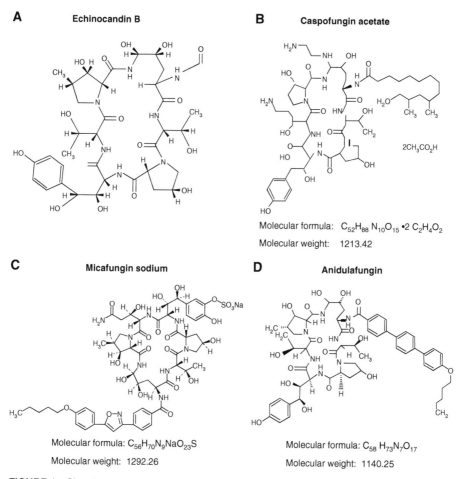

A Echinocandin B

B Caspofungin acetate

Molecular formula: $C_{52}H_{88}N_{10}O_{15} \cdot 2\,C_2H_4O_2$

Molecular weight: 1213.42

C Micafungin sodium

Molecular formula: $C_{56}H_{70}N_9NaO_{23}S$

Molecular weight: 1292.26

D Anidulafungin

Molecular formula: $C_{58}H_{73}N_7O_{17}$

Molecular weight: 1140.25

FIGURE 1 Chemical structure of echinocandins.

MECHANISM OF ACTION

The cell membranes of eukaryotic (human and fungal) organisms consist of a fluid mosaic bilayer of phospholipids in which proteins are embedded (17). In eukaryotes, complex lipids, called sterols, constitute up to 25% of the lipids in the cell membranes. Sterols give the eukaryotic cell membranes rigidity, important in withstanding physical stress. The predominant sterol in fungal cell membranes is ergosterol, while that in human cell membranes is cholesterol. This difference is what allowed for the development of systemic antifungal drugs such as azoles and polyene antifungal agents, which target the fungal cell membrane (Fig. 2). Azoles and polyene compounds have been the mainstay of antifungal therapy in humans for decades.

Another difference between human and fungal cells is the presence of a cell wall in fungi, and its complete absence in humans. The main components of the fungal cell wall are the polymers glucan, mannose, and chitin. These polymers are responsible for cell wall shape and strength. Glucan, which accounts for 30% to

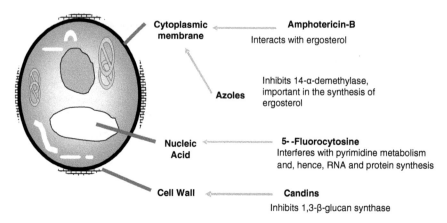

FIGURE 2 Targets for antifungal drugs.

60% of the cell wall of *Candida* spp., is made of three helically entwined glucose polymers (18–20). The glucose polymers are linked by β-1,3-, α-1,3, or β-1,6-bonds. 1,3-β-D-glucan synthesis is catalyzed by the enzyme complex 1,3-β-D-glucan synthase (Fig. 3), which is partially encoded by the genes *FKS1, FKS2, RHO1,* and *ETG1.* The enzyme complex (Fig. 3) consists of two main components, the first of which is a membrane-bound catalytic component that uses UDP-glucose as its substrate and catalyzes the formation of linear polymers of 1,3-β-D-glucan (19,21). There is a second soluble component that binds guanidine triphosphate (GTP) and links glucan synthesis to the cell cycle.

Echinocandins act by concentration-dependent, noncompetitive inhibition of 1,3-β-D-glucan synthesis, which leads to inhibition of 1,3-β-D-glucan formation, osmotic fragility, and fungal cell lysis (7,19,20). Because the cell wall is not found in mammalian cells, the echinocandins demonstrate an excellent safety profile in the clinic.

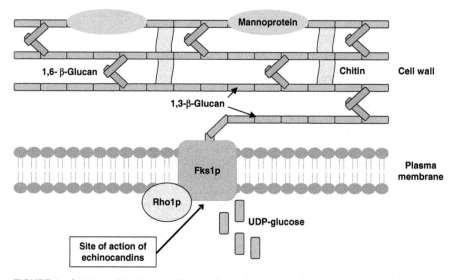

FIGURE 3 Structure of the fungal cell membrane and cell wall and the site of action of echinocandins.

SUSCEPTIBILITY OF FUNGI TO ECHINOCANDINS

Standard methods for in vitro susceptibility testing of the echinocandins have not been established. The most widely used procedure is the broth dilution method recommended by the Clinical and Laboratory Standards Institute [formerly the National Committee for Clinical Laboratory Standards (NCCLS)] (22,84) in which susceptibility testing is conducted in RPMI 1640 plus morpholinopropane sulfonic acid, and the minimum inhibitory concentration (MIC) is defined as the minimum concentration that results in 80% reduction of turbidity compared to the growth of controls after incubation at 35°C for 48 hours. Some laboratories have suggested variations in the test procedure, including supplementing RPMI 1640 with 2% glucose (3,23), substituting antibiotic medium 3 (AM3) for RPMI 1640, and reading the MIC at 24 hours instead of 48 hours (3). Unfortunately, addition of 2% glucose leads to wider variability in the MICs (3,23), and the manufacture of AM3 is not standardized. When susceptible fungi are exposed to echinocandins, there is a dose-dependent alteration of fungal cell morphology starting with shortening of hyphal elements, followed by swelling and terminal vacuolization (24) as shown in Figure 4. Because of this unique mechanism of action by echinocandins, some laboratories have proposed use of a susceptibility index for molds termed the minimal effective concentration (MEC). The MEC is the lowest concentration of echinocandin needed to produce abnormal hyphal growth after 24 or 48 hours of incubation (25,26).

A good laboratory method should be able to produce susceptibility patterns that have clinical relevance. If a fungal isolate from a patient is judged to be "susceptible" by virtue of an MIC value, the implication is that a favorable clinical

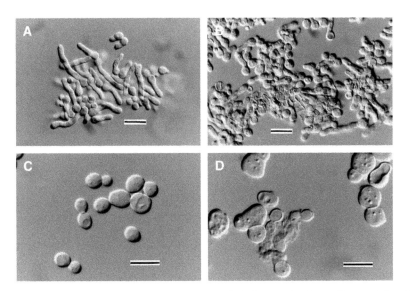

FIGURE 4 Echinocandin induced morphological changes in *Aspergillus* and *Candida* species. Differential interference contrast photomicrographs of drug-induced morphology changes in *Aspergillus fumigatus* TIMM3968 (**A** and **B**) and *Candida albicans* ATCC90028 (**C** and **D**). (**A**) saline control after challenge for five hours; (**B**) 0.01 µg/mL of micafungin after five-hour challenge; (**C**) saline control after challenge for three hours; (**D**) 0.1 µg/mL of micafungin after three-hour challenge. (Bar indicates 10 mcm). *Source*: From Ref. 24.

outcome is expected if that patient is adequately treated with the particular drug in question. Conversely, if the MIC is high and the isolate is classified as "resistant" the implication is that, *Ceteris paribus*, there would be an increased likelihood of poor outcome if the patient was treated with the drug to which the isolate is "resistant" (27). To date, MIC values for echinocandins, especially in the low ranges, have failed to discriminate between poor and good treatment outcome.

Hernandez et al. (28) reported on the case of a patient with AIDS-associated esophagitis caused by fluconazole-resistant *C. albicans*. *Candida* isolates were sequentially cultured from the patient prior to the initiation of successful intravenous caspofungin therapy (isolate 1) administered at a dose of 50 mg/day, during a relapse (isolate 2), and after failure of a second caspofungin treatment (isolate 3). The MICs of the three isogenic strains were performed using the CLSI method that employed either RPMI 1640 or AM3 and were read after 48 hours of incubation. The caspofungin MICs of isolates 1 and 2 were 0.25 mg/L in RPMI 1640, while that of isolate 3 was >64 mg/L. The MICs for isolates 1 and 2 were 0.125 mg/L in AM3, while that of isolate 3 was 0.5 mg/L. When the three isogenic strains were used to infect mice, and the mice were then treated with caspofungin, the minimum dose that reduced kidney fungal burden was 0.0625 mg/kg for isolate 1, 0.125 mg/kg for isolate 2, and 1 mg/kg for isolate 3. In this case, the results suggest that MICs with either method are predictive of in vivo success of caspofungin. On the other hand, Bartizal et al. (29) examined the MICs (CLSI macrobroth method in RPMI 1640) of 210 *Candida* isolates cultured from patients who had been treated with caspofungin. There were 14 patients with isolates that had MIC of $\geq$4 mg/L. However, all 14 patients were successfully treated with standard doses of caspofungin and had an outcome similar to patients infected with "susceptible" strains. In another study, Gonzalez et al. (30) compared the predictive value of *Coccidioides immitis* MICs (CLSI macrobroth method) to that of MECs. The two isolates had MICs of 8 and 64 mg/L by the CLSI method, but MECs of 0.125 mg/L. Despite the "high" MICs, which would have predicted failure, caspofungin therapy resulted in 100% survival in mice infected with either strain. Collectively, these studies indicate that the correlation of MICs to echinocandins and therapeutic success has not been established. It is in this context that the echinocandin susceptibility patterns of various fungi shown in Table 1 should be viewed. In order to provide a comparison to standard antifungal compounds, susceptibility patterns of the isolates to azoles and amphotericin B are also shown.

In general, as shown in Table 1, *Candida* and *Aspergillus* species are susceptible to echinocandins. However, the MIC_{90}s are high for *Fusarium* (>16 mg/L), *Trichosporon beigelii* (>16 mg/L), *Rhizopus* (>16 mg/L), and *Cryptococcus* (>16 mg/L) species (4–6). Echinocandins have low MICs (range: 0.0078–0.0625 mg/L) against mycelial forms of the endemic fungi *Histoplasma capsulatum*, *Blastomyces dermatitidis*, and C. *immitis*, but high MICs (range 32–>64 mg/L) against the yeast forms (31).

Since echinocandins are highly protein bound in human serum (32–34), it would be expected that MICs performed in media supplemented with either human or animal serum would be higher than those obtained without serum. Bartizal et al. (35) have demonstrated that the MIC of caspofungin to *C. albicans* increased from 0.06 μg/mL for susceptibility studies conducted in RPMI 1640 to 0.25 μg/mL when studies were performed with 50% RPMI/50% human serum. The MICs increased by more than eightfold when performed in 50% RPMI/50% mouse serum, compared to studies conducted in RPMI 1640 alone. Louie et al. (36) determined the MIC of caspofungin for a single *C. albicans* strain in RPMI 1640 and in 20% RPMI/

80% mouse serum and found the MIC for the fungal strain was 0.2 mg/L in either case, suggesting that protein binding had little effect on caspofungin activity. In contrast, some scientists demonstrated an "enhanced effect" for caspofungin against *Aspergillus fumigatus* by addition of as little as 5% human serum to RPMI 1640 (37). On the other hand, increases in MIC of micafungin of 128 times for *C. albicans*, 256-fold for *C. glabrata*, and 64-fold for *A. fumigatus* have been demonstrated when 4% human serum "albumin" was added to RPMI 1640 (38). Surprisingly, anidulafungin, which is apparently less protein bound than caspofungin (34), had increases in *C. albicans* MICs of 8- to 32-fold in 80% human serum/20% RPMI, and four- to eightfold in RPMI 1640 supplemented with 50 mg/L of bovine serum albumin when compared to susceptibility studies conducted using RPMI 1640 alone (39). The therapeutic and pharmacodynamic meaning, if any, of these findings is unclear.

MECHANISM OF RESISTANCE

Since 1,3-β-D-glucan synthesis inhibitors act by a mechanism distinct from other antifungal therapies, there appears to be a low potential for cross-resistance with other classes of antifungal agents (19). In a study evaluating in vitro susceptibility, fluconazole-resistant *Candida* isolates demonstrated no cross-resistance to micafungin (38,40). *Candida* strains resistant to pneumocandins have been generated in the laboratory. CAI4R1, NR2, NR3, and NR4 are pneumocandin-resistant mutant strains of *C. albicans* that were selected on agar plates containing the pneumocandin L-733,560. These isolates contain a mutation in *fks1*, a gene which encodes subunits of glucan synthase (19). Kurtz et al. (41) have generated four independent spontaneous *C. albicans* mutants also resistant to the pneumocandin L-733,560. These mutants have glucan synthase activity that was more resistant to the effect of echinocandin compared to that of the wild-type enzyme. The virulence of these spontaneous mutants was unimpaired in a mouse model of candidiasis; however, the spontaneous mutant CAI4R1 had a therapeutic response to pneumocandin L-733,560 at lower levels than would have been predicted based on in vitro susceptibility. Clinical isolates of *C. albicans* resistant to caspofungin have already been isolated in a clinical situation in which a patient who was successfully treated for esophageal candidiasis with this echinocandin developed relapse with an isolate resistant to caspofungin (28).

CASPOFUNGIN ACETATE
Animal Pharmacokinetics

The plasma pharmacokinetics of parenterally administered caspofungin in mice, rats, rabbits, and nonhuman primates, as reported by Sandhu et al., Hajdu et al., and Groll et al. (42–44), is summarized in Table 2. In mice, rats, and monkeys, caspofungin has a multiexponential distribution, with a short $t_{1/2\alpha}$ of 4.4 to 5.5 hours and a longer terminal $t_{1/2}$ of between 44.7 and 59.7 hours. Rabbit pharmacokinetics has produced results indicative of shorter $t_{1/2}$s, with one study reporting a $t_{1/2\alpha}$ of 1.2 ± 0.2 hour and a $t_{1/2\beta}$ of 11.7 ± 8.5 hour (42) while another study reported a $t_{1/2\alpha}$ of 0.1 ± 0.0 hour, a $t_{1/2\beta}$ of 3.5 ± 0.5 hour, and a $t_{1/2\gamma}$ of 30.9 hour (44). Hadju et al. (43) reported a terminal $t_{1/2}$ of 7.6 ± 1.0 hour in mice, somewhat similar to the terminal $t_{1/2}$ of 4.5 hours reported by Wiederhold et al. (45) in mice with aspergillosis. However, both studies only examined pharmacokinetics up to 24 hours after drug administration, thus ending their pharmacokinetic studies before the terminal phase had commenced (43). The terminal $t_{1/2}$s they reported mostly reflect a $t_{1/2\alpha}$.

TABLE 2 Pharmacokinetics of Caspofungin in Mice, Rats, Rabbits, Monkeys, and Chimpanzees

Ref.	Species	Dose (mg/kg)	Clearance (mL/min/kg)	V_{ss} (L/kg)	$t_{\frac{1}{2}\alpha}$ (hr)	$t_{\frac{1}{2}\beta}$ (hr)	$AUC_{0-\infty}$ mg*hr/L
42	Mouse	5	0.3 ± 0.1	0.5 ± 0.2	4.4 ± 1.2	46.9 ± 2.8	296.4 ± 67.8
42	Rat	2	0.4 ± 0.0	0.5 ± 0.1	5.9 ± 0.4	59.7 ± 14.5	77.1 ± 4.2
42	Rabbit	5	1.1 ± 0.3	0.3 ± 0.1	1.2 ± 0.2	11.7 ± 8.5	82.9 ± 21.3
44	Rabbit	6	0.7 ± 0.1	0.4 ± 0.0	0.1 ± 0.0	3.5 ± 0.5^{a}	158.4 ± 15.6
42	Monkey	5	0.3 ± 0.0	0.3 ± 0.0	5.5 ± 0.6	44.7 ± 7.2	278.3 ± 26.1
43	Chimpanzee	0.5	0.2 ± 0.0	0.1 ± 0.0	6.7 ± 2.1	–	–

[a]Reports a $t_{\frac{1}{2}\gamma}$ of 30.9 ± 1.0 hour.
Source: From Refs. 42–44.

We have determined the "serum and kidney" concentration–time profiles of caspofungin over a period of 96 hours in mice infected with *C. albicans* and analyzed the data using a "population pharmacokinetic" approach (36). The calculated terminal $t_{\frac{1}{2}}$ was 20.2 hours when only the serum pharmacokinetic data was considered but increased to 59.2 hours when we comodeled the serum and kidney concentration–time data, indicating the drug resides for a very long time in the peripheral tissues.

Caspofungin distributes extensively to peripheral tissues, but the equilibration process is slow (42,43). Compared to plasma concentrations, the C_{max} concentrations achieved in mice are 20 times higher in the liver, seven times higher in kidneys, and two to three times higher in lungs and spleen (43). This leads to achievement of higher area under the concentration–time curve (AUC_{0-24hr}) in the tissues, as exemplified by the liver in which the AUC_{0-24hr} is 16 times greater than in plasma (43). Caspofungin is taken up by hepatocytes, initially via rapid binding to the cell surface, later by slow transport processes into the intracellular compartment. It is then hydrolyzed to a number of inactive metabolites, designated M0-M6. These metabolites are excreted via the fecal and urinary routes (42).

Human Pharmacokinetics

The steady-state pharmacokinetic profile of caspofungin in humans given multiple daily doses of caspofungin (Table 3) closely follows that in animals. Caspofungin pharmacokinetics in humans is consistent with a linear, three-compartment model (48). Following a single intravenous infusion of a 70 mg dose to volunteers, there is a short α-phase of six hours. At this stage, caspofungin is confined to the plasma where 97% is bound to albumin. Therefore, its volume of distribution in the α-phase is roughly equal to that of the albumin space ($\sim$8 L). The drug is then gradually distributed to the extracellular fluid space, resulting in a β-phase ($t_{\frac{1}{2}} = $ 8–9 hour) lasting between 6 and 48 hours. The volume of distribution increases over the next two to three days approaching a plateau of greater than 23 L (48). Peak caspofungin concentrations are achieved in the tissues 36 to 48 hours after drug infusion, at which point less than 5% of the caspofungin remains in plasma. Less than 4% of drug is excreted in urine and feces during the first two days (48). Based on animal experiments, it is thought that extensive biotransformation starts at this time, which heralds the terminal γ-phase ($t_{\frac{1}{2}} = 27$ hour) of about one week in duration (documented until the lower limit of assay detection) (48). Metabolism of the caspofungin in humans is by peptide hydrolysis and *N*-acetylation to produce the biologically inactive open ring form, M0. By the fifth day, most caspofungin in plasma has been converted to M0. M1–6 is a hydrolysis product of M0 (48,49). Thus, the rates of urinary (2/3) and fecal (1/3) excretion of these

TABLE 3 Multiple Dose Caspofungin Pharmacokinetics in Humans

Patient population	Dose	C_{1hr} mg/L (90% CI)	C_{24hr} mg/L (90% CI)	AUC_{0-24hr} mg*hr/L (90% CI)	$t_{\frac{1}{2}\beta}$ (hr) (SD)	$t_{\frac{1}{2}\gamma}$ (hr) (SD)
Adults (healthy)	15 mg/day	2.8 (2.5–3.1)	0.4 (0.3–0.5)	24.4 (21.5–27.7)	8.6 (0.9)	40.9 (2.9)
	35 mg/day	6.0 (5.4–6.6)	0.9 (0.8–1.1)	54.9 (48.3–62.4)	9.5 (1.2)	46.7 (5.7)
	50 mg/day	8.7 (7.9–9.6)	1.6 (1.3–2.0)	86.9 (76.1–99.3)	10.1 (1.6)	42.1 (10.4)
	70 mg/day	14.0 (13.4–14.9)	2.4 (2.1–2.7)	129.6 (117.9–142.5)	11.2 (1.2)	45.5 (6.3)
	100 mg/day	21.5 (19.6–23.5)	4.3 (3.8–4.8)	218.9 (201.8–237.5)	12.6 (1.7)	49.4 (4.8)
Children						
Age 2–11 yr	50 mg/m²	15.4 (8.9–23.8)	1.3 (0.3–2.5)	113.6 (66.6–156.6)	8.0 (5.8–10.8)[a]	NR
Age 12–17 yr	1 mg/kg	7.5 (5.0–20.6)	0.7 (0.3–1.1)	54.0 (37.3–72.5)	11.7 (10.3–14.6)[a]	NR

All values reported for the adult pharmacokinetic data are means with 90% confidence intervals.
[a]The β-$t_{\frac{1}{2}}$ in children reported with 95% confidence intervals.
Abbreviations: AUC, area under the concentration–time curve; NR, not reported.
Source: Derived from Refs. 32,46,47.
Source: From Refs. 32,46,47.

products become marked after the first two days, peak at six to seven days, and then start declining, but are still detectable for up to three weeks (48).

When the "same" dose of an antifungal drug is given to a population of patients, it achieves "different" C_{max} and clearance (and therefore $t_{1/2}$) values in different people. Thus, while the pharmacokinetic data are often reported in the literature as mean values, by definition half of patients will not be able to achieve this mean value and may be at greater risk for therapeutic failure. A more useful data presentation is one that employs a population pharmacokinetic description. Population pharmacokinetics describes not only the mean parameter estimates, but also measures of dispersion, variability between- and within-subjects, and measurement error within a patient population that has been given a particular drug dose. This is more important in predicting therapeutic success in a population of patients. Unfortunately, there are no published human population pharmacokinetic studies for caspofungin.

Pharmacodynamics of Caspofungin: Preclinical

When different doses of a drug are given to infected animals (and to humans), there is a relationship between the dose administered and response (e.g., microbial kill) of the fungus. This dose–response relationship is mathematically described by the inhibitory sigmoid E_{max} model. The inhibitory sigmoid-E_{max} model is defined by the equation:

$$\text{Effect} = E_{con} - E_{max} \times [C]^{H} / ([C]^{H} + [EC\text{–}50]^{H})$$

where E is the residual fungal density (CFU/g) at the site of infection after drug exposure, E_{con} is the fungal burden (CFU/g) in untreated animals, E_{max} is the maximal reduction in fungal density (CFU/g) achievable with drug therapy, C is the antifungal drug exposure intensity [i.e., dose (mg/kg), AUC/MIC, C_{max}/MIC, or $T > $ MIC], EC–50 is the exposure intensity at which 50% of the maximal effect is observed, and H is the slope or Hill constant. An example of an inhibitory sigmoid E_{max} curve for caspofungin is shown in Figure 5, which shows the "dose-dependent response" of *C. albicans* in the kidneys of immunocompetent mice 96 hours after they were treated with a single injection of various doses of caspofungin. In this study, the E_{max} effect was a reduction of 2.25 $\log_{10}$ CFU of *C. albicans* per gram of kidney tissue compared with no therapy, and the EC–50 was a dose of 0.2 mg/kg. A dose of approximately 1 mg/kg was associated with 90% of maximal effect (i.e., EC–90). The EC–90 and E_{max} of caspofungin achieved in other mouse studies that examined different fungal pathogens are shown in Table 4.

Some investigators have described a paradoxical in vitro effect in which caspofungin concentrations much higher than the MIC and minimal fungicidal concentrations resulted in less killing of fungi compared to lower concentrations of drug (52). This paradoxical effect, analogous to the "Eagle effect" seen with penicillin (53), has also been noted in mice with aspergillosis treated with caspofungin (45). The therapeutic meaning, if any, in humans with fungal infection is unknown.

Dose-fractionation studies are used to determine the pharmacodynamic parameter (C_{max}/MIC ratio, AUC/MIC ratio, or Time above MIC) and hence dosing schedule (i.e., dosing frequency) associated with optimal microbial effect. Louie et al. (36) investigated the pharmacodynamic parameter linked with the efficacy of caspofungin in a mouse model of disseminated candidiasis. In these dose-fractionation

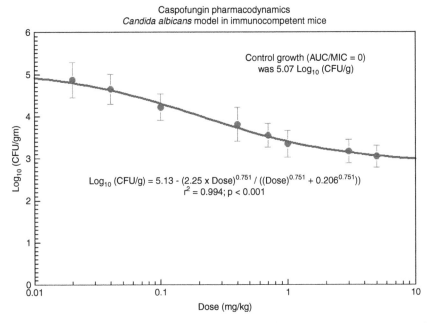

FIGURE 5 Dose–response relationship between caspofungin and kidney fungal burden. *Abbreviations*: AUC, area under the concentration–time curve; MIC, minimum inhibitory concentration.

studies, the reductions of the fungal densities in kidneys were similar in groups of mice that received each of the total doses of caspofungin investigated as 1, 2, or 4 equally divided doses over 96 hours. These results suggested that the pharmacodynamic parameter linked with efficacy was the AUC/MIC ratio. However, in dose-fractionation studies using a murine model of pulmonary aspergillosis, Wiederhold et al. (45) suggested that the C_{max}/MIC ratio was the pharmacodynamic parameter that best predicts the efficacy of caspofungin. Andes et al. (54) reported that the C_{max}/MIC ratio was the pharmacodynamic parameter linked with efficacy for the

TABLE 4 Fungal Reduction in Mice Kidneys After Treatment with Caspofungin

Mouse model	Pathogen (no. of strains)	MIC range (mg/L)	ED$_{90}$ (mg/kg)	E_{max} (log$_{10}$CFU/g)
DBA/2N[a]	*Candida albicans* (4)	0.125–0.25	0.003–0.02	3.66–4.69
–	*Candida tropicalis* (3)	0.125–0.25	0.03–0.06	3.51–3.97
–	*Candida glabrata* (2)	0.25–0.5	0.03–0.06	2.19–2.26
–	*Candida lusitaniae* (1)	0.5	0.16	2.8
–	*Candida parapsilosis* (1)	1.0	1.00	1.27
–	*Candida krusei* (1)	1.0	none	0.95
ICR	*Aspergillus fumigatus* (1)	0.25	0.49→1.0	N/A
–	*C. albicans* (1)	0.50	0.12	4.37–4.84

N/A, not available because end point was a 28-day survival after seven days of therapy.
The ED$_{90}$ were calculated using regression analysis.
[a]Day 7 sacrifice after daily intraperitoneal treatment for four days starting half an hour after infection.
Abbreviation: MIC, minimum inhibitory concentration.
Source: From Refs. 50,51.

Q6

echinocandin HMA 3270 in a neutropenic model of systemic candidiasis. Importantly, for antibiotics in which the AUC/MIC ratio is the pharmacodynamic parameter linked with efficacy, the antimicrobial effect is similar regardless of whether the cumulative dose is given as a single dose or as multiple equally divided doses per day. Thus, all of the cited dose-fractionation studies suggest that the administration of caspofungin as a single daily infusion (rather than multiple infusions each day) would optimize treatment efficacy while minimizing health provider manpower needs and patient inconvenience.

Pharmacokinetic studies in mice infected with *C. albicans* demonstrate that caspofungin remains in many tissues, including the kidney, longer than in serum (36). However, it has been suggested that measuring total concentrations of drug in tissue homogenates may have little meaning for extracellular pathogens (48) such as *Candida* species. To test this hypothesis, Louie et al. (55) developed an in vivo bioassay to determine the effect of caspofungin at the site of infection in the kidney of mice with systemic candidiasis. These investigators measured the concentrations of caspofungin in the serum and homogenates of kidney collected from mice that were treated with 0.4 mg/kg of caspofungin (which produced 50% of maximal microbial effect in a dose-range study) and found that the concentrations in the serum at 24, 48, and 72 hours after drug administration were 0.17, 0.05, and 0.03 mg/L, respectively. At these time points, the concentrations of caspofungin in the kidneys of these mice were 1.26, 0.58, and 0.13 mg/L, respectively. Mice inoculated with *C. albicans* at the 48-hour time point (>24 hour after serum concentrations fell below the fungal isolate's MIC of 0.2 mg/L) had a significant reduction in the fungal densities in their kidneys 24 hours later versus growth of fungi in the kidneys of untreated controls. The in vivo bioassay shows that "therapeutic" concentrations of caspofungin persist "at the site of infection" in kidney tissue well after serum concentrations fall below the MIC, underscoring the primacy of caspofungin levels in tissues in determining treatment outcome.

Important in determining the most appropriate frequency of administration of a drug is the duration of time a drug dose maintains therapeutic drug concentrations at the infection site. Another important factor is the duration of postantibiotic effect. Postantibiotic effect is the length of time that a drug maintains microbial growth inhibition after the drug falls below the MIC of the infecting pathogen. Ernst et al. (56) incubated two *C. albicans* isolates with different concentrations of caspofungin for one hour before washing away the drug and found that the postantifungal effect of caspofungin was concentration dependent. For each *Candida* strain, concentrations below the MICs had no postantifungal effect. However, the postantifungal effect for concentrations of $\geq 1 \times$ MIC was more than 12 hours. The postantifungal effect was maintained when the fungal isolates were incubated with caspofungin for as little as 0.25 hours. Louie et al. (36) also demonstrated that the postantifungal effect of caspofungin was concentration-dependent. The postantifungal effects after incubating a *C. albicans* isolate with 0.25×, 0.5×, 1×, and 5× MIC for three hours were 2, 4, 7, and 14 hours, respectively. In contrast, Manavathu et al. (57) reported that caspofungin did not exhibit a postantifungal effect against *A. fumigatus*.

Pharmacodynamics in Human Patients

Caspofungin pharmacodynamics was examined in a recent analysis that compared effect of different caspofungin concentrations to efficacy in human patients with candidiasis and aspergillosis (32). Exposures such as AUC_{0-24hr}, C_{1hr} (which approximate C_{max}), and C_{24hr} (which approximates trough in multiple dosing regimens) were examined in relationship to clinical response in patients who received 35, 50, and 70 mg/day of caspofungin for the treatment of esophageal candidiasis. The odds ratio for therapeutic success increased 3.48 times for every 2.72-fold increase in C_{24hr} (32). In addition, AUC_{0-24hr} was also marginally predictive of success. However, because of the design of the study, C_{1hr}, C_{24hr}, and AUC_{0-24hr} directly covaried, which meant that if C_{24hr} was associated with therapeutic success, C_{1hr}, and AUC_{0-24hr} would also be associated with success. Given that the MIC_{90} (CLSI method using RPMI 1640) of the Candida isolate cultured from the patients was 1 mg/L (58–60), we calculate (based on Table 3) that the corresponding AUC_{0-24hr}/MIC that were studied were approximately 55, 87, and 130. It is unclear where the doses that were studied are on the dose–response curve, although given the linear relationship between the odds ratio of success and C_{24hr} it would be reasonable to assume that the exposure range was on the steep portion of the sigmoid E_{max} curve. Nevertheless, it is intriguing that if the exposures studied in humans are loosely compared to those achieved in our murine dose–response study (Fig. 5), the same AUC/MIC ratios noted would also fall between EC_{20} and EC–90 on the steep portion of the dose–response curve. With regards to aspergillosis, however, there was no obvious dose–response relationship between caspofungin concentrations and clinical resolution of aspergillosis. The reasons for this are unclear. A potential explanation may be that the caspofungin doses studied achieve lung tissue concentrations on the maximal effect portion of the dose–response curve. Additional studies are needed.

MICAFUNGIN
Animal Pharmacokinetics
Similar to caspofungin, micafungin is poorly absorbed by the gastrointestinal tract. Single-dose pharmacokinetics of parenterally administered micafungin was determined in mice, rats, and dogs at doses of 0.32, 1.0, and 3.2 mg/kg (Table 5) (61). In all species, micafungin plasma concentrations declined bioexponentially with short α-phases of about 0.5 hours duration and longer β-phases, with a terminal elimination $t_{1/2}$ of approximately four to six hours. In all species, the $AUC_{0-\infty}$ of micafungin increased in proportion to the dose administered. Total clearance was approximately 1 mL/min/kg at all doses. A volume of distribution at a steady state of 0.38–0.56 L/kg was calculated for mice and rats, whereas it was 0.23–0.38 L/kg for dogs. No gender differences were observed. In rats, dosing rate had no effect on the plasma concentration–time profile of micafungin. The mean plasma concentration at five minutes after intravenous bolus injection was 2.92 mg/L while it was 2.10 mg/L at the end of the one-hour intravenous infusion. There were no differences in the mean $AUC_{0-\infty}$ or $t_{1/2}$ between the two methods of administration. In dogs, the plasma micafungin concentration–time profiles were also similar when administered by intravenous bolus or one-hour infusion. There were no substantial differences in calculated $AUC_{0-\infty}$ or $t_{1/2}$ between the two methods of administration. The mean plasma concentration at five minutes after intravenous bolus injection was 5.37 ± 0.21 mg/L and the mean plasma concentration following the one-hour infusion was 3.76 ± 0.47 mg/L. In our study of neutropenic mice with disseminated *C. glabrata* infection, population pharmacokinetic analysis for single doses of micafungin ranging between 0.3 and 3 mg/kg revealed a mean serum clearance of 0.7 ± 0.07 mL/min/kg, while the serum terminal $t_{1/2}$ was 6.7 hours (62).

TABLE 5 Pharmacokinetic Parameters (Mean ± Standard Error) After Single Administration of Micafungin to Mice, Rats, and Dogs: Intravenous Bolus Injection or Intravenous Infusion

Animal species	Dosing	Sex	Dose (mg/kg)	C_{max} (mg/L)	$AUC_{0-\infty}$ (mg*hr/L)	Terminal $t_{1/2}$ (hr)	V_{dss} (L/kg)	Cl (mL/min/kg)
Mice	Bolus	Male	0.32	0.809	4.9	5.71	0.49	1.01
			1.0	2.513	16.3	5.34	0.39	0.95
			3.2	8.835	54.0	5.63	0.38	0.92
Rats	Bolus	Male	0.32	0.656	3.3	3.96	0.53	1.50
			1.0	2.106	11.9	4.97	0.56	1.33
			3.2	7.612	48.3	5.05	0.42	1.04
		Female	1.0	2.590	11.3	4.53	0.53	1.40
	Infusion for 1 hr	Male	1.0	2.10 ± 0.13	16.4 ± 1.6	5.28 ± 0.21	0.40 ± 0.03	0.98 ± 0.10
	Bolus	Male	1.0	2.92 ± 0.08	15.7 ± 0.4	4.91 ± 0.14	0.38 ± 0.01	1.00 ± 0.03
			0.32	2.18 ± 0.22	7.2 ± 0.3	4.24 ± 0.06	0.23 ± 0.01	0.70 ± 0.03
	Bolus	Male	1.0	5.37 ± 0.21	21.2 ± 3.6	4.57 ± 0.51	0.25 ± 0.02	0.78 ± 0.15
Dogs	Infusion for 1 hr	Male	3.2	17.7 ± 2.0	84.8 ± 5.4	5.43 ± 0.15	0.23 ± 0.01	0.59 ± 0.04
			1.0	3.76 ± 0.47	20.0 ± 3.6	4.60 ± 0.41	0.27 ± 0.03	0.83 ± 0.14

Abbreviation: AUC, area under the concentration–time curve.
Source: From Ref. 61.

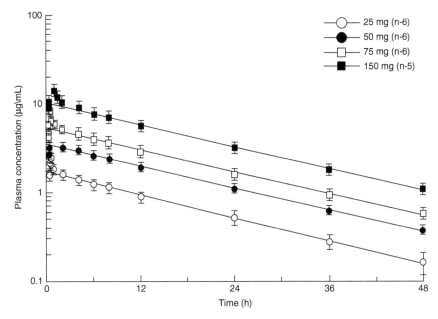

FIGURE 6 Plasma concentrations (mean ± standard deviation) after single intravenous infusion of 25, 50, 75, and 150 mg micafungin. *Source*: From Ref. 63.

Human Pharmacokinetics

When 25, 50, and 75 mg of micafungin were administered by intravenous infusion to healthy human adult volunteers over 30 minutes or 150 mg over one hour, the AUCs of the unchanged compound increased in proportion to the doses (Fig. 6 and Table 6) (63). Plasma concentrations reached a maximum at completion of administration and the mean elimination $t_{1/2}$ was 13.9 ± 1.0 hours. When 75 mg of micafungin was administered by intravenous infusion to healthy adult volunteers over 30 minutes once daily for seven days, plasma concentrations reached a steady state on day 4 (63). C_{max} and the elimination $t_{1/2}$ at the final administration were 10.87 mg/L and 14.0 hours, respectively.

No formal population pharmacokinetic study for micafungin administered to patients with fungal infections has been published. However, when 50 mg of micafungin was administered by intravenous infusion over an hour to elderly volunteers 66 to 78 years of age and nonelderly volunteers 20 to 24 years of age, plasma concentrations of the unchanged compound showed a similar time–course profile in both the elderly group and the nonelderly group. There were no differences in C_{max}, $AUC_{0-\infty}$, $t_{1/2}$, and protein binding rate in either group (64).

Pharmacodynamics of Micafungin: Preclinical

The efficacy of micafungin was evaluated in granulocytopenic mouse models of disseminated candidiasis and pulmonary aspergillosis (62,65–68). In candidiasis caused by *C. albicans*, *C. glabrata*, *Candida tropicalis*, *C. krusei*, *Candida parapsilosis*, *and C. guilliermondii*, micafungin exhibited 50% effective doses (ED_{50}s) in the range of 0.14 to 1.61 mg/kg (68). In pulmonary aspergillosis caused by *A. fumigatus*, micafungin exhibited ED_{50}s in the range of 0.26 to 0.45 mg/kg 15 days after

TABLE 6 Pharmacokinetic Parameters (Mean ± Standard Deviation) of Single-Dose Micafungin in Humans

Dose (mg)	No. of subjects	T_{max} (hr)	C_{max} (mg/L)	$AUC_{0-\infty}$ (mg*hr/L)	Terminal $t_{\frac{1}{2}}$ (hr)
25	6	0.5 ± 0	2.52 ± 0.28	34.3 ± 5.8	14.0 ± 1.2
50	6	0.5 ± 0	5.23 ± 0.38	74.3 ± 6.2	14.2 ± 1.2
75	6	0.5 ± 0	7.90 ± 1.35	106.5 ± 13.4	13.3 ± 0.7
150	5	1.0 ± 0	14.30 ± 1.31	216.6 ± 23.1	14.0 ± 0.9
25–150	23	NC	NC	NC	13.9 ± 1.0

Abbreviation: NC, not calculated.
Source: From Ref. 63.

infection (67). These results indicate that micafungin is a potent parenteral therapeutic agent for disseminated candidiasis and pulmonary aspergillosis in granulocytopenic mice.

The minimum effective plasma concentrations (MEPCs) of micafungin against mouse disseminated candidiasis and pulmonary aspergillosis were determined in mouse continuous infusion models using a miniosmotic pump (69). The activity of micafungin was evaluated in a disseminated candidiasis and pulmonary aspergillosis target organ assay, in order to determine the MEPC. Immunosuppressed mice were challenged intravenously with *C. albicans* or intranasally with *A. fumigatus* and treated with various concentrations of micafungin infused continuously, via a subcutaneously implanted pump. Efficacy was evaluated on the basis of a comparison of the mean $\log_{10}$ CFU/g in kidneys or lungs in groups treated with micafungin and in control groups, five days postinfection. These continuous infusion studies demonstrate an exposure-dependent effect for both *Candida* and *Aspergillus* infections (Tables 7 and 8). The estimated MEPCs of micafungin were 0.16 to 0.26 mg/L and 0.55 to 0.80 mg/L in mouse candidiasis and aspergillosis, respectively.

In vitro, micafungin demonstrates a concentration-dependent postantifungal effect against Candida species that ranged from 0 to ≥ 20.1 hours (70). The postantifungal effect was strain-dependent. Gumbo et al. (62) reported that micafungin has a long in vivo persistent effect against *C. glabrata*. Mice treated with a single 3 mg/kg micafungin dose showed persistent inhibition of kidney fungal growth for three days after micafungin kidney concentrations had fallen below the MIC of the *C. glabrata* isolate, while those treated with a single dose of 30 mg/kg revealed persistent inhibition of fungal growth in kidneys for five days after kidney drug concentrations had declined below the MIC of the fungal strain. This may allow for the study of intermittent micafungin dosing regimens in the treatment of disseminated candidiasis. The postantifungal effect of micafungin for *Aspergillus* species has not been reported.

Pharmacodynamics in Human Patients

In human clinical trials of HIV-infected patients with esophageal candidiasis, 12.5 mg/day (serum $AUC_{0-\infty}$ of 17 mg*hr/L) was associated with 80% clinical success, 25 mg/day (serum AUC 34 mg*hr/L) with 90% success, and 50 mg/day (AUC $\sim$ 74 mg*hr/L) with 100% success (71). This shows that micafungin has a dose-dependent effect in esophageal *Candida* infection in humans.

TABLE 7 Effect of Continuous Infusion of Micafungin on Kidney Fungal Titers After Five Days of Therapy in a Mouse Model of Disseminated Candidiasis

Concentration of micafungin injected into pump (mg/mL)	Plasma concentration (mg/L: mean ± SE)	Viable count ($\log_{10}$ CFU/kidney ± SE)
Control	–	6.36 ± 0.02
0.125	0.15	6.30 ± 0.13
0.25	0.16 ± 0.01	3.81 ± 0.34[a]
0.5	0.26 ± 0.04	2.77 ± 0.42[a]
1	0.37 ± 0.04	<1.00

Mice: ICR strain, male, five weeks old, five mice per group. Hydrocortisone was subcutaneously administered at 100 mg/kg at one day before and one day after infection.
Infection: *Candida albicans* FP633 was suspended in physiological saline and injected intravenously (9.2×10^4 CFU).
Treatment: ALZET miniosmotic pump filled with micafungin or saline (control group) was implanted subcutaneously in mice after infection.
Note: Values below the lower limit of quantification (0.05 mg/L) were recorded as 0.05 mg/L.
[a]Significantly ($p < 0.01$) different from the control (one way layout analysis of variance and Dunnett's multiple comparison).
Abbreviation: SE, standard error.
Source: From Ref. 69.

ANIDULAFUNGIN
Animal Pharmacokinetics

Pharmacokinetic studies of anidulafungin in animals have been published for rabbits (34,72,73), but not for other animals. In "noninfected" neutropenic rabbits, anidulafungin plasma concentration–time profiles exhibit multiexponential decline. There is a rapid α-phase (initial distribution) of less than two hours duration ($t_{1/2} = 0.2$–0.3 hour), a slower β-phase (distribution-elimination) of ~24 hours ($t_{1/2} = 3.0$–6.6 hour), and a γ-phase (duration dependent on dose) with a terminal $t_{1/2}$ of 9.8 to 17.6 hours (73). Clearance was constant and was between 0.07 and 0.12 L/hr/kg. With multiple daily dosing of anidulafungin, the trough concentrations in the liver, lung, spleen, and kidney were at least 32, 43, 21, and 17 times higher, respectively, than those measured in plasma. In noninfected rabbits, there was a linear relationship between

TABLE 8 Effect of Continuous Infusion of Micafungin on Mouse Lung Fungal Titers After Five Days of Therapy in a Model of Pulmonary Aspergillosis

Concentration of micafungin injected into pump (mg/mL)	Plasma concentration (mg/L: mean ± SE)	Viable count ($\log_{10}$ CFU/lung ± SE)
Control	–	3.99 ± 0.18
0.5	0.23 ± 0.04	3.71 ± 0.22
1	0.55 ± 0.13	2.82 ± 0.22
2	0.91 ± 0.19	2.14 ± 0.35[a]
4	1.55 ± 0.17	1.71 ± 0.29[a]

Mice: ICR strain, male, five weeks old, five mice per group. Hydrocortisone was subcutaneously administered at 100 mg/kg, one day before and one day after infection.
Infection: *Aspergillus fumigatus* TIMM0063 was suspended in physiological saline and injected intranasally (6.0×10^4 CFU).
Treatment: ALZET miniosmotic pump filled with micafungin or saline (control group) was implanted subcutaneously in mice after infection.
Note: Values below the lower limit of quantification (0.05 mg/L) were recorded as 0.05 mg/L.
[a]Significantly ($p < 0.05$) different from the control (one-ay layout analysis of variance and Dunnett's multiple comparison)
Abbreviation: SE, standard error.
Source: From Ref. 69.

plasma C_{max}, $AUC_{0-\infty}$, and dose. However, pharmacokinetic studies in rabbits with life-threatening disseminated candidiasis revealed a reduction in clearance of anidulafungin in these animals to between 0.05 and 0.08 L/hr/kg, and a smaller volume of distribution, which was significantly lower than in "healthy" animals. Consequently, infected animals achieved higher AUCs than noninfected animals receiving the same doses. Furthermore, doses between 0.1 and 1 mg/kg produced nonlinear kinetics in infected rabbits (73).

Human Pharmacokinetics

Data on human pharmacokinetics of anidulafungin are scant and have been mostly published in abstract form (74–78). The data are therefore incomplete and do not allow a proper examination of the methods employed. The data are summarized in Table 9. C_{max} and AUC increased linearly with dose (74,78). Metabolism of anidulafungin does not rely upon hepatic microsomal enzyme biotransformation. Nonenzymatic chemical degradation leads to opening of the ring to produce a linear peptide, which in turn is degraded in plasma by nonspecific peptidases. In fact, in patients with severe hepatic dysfunction, there is some increase in clearance of drug (Table 9) (76), although mechanisms to explain this are unclear. Renal dysfunction has no effect on metabolism or elimination of the drug. While the plasma $t_{1/2}$ of anidulafungin is approximately one day, that of its metabolites is five days (77). The entire dose of anidulafungin is eliminated via fecal excretion, 10% as parent drug and 90% as degradants (77). There is thus no need for dose adjustment in patients with renal failure, with hepatic dysfunction.

Human Population pharmacokinetics

The population pharmacokinetics of anidulafungin has been studied in 129 patients who were being treated for esophageal candidiasis, 87 patients being treated for invasive candidiasis, seven patients being treated for invasive aspergillosis, and two with azole refractory mucosal candidiasis (79). This represents a good mix of patients for whom anidulafungin is likely to be used. Anidulafungin was administered as daily infusions in three doses: 50, 75, and 100 mg. Steady-state concentrations were analyzed using a mixed effects model. The data best fit a two-compartment model with first-order elimination. Clearance of anidulafungin increased with body weight, diagnosis of invasive candidiasis, and male gender. The percentage relative standard error for weight on clearance was 27% while it was 25% for gender on clearance. In addition, central volume of distribution increased with weight. However, the covariates explained only 20% of the

TABLE 9 Pharmacokinetics of Single-Dose Anidulafungin in Humans

Patient population (no. of patients)	Intravenous dose (mg/day)	C_{max} (mg/L)	$AUC_{0-\infty}$ (mg*hr/L)	Clearance (L/hr)	Terminal $t_{1/2}$ (hr)
Mild hepatic impairment (6)	50	2.2 (0.3)	–	0.9 (0.2)	34.0(2.5)
Severe hepatic impairment (5)	50	1.6 (0.4)	42.4 (9.9)	1.23 (0.3)	–
Creatinine clearance <30 mL/min (6)	50	2.3 (0.5)	54.2 (10.9)	1.0 (0.2)	–
Healthy volunteers (8)	50	2.1 (0.2)	51.0 (5.0)	1.0 (0.1)	–
Healthy volunteers (9)	90	4.1 (14.1)	102.2 (13.6)	–	27.7 (8.4)

Abbreviation: AUC, area under the concentration–time curve.
Source: From Refs 74–78.

intersubject variability in clearance. The magnitude of intersubject variability in patients who receive 50, 75, and 100 mg of anidulafungin is therefore small and may thus be of limited clinical relevance.

Pharmacodynamics of Anidulafungin: Preclinical

Petraitis et al. (80) have examined the concentration–response relationships of anidulafungin in a rabbit model of esophageal candidiasis. The relationship between esophageal fungal burden and plasma concentration was described by an inhibitory sigmoid E_{max} equation. Inspection of their data reveals that concentrations between 8 and 10 mg/L were associated with maximal effect. Unfortunately, the drug concentrations that were measured were a single concentration two hours after administration of the anidulafungin, which makes it difficult to calculate the precise C_{max} and AUC achieved.

The pharmacodynamic parameters associated with optimal anidulafungin effect have been investigated in rabbits with disseminated candidiasis and invasive aspergillosis (73). Studies using a rabbit model of disseminated candidiasis revealed a dose–response relationship with maximal effect seen in animals that achieved a C_{max} of 1.95 mg/L and an AUC_{0-24hr} of 8.25 mg*hr/L. The Candida isolate studied had an MIC (CLSI method but using AM 3) of 0.015 mg/L and a minimum fungicidal concentration of 0.25 mg/L, which means the effective AUC_{0-24}/MIC ratio was 550. Unfortunately, the study could not identify the pharmacodynamic parameter [C_{max}/ minimum fungicidal concentration (MFC), AUC/MFC or T > MFC] linked to efficacy; the coefficient of determination between effect and pharmacodynamic parameter was similar and between 0.77 and 0.82 in the *Candida* infection model. Furthermore, the study in pulmonary aspergillosis failed to demonstrate a relationship between either plasma or tissue drug exposure and response as measured by quantitative cultures of lung fungal burden, measure of tissue injury, or survival (73). Gumbo et al. (81,82) examined the pharmacodynamics of anidulafungin in a neutropenic mouse model of disseminated candidiasis. The plasma pharmacokinetics in animals that we treated with escalating anidulafungin doses revealed that there was a linear relationship between C_{max} and $AUC_{0-\infty}$ and a mean serum terminal $t_{1/2}$ of 21.6 hours (± 4.6). Pharmacodynamic studies conducted over 96 hours revealed a dose–response effect defined by an inhibitory sigmoid E_{max} curve whose parameters included an EC–50 (AUC_{0-96hr}/MIC) of 642. Exposures beyond the EC–50 were associated with persistent antifungal effect that lasted more than four days. These in vivo results are consistent with the in vitro postantifungal effect of anidulafungin (56), although persistent tissue concentrations may also play an important role. Taken together, these pharmacodynamic studies, in conjunction with human pharmacokinetics of anidulafungin and the MIC_{90} of most *Candida* species isolates, suggest that doses of anidulafungin of above 50 mg a day are likely to achieve good microbiological outcome. However the dosing frequency associated with optimizing drug effect has not yet been established and await dose-fractionation studies.

Human Pharmacodynamics

Dowell et al. (83) examined clinical data from four studies of human patients with esophageal candidiasis who had been treated with a total of eight different daily doses of anidulafungin. Outcome was defined as clinical resolution of the esophageal candidiasis. There was a sigmoid E_{max} response when outcome was plotted against anidulafungin exposures. Good outcome was associated with a steady-state AUC of >35 mg*hr/L (equivalent to a trough concentration >1.5 mg/ L) (83). It is unclear what the anidulafungin MIC distribution was among the *Candida* isolates causing disease among these patients. Nevertheless, it is interesting to compare these results to findings in published animal studies (34,73), in which the maximal effect of anidulafungin in rabbits was associated with an $AUC_{0-\infty}$ (per our calculations) of between 29 and 36 mg*hr/L. We would like to point out that while the animal and human data show remarkable concordance, the final pharmacodynamic target value would need to take into account the MICs of *Candida* isolates in both studies, so that an AUC/MIC ratio that may be used as a therapeutic target can be calculated.

THERAPEUTIC USES AND APPROVED INDICATIONS FOR ECHINOCANDINS

Caspofungin has been approved by the Food and Drug Administration (FDA) as an empirical therapy for presumed fungal infections in febrile, neutropenic patients; treatment of candidemia; treatment of intra-abdominal abscesses, peritonitis, and pleural space infections due to *Candida* species; for esophageal candidiasis; and for treatment of invasive aspergillosis in patients who are refractory to or intolerant of polyene or azole therapy. It is administered as an intravenous dose of 70 mg on the first day, followed by 50 mg each day for the duration of therapy. Micafungin has been approved in Japan for the treatment of fungemia, respiratory mycosis, and gastrointestinal mycosis caused by *Aspergillus* spp. and *Candida* spp. For aspergillosis, the usual single daily dose is 50 to 150 mg of micafungin and this drug should be infused intravenously once daily. The dosage can be increased according to the patient's condition for severe or refractory aspergillosis up to 300 mg/day. For candidiasis, the standard single daily dose is 50 mg of micafungin and this drug should be infused intravenously once daily. The dosage can be increased according to the patient's condition for severe or refractory candidiasis up to 300 mg/day. Anidulafungin is being considered by the FDA for treatment of esophageal candidiasis. In the United States micafungin is FDA-approved for the treatment of esophageal candidasis and for prophylaxis of *Candida* infections in patients undergoing hematopoietic stem cell transplantation. For esophageal candidiasis, the FDA-approved dose is 150 mg/day. For prophylaxis of *Candida* infections, the FDA-approved dose is 50 mg/day. Anidulafungin is approved by the FDA for treatment of candidemia, esophageal candidiasis, and other forms of *Candida* infections (i.e., intraabdominal abscess and peritonitis. Anidulafungin is administered as a 200 mg loading dose followed by 100 mg/day thereafter.

REFERENCES

1. von Benz F, Knusel F, Nuesch J, Treichler H, Voser W, Nyfeler R. Echinocandin B, ein neuartiges polipeptide-antibiotikum aus *Aspergillus nidulans* var. *echinatus*: isolierung und Bausteine. Helvetica Chemica Acta 1974; 57:2459–2477.
2. Keller-Juslen C, Kuhn M, Loosli HR, Pechter TJ, von Weber HP, Wartburg A. Structure des cyclopeptide-antibiotikum SL 7810 (= Echinocandin B). Tetrahedron Letters 1976; 4147–4150.
3. Pfaller MA, Messer SA, Boyken L, et al. Further standardization of broth microdilution methodology for in vitro susceptibility testing of caspofungin against *Candida* species by use of an international collection of more than 3,000 clinical isolates. J Clin Microbiol 2004; 42:3117–3119.
4. Diekema DJ, Messer SA, Hollis RJ, Jones RN, Pfaller MA. Activities of caspofungin, itraconazole, posaconazole, ravuconazole, voriconazole, and amphotericin B against 448 recent clinical isolates of filamentous fungi. J Clin Microbiol 2003; 41:3623–3626.

5. Serrano MC, Valverde-Conde A, Chavez MM, et al. In vitro activity of voriconazole, itraconazole, caspofungin, anidulafungin (VER002, LY303366) and amphotericin B against *Aspergillus* spp. Diagn Microbiol Infect Dis 2003; 45:131–135.
6. Espinel-Ingroff A. Comparison of In vitro activities of the new triazole SCH56592 and the echinocandins MK-0991 (L-743,872) and LY303366 against opportunistic filamentous and dimorphic fungi and yeasts. J Clin Microbiol 1998; 36:2950–2956.
7. Onishi J, Meinz M, Thompson J, et al. Discovery of novel antifungal (1,3)-beta-D-glucan synthase inhibitors. Antimicrob Agents Chemother 2000; 44:368–377.
8. Abi-Said D, Anaissie E, Uzun O, Raad I, Pinzcowski H, Vartivarian S. The epidemiology of hematogenous candidiasis caused by different *Candida* species. Clin Infect Dis 1997; 24:1122–1128.
9. Nguyen MH, Peacock JE Jr, Morris AJ, et al. The changing face of candidemia: emergence of non-*Candida albicans* species and antifungal resistance. Am J Med 1996; 100:617–623.
10. Nguyen MH, Clancy CJ, Yu VL, et al. Do in vitro susceptibility data predict the microbiologic response to amphotericin B? Results of a prospective study of patients with *Candida* fungemia. J Infect Dis 1998; 177:425–430.
11. Gumbo T, Isada CM, Hall G, Karafa MT, Gordon SM. *Candida glabrata* Fungemia. Clinical features of 139 patients. Medicine (Baltimore) 1999; 78:220–227.
12. Gumbo T, Chemaly RF, Isada CM, Hall GS, Gordon SM. Late complications of *Candida* (*Torulopsis*) *glabrata* fungemia: description of a phenomenon. Scand J Infect Dis 2002; 34:817–818.
13. Clancy CJ, Barchiesi F, Falconi DL, et al. Clinical manifestations and molecular epidemiology of late recurrent candidemia, and implications for management. Eur J Clin Microbiol Infect Dis 2000; 19:585–592.
14. Wingard JR, Kubilis P, Lee L, et al. Clinical significance of nephrotoxicity in patients treated with amphotericin B for suspected or proven aspergillosis. Clin Infect Dis 1999; 29:1402–1407.
15. Bates DW, Su L, Yu DT, et al. Mortality and costs of acute renal failure associated with amphotericin B therapy. Clin Infect Dis 2001; 32:686–693.
16. Drusano GL. Antimicrobial pharmacodynamics: critical interactions of "bug and drug." Nat Rev Microbiol 2004; 2:289–300.
17. Singer SJ, Nicolson GL. The fluid mosaic model of the structure of cell membranes. Science 1972; 175:720–731.
18. Fleet GH. Composition and structure of yeast cell walls. Curr Top Med Mycol 1985; 1:24–56.
19. Douglas CM, D'Ippolito JA, Shei GJ, et al. Identification of the FKS1 gene of *Candida albicans* as the essential target of 1,3-beta-D-glucan synthase inhibitors. Antimicrob Agents Chemother 1997; 41:2471–2479.
20. Debono M, Gordee RS. Antibiotics that inhibit fungal cell wall development. Annu Rev Microbiol 1994; 48:471–497.
21. Douglas CM. Fungal β (1,3)-D-glucan synthesis. Med Mycol 2001; 39(suppl 1):55–66.
22. National Committee for Clinical Laboratory Standards Reference method for dilution antifungal susceptibility testing of yeast; approved standards. NCCLS document M27-A2. Wayne, PA: National Committee for Clinical Laboratory Standards, 2002.
23. Ostrosky-Zeichner L, Rex JH, Pappas PG, et al. Antifungal susceptibility survey of 2,000 bloodstream *Candida* isolates in the United States. Antimicrob Agents Chemother 2003; 47:3149–3154.
24. Yamaguchi H et al. Antifungal mechanisms of micafungin: enzymological and morphological studies of micafungin against *Candida albicans* and *Aspergillus fumigatus*. Chemotherapy (Tokyo) 2002; 50(S-1):20–29.
25. Espinel-Ingroff A. Evaluation of broth microdilution testing parameters and agar diffusion Etest procedure for testing susceptibilities of *Aspergillus* spp. to caspofungin acetate (MK-0991). J Clin Microbiol 2003; 41:403–409.
26. Arikan S, Lozano-Chiu M, Paetznick V, Rex JH. In vitro susceptibility testing methods for caspofungin against *Aspergillus* and *Fusarium* isolates. Antimicrob Agents Chemother 2001; 45:327–330.
27. Rex JH, Pfaller MA, Galgiani JN, et al. Development of interpretive breakpoints for antifungal susceptibility testing: conceptual framework and analysis of in vitro-in vivo correlation data

for fluconazole, itraconazole, and candida infections. Subcommittee on Antifungal Suscept-ibility Testing of the National Committee for Clinical Laboratory Standards. Clin Infect Dis 1997; 24:235–247.

28. Hernandez S, Lopez-Ribot JL, Najvar LK, McCarthy DI, Bocanegra R, Graybill JR. Caspofun-gin resistance in *Candida albicans*: correlating clinical outcome with laboratory susceptibility testing of three isogenic isolates serially obtained from a patient with progressive *Candida* esophagitis. Antimicrob Agents Chemother 2004; 48:1382–1383.

29. Bartizal K, Motyl M, Hicks P, Sable C, DiNubile M. In vitro susceptibility results and correlation with outcome: results from the randomized invasive candidiasis (IC) study of caspofungin (CAS) vs. Amphotericin B (AmB) [abstr] M-1240. 42nd Interscience Conference of Antimicrobial Agents and Chemotherapy, San Diego, CA, Sep 27–30, 2002.

30. Gonzalez GM, Tijerina R, Najvar LK, et al. Correlation between antifungal susceptibilities of *Coccidioides immitis* in vitro and antifungal treatment with caspofungin in a mouse model. Antimicrob Agents Chemother 2001; 45:1854–1859.

31. Nakai T, Uno J, Ikeda F, Tawara S, Nishimura K, Miyaji M. In vitro antifungal activity of Micafungin (FK463) against dimorphic fungi: comparison of yeast-like and mycelial forms. Antimicrob Agents Chemother 2003; 47:1376–1381.

32. Background Document for Antiviral Drug Products Advisory Committee Meeting January 10, 2001 CANCIDAS. www.fda.gov/ohrms/dockets/ac/01/briefing/3676b1_01.pdf. 2001.

33. Mukai T, Ohkuma T, Nakahara K, Takaya T, Uematsu T. Pharmacokinetics of FK463, a novel Echinocandin analogue, in elderly and non-elderly subjects [abstr A-30]. 41st Interscience Conference of Antimicrobial Agents and Chemotherapy, Chicago, IL, Sep 22–25, 2001.

34. Petraitiene R, Petraitis V, Groll AH, et al. Antifungal activity of LY303366, a novel echino-candin B, in experimental disseminated candidiasis in rabbits. Antimicrob Agents Chemother 1999; 43:2148–2155.

35. Bartizal K, Gill CJ, Abruzzo GK, et al. In vitro preclinical evaluation studies with the echino-candin antifungal MK-0991 (L-743,872). Antimicrob Agents Chemother 1997; 41:2326–2332.

36. Louie A, Deziel M, Liu W, Drusano M, Gumbo T, Drusano GL. Pharmacodynamics of caspofungin in a murine model of systemic candidiasis: importance of persistence of caspo-fungin in tissues to understanding drug activity. Antimicrob Agents Chemother 2005; 49:5058–5068.

37. Chiller T, Farrokhshad K, Brummer E, Stevens DA. Influence of human sera on the in vitro activity of the echinocandin caspofungin (MK-0991) against *Aspergillus fumigatus*. Antimicrob Agents Chemother 2000; 44:3302–3305.

38. Tawara S, Ikeda F, Maki K, et al. In vitro activities of a new lipopeptide antifungal agent, FK463, against a variety of clinically important fungi. Antimicrob Agents Chemother 2000; 44:57–62.

39. Zhanel GG, Saunders DG, Hoban DJ, Karlowsky JA. Influence of human serum on anti-fungal pharmacodynamics with *Candida albicans*. Antimicrob Agents Chemother 2001; 45:2018–2022.

40. Ikeda F. In vitro activity of a new lipopeptide antifungal agent, micafungin, against clinically important fungi. Chemotherapy (Tokyo) 2002; 50(S-1), 8–19.

41. Kurtz MB, Abruzzo G, Flattery A, et al. Characterization of echinocandin-resistant mutants of *Candida albicans*: genetic, biochemical, and virulence studies. Infect Immun 1996; 64:3244–3251.

42. Sandhu P, Xu X, Bondiskey PJ, et al. Disposition of caspofungin, a novel antifungal agent, in mice, rats, rabbits, and monkeys. Antimicrob Agents Chemother 2004; 48:1272–1280.

43. Hajdu R, Thompson R, Sundelof JG, et al. Preliminary animal pharmacokinetics of the parenteral antifungal agent MK-0991 (L-743,872). Antimicrob Agents Chemother 1997; 41:2339–2344.

44. Groll AH, Gullick BM, Petraitiene R, et al. Compartmental pharmacokinetics of the antifun-gal echinocandin caspofungin (MK-0991) in rabbits. Antimicrob Agents Chemother 2001; 45:596–600.

45. Wiederhold NP, Kontoyiannis DP, Chi J, Prince RA, Tam VH, Lewis RE. Pharmacodynamics of caspofungin in a murine model of invasive pulmonary aspergillosis: evidence of concen-tration-dependent activity. J Infect Dis 2004; 190:1464–1471.

46. Stone JA, Holland SD, Wickersham PJ, et al. Single- and multiple-dose pharmacokinetics of caspofungin in healthy men. Antimicrob Agents Chemother 2002; 46:739–745.

47. Walsh TJ, Adamson PC, Seibel NL, et al. Pharmacokinetics of caspofungin in pediatric patients [abstrt M-896]. 42nd Interscience Conference of Antimicrobial Agents and Chemotherapy, San Diego, CA, Sep 27–30, 2002.

48. Stone JA, Xu X, Winchell GA, et al. Disposition of caspofungin: role of distribution in determining pharmacokinetics in plasma. Antimicrob Agents Chemother 2004; 48:815–823.

49. Balani SK, Xu X, Arison BH, et al. Metabolites of caspofungin acetate, a potent antifungal agent, in human plasma and urine. Drug Metab Dispos 2000; 28:1274–1278.

50. Abruzzo GK, Flattery AM, Gill CJ, et al. Evaluation of the echinocandin antifungal MK-0991 (L-743,872): efficacies in mouse models of disseminated aspergillosis, candidiasis, and cryptococcosis. Antimicrob Agents Chemother 1997; 41:2333–2338.

51. Abruzzo GK, Gill CJ, Flattery AM, et al. Efficacy of the echinocandin caspofungin against disseminated aspergillosis and candidiasis in cyclophosphamide-induced immunosuppressed mice. Antimicrob Agents Chemother 2000; 44:2310–2318.

52. Stevens DA, Espiritu M, Parmar R. Paradoxical effect of caspofungin: reduced activity against *Candida albicans* at high drug concentrations. Antimicrob Agents Chemother 2004; 48:3407–3411.

53. Eagle H, Musselman AD. The rate of bactericidal action of penicillin in vitro as a function of its concentration, and its paradoxically reduced activity at high concentrations against certain organisms. J Exp Med 1948; 88:99–131.

54. Andes D, Marchillo K, Lowther J, Bryskier A, Stamstad T, Conklin R. In vivo pharmacodynamics of HMR 3270, a glucan synthase inhibitor, in a murine candidiasis model. Antimicrob Agents Chemother 2003; 47:1187–1192.

55. Louie A, Deziel M, Liu W, Drusano M, Gumbo T, Drusano GL. Kidney concentrations are critical in understanding the pharmacodynamics of caspofungin [abstr A-1573]. 43rd Interscience Conference of Antimicrobial Agents and Chemotherapy, Chicago, IL, Sep 14–17, 2003.

56. Ernst EJ, Klepser ME, Pfaller MA. Postantifungal effects of echinocandin, azole, and polyene antifungal agents against *Candida albicans* and *Cryptococcus neoformans*. Antimicrob Agents Chemother 2000; 44:1108–1111.

57. Manavathu EK, Ramesh MS, Baskaran I, Ganesan LT, Chandrasekar PH. A comparative study of the post-antifungal effect (PAFE) of amphotericin B, triazoles and echinocandins on Aspergillus fumigatus and Candida albicans. J Antimicrob Chemother 2004; 53:386–389.

58. Villanueva A, Arathoon EG, Gotuzzo E, Berman RS, DiNubile MJ, Sable CA. A randomized double-blind study of caspofungin versus amphotericin for the treatment of candidal esophagitis. Clin Infect Dis 2001; 33:1529–1535.

59. Villanueva A, Gotuzzo E, Arathoon EG, et al. A randomized double-blind study of caspofungin versus fluconazole for the treatment of esophageal candidiasis. Am J Med 2002; 113:294–299.

60. Arathoon EG, Gotuzzo E, Noriega LM, Berman RS, DiNubile MJ, Sable CA. Randomized, double-blind, multicenter study of caspofungin versus amphotericin B for treatment of oropharyngeal and esophageal candidiases. Antimicrob Agents Chemother 2002; 46:451–457.

61. Yamato Y. Pharmacokinetics of the antifungal drug micafungin in mice, rat and dogs, and its in vitro protein binding and distribution to blood cells. Chemotherapy (Tokyo) 2002; 50(S-1): 74–79.

62. Gumbo T, Drusano GL, Liu W, et al. Once-weekly micafungin therapy is as effective as daily therapy for disseminated candidiasis in mice with persistent neutropenia. Antimicrob Agents Chemother, 2007; 51: 968–974.

63. Azuma J. Pharmacokinetics study of micafungin. 2. Chemotherapy (Tokyo) 2002; 50(S-1): 155–184.

64. Azuma J. Pharmacokinetics study of micafungin in elderly subjects. Chemotherapy (Tokyo) 2002; 50(S-1):148–154.

65. Matsumoto S, Wakai Y, Nakai T, et al. Efficacy of FK463, a new lipopeptide antifungal agent, in mouse models of pulmonary aspergillosis. Antimicrob Agents Chemother 2000; 44:619–621.

66. Ikeda F, Wakai Y, Matsumoto S, et al. Efficacy of FK463, a new lipopeptide antifungal agent, in mouse models of disseminated candidiasis and aspergillosis. Antimicrob Agents Chemother 2000; 44:614–618.

67. Matsumoto F. Efficacy of micafungin, a new lipopeptide antifungal agent, in mouse models of pulmonary aspergillosis. Chemotherapy (Tokyo) 2002; 50(S-1):37–42.

68. Matsumoto F. Efficacy of micafungin, a new lipopeptide antifungal agent, in mouse models of disseminated candidiasis and aspergillosis. Chemotherapy (Tokyo) 2002; 50(S-1):30–36.

69. Wakai Y. Minimum effective concentrations of micafungin for the treatment of disseminated *Candida albicans* infection and pulmonary *Aspergillus fumigatus* infection in the mouse. Chemotherapy (Tokyo) 2002; 50(S-1):43–47.

70. Ernst EJ, Roling EE, Petzold CR, Keele DJ, Klepser ME. In vitro activity of micafungin (FK-463) against *Candida* spp.: microdilution, time-kill, and postantifungal-effect studies. Antimicrob Agents Chemother 2002; 46:3846–3853.

71. Pawlitz D, Young M, Klepser ME. Micafungin. A new echinocandin. Formulary 2003; 38:354–367.

72. Petraitis V, Petraitiene R, Groll AH, et al. Antifungal efficacy, safety, and single-dose pharmacokinetics of LY303366, a novel echinocandin B, in experimental pulmonary aspergillosis in persistently neutropenic rabbits. Antimicrob Agents Chemother 1998; 42:2898–2905.

73. Groll AH, Mickiene D, Petraitiene R, et al. Pharmacokinetic and pharmacodynamic modeling of anidulafungin (LY303366): reappraisal of its efficacy in neutropenic animal models of opportunistic mycoses using optimal plasma sampling. Antimicrob Agents Chemother 2001; 45:2845–2855.

74. Thye D, Kilfoil T, White RJ, Lasseter KC. Anidulafungin: pharmacokinetics in subjects with mild and moderate hepatic impairment [abstr A-34]. 41st Interscience Conference of Antimicrobial Agents and Chemotherapy, Chicago, IL, Sep 22–25, 2001.

75. Thye D, Marbury T, Kilfoil T, Henkel T. Anidulafungin pharmacokinetics in subjects with renal failure [abstr A-1391]. 42nd Interscience Conference of Antimicrobial Agents and Chemotherapy, San Diego, CA, Sep 27–30, 2002.

76. Thye D, Kilfoil T, Kilfoil G, Henkel T. Anidulafungin: Pharmacokinetics (PK) in subjects with severe hepatic impairment [abstr A-1392]. 42nd Interscience Conference of Antimicrobial Agents and Chemotherapy, San Diego, CA, Sep 27–30, 2002.

77. Dowell JA, Pu F, Lee J, Stogniew M, Krause D, Henkel T. A clinical mass balance study of anidulafungin (ANID) showing complete fecal elimination [abstr] A-1576. 43rd Interscience Conference of Antimicrobial Agents and Chemotherapy, Chicago, IL, Sep 14–17, 2003.

78. Brown GL, White RJ, Taubel J. Phase I dose optimization study for V-echinocandin. 40th Interscience Conference of Antimicrobial Agents and Chemotherapy, Toronto, Ontario, Sep 17–20, 2000.

79. Dowell JA, Knebel W, Ludden T, Stogniew M, Krause D, Henkel T. Population pharmacokinetic analysis of anidulafungin, an echinocandin antifungal. J Clin Pharmacol 2004; 44:590–598.

80. Petraitis V, Petraitiene R, Groll AH, et al. Dosage-dependent antifungal efficacy of V-echinocandin (LY303366) against experimental fluconazole-resistant oropharyngeal and esophageal candidiasis. Antimicrob Agents Chemother 2001; 45:471–479.

81. Gumbo T, Louie A, Liu W, Drusano M, Deziel MR, Drusano GL. Relative efficacies of anidulafungin, fluconazole, and amphotericin B for treatment of disseminated Candida glabrata in neutropenic mice [abstr A-1577]. 43rd Interscience Conference of Antimicrobial Agents and Chemotherapy, Chicago, IL, Sep 14–17, 2003.

82. Gumbo T, Drusano GL, Liu W, et al. Anidulafungin pharmacokinetics and microbial response in neutropenic mice with disseminated candidiasis. Antimicrob Agents Chemother, 2006; 50: 3695–3700.

83. Dowell JA, Stogniew M, Krause D, Henkel T. Anidulafungin (ANID) pharmacokinetic (PK)/ Pharmacodynamic (PD) Correlation: Treatment of esophageal candidiasis [abstr A-1578]. 43rd Interscience Conference of Antimicrobial Agents and Chemotherapy, Chicago, IL, Sep 14–17, 2003.

84. National Committee for Clinical Laboratory Standards Reference Method for Broth Dilution Antifungal Susceptibility Testing of Filamentous Fungi; Approved standard. NCCLS document M38-A. Wayne, PA: National Committee for Clinical Laboratory Standards, 2002.

19 Antimalarial Agents

Elizabeth A. Ashley
Shoklo Malaria Research Unit, Mae Sot, Thailand; Faculty of Tropical Medicine,
Mahidol University, Bangkok, Thailand; and Centre for Clinical Vaccinology and
Tropical Medicine, Churchill Hospital, Headington, Oxford, U.K.

Nicholas J. White
Faculty of Tropical Medicine, Mahidol University, Bangkok, Thailand and Centre
for Clinical Vaccinology and Tropical Medicine, Churchill Hospital, Headington,
Oxford, U.K.

INTRODUCTION/BRIEF HISTORY OF ANTIMALARIAL AGENTS

Malaria is the most important parasitic infection of man. *Plasmodium falciparum* is estimated to kill over a million people each year. Most of these deaths occur in African children. Mortality is rising—directly as a result of drug resistance. The naturally occurring ancient antimalarials, qinghaosu (artemisinin) and quinine (from Cinchona bark), have been used to treat fever since at least *AD* 340 and 1631, respectively (1,2). These two plant-derived medicines remain the mainstay of treatment for severe malaria to this day having survived over decades that have seen synthetic antimalarials come and then fall to resistance (Fig. 1). The first synthetic antimalarial, the 8-aminoquinoline pamaquine, was discovered in 1926, followed in the next decade by mepacrine (1932) and chloroquine (CQ) (1934). The discovery of the antifols, proguanil (1945), pyrimethamine (1952), and the 4-aminoquinoline amodiaquine (AQ) (1952) coincided with a global initiative by the newly formed World Health Organisation to eradicate malaria. Research stimulated by the conflict in Vietnam led to the discovery of mefloquine and halofantrine by the Walter Reed Army Institute of Research in the United States. Mefloquine was ready for use in 1977 but was not deployed in an endemic area until 1984 (3). This was followed by a hiatus in antimalarial drug development in the West despite the emergence and spread of CQ and antifol resistance in Asia and South America. In 1979, Chinese scientists reported the discovery, chemical structure, and antimalarial activity in vitro and in vivo of a plant-derived compound qinghaosu (artemisinin) with an entirely novel chemical structure. They then described the subsequent synthesis of more active derivatives artesunate, artemether, arteether (artemotil), and dihydroartemisinin (DHA) (4,5). Use of one of these compounds in combination with another antimalarial with a different mechanism of action [(artemisinin combination treatment (ACT)] is now the preferred treatment strategy in areas where multiple drug resistance is established (6). Indeed the World Health Organisation now recommends that any country changing antimalarial drug policy should switch to an ACT. In the last 10 years, there has been renewed interest in antimalarial drug development as part of global initiatives to alleviate the major diseases of poverty, and realization of the enormous health and economic consequences of antimalarial

AD340	Qinghaosu
1631	Quinine
1930s	Pamaquine, mepacrine, chloroquine
1940s	Proguanil
1960s	Sulfadoxine-pyrimethamine
1970s	Mefloquine, artemisinin derivatives
1980s	Halofantrine
1990s	Artemether-lumefantrine, atovaquone-proguanil
2000s	Chlorproguanil-dapsone, dihydroartemisinin-piperaquine

FIGURE 1 Antimalarial drug discovery/development.

drug resistance. New antimalarials developed include several compounds, e.g., lumefantrine, naphthoquine, and pyronaridine discovered and manufactured in China. Lumefantrine is available only in a fixed combination with artemether. New combinations of older drugs have also been introduced, e.g., atovaquone-proguanil, chlorproguanil-dapsone, and DHA-piperaquine.

Antimalarials may be categorized into five broad groups: the arylaminoalcohols and related compounds (quinine, quinidine, CQ, AQ, mefloquine, halofantrine, lumefantrine, piperaquine, pyronaridine, primaquine, and tafenoquine); the antifols (pyrimethamine, proguanil, chlorproguanil, and trimethoprim); the artemisinin derivatives (artemisinin, DHA, artemether, artemotil, and artesunate); hydroxynaphthaquinones (atovaquone); and antibacterial drugs with antimalarial activity (clindamycin, tetracyclines, and azithromycin). These drugs exhibit considerable differences in pharmacokinetic (PK) properties with terminal elimination half-lives which vary between less than one hour and more than one month. Several compounds are lipophilic and hydrophobic and very variably absorbed. All drugs exert their maximum effects on the blood stage parasite at the mature trophozoite stage, but this activity varies in terms of parasite killing from a fractional reduction in parasite densities of 10- to 10,000-fold per asexual cycle. The asexual cycle is one day for *Plasmodium knowlesi*, two days for *P. falciparum*, *Plasmodium vivax, and Plasmodium ovale*, and three days for *Plasmodium malariae*.

Mechanism of Action
The antimalarial drugs kill asexual stages of all parasites and the sexual stages of *P. vivax, P. malariae*, and *P. ovale*. With the exception of the 8-aminoquinolines and the artemisinin derivatives, they do not affect the sexual stages (gametocytes) of *P. falciparum*. The 8-aminoquinolines and hydroxynaphthaquinones inhibit the liver stages (hepatic schizonts) of the parasite life cycle. Only the 8-aminoquinolines kill the persistent liver stages (hypnozoites) of *P. vivax* and *P. ovale*.

Arylaminoalcohols
4-Aminoquinolines (CQ and AQ)
CQ is a weak base, which is concentrated within the parasite food vacuole. The parasite digests hemoglobin-producing toxic heme, which is polymerized (stacked in dimers) to form the relatively inert hemozoin or malaria pigment. CQ binds to ferriprotoporphyrin IX (ferric haem), a product of hemoglobin degradation,

thereby chemically inhibiting haem dimerization. Inhibition of this process provides a plausible although not proven explanation for the selective antimalarial action of CQ and probably the other aminoquinolines. CQ also competitively inhibits glutathione-mediated haem degradation, another parasite detoxification pathway. AQ and the closely related amopyroquine are Mannich base 4-aminoquinolines. AQ is almost entirely metabolized to a biologically active metabolite desethylamodiaquine (DAQ) (7–9).

Piperaquine is a bisquinoline compound related to CQ and other 4-aminoquinolines. It was synthesized and developed by Chinese scientists about 30 years ago and has been used as first-line treatment for CQ-resistant falciparum malaria in China although not elsewhere. The analogues hydroxypiperaquine and hydroxypiperaquine phosphate have also been tested in field studies in China with good results. Piperaquine base (phosphate) is now available in a fixed combination with DHA (and also sometimes trimethoprim and primaquine) (10). AQ is more active than CQ against resistant parasites, but piperaquine retains activity against highly CQ- and AQ-resistant *P. falciparum*. The reason for these differences in activity is not known. No major differences in pharmacodynamic properties among these compounds have been demonstrated.

8-Aminoquinolines (Primaquine and Tafenoquine)
Although structurally related to the 4-aminoquinolines, the activity of these drugs is different. They have very weak asexual stage activity against *P. falciparum*, but unlike other antimalarials they have potent sexual stage (gametocytocidal) activity. Against *P. vivax*, *P. malariae*, and *P. ovale*, these compounds have significant asexual stage activity (albeit weaker than CQ), and they also kill the hypnozoites of *P. vivax* and *P. ovale* (radical curative activity). The mechanism of action of primaquine (and metabolites) has not been characterized, although it may interfere with the function of plasmodium DNA rather than interference with hemoglobin degradation. Tafenoquine possesses greater activity against both the blood and liver stages of malaria than primaquine (11).

Cinchona Alkaloids (Quinine and Quinidine)
Quinine is derived from the bark of the Cinchona tree and usually formulated as the dihydrochloride salt for parenteral administration, and as the sulphate, bisulphate, dihydrochloride, ethylcarbonate, hydrochloride, or hydrobromide salts for oral administration. Quinine also inhibits haem detoxification in vitro suggesting it has a similar final mechanism of action to CQ. Quinidine is the dextrorotatory diastereoisomer of quinine. It is intrinsically more active as an antimalarial, but it is also more cardiotoxic (12).

Mefloquine
Mefloquine is a fluorinated 4-quinoline methanol compound used for the treatment of multidrug-resistant falciparum malaria. It has two asymmetric carbon atoms and is used clinically as a 50:50 racemic mixture of the erythroisomers. Mefloquine also seems to target the parasite food vacuole. The precise mechanism of action is unclear, although it probably also involves inhibition of haem detoxification. It has also been suggested that there is a hydrogen bond formation between mefloquine and a cellular effector or transport proteins (13).

Halofantrine

Halofantrine is a 9-phenanthrene methanol. It has one asymmetric carbon atom and is used as a racemate. The enantiomers have equal antimalarial activity. Halofantrine has a similar mechanism of action to but is intrinsically more potent than quinine or mefloquine. Unfortunately it is associated with rare but potentially lethal ventricular tachycardias, which have curtailed its use.

Lumefantrine

Formerly called benflumetol, lumefantrine was developed by Chinese scientists. It is a racemic 2,4,7,9-substituted fluorine derivative, thought to have a similar mechanism of action to the other arylaminoalcohols. Lumefantrine is available only in a fixed tablet combination with artemether. Each tablet contains artemether 20 mg and lumefantrine 120 mg. The two compounds are moderately synergistic in vitro against *P. falciparum* (14).

Quinine, mefloquine, halofantrine, and lumefantrine have very different PK properties but their pharmacodynamic properties are similar.

Pyronaridine

An acridine-type (benzonaphthyridine) Mannich base compound related to AQ, which was developed in China in 1970. It is more active than AQ against resistant parasites (15). It is being produced in a fixed combination with artesunate.

Antifols (Pyrimethamine, Cycloguanil, Chlorcycloguanil, and Trimethoprim)

These drugs exert their antimalarial effect via the inhibition of folate biosynthesis, essential for pyrimidine synthesis and consequently DNA replication. Pyrimethamine acts by inhibiting plasmodial dihydrofolate reductase (DHFR) while the sulpha drugs, with which they are combined, inhibit dihydropteroate synthase (DHPS), sequential key enzymes in the folate biosynthetic pathway. There is marked synergy between the two classes of compound used in combination. The biguanide antifols, proguanil and chlorproguanil, act as prodrugs for the active triazine metabolites cycloguanil and chlorcycloguanil, which also inhibit DHFR. Chlorproguanil-dapsone (*Lapdap*®) is the latest addition to this class and is more effective than the very widely used sulfadoxine-pyrimethamine (SP) against *P. falciparum* in Africa.

Qinghaosu

Qinghaosu or artemisinin is a sesquiterpene lactone peroxide extracted from the leaves of the sweet wormwood plant *Artemisia annua* (Qinghao). Four derivatives are used widely: the water-soluble hemisuccinate derivative artesunate, the oil-soluble ethers; the methyl ether artemether, the ethyl ether artemotil (arteether), and their common main metabolite DHA. DHA is 5 to 10 times more potent as an antimalarial compared with artemisinin. How these antimalarials exert their effect is not entirely clear. Parasiticidal activity is dependent on the integrity of the peroxide bridge. Carbon-centered free radicals are produced following the non-haem iron catalyzed cleavage of the endoperoxide bridge in the parasite food vacuole. It was thought that these might alkylate critical proteins (16). Recently, the artemisinin compounds have been shown to be specific and potent inhibitors of the sarcoplasmic reticulum Ca^{2+}-ATPase (SERCA) or PfATPase 6 outside the parasite food vacuole following activation by iron (17).

Atovaquone-Proguanil (Malarone™)

Atovaquone is a hydroxynaphthoquinone compound, which acts on the ubiquinone metabolic pathway thereby inhibiting mitochondrial cytochrome electron transport and thus cellular respiration (18). Proguanil may have a similar site of action. The two compounds are synergistic in vitro. Originally atovaquone was developed to be used alone, but high-level resistance developed rapidly in approximately one-third of treated patients. Interestingly it is the parent compound proguanil, which is the important synergistic contributor to antimalarial efficacy in this fixed combination, rather than the antifol triazine metabolite, as atovaquone-proguanil is equally effective against highly antifol-resistant parasites, and also in individuals with low CYP2C19 activity who are unable to convert proguanil to cycloguanil.

Antibacterial Agents

The antibacterials, which act on protein or nucleic acid synthesis, often have significant antimalarial activity, but low parasite-killing rates. The antifol trimethoprim has good antimalarial activity and shares resistance profiles with pyrimethamine. The tetracyclines are consistently active against all species of malaria, with doxycycline being the most widely used. Clindamycin is equally active. The macrolides are active in vitro but are generally disappointing in vivo, with the exception of azithromycin, which has been evaluated in prophylaxis and treatment. Other drugs affecting protein or nucleic acid synthesis, e.g., chloramphenicol and rifampicin exhibit weak antimalarial activity. These drugs all act relatively slowly and are therefore used in combination with more rapidly acting agents, e.g., clindamycin or doxycycline is used in combination with quinine or artesunate in the treatment of resistant falciparum malaria. They should not be used alone. The main disadvantage of drug combinations containing an antibiotic is the need to give seven days of treatment (19).

New Drug Targets

Several new drug targets are under investigation. These include *P. falciparum* protein farnesyltransferase, a key enzyme in parasite isoprenoid biosynthesis and 1-deoxy-d-xylulose-5-phosphate (DOXP) reductoisomerase, which is inhibited by the antibiotic fosmidomycin (20).

SUMMARY OF SPECTRUM OF MICROBIOLOGICAL ACTIVITY

There are two discrete components of antimalarial activity, one of which is analogous to the spectrum of antibiotic activity; this is the variation in activity against different parasites, particularly *P. falciparum* from different geographic regions. The other does not have an antibacterial correlate; this is the relative activity against the different stages of development in the asexual life cycle and is of importance in the treatment of severe malaria where prevention of parasite development from circulating ring stages to more pathological sequestered cytoadherent stages may prevent death.

Parasite Life Cycle

The infected female Anopheline mosquito injects up to 100 (median circa 10) motile sporozoites into the circulation while taking a blood meal. The sporozoites find their way rapidly to the liver where they invade individual hepatocytes. They then develop into hepatic schizonts, which mature over five to seven days before rupturing to release thousands of merozoites into the blood. These immediately invade red blood cells. This marks the beginning of the asexual life cycle of the "blood stage" parasite. The young trophozoites start as tiny "rings," which look like signet rings or stereo headphones under light microscopy. This circulating stage, which lasts 12 to 18 hours, is less pathogenic than the more mature trophozoite stages, which in *P. falciparum* infections cause cytoadherence of the infected erythrocytes. Sequestration of the parasitized erythrocytes leads to microvascular obstruction and is an important mechanism in the pathogenesis of severe disease. Mature trophozoites develop into schizonts, which will rupture at the end of the 48-hour asexual cycle releasing more merozoites. These invade more red cells starting a new cycle and causing the infection to expand exponentially. The sexual life cycle, which in *P. falciparum* starts slightly later than the asexual cycle, results from the production of male and female gametocytes from asexual parasites. As the switch from asexual to sexual development is density dependent, early effective treatment of falciparum malaria prevents significant gametocytogenesis and therefore interrupts transmission. The gametocytes may persist in the circulation for days or weeks. They are not pathogenic to their host, but they are the source of transmission of malaria, and thus gametocytocidal activity is of public health importance. In *P. vivax* and *P. ovale* infections, some of the sporozoites inoculated by the mosquito remain dormant in the liver as hypnozoites, which may cause relapses weeks or months later—even after an apparently successful treatment of the blood stage infection.

Antimalarial Stage Specificity

The term blood schizontocide is widely used to describe the action of antimalarials on blood stage parasites, although it is a slight misnomer as mature trophozoites are more susceptible to the antimalarial drugs and formed schizonts are relatively resistant. Young ring trophozoites are also relatively drug resistant (particularly to quinine and pyrimethamine). CQ acts mainly on the large ring form and mature trophozoite stages of the parasite. Quinine acts principally on the mature trophozoite stage of parasite development while the antifols act a little later. The artemisinin derivatives are the most rapidly acting of the known antimalarials, and they have the broadest time window of antimalarial effect (from ring forms to early schizonts). These compounds prevent maturation of ring stages, thus reducing subsequent cytoadherence and by inference, disease severity. Only the 8-aminoquinolines, primaquine, and tafenoquine are capable of killing the hypnozoites of *P. vivax* and *P. ovale*, and these are also the only compounds, which kill the mature (stage 5) gametocytes of *P. falciparum*. The artemisinin derivatives reduce gametocytemia by rapidly reducing the parasite biomass and also killing immature gametocytes (stages 1 to 4). In contrast, in the treatment of the other three malarias, all the antimalarial drugs are gametocytocidal (21,22).

MECHANISMS OF RESISTANCE

Stable resistance arises from the selection and spread of parasites with spontaneous chromosomal point mutations or gene duplications, which are independent of drug selection pressure (Table 1). The frequency of naturally occurring viable mutants

TABLE 1 Single Nucleotide Polymorphisms Related to Antimalarial Drug Resistance

Protein/enzyme	Resistant mutation[a]	Antimalarial drugs affected
PfCRT (Chromosome 7)	D/N75E K76T A220S Q271E N326S I356T R371	Chloroquine, amodiaquine
PfMDR1 (Chromosome 5)	N86Y Y184F S1034C N1042D D1246Y	Chloroquine, amodiaquine
PfDHFR	A16V C50R N51I C59R S108N/T I164L	Pyrimethamine, proguanil, chlorproguanil
PfDHPS	S436A/F/C A437G K540E A581G A613S/T	Sulfadoxine, dapsone
PvDHFR	N50K F57L S58R T61M S/I117N S/I117T I173L	Pyrimethamine, proguanil, chlorproguanil

[a] A, alanine; C, cysteine; D, aspartate; E, glutamate; F, phenylalanine; G, glycine; I, isoleucine; K, lysine; L, leucine; M, methionine; N, asparagine; Q, glutamine; R, arginine; S, serine; T, threonine; V, valine; Y, tyrosine.

resistant to the different drug classes varies, as does the degree of resistance to that class conferred by each mutation or duplication. After the initial de novo occurrence (which is a relatively rare event), the subsequent spread of resistant parasites is facilitated by the use of drugs with long elimination phases. These provide a "selective filter," allowing infection by the resistant parasites while the residual antimalarial activity prevents infection by sensitive parasites.

Despite extensive investigations, the precise mechanism of resistance to many of the drug classes remains unclear, with the exception of the antifols and atovaquone.

Arylaminoalcohols
Resistance to these antimalarials results from altered intraparasitic drug transport.

Aminoquinolines

Concentration of CQ in the acidic food vacuole compartment of the intraerythrocytic parasite is essential for it to exert its pharmacological activity. CQ-resistant parasites show decreased accumulation of the drug in the food vacuole. Both reduced influx and increased efflux have been implicated. Resistance is linked mainly to multiple mutations in a transporter protein in the vacuolar membrane, PfCRT (*P. falciparum* CQ resistance transporter). The *Pfcrt* gene is on chromosome 7 and there are a number of polymorphisms in this gene, which have been associated with CQ resistance. The principal correlate is a mutation giving a codon change, which results in substitution of lysine by threonine at position 76 (23,24). Resistance to CQ does not equate to resistance to all aminoquinolines. It has been proposed that 4-aminoquinolines, which are more polar at lysosomal pH (such as CQ), are more likely to pass through the mutated PfCRT channel and escape from the lysosome into the parasite cytoplasm, while those which are more hydrophobic (AQ and its active metabolite DAQ) are likely to bind to the hydrophobic channel lining and stay inside. Their positive charges repel further access of 4-aminoquinoline to the channel (25).

The ATP-requiring transmembrane pump, P-glycoprotein, originally identified in mammalian tumor cells, is thought to contribute to CQ efflux. The gene *pfmdr1* encodes for a 162 kDa homologue of the P-glycoprotein pump. Transfection studies confirm a role for *pfmdr* in mediating resistance to CQ and mefloquine. These MDR genes are found in increased copy numbers in most quinine- and mefloquine-resistant parasites, and point mutations (notably asparagine to tyrosine at position 86) are associated with CQ resistance. Mutant MDR genes are much less likely to amplify than the wild type so point mutations are negatively associated epidemiologically with mefloquine resistance. CQ resistance can be reversed in vitro by a number of structurally unrelated pump inhibitors such as verapamil, fluoxetine, amlodipine, several antihistamines, and phenothiazines. It is likely that multiple unlinked mutations are required for the development of CQ resistance, and that other contributors to quinoline resistance remain to be discovered. Clinical resistance to piperaquine developed in Southern China after years of widespread and intensive use, including mass prophylaxis. The mechanism is unclear as there is little evidence of cross-resistance with CQ in vitro (26–28). The mechanism of chloroquine resistance in *P. vivax* has not been elucidated.

Quinine

Quinine resistance is also associated with reduced drug uptake by the parasite but the precise mechanism by which this occurs is unclear. There is a weak association with the *pfmdr* genotype (PfMDR1 amplification reduces quinine susceptibility), however, it is probable that resistance to quinine is also the result of mutations in multiple genes. There have been occasional case reports of high-grade clinical resistance to quinine, some of which can be attributed to inadequate plasma concentrations, but despite a moderate reduction in sensitivity in Southeast Asia, quinine is still effective.

Mefloquine

Increased *pfmdr1* copy number is the principal correlate of mefloquine resistance, accounting for over 60% of the variance in mefloquine susceptibility in field isolates. However it is not the only determinant. In vitro experiments in certain mefloquine-resistant clonal infections with known *pfmdr1* copy numbers have shown no relationship, which suggests that mechanisms apart from amplification and over-expression of the gene are also important. Mefloquine resistance is reversed in vitro by penfluridol, which does not reduce CQ efflux (29–31).

Antifols

The mechanism of *P. falciparum* resistance to antifols has been characterized in detail. Comparison of the *pfdhfr* gene sequence from sensitive and resistant parasites shows that resistance results from the sequential accumulation of point mutations. The initial mutation conferring DHFR resistance is usually serine to asparagine at position 108 (S108N). This appears to be the key mutation. Further mutations arise at positions 51 (N51I) and 59 (C59R), conferring further increases in resistance to the drugs; however, some therapeutic response is usually seen, and in the presence of background immunity (i.e., older children and adults in endemic areas), cure rates may still be high. Acquisition of a fourth mutation at position 164 (I164L), as has happened in parts of Asia and South America, renders all available antifols completely ineffective (32,33). Pyrimethamine resistance is generally assumed to have evolved multiple times because the genetic basis is simple and resistance can be selected easily in the laboratory; however, data from Southeast Asia, Southern Africa, and South America indicate remarkable clonal spread of SP-resistant parasites carrying resistant DHFR alleles (34). The initial mutation encoding for cycloguanil resistance is different to that found in pyrimethamine resistance (35). Serine at position 108 is converted to threonine and there is a second mutation at position 16 (alanine to valine). In general, the biguanides are more active than pyrimethamine against resistant mutants, except against parasites with the DHFR 164 mutation. *P. vivax* shares similar antifol resistance mechanisms with *P. falciparum* through serial acquisition of mutations in *Pvdhfr*, although the sequence of acquisition and levels of resistance conferred differ (Table 1).

Five point mutations in PfDHPS conferring resistance to sulfonamides and sulfones have been identified. These are S436A/F, A437G, K540E, A581G, and A613S/T. Of these, the A437G mutation is found with the highest frequency in field isolates and so may be the key initial mutation for sulpha resistance. There has been considerable debate on the contribution of DHPS mutations to resistance to antifol-sulpha combinations. Current evidence suggests that DHFR mutations play a primary role and DHPS mutations an important supporting role.

Artemisinin Derivatives (Qinghaosu)

Clinical resistance to these drugs has not been documented. In vitro multidrug-resistant parasites are more artemisinin resistant, and reduced susceptibility can be selected in the laboratory (up to a factor of approximately 10), but these parasites are still fully susceptible in vivo to therapeutic concentrations of artemisinin derivatives. In the Xenopus oocyte expression system, replacement of leucine by glutamic acid at PfATPase 6 position 263 abolished SERCA inhibition by artemisinins. To date, with one controversial exception, polymorphisms in PfATPase 6 in field isolates have not been correlated with susceptibility. Although significant resistance to this class of drugs has not been reported, cure rates with artemisinin derivatives are not 100%. This is attributed to the persistence of a few parasites, probably at the merozoite/very young ring stage (i.e., just after red cell invasion), which show maturation arrest and may persist for many days despite therapeutic antimalarial drug levels. These dormant forms may "awaken" to cause recrudescence of the infection with parasites, which remain fully susceptible to artemisinin (36,37).

Atovaquone-Proguanil

Atovaquone resistance is associated with single point mutations in the *cytochrome b* gene of the parasite encoding the cytochrome bc(1) complex of the parasite inner mitochondrial membrane. Resistance mutations arise commonly, at an approximate in vivo frequency of 1 in 10^{12} parasites. They emerge in one-third of patients receiving atovaquone alone and a much lower (100 times) frequency following atovaquone-proguanil (38).

PHARMACOKINETICS

The PK properties of many antimalarial agents have been characterized in detail. These PK properties often vary depending on whether the drug is given in health, disease, pregnancy, or in combination with another drug (Table 2).

Arylaminoalcohols
Aminoquinolines
Chloroquine

CQ is rapidly absorbed from the gastrointestinal tract, reaching maximum plasma concentrations in approximately five hours. Because of extensive tissue binding CQ and its metabolite desethyl CQ has a very large apparent volume of distribution (V_d) of approximately 500 L/kg. CQ is metabolized slowly to desethyl CQ, which has approximately equivalent antimalarial activity. The decline in blood concentrations of CQ is multiexponential with a terminal elimination half-life of one to two months. Thus the blood concentration profile during malaria is determined mainly by distribution rather than elimination processes. The PK properties of CQ are not significantly altered by disease severity, pregnancy, or age. As approximately 50% of systemic clearance is renal, dose adjustment is needed in renal failure with prophylactic administration but not for treatment. As the volume of the central compartment is some 1000 times smaller than the total apparent V_d, the rapid absorption of CQ following subcutaneous or intramuscular injection may outpace distribution, and transiently toxic concentrations may occur with individual doses over 3.5 mg base/kg. Intravenous CQ should be given by continuous infusion—never by bolus injection. A suppository formulation has been developed, which also has good bioavailability. CQ is approximately 55% bound to plasma proteins. CQ is taken up by platelets and leukocytes, so plasma levels are lower than corresponding serum values. Concentrations are higher in whole blood than in plasma. Whole blood measurements are simple and reliable and can be performed on small volumes or filter paper samples (40–43).

Amodiaquine

Oral AQ is absorbed rapidly from the gastrointestinal tract. Peak plasma concentrations are reached after a mean of 30 minutes (healthy volunteers) and 1.75 hours in malaria patients. It is extensively metabolized in the liver to desethylamodiaquine (DAQ); the main antimalarial compound, and 2-hydroxyamodiaquine. Peak plasma concentrations (C_{max}) of these metabolites are reached after a mean of 3.4 hours (healthy volunteers). There is a very large first-pass effect. The mean plasma concentration of DAQ is six to seven times greater than that of AQ (healthy volunteers), and it is not uncommon for the parent compound to be undetectable. DAQ accumulates in red cells to give a red cell:plasma ratio of 3:1.

Drug	Absorption: Time to peak (h) p.o.	i.m.	Oral dose (mg/kg)	Peak level (mg/L)	Plasma Binding (%)	V_d/f (L/kg)	Clearance/f (ml/kg/min)	$t_{1/2\beta}$ (hours unless otherwise stated)
Uncomplicated malaria:								
Quinine[a]	6	1	10	8	90	0.8	1.5	16
Quinidine	1	—	10	5	85	1.3	1.7	10
Chloroquine[b]	5	0.5	10	0.12	55	500	2.0	30–60 days
Desethylamodiaquine[c]	—		30	—	—	—	—	11 days
Piperaquine[d]	6	—	55	0.15	—	728	23	28 days
Mefloquine[e]	17	—	25	0.9	>98	20	0.35	14 days
Halofantrine[d,f]	15	—	8	3.5	>98	—	7.5	113
Lumefantrine[d,f]	6	—	9	0.5	>98	2.7	3.0	86
Pyrimethamine	4	41	1.25	0.5	94	30	0.33	87
Chlorproguanil[f]	4	—	2.0	0.1	75	6	20	35
Atovaquone[d]	6	—	15	5	99.5	—	2.5	30
Artesunate[g]	1.5	0.5	4	0.5	—	0.15	50	0.75
Artemether[h]	2	3–18	4	1.5	95	2.7	54	1
Dihydroartemisinin	4	—	4	—	70	—	—	1
Healthy subjects								
Primaquine	3	—	0.6	0.15	—	3	6	6
Proguanil[i] (Chloroguanide)	3	—	3.5	0.17	75	24	19	16
Pyrimethamine	4	—	0.3	0.35	—	2.9	0.4	85

[a] Protein binding increased. V_d and clearance further reduced in severe malaria. Rate of i.m. absorption proportional to concentration of injectate.

[b] Concentrated in red cells, white cells and platelets. Kinetics unaffected by disease severity.

[c] Almost all antimalarial activity of amodiaquine after oral administration is provided by the metabolite.

[d] Absorption increased by fats.

[e] (+) RS enantiomer concentration higher, and (–)SR enantiomer lower, in whole blood than plasma. Metabolized to active desbutyl metabolite which is eliminated more slowly.

[f] Mainly a prodrug for active triazine metabolite chlorcycloguanil which is eliminated more rapidly.

[g] Rapidly hydrolyzed in the stomach to dihydroartemisinin (DHA). Very little artesunate in blood after oral administration.

[h] Rapidly metabolised to DHA which predominates in plasma.

[i] Mainly a prodrug for active triazine metabolite cycloguanil which is eliminated more rapidly.

Note: V_d/f is the total apparent volume of distribution; i.e., V_d divided by f, the fraction of drug absorbed.

Clearance/f is the apparent clearance.

The pharmacokinetic properties of chloroquine, lumefantrine, artemether, artesunate, atovaquone and proguanil in late pregnancy are significantly altered so that plasma concentrations are approximately half those in non-pregnant adults. Pyrimethamine and sulfadoxine plasma concentrations in young children are approximately half those in older children and adults for the same administered dose.

Source: Adapted from Ref. 17.

AQ and its metabolites are >90% protein-bound and are eliminated by renal excretion. The mean terminal half-life of AQ is 5.2 (±1.7) hours. DAQ has a much longer terminal $t_{\frac{1}{2}}$ than AQ. For such a widely used drug, there are remarkably few reliable data on the elimination kinetics of the active metabolite estimated as approximately 11 days. There are no parenteral formulations commercially available, although a structurally similar compound, amopyroquine, is available for intramuscular administration in some countries (44,45).

Piperaquine

Available data on the PK properties of piperaquine are recent and come from estimates in small numbers of patients (46). As expected, piperaquine has a large apparent V_d of >500 L/kg and a terminal elimination half-life of approximately 28 days in adults and 14 days in children (Tarning, J. Personal communication, 2007). Given the structural similarities with CQ, it is likely that the PK properties of the two drugs will be similar. A study in healthy volunteers has shown oral bioavailability increases with coadministration with fat (47). There are no published data yet on protein binding (which is very high), partition with blood cells, or metabolism.

Primaquine

Commercial primaquine is usually contaminated with another 8-aminoquinoline antimalarial compound; quinocide. Primaquine is well absorbed reaching peak plasma levels at three hours. It is widely distributed in tissues but not extensively bound. It is cleared by hepatic biotransformation to the more polar metabolite carboxyprimaquine and several other metabolites (including the oxidant 5-hydroxypremaquine), with an elimination half-life of six hours and renally excreted. It is not known whether primaquine itself or its metabolites are responsible for the action against *P. vivax* hypnozoites.

Tafenoquine

Tafenoquine is also well absorbed and widely distributed with longer absorption and elimination half-lives than primaquine of 1.0 hour and 16.4 days, respectively (48).

Quinine/Quinidine

Quinine is well absorbed after oral or intramuscular administration both in adults and in children. Oral bioavailability is approximately 70%. Peak levels are usually reached within six hours (more rapidly if the intramuscular injections are diluted). Intramuscular and intravenous infusions give similar peak levels. Quinine should never be given by bolus intravenous injection as lethal hypotension may ensue. In acute malaria, the total apparent V_d is contracted and systemic clearance reduced in proportion to disease severity. As a result, blood concentrations are higher in uncomplicated malaria than in healthy subjects and highest in patients with severe malaria (Fig. 2). The elimination half-life is approximately 18 to 20 hours in cerebral malaria, 16 hours in uncomplicated malaria, and 11 hours in health. In children and pregnant women, the apparent V_d is relatively smaller and elimination is more rapid (49–58). Quinine is a base and is bound principally to the acute-phase plasma protein α_1-acid glycoprotein. Plasma protein binding is increased from approximately 75% to 80% in healthy subjects to over 90% in

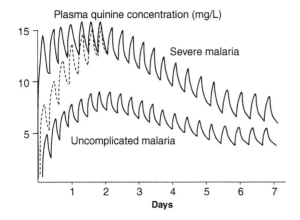

FIGURE 2 Plasma quinine concentrations during the treatment of malaria.

patients with severe malaria (59–61). Approximately 80% of the administered drug is eliminated by hepatic biotransformation, and the remaining 20% is excreted unchanged by the kidney. Although systemic clearance is reduced in severe malaria, this 80:20 proportion is preserved. The CYP 3A4 subfamily of the cytochrome P450 mixed-function oxidase system is the main metabolizing enzyme. The principal metabolite 3-hydroxyquinine is biologically active, contributing approximately 10% to antimalarial activity, but more in renal failure where it accumulates. The other more polar metabolites are either much less active or inactive as antimalarials. Oral quinidine is well absorbed in patients with malaria. The V_d and systemic clearance are significantly greater than for quinine, and the free fraction in plasma is approximately twice that of quinine. As a result, it has a shorter $t_{1/2}$. As for quinine, systemic clearance and V_d are reduced in malaria in proportion to disease severity. Quinidine, like quinine, is metabolized to hydroxylated more polar metabolites. CYP 3A4 and CYP2D6 are the main metabolizing enzymes (63,64).

Mefloquine
The PKs of mefloquine are highly stereo-specific and all PK parameters except T_{max} are significantly different for the (+) and (−) enantiomers. Mefloquine is moderately well absorbed, extensively distributed, and slowly eliminated. In healthy volunteers, absorption is biphasic and peak levels are reached between 8 and 24 hours after administration. It is highly (>98%) bound to plasma proteins. Mefloquine is excreted largely unchanged in feces. There is enterohepatic recycling. The apparent V_d and clearance of the (+)RS enantiomer is approximately four times higher than for the (−)SR enantiomer. The terminal elimination half-life is approximately three weeks in healthy subjects and two weeks in patients with malaria. In patients with malaria, absorption is reduced in the acute phase of illness, so splitting the dose or administering mefloquine after an artemisinin derivative (which gives rapid recovery) increases absorption and peak blood concentrations. The PKs in adults and children is similar (65,66).

Halofantrine
Halofantrine is intrinsically more active than mefloquine. This drug is poorly and erratically absorbed. Furthermore, absorption appears to be "saturable," i.e., with

individual doses over 8 mg/kg, no increment in blood concentrations occurs. Absorption is increased markedly by coadministration with fats. Halofantrine is extensively distributed and cleared largely by hepatic biotransformation. It is bound principally to lipoproteins in the plasma. The terminal elimination half-life is about one to three days in healthy subjects and approximately five days in patients with malaria. There is significant first-pass metabolism to a biologically active desbutyl metabolite. This is eliminated more slowly ($t_{1/2}$ 3–7 days) than the parent compound and undoubtedly contributes significantly to antimalarial activity (67,68).

Lumefantrine

Lumefantrine is absorbed more slowly than its partner drug artemether with peak concentrations occurring at six hours. Absorption is dose-limited, so this drug needs to be given twice daily (99). There is no apparent PK interaction between the constituent drugs in the combination. Lumefantrine is lipophilic and hydrophobic. Taking the drug with a fatty meal increases its relative bioavailability up to 16-fold. Absorption is reduced in the acute phase of malaria, but then increases considerably as symptoms resolve and the patient starts to eat. Lumefantrine is bound principally to lipoproteins in plasma. The principal metabolite is desbutyl lumefantrine, which also has antimalarial activity. CYP 3A4 is the main metabolizing enzyme. The elimination half-life is three to four days. The PK properties of lumefantrine are similar in adults and children (14). Lumefantrene concentrations are reduced by approximately half in late pregnancy (100).

Antifols
Sulfadoxine-Pyrimethamine

Pyrimethamine and sulfadoxine are well absorbed orally reaching peak plasma levels after about four hours. The estimated V_d for sulfadoxine is 0.14 L/kg and for pyrimethamine is 2.3 L/kg. Both are extensively protein-bound (90%). They are able to cross the placental barrier and pass into breast milk. About 5% of sulfadoxine appears in the plasma as the acetylated metabolite and about 2% to 3% as the glucuronide. The metabolites of pyrimethamine have not been characterized. The treatment is administered as a single dose. The terminal elimination half-lives are about 85 hours for pyrimethamine and 200 hours for sulfadoxine. Both are eliminated mainly via the kidneys. Following intramuscular injection, absorption is as rapid as after oral administration but blood concentrations are lower and more variable, which suggest incomplete intramuscular bioavailability (69). A recent very large study has shown that following standard dosing plasma concentrations of pyrimethamine and sulfadoxine in African children aged 2–5 years with falciparum malaria are approximately half those in older children and adults (101).

Proguanil and Chlorproguanil

Proguanil and chlorproguanil are well absorbed by mouth, and converted rapidly to the antifol triazine metabolites. These in turn are metabolized to the inactive metabolites chloro- and dichlorophenylbiguanide, respectively. As the parent compounds are eliminated more slowly than the metabolites, the profile of antimalarial activity resulting from the cyclic metabolites is determined by the parent drug distribution and elimination. The $t_{1/2}$ chlorproguanil is approximately 16 hours in healthy subjects and 13 hours in patients with malaria, although recent

population kinetic studies in malaria indicate a much longer half life of chlorproguanil of 35 hours (102). Approximately 3% of Caucasian and African populations,
but up to 20% of Orientals, fail to convert the parent compounds to their active
metabolites. In some parts of Micronesia, the prevalence is even higher. This is
related to a genetic polymorphism in CYP 2C19. The conversion of proguanil to
the active metabolite is reduced in pregnancy and also in women taking the oral
contraceptive pill (estrogens inhibit CYP2C19 activity) (70–74).

Artemisinin Derivatives

The artemisinin derivatives are rapidly absorbed and eliminated. Artesunate,
artemether, and artemotil are all hydrolyzed to the active metabolite DHA, which
has an elimination half-life of approximately 45 minutes. They are by far the most
rapidly eliminated of the antimalarial drugs. Despite this, they are highly effective
when given once daily. After oral or parenteral administration, artesunate is hydrolyzed rapidly (by stomach acid, and esterases in plasma and erythrocytes) and most
of the antimalarial activity results from the DHA metabolite. Oral absorption is
rapid and bioavailability is approximately 60%. Rectal bioavailability is more
variable; following administration of a rectal formulation, the Rectocap^{TM}, bioavailability averages 50% (although absorption is more variable). After oral administration, artemether is absorbed rapidly, but is converted more slowly (via CYP 3A4)
to DHA, although the metabolite still accounts for the majority of antimalarial
activity. In contrast after intramuscular administration, absorption of artemether
and artemotil is slow and erratic. Peak concentrations are often not reached for
many hours and concentrations of the parent compound exceed those of the active
DHA metabolite. Oral formulations of DHA contain excipients, which promote
absorption and give bioavailability comparable to that of artesunate. Elimination of
DHA is largely by conversion to inactive glucuronides. As for quinine, there is a
contraction in the V_d and reduced clearance in acute malaria, which increases blood
concentrations. There may also be a malaria-related inhibition of intestinal and
hepatic first-pass metabolism, which improves oral bioavailability (75,76).

Atovaquone

Atovaquone is rapidly absorbed reaching C_{max} in approximately six hours. It is a
lipophilic compound that is highly protein-bound (99%) with an apparent V_d of
6 L/kg. The majority of atovaquone is excreted unchanged in the feces. Elimination
is slower in patients of African origin ($t_{1/2}$ 70 hours) than in Oriental patients ($t_{1/2}$
30 hours) and faster in pediatric patients (1–2 days) than in adult patients (2–3
days). Oral absorption is augmented considerably by fats. There are no significant
interactions with proguanil or artesunate. There is limited experience of use of
atovaquone-proguanil in pregnancy where V_d and clearance of both drugs are
increased and as a consequence plasma concentrations are relatively low (77,78).

PHARMACODYNAMICS

Terminology

Malaria parasites are eukaryotes, which divide asexually in the red blood cells.
Multiplication rates can reach 20/cycle. For the main human parasites, the cycle
length is two days. Some familiar terms can usefully be borrowed from antibacterial pharmacodynamics, and some cannot. The minimum inhibitory concentration (MPC) is the lowest plasma concentration giving maximum parasite-killing

activity. This is a useful term as it represents a target concentration. The minimum inhibitory concentration (MIC) is not so useful as it would refer to a killing rate of one per cycle—resulting in no net change in numbers—or an effect considerably below maximum. It therefore has clinical relevance only in determining the time to recrudescence.

General Principles

The principal effect of antimalarial drugs in the treatment of uncomplicated malaria is to inhibit parasite multiplication (by stopping parasite development). The untreated infection can multiply at a maximum rate given by the average number of viable merozoites per mature schizont (100% efficiency). In nonimmunes, multiplication is often relatively efficient with multiplication rates of 6 to 20/cycle (30–90% efficiency). Antimalarials exerting their maximum effects (E_{max}) will convert this to a negative figure from –10 to –10,000, thus reducing parasite numbers by between 10- and 10,000-fold per cycle. The E_{max} is the effect represented at the top of the sigmoid dose–response or concentration-effect relationship (Fig. 4). Patients with acute malaria may have up to 10^{12} parasites in the circulation. Even with killing rates per cycle of 99.99%, it will take at least three life cycles (six days) to eradicate all the parasites from the body. Thus antimalarial treatment must usually provide therapeutic drug concentrations for seven days (covering four cycles) to effect a cure reliably. For rapidly eliminated drugs, this means the course of treatment must be seven days. Drugs differ in their E_{max}; for example, the artemisinins often produce a 10,000-fold reduction per asexual cycle, whereas antimalarial antibiotics such as tetracycline or

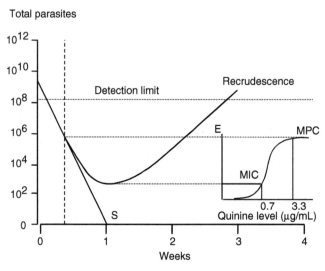

FIGURE 3 In vivo pharmacodynamics of quinine in falciparum malaria. The mean profile of the parasite burden in successfully treated patients declines logarithmically until total eradication(s) because the concentrations in plasma remain above the MPC (approximately 3.3 μg/ml). In patients with recrudescent infections, parasite killing (E) falls below maximal values after 48 h of treatment. A parasite multiplication rate of 1 per cycle results from average MICs of 0.7 μg/ml at the end of the first week, and thereafter, the parasite multiplication rate (10 per cycle) is unrestrained. *Source:* From Ref. 62.

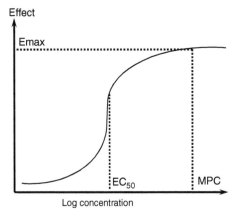

Effect

Emax

EC_{50} MPC

Log concentration

FIGURE 4 Sigmoid concentration-effect relationship for antimalarial drugs. *Abbreviations*: EC-50, concentration at which 50% maximal parasite killing effect is achieved; MPC, minimum parasitocidal concentration.

clindamycin may achieve only at most a 10-fold parasite reduction per cycle (Fig. 5). Parasite reduction appears to be a first-order process throughout. This means that provided the MPC is exceeded for a critical period of each asexual cycle, then a fixed fraction of the population is removed in each successive cycle. This simple model generally fits observed effects (although the persistence of a few blood stages following artemisinin treatment is an important exception). The concentration-effect relationship of antimalarials in vitro varies depending on the stage of parasite development. In contrast to antibiotics, where dose intervals exceed the life of the organism, in malaria the organism lives for two to three days. In the treatment of malaria, it is clearly not necessary to exceed the MPC throughout the dose interval—maximum effects on the mid-trophozoite (the most sensitive stage) are achieved in approximately four hours. Thus the artemisinin derivatives, which are rapidly eliminated ($t_{1/2}$ <1 hour), are effective when given only once per day (79).

Antimalarial Drug Susceptibility Testing

Antimalarial activity in vitro is assessed in terms of inhibition of morphological development and parasite multiplication or inhibition of glucose, amino acid, and hypoxanthine uptake (80). A newer technique utilizes the quantification of parasite

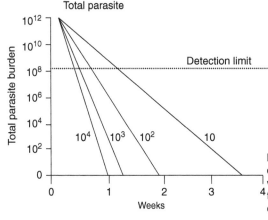

Total parasite

10^{12}

10^{10}

10^8 .... Detection limit

10^6

10^4

10^4 10^3 10^2 10

10^2

0

| | | | | |
| 0 | 1 | 2 | 3 | 4 |

Weeks

Total parasite burden

FIGURE 5 Parasite clearance following different antimalarial treatments. Provided that the MPC is exceeded, then a fixed fraction of the population is removed each successive asexual cycle.

lactate dehydrogenase production as an indicator of growth (81). Data from in vitro sensitivity testing are useful in screening new compounds for antimalarial activity and mapping the epidemiology of drug resistance patterns in different geographical areas, but they are poor predictors of individual treatment responses. This is because there are no antiparasitic factors such as antibodies in vitro; there is little protein or lipoprotein binding; there may be variable solubility and adsorption to laboratory materials; and concentrations are constant. Until recently, susceptibility testing was confined to *P. falciparum*, but now accurate reproducible methods are available for *P. vivax*.

As the mechanisms of resistance become dissected, parasite genotyping of known resistance markers is increasingly being validated as an alternative to in vitro susceptibility testing. This is best established for *Pfdhfr* where there are good correlates between the genotype and the therapeutic response to SP. The PfCRT (K76T mutation) is also a good marker of low-level CQ resistance. *Pfmdr* amplification identifies mefloquine-resistant parasites, and the single point mutations in cytochrome b indicate high-level atovaquone resistance.

Pharmacodynamics in Animal Models

In general, animal models do not reliably reproduce disease or resistance findings in human malarias. Rodent models have been used to screen new drugs and assess the potential for resistance to be selected, and a primate model (*Plasmodium cynomolgus*) has been used to assess radical curative activities as it also has persistent liver forms. To study the human parasites, primates (Aotus or Saimiri monkeys for *P. falciparum* and chimpanzees for *P. vivax*) are required. Preclinical testing of antimalarials in animals has revealed certain patterns of toxicity, which have had to be refuted by careful studies in humans. The artemisinin derivatives cause an unusual dose-related selective pattern of neuronal cell damage affecting certain brain stem nuclei in animals related to the PK properties of the drug. Neurotoxicity results from protracted exposure to sustained blood concentrations, as follows intramuscular administration of the oil-based artemether and artemether (artemotil). Neurotoxicity is much less in these models following oral administration or intravenous artesunate because the drugs levels are not sustained, even though bioavailability is better and peak levels considerably higher (82). Careful clinical audiometric and auditory-evoked potential measurement in clinical studies and some pathological studies have not shown similar findings in man. Arteether has been shown to prolong the QT interval in beagle dogs but not in man. The artemisinin derivatives cause fetal resorption in rodents and rabbits. SP has been shown to be teratogenic in rats with an estimated minimum oral teratogenic dose of approximately 0.9 mg/kg pyrimethamine plus 18 mg/kg sulfadoxine (83).

Pharmacodynamics (PD) in Humans

In the treatment of bacterial and viral infections, the principal PK determinant of therapeutic response [C_{max}, area under the (AUC), time above MIC, etc.] is usually determined first in animals, and then in man. These relationships have not been determined for most antimalarial drugs, although in general it is the time above the MPC that determines the probability of cure in uncomplicated malaria.

Malaria has the advantage over many other infections in terms of PK-PD assessment in that the infection is in red blood cells, so extravascular concentrations are not important for the therapeutic response, and the number of organisms can be quantitated. Speed of initial response is determined by the initial killing rate (fractional reduction in parasite numbers per asexual cycle; PRR). Cure rate is determined by the probability that drug levels and host immunity will combine to eradicate the infection from the body. If the MPC is exceeded for long enough, then the infection will usually be cured. The minimum time to ensure cure is four asexual cycles (eight days). A population-based approach has frequently been employed to distinguish and characterize patient and disease contributors to interindividual variance in antimalarial drug PKs. This information is generated from sparse PK data in large numbers of patients using statistical methods and is a more practical and relevant approach to the study of PK-PD in malaria-endemic areas (84). There have been relatively few studies of antimalarial dose–response or concentration–effect relationships in vivo. Most antimalarial drugs given at currently recommended doses provide concentrations well above the MPC. The only exceptions are poorly absorbed lipophilic drugs given orally in uncomplicated malaria (e.g., initial doses of lumefantrine), or intramuscularly in severe malaria (e.g., artemether). Thus the therapeutic response in uncomplicated malaria is determined by the intrinsic activity (reflected in the fractional kill rate per cycle or parasite reduction ratio; PRR) and the length of time blood concentrations exceed the MPC. Infections are often with more than one parasite clone (i.e., genotype), and these may have different drug susceptibilities. Thus investigations of antimalarial drug efficacy must take into account host immunity, which may clear drug-resistant infections, interindividual variation in PKs, and also the heterogeneity of parasite susceptibility. Pharmacodynamic effects can be assessed in terms of parasite reduction as parasite counts fall over the countable range from >100,000 to 50/μL. If recrudescence occurs later then the parasite counts below detectable levels can be modeled based on in vitro susceptibility (assuming that the slope of the concentration-effect relationship is the same in vitro and in vivo even if absolute values are not), antimalarial blood concentration profile, and time to recrudescence. For slowly eliminated drugs, measurement of residual drug levels at the time of recrudescence is important as this provides a concentration below the in vivo MIC—otherwise parasite numbers could not have increased (with the caveat that illness may affect V_d and thus blood concentrations). In *P. falciparum* and *P. malariae* infections, prevention of recrudescence is the object of treatment, and this is easily defined in a recurrent infection by a comparison of genotypes to distinguish recrudescence from a newly acquired infection. In *P. vivax* and *P. ovale* infections, relapses may occur three weeks after primary treatment, but these do not represent a failure of treatment of the asexual stages in the initial infection. This complicates interpretation of clinical trial results. Interestingly, CQ, mefloquine, and piperraquine suppress the first relapse of tropical *P. vivax* (but not the second at six weeks), so appearance of parasites within one month of treatment does reflect resistance (provided drug levels are adequate), whether this is a recrudescence or a relapse breakthrough, or even a new infection.

PK-PD Determinants of the Therapeutic Response
Uncomplicated Falciparum Malaria
Chloroquine
As resistance worsens the blood concentrations following the 25 mg base/kg dose are insufficient to suppress parasite multiplication for long enough to eradicate the infection—but they do act as a "brake" on multiplication. Initially the treatment

responses appear normal and the recrudescences do not present until weeks after drug administration. As resistance worsens, the recrudescences appear earlier and earlier until parasitemia no longer clears. Eventually some patients fail to respond at all—at this stage, mortality rises steeply. Because of the peculiar PK properties of CQ, the very high concentrations that occur initially may still temporarily suppress even highly resistant infections. Measurement of blood concentrations at the time of recrudescence or at day 7 or 14 post-treatment is useful in assessing treatment responses and assessing resistance.

Sulfadoxine-Pyrimethamine
Treatment failures following SP follow a similar pattern to CQ with the important exception that SP does not appear to provide temporary suppression of highly resistant parasites. The levels required for synergistic activity differ greatly depending on the parasite *dhfr* and *dhps* genotype. Resistance appears to have a differential effect on gametocytogenesis, in that a small increment in failure rate is associated with a large increase in gametocyte carriage (which fuels the spread of resistance).

Mefloquine
Mefloquine was originally prescribed as monotherapy for the treatment of resistant malaria at a dose of 15 mg base/kg. Initial high cure rates declined rapidly as resistance emerged leading to an increase in dose to 25 mg/kg. There is theoretical evidence that starting with this lower dose encouraged the selection of resistance (Figs. 6–10). Thus 25 mg base/kg should be the only recommended dose. The absorption of mefloquine is reduced in the acute phase of illness and bioavailability of the higher 25 mg/kg dose is improved by dividing it (e.g., giving 15 mg/kg initially and 10 mg/kg 8–24 hours later, or 8mg/kg/day for 3 days). Splitting the dose also reduces the incidence of acute adverse, effects, e.g., early vomiting. Blood concentrations are higher in patients with malaria than in healthy subjects (85,86). As resistance emerges, it follows a similar pattern of evolution to CQ.

Quinine
The therapeutic range has not been well defined but total plasma concentrations of between 8 and 15 mg/L are certainly safe and effective. Recent studies in Thailand suggested an in vivo MIC of 0.7 μg/mL and MPC of 3.3 μg/mL (Fig. 11) (62). Toxicity is increasingly likely with plasma concentrations over 20 mg/L (free quinine >2 mg/L). There is no established high-level quinine resistance. Treatment failure is associated particularly with poor adherence to seven-day treatment regimens or a large

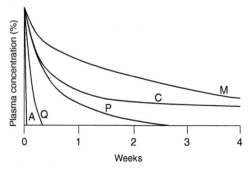

FIGURE 6 Plasma concentration profiles of different antimalarial drugs following a single treatment dose. *Abbreviations*: A, artesunate; Q, quinine; P, pyrimethamine; C, chloroquine; M, mefloquine.

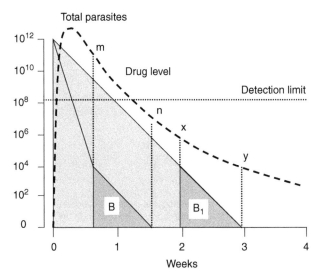

FIGURE 7 Artemisinin combination therapy. The impact of adding a 3 day course of artesunate (4 mg/kg/day) to mefloquine (25 mg/kg) on parasite killing in an area of mefloquine resistance. Without artesunate the parasitaemia declines 100-fold per asexual cycle and is eliminated in 3 weeks. Addition of artesunate for 3 days, covering 2 asexual cycles reduces the parasite biomass by a factor of 10^8 leaving a smaller residuum of parasites (B) for the mefloquine to remove while plasma concentrations are high. This lowers the chance of selecting a resistant parasite 10^7 fold. Without the artesunate the number of parasites corresponding to B, i.e. B1 are exposed to a much lower concentration of mefloquine (from x to y, compared with m to n) thus increasing the risk of recrudescence. *Source*: From Ref. 19.

V_d and low blood levels. Treatment failure characteristically manifests by recrudescence three weeks after starting treatment.

Artemisinin-Based Combinations (ACTs)

The artemisinin component reliably accelerates the clinical and parasitological response, increases overall cure rates, and reduces gametocyte carriage and thereby transmissibility. The combination should also delay the emergence of resistance (Fig. 7).

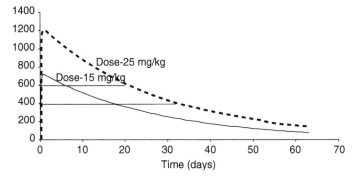

FIGURE 8 Simulated pharmacokinetic profiles for 15 mg/kg and 25 mg/kg doses of mefloquine.

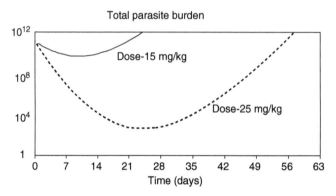

FIGURE 9 Total malaria parasite burden over time in a mefloquine-resistant infection for doses based on PK/PD parameter estimates. The initial parasite burden corresponds to an initial parasitaemia of 2% in an adult with falciparum malaria.

Mefloquine + Artesunate

When used in combination with artesunate, delaying the administration of mefloquine has been found to be associated with improved bioavailability. Coadministration with artesunate results in a more rapid recovery from malaria, which enhances oral bioavailability if the mefloquine dose is split (87). This and the added antimalarial effects of artesunate augment curative activity.

Artemether-Lumefantrine

There are probably more data on the PK determinants of cure for this compound than any other. The AUC of lumefantrine correlates with treatment response. The plasma level on day 7 after starting treatment is a good surrogate of the AUC; plasma levels of lumefantrine above 500 ng/mL are associated with a >90% cure rate (14,88). A plasma lumefantrine level of 280 µL on day 7 has been found to be a useful predictor of risk of subsequent recrudescence. In one study, 75% of

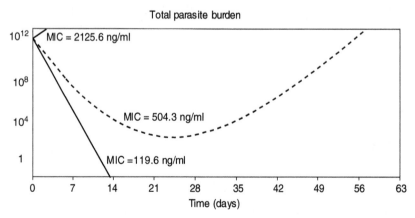

FIGURE 10 Modeling mefloquine resistance (86); relationship between parasite clearance over time and the MIC in vivo. In this example P0 is 10^{12}, a is 1.15/day, k is 0.036/day, C_0 is 1200 ng/ml, k_1 is 3.45/day and γ is 2.5.

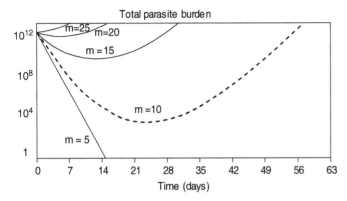

FIGURE 11 Modeling mefloquine resistance (86); relationship between parasite clearance over time and m (scalar value relating EC_{90} in vitro to MIC in vivo). In this example P_0 is 10_{12}, a is 1.15/day, k is 0.036/day, C_0 is 1200 ng/ml, k_1 is 3.45/day, γ is 2.5, and EC_{90} in vitro is 50.43 ng/ml.

patients with plasma lumefantrine levels above 280 µg/L were cured compared with only 51% of patients with lower concentrations.

Severe Malaria

The primary objective of the treatment of severe malaria is to save life. Prevention of recrudescence is of lesser concern. Thus the first parenteral dose is the most important, and the speed of effect is critical. To ensure the MPC is achieved reliably as soon as is compatible with safety, quinine or quinidine is given in an initial loading dose (usually twice the maintenance dose) (89). The initial dose of intramuscular artemether is also twice the maintenance dose. The bioavailability of intramuscular artemether in severe malaria is variable and may be associated with an inadequate therapeutic response in some shocked patients (90,91). However large randomized trials show that artemether treatment is associated with a similar mortality to quinine in African children, and a lower mortality in Southeast Asian adults (92). Thus it seems that the intrinsic advantages of the artemisinin derivative, in terms of earlier stage specificity of action, may have been offset by variable absorption. For these reasons, attention has shifted to the use of the water-soluble artesunate, which may be given intravenously or intramuscularly with excellent bioavailability. Recently the largest prospective trial ever conducted in severe malaria showed that artesunate reduced the mortality by 35% compared with quinine (93).

THERAPEUTIC USES AND DOSING REGIMENS

The recommended treatment options for uncomplicated malaria are shown in Table 3. The last bastions of CQ-sensitive falciparum malaria are The Caribbean, Central America west of the Panama Canal, and the Middle East. CQ can no longer be recommended for the treatment of falciparum malaria elsewhere, but it is still used widely and effectively for the treatment of *P. vivax*, *P. malariae*, and *P. ovale*. The preferred approach to the treatment of uncomplicated malaria is the use of combinations of drugs with different mechanisms of action, and therefore different drug targets to prevent the emergence of resistance. If two drugs are used, which do not share a common mode of action and therefore the parasite develops different mechanisms of resistance to them, the probability of a single

TABLE 3 Treatment of Uncomplicated Malaria

Malaria	Recommended drug treatment
P. vivax, P. malariae, P. ovale	**Chloroquine** phosphate (1 tablet contains 250 mg salt, equivalent to 155.3 mg base); 10mg/kg base at 0, 24 h followed by 5 mg/kg base at 48 h or **amodiaquine** 10 mg base/kg/day for 3 days
Uncomplicated *P. falciparum*	**Artemether-lumefantrine (Coartem™):** One dose at hours 0, 8, 24, 36, 48, 60 according to weight:

kg	Tablets per dose
5–14	1
15–24	2
25–34	3
>34	4

Artesunate-amodiaquine:
4 mg/kg of body weight artesunate and 10 mg base/kg amodiaquine once a day for 3 days

Artesunate-sulfadoxine-pyrimethamine:
4 mg/kg of body weight artesunate once a day for 3 days and a single dose of sulfadoxine-pyrimethamine (25/1.25 mg base/kg of body weight) on day 1

Artesunate-mefloquine:
4 mg/kg of body weight artesunate once a day for 3 days and mefloquine 25 mg base/kg of body weight split over 2 or 3 days

Quinine:
10 mg salt/kg three times daily plus tetracycline 4 mg/kg four times daily or **doxycycline** 4 mg/kg once daily or **clindamycin** 5 mg/kg three times per day for 7 days

Atovaquone:
20 mg/kg/day, **proguanil** 8 mg/kg/day for 3 days

Note:
- Pregnancy: Mefloquine and artesunate should not be given in the first trimester. Primaquine and tetracycline should not be used at any time in pregnancy.
- Vomiting is less likely if the patient's temperature is lowered before oral drug administration.
- Contraindications to mefloquine treatment include treatment with the drug in the previous 63 days, epilepsy or neuropsychiatric disorder, history of allergy.
- Short courses of artesunate or quinine (< 7 days) alone are not recommended.
- In renal failure the dose of quinine should be reduced by one-third to one-half after 48 hours, and doxycycline but *not* tetracycline should be prescribed.
- The doses of all drugs are unchanged in children and pregnant women.
- Oral treatment of uncomplicated hyperparasitaemic infections should include an artemisinin derivative and be prolonged to minimize the chance of recrudescence e.g. artesunate loading dose of 4mg/kg initially followed by 2 mg/kg/day on the following 6 days, in combination with mefloquine or doxycycline as in the table.
- Patients with *P. vivax* and *P. ovale* infections should also be given primaquine 0.25 mg base/kg daily (0.375–0.5 mg base/kg in Oceania) for 14 days to prevent relapse. In mild G6PD deficiency 0.75 mg base/kg should be given once weekly for 6 weeks.
- Use of tetracyclines in pregnant women or children under 8 years of age is contraindicated.

Source: Adapted from WHO Malaria Treatment Guidelines, World Health Organization, Geneva, Switzerland, 2006.

parasite developing simultaneous resistance to both drugs is of the product of the probabilities of developing resistance to the individual drugs (94). This is the same rationale underlying antituberculosis drug treatment with combinations. The lower the de novo per parasite probability of developing resistance, the greater the delay in the emergence of resistance. In order to succeed this approach requires good levels of population coverage and affordable simple drug regimens, which are adhered to. From the drugs that are available, the properties of the artemisinin derivatives make them the preferred choice as one of the drugs in such a combination (Fig. 12). Even though the drug of choice for treating severe malaria is intravenous artesunate, this drug is not widely available outside Asia. Alternative treatments are listed in Table 4. Artesunate gelatin–coated suppositories (Rectocap®) may be used as a holding measure in rural settings before a patient may be transferred for parenteral treatment.

There are two strategies for prevention of malaria: chemoprophylaxis in travelers (Table 5) and intermittent presumptive treatment. Studies from Africa indicate that administration of a treatment dose of SP twice during pregnancy (once in the second and once in the third trimester) has a beneficial effect on maternal anemia and pregnancy outcome (birth weight). HIV-positive women need monthly SP for the same effect. This approach has now been extended to infancy, with a view to possible incorporation in EPI programs in areas where SP retains good efficacy. Unfortunately the rapid spread of resistance is compromising this approach (95–97) and there have been no pharmacokinetic studies in the target groups (pregnant women and infants).

The majority of the most effective treatments for severe malaria or CQ-resistant malaria are not approved by the Food and Drug Administration. Quinidine and quinine are the only drugs approved for the treatment of severe malaria. Atovaquone-proguanil, mefloquine, SP, and CQ are approved for the treatment and prophylaxis of falciparum malaria. CQ and mefloquine are also approved for

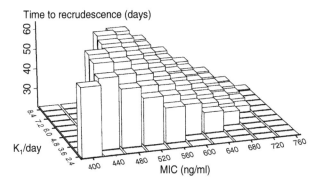

FIGURE 12 Time to recrudescence following treatment with 25 mg/kg mefloquine as a function of MIC in vivo and killing rates (k1). The z axis is the time to recrudescence, the y axis is the killing rate of mefloquine (k_1), and the X axis is the MIC in vivo. The pharmacokinetic parameters used in the simulation were the population mean values from studies in northwestern Thailand (ref). Nonraised rectangles represent 2 possible scenarios: either the patient is cured or at day 7 the parasites are still detectable. This illustrates that for relatively drug-sensitive parasites (MIC < 500 ng/ml) the infections are all cured with high killing rates and that with low killing rates recrudescences occur long after the conventional follow-up period of 28 days.

TABLE 4   Treatment of Severe Malaria

Artesunate i.v.: 2.4 mg/kg given on admission, at 12 hr, 24 hr, and then every 24 hr
Alternative treatments:
Artemether i.m.: initial dose of 3.2 mg/kg followed by 1.6 mg/kg every 24 hr[a]
or
Quinine i.v.: LD 20 mg/kg given over 4 hr, then 10 mg given 8 hr after the LD was started,
   followed by 10 mg/kg every 8 hr
or
Quinine i.m.: LD (20 mg/kg) is given as 2 simultaneous injections (each 10 mg/kg) in the
   anterior thigh after 50% dilution of the quinine in sterile water. The maintenance dose (10
   mg/kg) is given as
   1 i.m. injection every 8 hr using the same dilution
or
Quinidine i.v.: 10 mg base/kg infused over 1–2 hr followed by 1.2 mg base/kg/hr.
   Electrocardiographic monitoring advisable[b]
Total treatment duration for all regimens = 7 day
Once the patient has recovered sufficiently to tolerate oral medication, a second drug should
   be added such as doxycycline 4 mg/kg for 7 day, clindamycin 5 mg/kg t.i.d. for 7 day or
   atovaquone 20 mg/kg/day + proguanil 8 mg/kg/day for 3 day[c]

[a]Absorption of i.m. artemether may be inadequate in a subgroup of patients with poor peripheral perfusion.
[b]Some authorities recommend a lower dose of 6.2 mg base/kg initially over one hour followed by 1.2 mg
base/kg/hr.
[c]Mefloquine should not be used because of the increased risk of postmalaria neurological syndrome.
Abbreviations: LD, loading dose; i.v., intravenous; i.m., intramuscular; t.i.d., thrice a day.

TABLE 5   Antimalarial Chemoprophylaxis[a]

–	Weight adjusted dose for children	Adult dose
Chloroquine-sensitive malaria		
Chloroquine[b]	5 mg base/kg weekly, or	300 mg base
and/or	1.6 mg base/kg daily	100 mg base
Proguanil	3.5 mg/kg daily	200 mg base
Chloroquine-resistant malaria		
Mefloquine[c]	5 mg base/kg weekly	250 mg base
or		
Doxycycline[d]	1.5 mg/kg daily	100 mg
or		
Primaquine[e]	0.5 mg base/kg daily with food	30 mg base
or		
Atovaquone-proguanil	4/1.6 mg/kg daily	250/100 mg

[a]Detailed local knowledge of *Plasmodium falciparum* antimalarial susceptibility and malaria risk should always
be obtained.
[b]Chloroquine should not be taken by people with a history of seizures, generalized psoriasis, or pruritus
previously on chloroquine.
[c]Mefloquine is not recommended for babies less than three months of age. Mefloquine should not be taken by
people with psychiatric disorders, epilepsy, or those driving heavy vehicles, trains, aeroplanes, etc., or deep sea
diving.
[d]Doxycycline may cause photosensitivity. Use of sunscreens is recommended.
[e]Primaquine is contraindicated in pregnancy and glucose-6-phosphate dehydrogenase deficiency.

treatment or prophylaxis of vivax malaria. There are no drugs effective against mdr falciparum malaria approved by the USFDA for use in pregnancy.

Antimalarial drugs have other therapeutic uses outside the treatment of malaria, e.g., hydroxychloroquine is used as an antirheumatic, and in the management of systemic lupus erythematosus and porphyria cutanea tarda. Quinine is also used to treat babesiosis and night cramps. Pyrimethamine is used with sulphadoxine to treat cerebral toxoplasmosis. Atovaquone is a broad-spectrum antiparasitic agent active against *Pneumocystis jiroveci* pneumonia (PCP), toxoplasmosis, and babesiosis. Primaquine has also been used to treat PCP in combination with clindamycin. The artemisinin derivatives have been shown to have activity against a number of organisms in vitro and are effective in vivo against Schistosoma and other trematodes; however, careless deployment of these drugs as monotherapies to treat other diseases in malaria-endemic countries should be avoided (98).

REFERENCES

1. Editorial. Rediscovering wormwood, qinghaosu, for malaria. Lancet 1992; 339:649–651.
2. Duran-Reynolds ML. The Fever Bark Tree. New York: Doubleday, 1946.
3. Rozman RS, Canfield CJ. New experimental antimalarial drugs. Adv Pharmacol Chemother 1979; 16:1–43.
4. Qinghaosu Antimalarial Coordinating Research Group. Antimalarial studies on qinghaosu. Chin Med J 1979; 92:811–816.
5. Li Y, Wu YL. How Chinese scientists discovered qinghaosu (artemisinin) and developed its derivatives? What are the future perspectives? Med Trop (Mars) 1998; 58(3 suppl):9–12.
6. White NJ. Antimalarial drug resistance and combination chemotherapy. Phil Trans R Soc Lond B 1999; 354:739–749.
7. Krugliak M, Ginsburg H. Studies on the antimalarial mode of action of quinoline-containing drugs: time dependence and irreversibility of drug action, and interactions with compounds that alter the function of the parasite's food vacuole. Life Sci 1991; 49:1213–1219.
8. Chou AC, Chevli R, Fitch CD. Ferriprotoporphyrin IX fulfills the criteria for identification as the chloroquine receptor of malaria parasites. Biochemistry 1980; 19:1543–1549.
9. Mungthin, Mathirut, Bray, et al. Central role of hemoglobin degradation in mechanisms of action of 4-aminoquinolines, quinoline methanols, and phenanthrene methanols. Antimicrob Agents Chemother 1998; 42:2973–2977.
10. Lindegårdh N, Ashton M, Bergqvist Y. Automated solid-phase extraction method for the determination of piperaquine in plasma by peak compression liquid chromatography. J Chromatogr Sci 2003; 41(1):44–49.
11. Brueckner RP, Lasseter KC, Lin ET, Schuster BG. First-time-in-humans safety and pharmacokinetics of WR 238605, a new antimalarial. Am J Trop Med Hyg 1998; 58:645–649.
12. Foote SJ, Cowman AF. The mode of action and the mechanism of resistance to antimalarial drugs. Acta Trop 1994; 56:157–171.
13. Karle JM, Karle IL. Crystal structure and molecular structure of mefloquine methylsulfonate monohydrate: implications for a malaria receptor. Antimicrob Agents Chemother 1991; 35(11):2238–2245.
14. White NJ, van Vugt M, Ezzet F. Clinical pharmacokinetics and pharmacodynamics of artemether-lumefantrine. Clin Pharmacokinet 1999; 37:105–125.
15. Chang C, Lin-Hua T, Jantanavivat C. Studies on a new antimalarial compound: pyronaridine. Trans R Soc Trop Med Hyg 1992; 86(1):7–10.
16. Meshnick SR. Artemisinin: mechanisms of action, resistance and toxicity. Int J Parasitol 2002; 32(13):1655–1660.
17. Eckstein-Ludwig U, Webb RJ, Van Goethem ID, et al. Artemisinins target the SERCA of *Plasmodium falciparum*. Nature 2003; 424:957–961.
18. Srivastava IK, Rottenberg H, Vaidya AB. Atovaquone, a broad spectrum antiparasitic drug, collapses mitochondrial membrane potential in a malarial parasite. J Biol Chem 1997; 272:3961–3966.

19. White N. Malaria. In: Cook G, Zumla A, eds. Manson's Tropical Diseases. 21st ed. Edinburgh: W. B. Saunders, 2003:1205–1296.
20. Medicines for Malaria Venture Drug Discovery/Development Portfolio URL: http://www.mmv.org/FilesUpld/180.pdf [April 2007].
21. Watkins WM, Woodrow C, Marsh K. Falciparum malaria: differential effects of antimalarial drugs on ex vivo parasite viability during the critical early phase of therapy. Am J Trop Med Hyg 1993; 49:106–112.
22. Udomsangpetch R, Pipitaporn B, Krishna S, et al. Antimalarial drugs reduce cytoadherence and rosetting of Plasmodium falciparum. J Infect Dis 1996; 173:691–698.
23. Djimde A, Doumbo OK, Cortese JF, et al. A molecular marker for chloroquine-resistant falciparum malaria. N Engl J Med 2001; 344:257–263.
24. Durand R, Jafari S, Vauzelle J, Delabre J, Jesic Z, Le Bras J. Analysis of pfcrt point mutations and chloroquine susceptibility in isolates of Plasmodium falciparum. Mol Biochem Parasitol 2001; 114:95–102.
25. Warhurst DC, Craig JC, Adagu IS, Meyer DJ, Lee SY. The relationship of physico-chemical properties and structure to the differential antiplasmodial activity of the cinchona alkaloids. Malar J 2003; 2(1):26.
26. Fidock DA, Nomura T, Talley AK, et al. Mutations in the P. falciparum digestive vacuole transmembrane protein PfCRT and evidence for their role in chloroquine resistance. Mol Cell 2000; 6:861–871.
27. Foote SJ, Thompson JK, Courman AF, Kemp DJ. Amplification of the multidrug resistance gene in some chloroquine-resistant isolates of Plasmodium falciparum. Cell 1989; 57:921–930.
28. Martin SK, Oduola AMJ, Milhous WK. Reversal of chloroquine resistance in Plasmodium falciparum by verapamil. Science 1987; 235:899–901.
29. Price RN, Cassar C, Brockman A, et al. The pfmdr1 gene is associated with a multidrug resistance phenotype in Plasmodium falciparum from the western border of Thailand. Antimicrob Agents Chemother 1999; 43:2943–2949.
30. Lim AS, Galatis D, Cowman AF. Plasmodium falciparum: amplification and overexpression of pfmdr1 is not necessary for increased mefloquine resistance. Exp Parasitol 1996; 83(3):295–303.
31. Oduola AM, Omitowoju GO, Gerena L, et al. Reversal of mefloquine resistance with penfluridol in isolates of Plasmodium falciparum from south-west Nigeria. Trans R Soc Trop Med Hyg 1993; 87:81–83.
32. Peterson DS, Walliker D, Wellems T. Evidence that a point mutation in dihydrofolate reductase-thymidylate synthase confers resistance to pyrimethamine in falciparum malaria. Proc Natl Acad Sci USA 1988; 85:9114–9118.
33. Sibley CH, Hyde JE, Sims PF, et al. Pyrimethamine-sulfadoxine resistance in Plasmodium falciparum: what next? Trends Parasitol 2001; 17(12):582–588.
34. Nair S, Williams JT, Brockman A, et al. A selective sweep driven by pyrimethamine treatment in South East Asian malaria parasites. Mol Biol Evol 2003; 20(9):1526–1536. Epub 2003 Jun 27.
35. Peterson DA, Milhous WK, Wellems TE. Molecular basis of differential resistance to cycloguanil and pyrimethamine in Plasmodium falciparum malaria. Proc Natl Acad Sci USA 1990; 87:3018–3022.
36. Oduola AM, Sowunmi A, Milhous WK, et al. Innate resistance to new antimalarial drugs in Plasmodium falciparum from Nigeria. Trans R Soc Trop Med Hyg 1992; 86(2):123–126.
37. White NJ. Why is it that antimalarial drug treatments do not always work? Ann Trop Med Parasitol 1998; 92(4):449–458.
38. Korsinczky M, Chen N, Kotecka B, Saul A, Rieckmann K, Cheng Q. Mutations in Plasmodium falciparum cytochrome b that are associated with atovaquone resistance are located at a putative drug-binding site. Antimicrob Agents Chemother 2000; 44:2100–2108.
39. Gustafsson LL, Walker O, Alvan G, et al. Disposition of chloroquine in man after single intravenous and oral doses. Br J Clin Pharmacol 1983; 15:471–479.
40. Frisk-Holmberg M, Bergqvist Y, Termond E, Domeij-Nyberg B. The single dose kinetics of chloroquine and its major metabolite desethylchloroquine in healthy subjects. Eur J Clin Pharmacol 1984; 26:521–530.

41. White NJ, Watt G, Bergqvist Y, Njelesani E. Parenteral chloroquine in the treatment of falciparum malaria. J Infect Dis 1987; 155:192–201.
42. Walker O, Daurodu AH, Adeyokunnu AA, et al. Plasma chloroquine and desethylchloroquine concentrations in children during and after chloroquine treatment for malaria. Br J Clin Pharmacol 1983; 16:701–705.
43. Minker F, Iran J. Experimental and clinicopharmacological study of rectal absorption of chloroquine. Acta Physiol Hung 1991; 77:237–248.
44. Winstanley P, Edwards G, Orme M Breckenridge AM. The disposition of amodiaquine in man after oral administration. Br J Clin Pharmacol 1987; 23:1–7.
45. White NJ, Looareesuwan S, Edwards G, et al. Pharmacokinetics of intravenous amodiaquine. Br J Clin Pharmacol 1987; 23:127–135.
46. Hung TY, Davis TM, Ilett KF, et al. Population pharmacokinetics of piperaquine in adults and children with uncomplicated falciparum or vivax malaria. Br J Clin Pharmacol 2004; 57(3):253–262.
47. Sim IK, Davis TM, Ilett KF. Effects of a high-fat meal on the relative oral bioavailability of piperaquine. Antimicrob Agents Chemother 2005; 49(6):2407–2411.
48. Edstein MD, Kocisko DA, Brewer TG, Walsh DS, Eamsila C, Charles BG. Population pharmacokinetics of the new antimalarial agent tafenoquine in Thai soldiers. Br J Clin Pharmacol 2001; 52(6):663–670.
49. Supanaranond W, Davis TME, Pukrittayakamee S, et al. Disposition of oral quinine in acute falciparum malaria. Eur J Clin Pharmacol 1991; 40:49–52.
50. Waller D, Krishna S, Craddock C, et al. The pharmacokinetic properties of intramuscular quinine in Gambian children with severe falciparum malaria. Trans R Soc Trop Med Hyg 1990; 84:488–491.
51. Mansor SM, Taylor TE, McGrath CS, et al. The safety and kinetics of intramuscular quinine in Malawian children with moderately severe falciparum malaria. Trans R Soc Trop Med Hyg 1990; 84:482–487.
52. Sabcharoen A, Chongsuphajaisiddhi T, Attanath P. Serum quinine concentrations following the initial dose in children with falciparum malaria. SE Asian J Trop Med Public Health 1989; 13:689–692.
53. White NJ, Looareesuwan S, Warrell DA, et al. Quinine pharmacokinetics and toxicity in cerebral and uncomplicated falciparum malaria. Am J Med 1982; 73:564–572.
54. Krishna S, Nagaraja NV, Planche T, et al. Population pharmacokinetics of intramuscular quinine in children with severe malaria. Antimicrob Agents Chemother 2001; 45: 1803–1809.
55. White NJ, Chanthavanich P, Krishna S, et al. Quinine disposition kinetics. Br J Clin Pharmacol 1983; 16:399–404.
56. Krishna S, White NJ. Pharmacokinetics of quinine, chloroquine and amodiaquine. Clinical implications. Clin Pharmacokinet 1996; 30:263–299.
57. Phillips RE, Looareesuwan S, White NJ, et al. Quinine pharmacokinetics and toxicity in pregnant and lactating women with falciparum malaria. Br J Clin Pharmacol 1986; 21: 677–683.
58. van Hensbroek MB, Kwiatkowski D, van den Berg B, Hoek FJ, van Boxtel CJ, Kager PA. Quinine pharmacokinetics in young children with severe malaria. Am J Trop Med Hyg 1996; 54:237–242.
59. Silamut K, Molunto P, Ho M, et al. Alpha-one acid glycoprotein (orosomucoid) and plasma protein binding of quinine in falciparum malaria. Br J Clin Pharmacol 1991; 32:311–315.
60. Silamut K, White NJ, Warrell DA, Looareesuwan S. Binding of quinine to plasma proteins in falciparum malaria. Am J Trop Med Hyg 1985; 34:681–686.
61. Mansor SM, Molyneux ME, Taylor TE, et al. Effect of *Plasmodium falciparum* malaria infection as the plasma concentration of alpha acid glycoprotein and the binding of quinine in Malawian children. Br J Clin Pharmacol 1991; 32:317–325.
62. Pukrittayakamee S, Wanwimolruk S, Stepniewska K, et al. Quinine pharmacokinetic-pharmacodynamic relationships in uncomplicated falciparum malaria. Antimicrob Agents Chemother 2003; 47(11):3458–3463.
63. White NJ, Looareesuwan S, Warrell DA, et al. Quinidine in falciparum malaria. Lancet 1981; ii:1069–1072.

64. Phillips RE, Warrell DA, White NJ, et al. Intravenous quinidine for the treatment of severe falciparum malaria. Clinical and pharmacokinetic studies. N Engl J Med 1985; 312:1273–1278.

65. Gimenez F, Pennie RA, Koren G, Crevoisier C, Wainer IW, Farinotti R. Stereoselective pharmacokinetics of mefloquine in healthy Caucasians after multiple doses. J Pharm Sci 1994; 83:824–827.

66. Karbwang J, White NJ. Clinical pharmacokinetics of mefloquine. Clin Pharmacokinet 1990; 19:264–279.

67. Milton KA, Edwards G, Ward SA, et al. Pharmacokinetics of halofantrine in man: effects of food and dose size. Br J Clin Pharmacol 1989; 28:71–77.

68. Veenendaal JR, Parkinson AD, Kere N, et al. Pharmacokinetics of halofantrine and n-desbutylhalofantrine in patients with falciparum malaria following a multiple dose regimen of halofantrine. Eur J Clin Pharmacol 1991; 41:161–164.

69. Winstanley P, Watkins WM, Newton CRJC, et al. The disposition of oral and intramuscular Pyrimethamine/sulphadoxine in Kenyan children with high parasitaemia but clinically non-severe falciparum malaria. Br J Clin Pharmacol 1992; 33:143–148.

70. Wattanagoon Y, Taylor RB, Moody RR, et al. Single dose pharmacokinetics of proguanil and its metabolites in healthy adult volunteers. Br J Clin Pharmacol 1987; 24:775–780.

71. Winstanley P, Watkins W, Muhia D, Szwandt S, Amukoye E, Marsh K. Chlorproguanil/dapsone for uncomplicated *Plasmodium falciparum* malaria in young children: pharmacokinetics and therapeutic range. Trans R Soc Trop Med Hyg 1997; 91:322–327.

72. Helsby NA, Ward SA, Edwards C, et al. The pharmacokinetics and activation of proguanil in man: consequences of variability in drug metabolism. Br J Clin Pharmacol 1990; 30:593–598.

73. Kaneko A, Kaneko O, Taleo G, Bjorkman A, Kobayakawa T. High frequencies of CYP2C19 mutations and poor metabolism of proguanil in Vanuatu. Lancet 1997; 349:921–922.

74. Wangboonskul J, White NJ, Nosten F, et al. Single dose pharmacokinetics of proguanil and its metabolites in pregnancy. Eur J Clin Pharmacol 1993; 44:247–251.

75. Navaratnam V, Mansor SM, Sit NW, Grace J, Li Q, Olliaro P. Pharmacokinetics of artemisinin-type compounds. Clin Pharmacokinet 2000; 39:255–270.

76. Teja-Isavadharm P, Nosten F, Kyle DE, et al. Comparative bioavailability of oral, rectal, and intramuscular artemether in healthy subjects: use of simultaneous measurement by high performance liquid chromatography and bioassay. Br J Clin Pharmacol 1996; 42(5): 599–604.

77. Hussein Z, Eaves J, Hutchinson DB, Canfield CJ. Population pharmacokinetics of atovaquone in patients with acute malaria caused by *Plasmodium falciparum*. Clin Pharmacol Ther 1997; 61(5):518–530.

78. McGready R, Stepniewska K, Edstein MD, et al. The pharmacokinetics of atovaquone and proguanil in pregnant women with acute falciparum malaria. Eur J Clin Pharmacol 2003; 59(7):545–552. Epub 2003 Aug 30.

79. White NJ. Assessment of the pharmacodynamic properties of antimalarial drugs in vivo. Antimicrob Agents Chemother 1997; 41:1413–1422.

80. Webster HK, Boudreau EF, Pavanand K, Yongvanitchit K, Pang LW. Antimalarial drug susceptibility testing of *Plasmodium falciparum* in Thailand using a microdilution radio-isotope method. Am J Trop Med Hyg 1985; 34(2):228–235.

81. Druilhe P, Moreno A, Blanc C, Brasseur PH, Jacquier P. A colorimetric in vitro drug sensitivity assay for *Plasmodium falciparum* based on a highly sensitive double-site lactate dehydrogenase antigen-capture enzyme-linked immunosorbent assay. Am J Trop Med Hyg 2001; 64:233–241.

82. Nontprasert A, Pukrittayakamee S, Nosten-Bertrand M, Vanijanonta S, White NJ. Studies of the neurotoxicity of oral artemisinin derivatives in mice. Am J Trop Med Hyg 2000; 62:409–412.

83. Phillips-Howard PA, Wood D. The safety of antimalarial drugs in pregnancy. Drug Saf 1996; 14(3):131–145.

84. Simpson JA, Aarons L, White NJ. How can we do pharmacokinetic studies in the tropics? Trans R Soc Trop Med Hyg 2001; 95(4):347–351.

85. Simpson JA, Price RN, ter Kuile FO, et al. Population pharmacokinetics of mefloquine in patients with acute falciparum malaria. Clin Pharmac Ther 1999; 66:472–484.
86. Simpson JA, Watkins ER, Price RN, Aarons L, Kyle DE, White NJ. Mefloquine pharmacokinetic-pharmacodynamic models: implications for dosing and resistance. Antimicrob Agents Chemother 2000; 44:3414–3424.
87. Price RN, Simpson JA, Teja-Isavatharm P, et al. Pharmacokinetics of mefloquine combined with artesunate in children with acute falciparum malaria. Antimicrob Agents Chemother 1999; 43:341–346.
88. Ezzet F, van Vugt M, Nosten F, Looareesuwan S, White NJ. The pharmacokinetics and pharmacodynamics of lumefantrine (benflumetol) in acute uncomplicated falciparum malaria. Antimicrob Agents Chemother 2000; 44:697–704.
89. White NJ, Warrell DA, Looareesuwan S, et al. Quinine loading dose in cerebral malaria. Am J Trop Med Hyg 1983; 32:1–5.
90. Murphy SA, Mberu E, Muhia D, et al. The disposition of intramuscular artemether in children with cerebral malaria; a preliminary study. Trans R Soc Trop Med Hyg 1997; 91(3):331–334.
91. Mithwani S, Aarons L, Kokwaro GO, et al. Population pharmacokinetics of artemether and dihydroartemisinin following single intramuscular dosing of artemether in African children with severe falciparum malaria. Br J Clin Pharmacol 2004; 57(2):146–152.
92. Artemether-Quinine Meta-analysis Study Group. A meta-analysis using individual patient data of trials comparing artemether with quinine in the treatment of severe falciparum malaria. Trans R Soc Trop Med Hyg 2001; 95(6):637–650.
93. Dondorp A, Nosten F, Stepniewska K, Day N, White N; South East Asian Quinine Artesunate Malaria Trial (SEAQUAMAT) group. Artesunate versus quinine for treatment of severe falciparum malaria: a randomised trial. Lancet 2005; 366(9487):717–725.
94. White NJ. Preventing antimalarial drug resistance through combinations. Drug Res Updates 1998; 1:3–9.
95. Rogerson SJ, Chululuka E, Kanjala M, Mkundika P, Mhango C, Molyneux ME. Intermittent sulfadoxine-pyrimethamine in pregnancy: effectiveness against malaria morbidity in Blantyre, Malawi, in 1997–99. Trans R Soc Trop Med Hyg 2000; 94:549–553.
96. Wolfe EB, Parise ME, Haddix AC, et al. Cost-effectiveness of sulfadoxine-pyrimethamine for the prevention of malaria-associated low birth weight. Am J Trop Med Hyg 2001; 64:178–186.
97. Schellenberg D, Menendez C, Kahigwa E, et al. Intermittent treatment for malaria and anaemia control at time of routine vaccinations in Tanzanian infants: a randomised, placebo-controlled trial. Lancet 2001; 357:1471–1477.
98. Utzinger J, Xiao S, N'Goran EK, Bergquist R, Tanner M. The potential of artemether for the control of schistosomiasis. Int J Parasitol 2001; 31(14):1549–1562.
99. Ashley EA, Stepniewska K, Lindegardh N, et al. Pharmacokinetic study of artemether-lumefantrine given once daily for the treatment of uncomplicated multidrug-resistant falciparum malaria. Trop Med Int Health 2007; 12(2):201–208.
100. McGready R, Stepniewska K, Lindegardh N, et al. The pharmacokinetics of artemether and lumefantrine in pregnant women with uncomplicated falciparum malaria. Eur J Clin Pharmacol 2006; 62(12):1021–1031. (Epub 2006 Oct 20)
101. Barnes KI, Little F, Smith PJ, et al. Sulfadoxine-pyrimethamine pharmacokinetics in malaria: pediatric dosing implications. Clin Pharmacol Ther 2006; 80:582–596.
102. Simpson JA, Hughes D, Manyando C, et al. Population pharmacokinetic and pharmacodynamic modelling of the antimalarial chemotherapy chlorproguanil/dapsone. Br J Clin Pharmacol 2006; 61:289–300.

20 | Human Pharmacodynamics of Anti-infectives: Determination from Clinical Trial Data

George L. Drusano
Ordway Research Institute, Albany, New York, U.S.A.

INTRODUCTION

Determination of the relationship between drug exposure and response or between drug exposure and toxicity is key to achieving the ultimate aim of chemotherapy—obtaining the maximal probability of a good therapeutic response while engendering the smallest possible probability of toxicity.

In the area of anti-infective chemotherapy, there is a single difference from other areas of clinical pharmacological investigation. In other areas, we deal with receptors for the drug that are human in origin. There are true between-patient differences in receptor affinity for the drug that are, currently, not measurable. This unmeasured variance leads to difficulty in generating pharmacodynamic relationships.

In anti-infectives, however, we are dealing with an external invader. Whether we are dealing with bacteria, viruses, or fungi, we can, with few exceptions (e.g., hepatitis C virus), grow the offending pathogen and obtain a measure of the drug's potency for that particular pathogen. These measures have different names, depending on the pathogen (e.g., minimal inhibitory concentration (MIC), effective concentration that reduces growth by half (EC-50), and minimal fungicidal (static) concentration (MFC). These measures of pathogen sensitivity to the drug can then be used to normalize the drug exposure in the patient relative to the invading pathogen. This markedly reduces the observed variability and improves the ability to define a relationship between exposure and response.

DETERMINANTS OF A PHARMACODYNAMIC RELATIONSHIP FOR ANTI-INFECTIVES
End Points

In order to determine a relationship between exposure and response or between exposure and toxicity, the first step is to identify an end point. Such end points differ according to what is being studied. End points may be continuous in nature (e.g., change in glomerular filtration rate after drug exposure), dichotomous or polytomous (e.g., success vs. failure; survival vs. death, and a three-point scale for cure-improvement-failure), or time to an event (time to death, time to relapse, and time to viral clearance). As will be discussed later, the end point chosen will in many ways determine a good deal of the rest of the analysis.

The first and most important step in defining a pharmacodynamic relationship is to obtain solid response end point data. If all other steps are performed well but the end point data are poorly defined, then the relationship will be poor at best and possibly misleading.

Pathogen Identification and Susceptibility Determination

If one is attempting to construct an effect relationship (i.e., between drug exposure and response), it is imperative that the offending pathogen be isolated and identified and that the susceptibility of that specific pathogen to the drug being used for therapy be measured. This is straightforwardly performed in many clinical antibacterial trials, because pathogen identification and MIC determination are integral parts of making a clinical study case both clinically and microbiologically evaluable for the Food and Drug Administration (FDA).

Amazingly little has been done regarding the EC-50 of viruses and their influence on outcome in clinical trials of antiviral chemotherapy. It has only been recently, with the availability of commercial homologous recombination assays or rapid sequencing assays for HIV, that determination of drug susceptibility has become a part of the clinical trial arena. Nonetheless, EC-50 has an important role to play, as demonstrated by the data in Figure 1A to F. In this evaluation, the HIV protease inhibitor indinavir was administered as a single agent. Plasma concentrations were measured by high powered liquid chromatography (HPLC), and EC-50 values were determined for indinavir by the ACTG/DOD consensus assay. A sigmoidal E_{max} effect model was fit to the data with area under the plasma concentration–time curve (AUC), peak concentration, and trough concentration each as the independent variable (Fig. 1A–C). In Figure 1D to F, the drug exposures were normalized to the EC-50 of that patient's virus (1). It can be seen that the normalization improves the fit of the model to the data. It should be noted that the normalization transforms several points from well off the best-fit curve to an area where the fit improves. This is because of the added information gained from treating a very susceptible viral strain.

It should also be noted that because the drug was administered at essentially the same dose and schedule in all the patients, there is significant colinearity. That is, one cannot make the peak rise without also raising the trough and without increasing the AUC. Therefore one should not draw the inference from these data that the AUC/EC-50 ratio is the pharmacodynamically linked variable for indinavir. Indeed, from other sources of data, the trough/EC-50 ratio or (perhaps preferably) the time >EC–95 is the linked variable for protease inhibitors (2). Clearly, normalization to a measure of potency for the viral isolate to indinavir improves the relationship between exposure and response.

Much the same is true for any type of pathogen. Our ability to grow the organism and identify its sensitivity to the therapeutic agent is key to our ability to formulate an exposure-response relationship.

However, the measure of sensitivity of the pathogen to the drug, although important, is not a sufficient condition for the development of a dynamics relationship. In order to have the highest probability of attaining a robust dynamics relationship, obtaining a good estimate of drug exposure for the individual patient is also critical. This was seen in a neutropenic rat model of fluoroquinolone pharmacodynamics (3). Two stable mutants of a parental strain of *Pseudomonas aeruginosa* (MICs to the test fluoroquinolone of 1, 4, and 8 mg/L) were derived. Therapy with the same dose of drug produced a clear difference in response by MIC (80 mg/kg once daily as therapy with survivorships of 70%, 15%, and 0% for

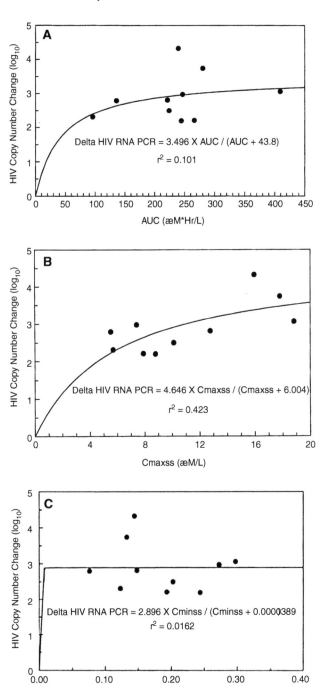

FIGURE 1 Pharmacodynamics of indinavir. E_{max} model of plasma copy number change. (For explanations, see text.) (*Continued*)

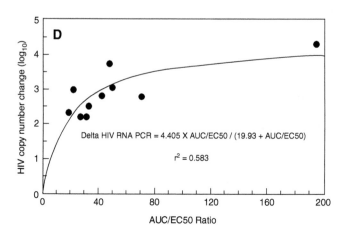

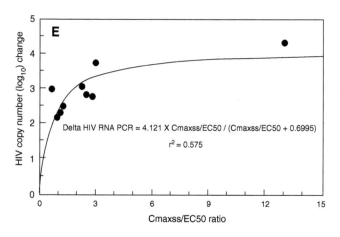

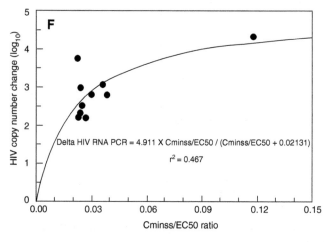

FIGURE 1 (*Continued*)

the groups challenged with MICs of 1, 4, and 8 mg/L, respectively). However, when the dose was altered (20 mg/kg once daily for the MIC of 1.0 mg/L challenge strain) so that the peak/MIC ratio and AUC/MIC ratio were the same as those seen for the challenge group with an MIC of 4.0 treated with 80 mg/kg once daily, the survivorship curves were identical. Because isogenic mutants were employed, this demonstrates that both pieces of information (drug exposure plus a measure of drug susceptibility) are necessary for the best pharmacodynamic relationships to be developed.

Drug Exposure in Clinical Trial Patients

In the past, amazingly few pharmacodynamic relationships have been derived in the anti-infective arena. Part of the reason for this is that patients being treated for infections are often quite ill and unwilling or unable to undergo the rigors of a traditional pharmacokinetic evaluation. Often, dose has been employed as a surrogate for actual exposure estimates. This has proven to be a failed strategy. Dose is a poor measure of exposure. There are true between-patient differences in the pharmacokinetic parameter values such as clearance and volume of distribution. Such true differences (but unmeasured, when dose is used as a measure of exposure) translate into large differences in peak concentration, trough concentration, and AUC in a population of patients receiving the same dose. It should not be surprising that dose is a particularly poor measure of drug exposure and a poor exposure variable to employ in developing pharmacodynamic relationships.

Figure 2 demonstrates the inadequacy of examining just dose as a measure of drug exposure. This is the marginal density plot for clearance for levofloxacin. This drug was studied in 272 patients enrolled in the first study to prospectively develop a relationship between exposure and response (4). This was done in a multicenter study that included 22 centers in the United States. In the study protocol, patients with serum creatinine values in excess of 2.0 mg/dL were excluded. Nonetheless, by inspection, the range of clearance exceeded tenfold. This

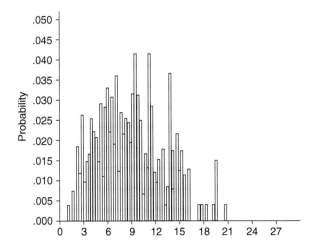

FIGURE 2   Approximate marginal density for clearance of levofloxacin.

also indicates that the range of AUC for a fixed dose would exceed tenfold. Obviously, any attempt to link exposure to outcome employing dose as the measure of exposure would be doomed to failure.

Over the past decade, a number of mathematical techniques have found their way into the toolbox of the kineticist or clinician wishing to construct such relationships.

Optimal Sampling Theory

The first is optimal sampling theory. This technique allows identification of sample times that are laden with "information." The definition of "information" is dependent upon the measure that is defined by the user. For instance, the most commonly employed measure is the determinant of the inverse Fisher information matrix. This is referred to as D-optimality. It has several properties that are desirable. The answers obtained are independent of how the system is parameterized and are also independent of units. This measure also has the remarkable property of replicativeness. That is, if one defines a four-parameter system, there will be exactly four optimal sampling times. If the investigator wishes to make the sampling scheme more robust to errors, D-optimality will tell the investigator to repeat one of the optimal sampling times. This is because D-optimality is deterministic and is based upon the (incorrect) assumption that there is only one true parameter vector, without true between-patient variability. Most other measures of information content (e.g., C-optimality and A-optimality) also suffer from being deterministic. Publications by D'Argenio (5) and Retout and Mentre (6) extended optimal sampling into the stochastic framework and allowed true between-subject variability in the parameter values. This allows the investigator to increase the number of samples and to have increasing amounts of information in the sampling scheme for patients whose values are more removed from the mean values.

Traditional (deterministic) optimal sampling has been well validated. Further, it is possible to employ traditional optimal sampling and still obtain sampling schedule designs robust for a large portion of the population.

One problem with optimal sampling strategy is that it assumes that the answer is already known, that is, one knows the true mean parameter vector for the model system. This obviously places limitations on the use of optimal sampling strategy in the early phases of drug development when little is known regarding the "true" model to be employed for a specific drug and less is known regarding the true mean parameter vector. Nonetheless, with only a little information regarding these issues, optimal sampling has been employed successfully.

There was no validation of this technique in patients until a series of studies were published by Drusano and coworkers (7–10). In what was, to our knowledge, the first clinical validation of optimal sampling theory, the drug ceftazidime was examined in young patients with cystic fibrosis receiving a single dose (7). With the use of a Bayesian estimator, the optimal sampling subset of the full sampling set produced precise and unbiased estimates of the important pharmacokinetic parameter values.

This group then examined the drug piperacillin in a population of septic, neutropenic cancer patients (8). Whereas the study with ceftazidime was performed with a single dose of drug, the study with piperacillin examined two issues: (*i*) whether optimal sampling would provide precise and unbiased estimates of parameter values in the steady-state situation and (*ii*) whether obtaining duplicate samples at the specified sample times would improve the precision of parameter estimation.

The results demonstrated that optimal sampling would, as expected, provide reasonably precise and unbiased parameter estimates. Further, this study also showed that resampling at the designated optimal times did *not* improve the precision of parameter estimation. The latter result was a bit of a surprise and flew in the face of the then-accepted theory regarding optimal sampling. Optimal sampling theory assumes that the mean parameter vector is known without error. Further, true between-patient variance is not incorporated into the optimal sampling time calculation. Given these limitations, it is not surprising that when queried regarding the next most optimal time to obtain a sample after the original optimal times have all been obtained, the theory forces one of the optimal times to be repeated (property of replication). This strategy may improve the precision for the mean patient, but in the clinical situation, where one is trying to construct a population model (part of the creation of a pharmacodynamic model), it is important to recognize that true between-patient variance exists for the parameter values.

If an investigator is to limit the number of plasma samples obtained to an optimal sampling set, it is important to know how robust optimal sampling is with regard to errors in nominal parameter values. This group also addressed this issue (10). Theophylline has been demonstrated to have its clearance altered by smoking cigarettes. The degree of this alteration has been in the order of a 50% increase in the mean clearance of the population. It was felt that by studying a population of smokers as well as a population of nonsmokers and employing optimal sampling strategies for both smokers and nonsmokers, they could examine how badly optimal sampling sets performed when systematic errors on the order of 50% (either high or low) were introduced into the nominal value for clearance. This study demonstrated that errors of this magnitude did not introduce significant bias or imprecision into the overall estimation of theophylline clearance. Further, because this study was performed in two stages, after the first stage they embedded a sampling set that was calculated by employing the patient's initial parameter values estimated from the full sample set obtained during the first stage. They demonstrated that the patient's own optimal samples provide excellent precision and minimal bias for the second stage of the study (patient by patient). Such a finding is important in that it means that toxic drugs can be adequately controlled with minimal sample acquisition, if patients are to be dosed over a relatively long period of time (as is the case in antiretroviral chemotherapy). Likewise, obtaining information about the patient's parameter values for effect control with limited sample acquisition also becomes possible in the routine clinical situation.

Others have recognized the importance of optimal sampling theory in guiding the acquisition of plasma samples in the clinical trial setting for the development of exposure-response relationships. Forrest's group (11) developed an optimal sampling strategy for ciprofloxacin that is useful in the environment of seriously ill hospitalized patients with lower respiratory tract infections. Fletcher and coworkers (12) adapted optimal sampling strategy to the AIDS arena for the development of concentration-controlled trials.

Population Pharmacokinetic Modeling

The second technique is population pharmacokinetic modeling. Credit for the initial development of this technique reflects to Sheiner and coworkers (13–15). After the initial development of the NONMEM system, other groups developed population modeling programs—Mallet (NPML) (16), Schumitzky et al. (NPEM) (17), S-ADAPT of Bauer (18), Davidian and Gallant (19), Lindstrom and Bates (20), and Forrest et al. (21), among others. Population modeling allows the development of a mean parameter vector for the model without requiring that every patient have a robust sampling set. It also provides an estimate of the covariance matrix, allowing construction of parameter distributions and also allowing Monte Carlo simulations, which has recently been shown to be useful in evaluation of doses and schedules. Of course, the important issue with population modeling is that the data must be well timed. The looseness of execution often associated with performing population pharmacokinetics modeling is not an excuse for poor timing of samples collection. Such poor attention to detail can have a severe adverse impact upon the estimates, rendering them either biased or imprecise. Nonetheless, it should be recognized that the ability to perform population modeling has resulted in nothing short of a revolution in our ability to obtain information about drug disposition in ill-target patients. The data presented in Figure 2 are from an analysis employing NPEM (4). Ill patients with community-acquired infections were studied, with each patient having an optimal sampling set of seven plasma determinations, each guided by stochastic design theory.

Human Pharmacodynamics of Anti-infectives

Once population modeling has been performed, it is then useful to perform Bayesian estimation. This allows point estimates of the model parameters to be obtained for all the patients in the populations. Measures of drug exposure [peak concentration, trough concentration, and AUC can be calculated and then normalized to the potency parameter (e.g., peak/MIC ratio, AUC/MIC ratio, and

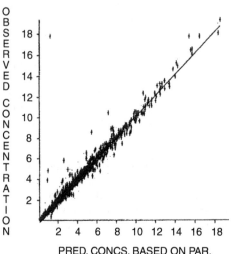

PRED. CONCS. BASED ON PAR.
MEDIANS FROM POST. DISTRIBUTION

FIGURE 3 Scatterplot and least squares line for the entire population.

time > MIC). It is now possible to examine the relationship between exposure and response and/or toxicity.

In the study cited above, a parameter vector was calculated for each patient by Bayesian estimation. The plasma drug concentrations were then simulated for the specific times they were obtained, and a predicted versus observed plot was produced. Figure 3 displays this analysis. The best-fit line was:

$$\text{Observed} = 1.001 \times \text{predicted} + 0.0054, \, r^2 = 0.966; \, p \ll 0.001$$

Once robust estimates of parameter values are obtained for each patient, it is straightforward to attempt to link measures of exposure (peak concentration/MIC or AUC/MIC ratio, time > MIC, etc.) to outcomes. For continuous outcome variables (e.g., viral copy number and CD4 counts), continuous functions such as a traditional sigmoidal E_{max} effect function would be a natural choice (Fig. 1A–F).

However, clinical trials frequently have either dichotomous outcome variables (e.g., success/failure and eradication/persistence) or time-to-event end points (e.g., time to death, time to opportunistic infection, and time-to-lesion change in cytomegalovirus retinitis). For dichotomous outcome variables, logistic regression analysis is a natural choice. For the prospective study examining levofloxacin cited above, we had an analysis plan that tested 13 covariates univariately (4,22). Model building then ensued from the covariates that significantly altered the probability of a good clinical or microbiological outcome (separate sets of analyses). The final models for clinical and microbiological outcomes are displayed graphically in Figures 4 and 5, respectively.

It should also be noted that small boxes on the probability curve denote independent variable "breakpoints." These are arrived at through classification and regression tree (CART) analysis. These merely indicate that patients whose

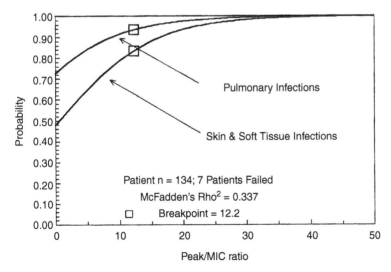

FIGURE 4 Logistic regression relationship between levofloxacin peak/MIC ratio and the probability of a good clinical outcome. *Abbreviation*: MIC, minimum inhibitory concentration.

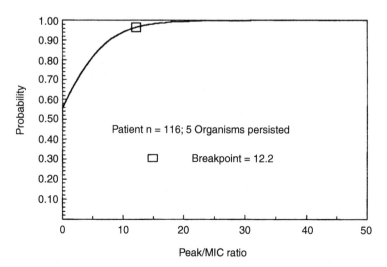

FIGURE 5 Logistic regression relationship between levofloxacin peak/MIC ratio and the probability of organism eradication. *Abbreviation*: MIC, minimum inhibitory concentration.

independent variable (here, peak concentration/MIC ratio) has a value equal to or greater than the breakpoint value have a significantly higher probability of obtaining a good outcome. CART is a useful adjunctive technique in pharmacodynamic analyses but should probably be seen as an exploratory tool and one for rational setting of breakpoints. Logistic regression should be seen as a primary tool for analysis with dichotomous end points.

In addition to modeling success/failure, logistic regression can also be employed to model the probability of occurrence of toxicity. An example can be seen in the analysis of aminoglycoside-related nephrotoxicity published by Rybak et al. (23). These authors performed a prospective, randomized, double-blind trial in which patients received their aminoglycoside either once daily or twice daily.

In the final model, the schedule of administration, the daily AUC of aminoglycoside, and the concurrent use of vancomycin all independently influenced the probability of occurrences of aminoglycoside-related nephrotoxicity.

Sometimes, as with the therapy of cytomegalovirus retinitis, the end point examined is the time to an event, here the time to CMV lesion progression. In this circumstance, after having performed the Bayesian estimation, the measures of exposure may be employed as covariates in a Cox proportional hazards model analysis. This semiparametric approach is a useful way to approach such analyses. For those instances where fuller knowledge of the shape of the hazard function is available, fully parametric analyses (e.g., Weibull analysis) can be performed.

In an analysis of the use of foscarnet for the therapy of cytomegalovirus retinitis, Drusano et al. (24) performed a population pharmacokinetic analysis followed by Bayesian estimation. The exposures then became part of the pharmacodynamic analysis. Five covariates were examined: (*i*) baseline CD4 count, (*ii*) peak CD4 count during therapy, (*iii*) whether or not the patient had a baseline

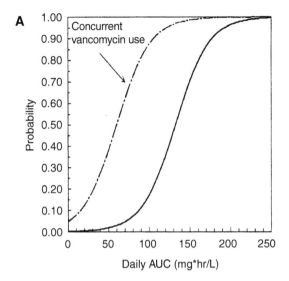

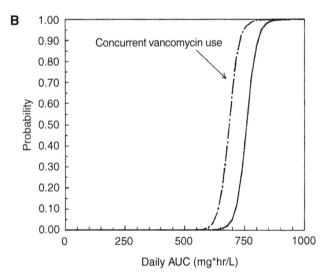

FIGURE 6 Probability of aminoglycoside toxicity. (**A**) Twice daily dosing—vancomycin use, and daily AUG. (**B**) Once daily dosing—vancomycin use and daily AUC. *Abbreviation*: AUC, area under the curve.

blood culture positive for CMV, (*iv*) the peak concentration achieved, and (*v*) the AUC achieved. Trough concentrations were not considered, because they would generally be below the level of assay detection. In fact, all five covariates significantly shifted the hazard function. With model building, only AUC and the baseline CMV blood culture status remained in the final model. Figure 7 demonstrates the exposure response from the final Cox model.

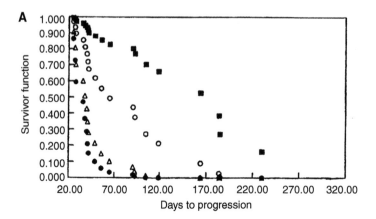

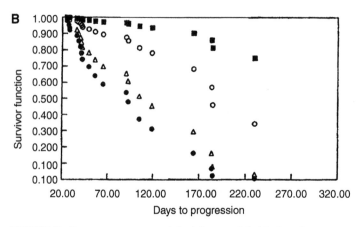

FIGURE 7 Exposure response of final Cox model. (**A**) Baseline cytomegalovirus blood culture positive. (**B**) Baseline cytomegalovirus blood culture negative. (●) Lowest AUC observed in the population; (△, □, ■) 20th, 50th, and 80th percentiles, respectively, of AUC in the population. *Abbreviation*: AUC, area under the curve.

Evaluation of Dose and Schedule by Monte Carlo Simulation

Monte Carlo simulation has recently been demonstrated to be useful for the evaluation of doses and schedules for anti-infective agents. This technique was first applied for this purpose by Drusano at a meeting of the FDA Anti-Infective Drug Products Advisory Committee (25). Two applications are demonstrated here. The first is for dose adequacy and for preclinical MIC breakpoint determination. The second is for the evaluation of the dosing schedule.

To evaluate the adequacy of a 500 mg dose of the fluoroquinolone levoflox-acin, Drusano and Craig collaborated for the following analysis. The mean parameter vector and covariance matrix from the levofloxacin study cited earlier were employed to create a 10,000 subject Monte Carlo simulation. The AUC distribution for a 500 mg i.v. dose for these subjects was generated. The data from the levofloxacin Tracking Resistance in the United States Today (TRUST) study were employed for the MIC distribution for *Streptococcus pneumoniae*. The key for this analysis was to set a "target goal." Craig's mouse thigh model allowed setting the AUC/MIC target goal of 27.0 (total drug) associated with stasis and 34.5 (total drug) for a drop in the CFU of one $\log_{10}$ unit (associated with the shutoff of bacteremia) for levofloxacin (Craig WA. Personal communication). Each of the 10,000 AUCs in the distribution was divided by the MIC range from 0.125 to 4.0 (in a twofold dilution series). The resultant values were compared to the target goals, and the frequency with which the target was achieved was ascertained. The outcome of this analysis is displayed in Figure 8A.

It is obvious that the goal-attainment rate is 100% for both targets until an MIC of 0.5 mg/L is reached. At 1.0 mg/L, both target attainments decline, but both are in excess of 90%. Only after this do we see a large decline in target attainment.

Much of the outcome observed depends upon the MIC distribution. If, instead of employing the TRUST data from 1998–1999 (Trust IV), one employs TRUST VI, a different MIC distribution is seen. Furthermore, Ambrose et al. (27) demonstrated that a free drug AUC/MIC ratio of 30 is associated with a good clinical outcome in patients. These two changes are displayed in Figure 8B. It is clear by inspection that the target-attainment rate at 1.0 mg/L, where the bulk of the clinical isolates reside, is below 80%.

If one uses a 750 mg levofloxacin dose, the resultant target attainment at an MIC of 1.0 mg/L for the TRUST VII distribution approximates 95% (Fig. 8C).

It is possible to remove the variability in the MIC by performing an expectation over the MIC distribution. In essence, we can multiply the target-attainment rate by the fraction of the strains of pneumococcus represented at each levofloxacin MIC value. This gives us an estimate of the target-attainment rate in a clinical trial, subject to the assumptions that the MIC distribution is representative

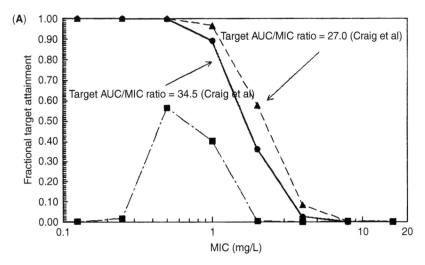

FIGURE 8 (A–C) Levofloxacin 10,000-subject Monte Carlo simulation. Pneumococcal target attainment with a 500 mg qid dose. Key: (●) 1 Log drop target; (▲) stasis target; (■) MIC distribution. (*Continued*)

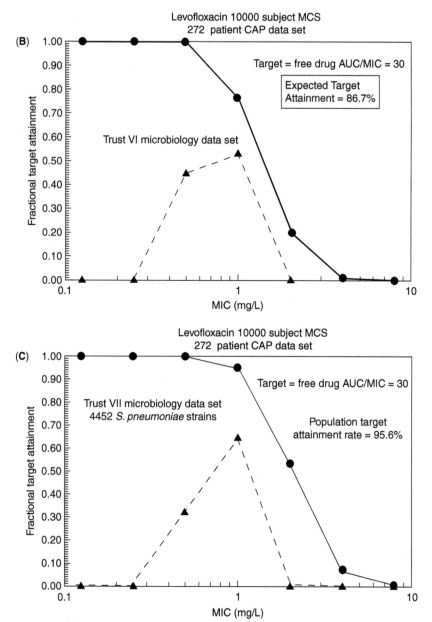

FIGURE 8 *(Continued)* Key: (●) 1 Log drop target; (▲) stasis target; (■) MIC distribution.

of that seen in clinical trials and that the AUC distribution likewise representative of a clinical Trial. The target-attainment rates are shown in Table 2.

This example demonstrates that a 500 mg dose of levofloxacin would likely be adequate for pneumococcal infections, given the distribution of the AUCs for the drug and the distribution of the MICs in TRUST IV. By Trust VII, a 750 mg

TABLE 1 Precision (%) of Kinetic Parameters of Theophylline as Determined from Different Optimal Sampling Strategies Relative to Those Determined from the Full Sampling Strategy[a]

–	V_c	V_{SS}	V_{area}	S_{cl}	$T_{1/2}$*
Correct7	2.20	1.26	1.30	2.97	2.99
Wrong7	1.66	1.01	1.04	3.56	3.98
Patient's7	2.28	1.34	1.30	2.98	3.66
Patient's4	2.60	2.20	2.28	2.99	3.77

[a]Correct7 represents the seven sample times derived from the "correct" prior population. Wrong7 represents the seven sample times derived from the "wrong" prior population. Patient's7 and Patient's4 represent the seven and four sample times derived from the patient's own prior parameter values.

levofloxacin dose would be required for robust activity with the new target value and with the changed MIC distribution. This has been demonstrated in clinical trials of levofloxacin in community-acquired pneumonia (22,28,29).

It is also possible to examine schedule with this technique. Drusano et al. (30) examined the combination of abacavir plus amprenavir for HIV. The interaction of the two agents was quantitated in the presence of human binding proteins in vitro using the Greco interaction equation (31). Population pharmacokinetic models were then derived from clinical trial data for both drugs. Monte Carlo simulations were derived of the effect-time curves for 500 subjects. In the simulations, doses of 300 mg of abacavir every 12 hours (q12h) plus 800 mg of amprenavir every eight hours (q8h) were simulated, as well as doses of abacavir 300 mg every 12 hours plus 1200 mg of amprenavir every 12 hours. In Figure 9A and B, the mean concentration-time profiles are shown for the various simulations for 500 subjects. Figure 9C and D show one subject selected from the population. Figure 9E and F show the effect versus time curves derived from that specific patient at steady state for the different schedules of administration.

The effect-time curves can be integrated over a 24-hour steady-state interval and divided by the interval length (24 hours). An average percent of maximal effect results from the calculation. These are plotted in Figure 10 for the two schedules of administration for all 500 simulated subjects.

It is obvious from inspection that the schedule of administration that is more fractionated for the protease inhibitor (amprenavir q8h) is providing greater effects. This can clearly be seen in the frequency histograms presented in Figure 11. Irrespective of how one tests the differences between regimes (frequency > 90% maximal effect, frequency > 70% maximal effect, difference between mean percent maximal effects), the more fractionated regimen is always statistically significantly superior.

TABLE 2 Levofloxacin 10,000-Subject Monte Carlo Simulation: Target Attainment Over a 4296 Isolate Database of *Streptococcus pneumoniae*

Target Attainment	1 Log drop (34.5 AUC/MIC ratio) 94.7	Stasis (27 AUC/MIC ratio) 97.8
Target Attainment for a *f*AUC/MIC = 30		
500 mg Levofloxacin 86.7%	750 mg Levofloxacin 95.6%	

Abbreviations: AUC, area under the plasma concentration–time curve; MIC, minimal inhibitory concentration.
Source: PK parameters, from Ref. 26; Isolate MICs from the 1998–1999 TRUST study; target-attainment data from Craig WA. Personal communication; Ambrose target from Ref. 27.

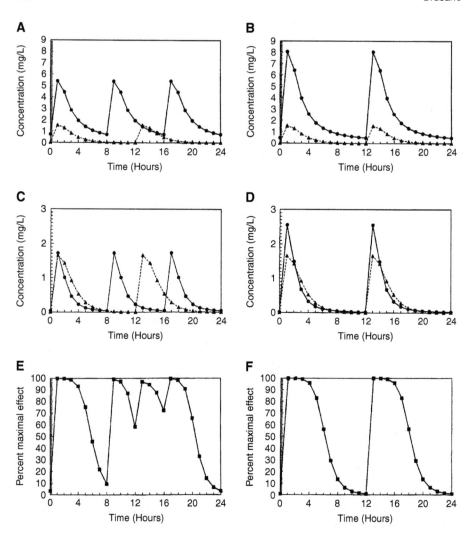

FIGURE 9 (**A**) Mean concentration-time curve for a steady-state dosing interval derived from a 500-subject Monte Carlo simulation with abacavir (300 mg q12h) and amprenavir (800 mg q8h). (**B**) Mean concentration-time curve for a steady-state dosing interval derived from a 500-subject Monte Carlo simulation with abacavir (300 mg q12h) and amprenavir (1,200 mg q12h). (**C**) Concentration-time curve for a steady-state dosing interval simulated for one of 500 simulated subjects using abacavir (300 mg q12h) and amprenavir (800 mg q8h). (**D**) Concentration-time curve for a steady-state dosing interval simulated for one of 500 simulated subjects using abacavir (300 mg q12h) and amprenavir (1,200 mg q12h). (**E**) Steady-state effect-time curve as calculated from the drug interaction parameters (3) for abacavir plus amprenavir at the concentrations displayed in panel C. (**F**) Steady-state effect-time curve as calculated from the drug interaction parameters (3) for abacavir plus amprenavir at the concentrations displayed in panel D.

Consequently, Monte Carlo simulation can be used for both dose and regimen evaluations and can also set preclinical and, after the human pharmacodynamic trials have been performed, clinical MIC breakpoints. It is obvious that

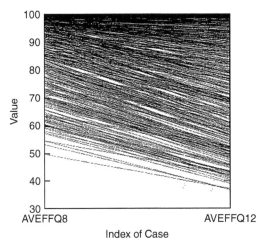

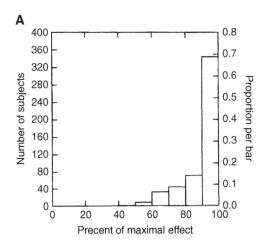

FIGURE 10 Average percent of maximal antiretroviral effect for combination therapy with abacavir plus amprenavir. Amprenavir schedule of administration was 800 mg (q8h) or 1200 mg (q12h). Abacavir schedule was 300 mg (q12h) in both groups. Results were calculated from a Greco interaction model. Drug concentrations were from Monte Carlo simulation.

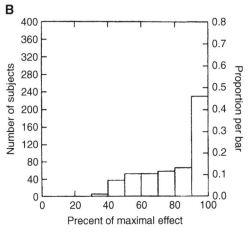

FIGURE 11 Frequency histograms for average percent of maximal effect for abacavir/amprenavir combination therapy. In (A), abacavir was 300 mg (q12h) and amprenavir dose was 800 mg (q8h). In (B), abacavir was 300 mg (q12h) and amprenavir was 1200 mg (q12h). Values were determined as in Figure 10.

TABLE 3 Paradigm for the Development of Exposure-Response Relationships for Anti-Infective Agents

Decide on an end point
Make potency measurements on pathogens from trials (MIC, EC-50, etc.)
Obtain exposure estimates for patients from those trials
Stochastic design for sampling scheme
Population pharmacokinetic modeling
Bayesian estimation for individual-patient exposure estimates
Decide on an end point analysis (the following are examples only)
Sigmoidal E_{max} analysis for a continuous variable
Logistic regression for dichotomous/polytomous outcomes
Cox proportional hazards modeling (or a variant) for time-to-event data
Classification and regression tree analysis for breakpoint determination
Stochastic control when effect-toxicity relationships are available

Abbreviations: MIC, minimal inhibitory concentration; EC–50, effective concentration that reduces growth by half.

there are other uses for this flexible and powerful technique for the clinical arena [e.g., drug penetration to the site of infection (32)].

DEVELOPING RELATIONSHIPS BETWEEN ANTI-INFECTIVE EXPOSURE VS. RESPONSE—HOW TO DO IT?

Each of the foregoing examples leads to a simple paradigm. It is straightforward to attempt to define pharmacodynamic relationships in the clinical trial setting. The approach is set forth in Table 3. Once such relationships are defined, particularly when relationships are available for efficacy and toxicity, as is the case for aminoglycoside antibiotics (23,33), the true goal of anti-infective chemotherapy (maximal effect with minimal toxicity) can be sought by using stochastic control techniques (34). As part of the approach, it is clear that plasma concentration determination is a requirement. It should be brought home forcefully to pharmaceutical sponsors that the paradigm has shifted. We can successfully seek and generate such relationships. In the "old days," the necessity for concentration determination was the death knell for the development of a compound. Now it is clear that for the patient's sake we can achieve maximal probability of a good clinical and microbiological outcome coupled with the minimal probability of toxicity and (perhaps) emergence of resistance by measuring drug concentrations in plasma. This should be the new (and positive) way of differentiating drugs. We should prefer those that provide us the possibility of having a rational basis for producing the best patient outcomes. Again, third party payors also need to understand that drugs with such relationships developed should have priority on clinical pathways, because they provide maximal probability of response with minimal probability of toxicity.

Finally, the approach set forth in Table 3 works well. Ambrose et al. (35) have recently reviewed generation of exposure-response relationships in man for anti-infectives. They reviewed dynamic relationships developed for hospital-acquired pneumonia with fluoroquinolones (36,37), community-acquired respiratory tract infections with fluoroquinolones, β-lactams and telithromycin (38–40), and bacteremia with oritavancin and linezolid (41,42), as well as complicated skin and skin-structure infections with linezolid and tigecycline (42,43). There is no excuse now for new clinical trials not to include aspects where the relationship between exposure and response as well as exposure and toxicity can be elucidated. Indeed, in an era where regulatory requirements are tightening and where superiority trials are being either demanded or heavily encouraged, this type of relationship delineation provides the highest likelihood of meeting the regulatory requirements with a modest number of well-studied patients. This is in everyone's best interest.

REFERENCES

1. Drusano GL. Antiviral therapy of HIV and cytomegalovirus. Symposium 130-A, S-121. 37th Interscience Conference on Antimicrobial Agents and Chemotherapy, Toronto, Canada, Sept 28–Oct 1, 1997. Am Soc Microbiol, Washington, DC.
2. Drusano GL, Bilello JA, Preston SL, et al. Hollow fiber unit evaluation of BMS232632, a new HIV-1 protease inhibitor, for the linked pharmacodynamic variable. Session 171, Presentation 1662, 40th Interscience Conference on Antimicrobial Agents and Chemotherapy, Toronto, Canada, Sept 17–20, 2000. Am Soc Microbiol, Washington, DC.
3. Drusano GL, Johnson DE, Rose M, Standiford HC. Pharmacodynamics of a fluoroquinolone antimicrobial in a neutropenic rat model of Pseudomonas sepsis. Antimicrob Agents Chemother 1993; 37:483–490.
4. Preston SL, Drusano GL, Berman AL, et al. Levofloxacin population pharmacokinetics in hospitalized patients with serious community-acquired infection and creation of a demographics model for prediction of individual drug clearance. Antimicrob Agents Chemother 1998; 42:631–639.
5. D'Argenio DZ. Incorporating prior parameter uncertainty in the design of sampling schedules for pharmacokinetic parameter estimation experiments. Math Biosci 1990; 99:105–118.
6. Retout S, Mentre F. Optimization of individual and population designs using S-plus. Pharmacokinet Pharmacodynam 2003; 30:417–443.
7. Drusano GL, Forrest A, Snyder MJ, Reed MD, Blumer JL. An evaluation of optimal sampling strategy and adaptive study design. Clin Pharmacol Ther 1988; 44:232–238.
8. Drusano GL, Forrest A, Plaisance KI, Wade JC. A prospective evaluation of optimal sampling theory in the determination of the steady state pharmacokinetics of piperacillin in febrile neutropenic cancer patients. Clin Pharmacol Ther 1989; 45:635–641.
9. Yeun GJ, Drusano GL, Forrest A, Plaisance KI, Caplan ES. Prospective use of optimal sampling theory: steady-state ciprofloxacin pharmacokinetics in critically ill trauma patients. Clin Pharmacol Ther 1989; 46(4):451–457.
10. Drusano GL, Forrest A, Yuen JG, Plaisance KI. Optimal sampling theory: effect of error in a nominal parameter value on bias and precision of parameter estimation. J Clin Pharmacol 1994; 34:967–974.
11. Kashuba AD, Ballow CH, Forrest A. Development and evaluation of a Bayesian pharmacokinetic estimator and optimal, sparse sampling strategies for ceftazidime. Antimicrob Agents Chemother 1996; 40:1860–1865.
12. Noormohamed SE, Henry WK, Rhames FS, Balfour HH Jr, Fletcher CV. Strategies for control of zidovudine concentrations in serum. Antimicrob Agents Chemother 1995; 39:2792–2797.
13. Grasela TH, Sheiner LB. Population pharmacokinetics of procainamide from routine clinical data. Clin Pharmacol 1984; 9:545–554.
14. Beal SL, Sheiner LB. Estimating population kinetics. CRC Crit Rev, Bioeng 1982; 8:195–222.
15. Sheiner LB, Rosenberg B, Marathe VV. Estimation of population characteristics of pharmacokinetic parameters from routine clinical data. J Pharmacokinet Biopharm 1977; 5:445–479.
16. Mallet A. A maximum likelihood estimation method for random coefficient regression models. Biometrika 1986; 73:645–656.
17. Schumitzky A, Jelliffe R, Van Guilder M. NPEM2: a program for pharmacokinetic population analysis. Clin Pharmacol Ther 1994; 55:163.
18. Bauer RJ. S-ADAPT. Biomedical Simulations Resource. University of Southern California 2006.
19. Davidian M, Gallant AR. Smooth nonparametric maximum likelihood estimation for population pharmacokinetics, with application to quinidine. J Pharmacokinet Biopharm 1992; 20:529–556.

20. Lindstrom M, Bates D. Nonlinear mixed effects models for repeated measures data. Biometrics 1990; 46:673–687.

21. Forrest A, Ballow CH, Nix DE, Birmingham MC, Schentag JJ. Development of a population pharmacokinetic model and optimal sampling strategies for intravenous ciprofloxacin. Antimicrob Agents Chemother 1993; 37:1065–1072.

22. Preston SL, Drusano GL, Berman AL, et al. Prospective development of pharmacodynamic relationships between measures of levofloxacin exposure and measures of patient outcome: a new paradigm for early clinical trials. J Am Med Assoc 1998; 279:125–129.

23. Rybak MJ, Abate BJ, Kang SL, Ruffing MJ, Lerner SA, Drusano GL. Prospective evaluation of the effect of an aminoglycoside dosing regimen on rates of observed nephrotoxicity and ototoxicity. Antimicrob Agents Chemother 1999; 43:1549–1555.

24. Drusano GL, Aweeka F, Gambertoglio J, et al. Relationship between foscarnet exposure, baseline cytomegalovirus blood culture and the time to progression of cytomegalovirus retinitis in HIV-positive patients. AIDS 1996; 10:1113–1119.

25. Drusano GL, Preston SL, Hardalo C, et al. Use of pre-clinical data for the choice of a phase II/III dose for SCH27899 with application to identification of a pre-clinical MIC breakpoint. Antimicrob Agents Chemother 2001; 45:13–22.

26. Drusano GL, Preston SL, Gottfried MH, Danziger LH, Rodvold KA. Levofloxacin penetration into epithelial lining fluid as determined by population pharmacokinetic modeling and Monte Carlo simulation. Antimicrob Agents Chemother 2002; 46:586–589.

27. Ambrose PG, Grasela DM, Grasela TH, Passarell J, Mayer HB, Pierce PF. Pharmacodynamics of fluoroquinolones against Streptococcus pneumoniae in patients with community-acquired respiratory tract infections. Antimicrob Agents Chemother 2001; 45:2793–2797.

28. File TM Jr, Segreti J, Dunbar L, et al. A multicenter, randomized study comparing the efficacy and safety of intravenous and/or oral levofloxacin versus ceftriaxone and/or cefuroxime axetil in treatment of adults with community-acquired pneumonia. Antimicrob Agents Chemother 1997; 41:1964–1972.

29. Dunbar LM, Wunderink RG, Habib MP, et al. High dose, short course levofloxacin for community-acquired pneumonia: a new treatment paradigm. Clin Infect Dis 2003; 37: 752–760.

30. Drusano GL, D'Argenio DZ, Preston SL, et al. Use of drug effect interaction modeling with Monte Carlo simulation to examine the impact of dosing interval on the projected antiviral activity of the combination of abacavir and amprenavir. Antimicrob Agents Chemother 2000; 44:1655–1659.

31. Drusano GL, D'Argeno DZ, Symonds W, et al. Nucleoside analog 1592U89 and human immunodeficiency virus protease inhibitor 141W94 are synergistic in vitro. Antimicrob Agents Chemother 1998; 42:2153–2159.

32. Drusano GL, Preston SL, Van Guilder M, et al. A population pharmacokinetic analysis of the prostate penetration of levofloxacin. Antimicrob Agents Chemother 2000; 44:2046–2051.

33. Kashuba AD, Nafziger AN, Drusano GL, Bertino JS Jr. Optimizing aminoglycoside therapy for nosocomial pneumonia caused by gram-negative bacteria. Antimicrob Agents Chemother 1999; 43:623–629.

34. Schumitzky A. Applications of stochastic control theory to optimal design of dosage regimens. In: D'Argenio DZ, ed. Advanced Methods of Pharmacokinetic and Pharmacodynamic Systems Analysis. Plenum, New York, 1991:137–152.

35. Ambrose PG, Bhavnani SM, Rubino CM, et al. Pharmacokinetics-pharmacodynamics of antimicrobial therapy: it's not just for mice anymore. Clin Infect Dis 2007; 44:79–86.

36. Forrest A, Nix DE, Ballow CH, Goss TF, Birmingham MC, Schentag JJ. Pharmacodynamics of intravenous ciprofloxacin in seriously ill patients. Antimicrob Agents Chemother 1993; 37:1073–1081.

37. Drusano GL, Preston SL, Fowler C, Corrado M, Weisinger B, Kahn J. The relationship between fluoroquinolone AUC/MIC Ratio and the probability of eradication in patients with nosocomial pneumonia. J Infect Dis 2004; 189:1590–1597.

38. Ambrose PG, Bhavnani SM, Owens RC. Clinical pharmacodynamics of quinolones. Infect Dis Clin North Amer 2003; 17:529–543.

39. Craig WA. Pharmacodynamics of antimicrobials: general concepts and applications. In: Nightingale CH, Murakawa T, Ambrose PG, eds. Antimicrobial pharmacodynamics in theory and in practice. New York: Marcel Dekker, 2002:1–22.
40. Lodise TP, Preston SL, Barghava V, et al. Pharmacodynamics of an 800 mg dose of telithromycin in patients with community-acquired pneumonia caused by extracellular pathogens. Diagn Microbiol Infect Dis 2005; 52:45–52.
41. Bhavnani SM, Passarrell JA, Owen JS, Loutit JS, Porter SB, Ambrose PG. Pharmacokinetic-pharmacodynamic relationships describing the efficacy of oritavancin in patients with *Staphylococcus aureus* bacteremia. Antimicrob Agents Chemother 2006; 50:994–1000.
42. Rayner CR, Forrest A, Meagher AK, Birmingham MC, Schentag JJ. Clinical pharmacodynamics of linezolid in seriously ill patients treated in a compassionate-use program. Clin Pharmacokinet 2003; 42:1411–1423.
43. Meagher A, Passarrell J, Cirincione B, et al. Exposure-response analysis of the efficacy of tigecycline in patients with complicated skin and skin structure infections [abstr P-1184]. In: Program and abstracts of the 15th European Congress of Clinical Microbiology and Infectious Diseases (Copenhagen). Basel, Switzerland: European Society of Clinical Microbiology and Infectious Diseases, 2005:373–374.

21 Application of Pharmacokinetics and Pharmacodynamics in Antimicrobial Global Drug Development

Sujata M. Bhavnani
Institute for Clinical Pharmacodynamics, Ordway Research Institute, Inc., Albany, New York, U.S.A.

INTRODUCTION

The implementation of a global drug development strategy is a challenging, time-consuming, and expensive process. To gain approval to market a drug, sponsors are required to provide rigorous evidence of efficacy and safety using data from large multicenter, randomized, controlled clinical trials. In the United States (US), for example, the Food and Drug Administration (FDA) requires that at least two clinical studies in the same targeted patient population be performed to confirm the reproducibility of the evidence demonstrating efficacy, safety, and dose–response of the investigational agent. Data used to obtain regulatory approval in one region (e.g., the US), however, often prove insufficient to obtain approval in another region (e.g., Japan), thereby necessitating the further collection of clinical trial data in the new region of interest. Such duplication of previously collected registrational clinical trial data not only requires additional resources, but may also result in a significant delay in bringing the drug to market in the new region.

The landscape of global drug development has been, however, changing over the last two decades. In April 1990, drug-regulatory authorities and pharmaceutical industry associations from Japan, the European Union (EU), and the US assembled to develop standardized or "harmonized" drug-regulatory requirements for these three regions (1). This unified effort resulted in the establishment of "The International Conference on Harmonisation of Technical Requirements for the Registration of Pharmaceuticals for Human Use" (ICH). One set of ICH guidelines, "E5, Ethnic Factors in the Acceptability of Foreign Clinical Data" (2), has been especially useful in providing a framework for evaluating (*i*) the likelihood of the impact of ethnic factors upon the safety and efficacy of a particular dosing regimen and (*ii*) the appropriateness of "bridging" data from two ethnically distinct populations. When appropriate, bridging strategies may be employed to maximize the utility of previously collected data by allowing for an extrapolation of data from one region or ethnic group to another. The application of pharmacokinetic–pharmacodynamic (PK–PD) principles to bridging strategies for the development of antimicrobial agents provides the opportunity to optimize dose selection and thereby the likelihood of regulatory success in multiple regions. In this chapter, the ICH E5 guidelines are reviewed together with the application of PK–PD principles to the development of antimicrobial agents.

BRIDGING DATA AND BRIDGING STUDIES

As part of a global drug development plan, bridging strategies designed to maximize the utility of existing data can include at least two distinct components: bridging of data from a preexisting clinical data package and actual bridging studies carried out in the new region of interest. The former utilizes selected data from a complete clinical data package that are relevant to the ethnic population of the new region, including PK, PD, and dose–response data. In the latter case, a bridging study may be performed in the new region to provide PK, PD, or clinical data on efficacy, safety, dosage, and dose regimen for that same region, the results of which may justify the extrapolation of the existing clinical data to the population in the new region (2).

The exact nature of the bridging study is dependent upon prior experience with the drug class, especially as it relates to the likelihood that ethnic factors could affect the nature of the drug safety, efficacy, and dose–response relationship. Extrapolation of clinical data may be feasible without a bridging study if the drug has minimal potential for ethnic sensitivity and if other regional factors (such as medical practice and conduct of clinical trials) of the two regions are similar. Even in the case where the drug has the potential for ethnic sensitivity, extrapolation of clinical data may still be possible without a bridging study if there is sufficient clinical experience with agents from the same class.

As outlined in the ICH E5 guidelines (2), a controlled PD bridging study may be required if two regions are ethnically dissimilar and the drug has the potential for ethnic sensitivity. In these studies, a pharmacologic end point that is reflective of relevant drug activity (e.g., a well-established surrogate end point) may be evaluated and these data may be used to support the extrapolation of preexisting efficacy, safety, dose, and dose regimen data to the new region. However, the availability of fully validated surrogate end points is limited (3,4).

When ethnic variability in pharmacodynamics is observed, the cause is usually related to variability in PKs. Thus, the evaluation of PKs in the above-described studies can further enhance the value of the study. In cases where ethnic variability in PKs suggests that there will be differences in response, adjustment of dosage regimens may be adequate without the need for a new trial (2). Data supporting ethnic differences in PDs among patients with similar exposures are limited. However, the literature for propranolol provides data describing the influence of race/ethnicity on factors affecting drug response including PKs, receptor subtype distribution, and sensitivity (5). In such cases where PD data suggest that there are interregional differences in response that are independent of PKs, a trial with clinical end points will likely be required (2).

Unlike drugs such as propranolol, the PD end points for antimicrobial agents, for example microbiological response, are primarily influenced by the PKs and potency of the agent to the bacterial organism, the latter of which is described by the minimum inhibitory concentration (MIC). In contrast to other classes of agents, the pharmacological effect of antimicrobial agents is driven by the binding of the drug to the bacterial receptor site. Thus, interregional differences in PDs will be a function of the differences in PKs and MIC distributions for the pathogens of interest, rather than differences in drug affinities for human receptors.

Despite the demonstration of safety in the original region, region-specific factors may elevate the concern for safety in the new region. Depending upon the nature of the safety concern, data could be obtained from a bridging study primarily evaluating efficacy but with a sufficient sample size to monitor for the safety event of interest (2). Given that adverse events are often a function of elevated exposure, the evaluation of PKs in a bridging study designed to also evaluate for safety end points is crucial. If a bridging study for efficacy is not required or is of insufficient duration to evaluate for safety information, a separate safety study may be needed (2). For rare but important adverse events, a larger study may be requested. In some cases, this could represent a Phase 4 commitment.

Evaluation of Ethnic Factors

The success of bridging data from a complete clinical data package to a new region or extrapolating data through bridging studies is dependent upon the degree of similarity between populations. The ICH E5 guidelines suggest that data from two populations of different ethnicity may be appropriately bridged if the data from the two populations are sufficiently "similar" (2). Although not explicitly defined, two populations may be considered "similar" for a specific parameter if the difference in that parameter between populations is not likely to influence clinical efficacy or safety. The likelihood that two populations will be similar in clinical efficacy or safety end points may be inferred by assessing both the PK and the PD properties of an agent and the potential for impact of ethnic factors on these properties. Similarity between populations may be anticipated if the agent possesses PK and PD properties with minimum potential for ethnic sensitivity.

Extensive reviews of the literature have served to show differences in PKs by ethnic groups for many different classes of drugs (5,6). While differences in absorption, distribution, metabolism, and excretion have been observed among ethnic groups, interethnic variability in PKs has largely been associated with differences in metabolic handling. The ICH E5 guidelines suggest that minimal differences between populations may perhaps be anticipated for drugs demonstrating linear rather than nonlinear PKs (2). However, isoniazid represents an example of an antimicrobial agent with linear PKs (7) and impressive ethnic sensitivity (7,8). For drugs with nonlinear PKs and ethnic sensitivity, concentrations at which saturable clearance is evident may be different among subpopulations. Greater differences among populations may be anticipated if an enzyme with the potential for genetic polymorphism represents a predominant route of drug elimination. Thus, a lower likelihood for interpopulation variability may be expected for an agent with a minimal degree of drug metabolism or if metabolism is accomplished through multiple pathways. The following represent additional PK characteristics, which may favor successful bridging of data among populations: a low potential for drug–drug, drug–food, and drug–disease interactions, high bioavailability with little-to-no first-pass effect, and a low degree of protein binding (2).

While the nature of PK–PD relationships is not likely to change among populations, factors affecting PKs will have an impact upon exposure and thus, the risk for a subtherapeutic effect or a safety event at a given dose. For example, if weight distributions are known to differ between two populations, the area under the concentration–time curve (AUC) at a given dose would be expected also to differ for a drug for which weight is an important determinant of clearance. Higher exposures at a given dose would be expected in the population with lower weight distributions, and accordingly, the risk for an exposure-related adverse event may also be higher in this population. Although less frequent, examples of ethnic differences in PDs have been observed that appear to be independent of differences in PKs (5). As with polymorphisms of drug-metabolizing enzymes, the nature of the binding to a pharmacologic drug target, and hence drug response, is also influenced by genetic polymorphisms (9). In addition, other non-PK factors may affect the probability of clinical efficacy or a safety event such as the prevalence of comorbid conditions, and differences in medical practices including diagnosis and assessment of response.

Factors with the potential for ethnic sensitivity and for having an impact on PKs and/or PDs may be grouped into two categories: intrinsic and extrinsic. Intrinsic factors are those that describe genetic and physiologic characteristics, and extrinsic factors are those that describe the cultural and environmental characteristics of a population. Examples of intrinsic factors include genetic polymorphisms of drug metabolism, height, weight, body composition, or differences in disease pathophysiology. Extrinsic factors tend to be less genetically and more culturally and behaviorally determined. Examples of extrinsic factors include the social and cultural aspects of a region such as diet, use of tobacco, use of alcohol, exposure to pollution and sunshine, and socioeconomic status. Medical training, clinical practice guidelines, and practices in clinical trial conduct and assessment of outcome measures are also important examples of extrinsic factors, which influence the value of clinical studies conducted in different regions (2). The classification of intrinsic and extrinsic factors is shown in Figure 1.

In addition to the extrinsic factors described, cultural perceptions are also a very important determinant of regional variability in medical practices and patient behavior (10). Understanding region-specific cultural perspectives may help to predict how a patient may perceive illness and death and how likely a patient will be to communicate, disclose information, report adverse events, and seek treatment.

Comparisons by relevant intrinsic or extrinsic ethnic factors may be useful to define and identify important differences between populations, as such differences may influence the ability to extrapolate clinical data between regions. Thus, understanding variability in the intrinsic and extrinsic factors of interest and the potential

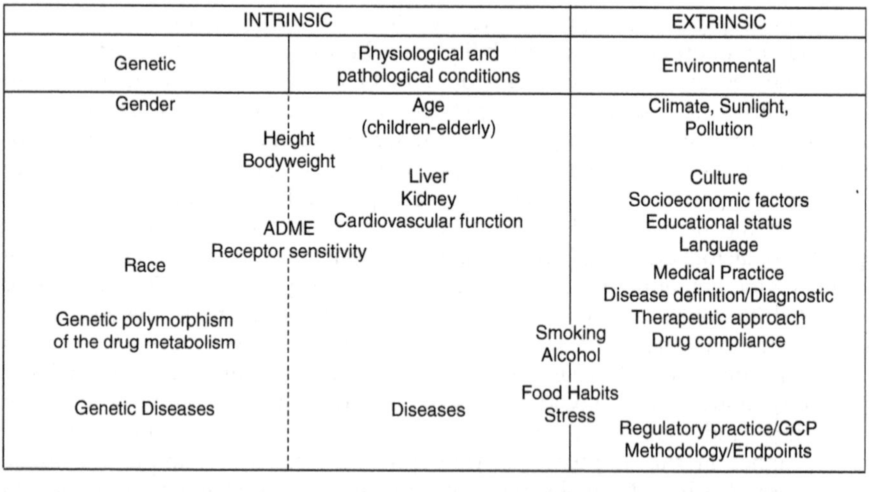

INTRINSIC			EXTRINSIC
Genetic	Physiological and pathological conditions		Environmental
Gender		Age (children-elderly)	Climate, Sunlight, Pollution
	Height Bodyweight		
		Liver	Culture
		Kidney	Socioeconomic factors
	ADME Receptor sensitivity	Cardiovascular function	Educational status Language
Race			Medical Practice Disease definition/Diagnostic
Genetic polymorphism of the drug metabolism		Smoking Alcohol	Therapeutic approach Drug compliance
Genetic Diseases		Diseases	Food Habits Stress
			Regulatory practice/GCP Methodology/Endpoints

FIGURE 1 Classification of intrinsic and extrinsic ethnic factors. *Source*: From Ref. 2.

impact of these factors on PKs and PDs is an essential element in designing and implementing multinational clinical studies and interpreting the results of such studies.

USE OF PK–PD PRINCIPLES FOR ANTIMICROBIAL DRUG DEVELOPMENT
Background
Over the last few decades, our understanding of the relationship between the PKs and PDs of antimicrobial agents has grown exponentially. This has been due in great part to the use of in vitro and animal models of infection, both of which have permitted the exploration of exposure–response relationships for efficacy for most antimicrobial agents against a multitude of microorganisms. As a result of this collective body of work, the PK–PD measure most closely associated with in vivo efficacy and the magnitude of the measures predictive of efficacy have been identified for many classes of agents (11,12). PK–PD data from infected patients have provided the opportunity to evaluate the degree of concordance between these data and those from in vitro and animal PK–PD studies. In multiple clinical indications and across different drug classes, the magnitudes of PK–PD measures necessary for clinical effectiveness in patients were shown to be similar to those identified from animal data (13).

Given these findings, there has been a recent and growing appreciation of the potential value of incorporating PK–PD principles gained from nonclinical models of infection into the early stages of clinical drug development for antimicrobial agents. The integration of PK–PD relationships derived from nonclinical infection models with Phase 1 PK data can be used to optimize antimicrobial dosing regimens for Phase 2 and 3 studies (14,15). The early integration of such knowledge has been advocated by the US FDA to increase the probability of selecting clinically efficacious antimicrobial dosing regimens for Phase 2/3 development (16). In addition to using PK–PD analyses to support early dose selection, sponsors are encouraged to compare early predictions to subsequent clinical trial results. In this regard, PK–PD analyses of Phase 2/3 studies allow for the validation of dose regimen selection decisions and provide the opportunity to "close the loop" between decisions made during the early and late clinical development of an agent (17).

In a recent review, approximately 50% of Phase 3 clinical trials across many classes of drugs were reported to have failed, with poor dose selection representing an important factor (18). As described above, the use of preclinical PK–PD infection models have been highly encouraged by the FDA as a guide to early dose selection decisions for antimicrobial agents (16,17). Thus, early consideration of PK–PD principles for dose selection along with evaluation of potential ethnic or regional factors that will have an impact upon the PKs or PDs of an investigational agent and the planning of efficient bridging studies will increase the likelihood of successfully obtaining simultaneous registration in multiple regions.

Sources of PK and PD Variability
Drusano et al. first described the paradigm whereby dose selection decisions for antimicrobial agents could be supported on the basis of PK–PD (14). Using mathematical modeling and Monte Carlo simulation, the following variables are considered to identify doses for further clinical study: (*i*) the variability in the in vitro activity of an antimicrobial agent against clinical isolates, (*ii*) the variability in antimicrobial exposure, (*iii*) the PK–PD target measures developed from animal

infection models, and (*iv*) the protein-binding characteristics of the investigational agent. Given that data for the latter two variables are based on animal infection models and in vitro determinations, respectively, the potential for the influence of regional differences in these data is minimal. However, the potential for tremendous differences in regional susceptibility patterns for an investigational agent has been demonstrated by the large surveillance systems, which collect and test the susceptibility of isolates collected globally (19–21).

For example, the rate of antimicrobial resistance for microorganisms such as *Pseudomonas aeruginosa* has been shown to vary tremendously by different regions of the world. In Latin America, the rate of resistance to many agents is higher than in other regions of the world, including the United States. Using meropenem as an example and examining data from the SENTRY Antimicrobial Surveillance Program between 1997 and 2000, the rate of resistance of *P. aeruginosa* to this agent was 15% in Latin America whereas in North America, the rate was only 4.8% (19). Given such differences in MIC distributions, the proportion of clinical successes across the MIC distribution would differ from region to region. Thus, the overall performance of an individual agent has the potential to differ immensely from region to region.

As described above, the variability in antimicrobial exposure among populations will be directly related to how an individual agent is handled and the potential for impact of intrinsic factors on exposure including body size, degree of kidney and liver function, and genetic polymorphisms of drug metabolism. Nonclinical and clinical PK studies have served to demonstrate the impact of body size on different physiologic functions related to drug elimination, including liver blood flow, metabolic rate, and glomerular filtration rate (22–27). For drugs primarily eliminated by glomerular filtration rate, evaluations for a number of drugs across therapeutic classes have demonstrated a significant relationship between body size and drug clearance. For example, 57% of the variability in amikacin clearance can be explained by variability in estimated creatinine clearance, for which body weight is an input (28). Not surprisingly, amikacin is also an example of an agent for which dosing recommendations are based on body weight and creatinine clearance.

Since drug clearance is a function of both rate of elimination and volume of distribution, it is important to understand the impact of body size on both the clearance and the volume of distribution of an investigational agent. In a review by Green and Duffull (29), studies among obese and nonobese subjects, which assessed the quantitative relationship between drug clearance or volume and various descriptors of body size were evaluated. These authors found that few studies evaluated the relationship between body size and either body clearance or volume of distribution in a quantitative manner. Despite the lack of quantitative studies, most studies found total body weight to be the best descriptor of volume and lean body mass to be the best descriptor of clearance. Even when a statistically significant relationship is found between body size and clearance or volume, however, it is important to evaluate the magnitude of impact of such a covariate on exposure. In the case of oritavancin, an investigational glycopeptide, body weight was found to be a significant covariate of clearance and central volume. However, with the exception of patients in the highest weight category (≥ 140 kg), the actual magnitude of impact on the maximum concentration (Cmax) or daily AUC value was generally modest (an absolute difference of 10–20%) for patients in weight groups above or below 70 to 90 kg (30). A subsequent exposure–response analysis of efficacy in patients with bacteremia failed to reveal a relationship between weight (as represented by ideal body weight) and microbiological response (31).

Examination of body weight of subjects across Japan versus the US has demonstrated lower measures of body mass index (BMI) and a lower prevalence of obesity among subjects from Japan (32) compared to those from the US (33). Additionally, body compositions studies have shown similar such patterns for BMI values between Caucasians and Asian populations. These studies have also served to show that the relationship between BMI and percent body fat is influenced by age, sex, and ethnic group (34). Such observations may be explained by differences in frame size and relative leg length, physical activity, and/or dietary practices among different ethnic groups. For example, the protein source in Japanese diets has traditionally been fish and vegetables, with minimum consumption of animal fat (35). However, as Western influences have become more prevalent in Japan and Asia and as urbanization has increased and transaction costs have decreased, there has been a growing intake of meat among individuals in these populations (36,37). As has been seen in Caucasian populations, the pattern of BMI and body fat is thus expected to increase more rapidly from the second decade of life to middle age among future Japanese and Asian populations.

When one considers the number of drugs that have been approved in Japan at lower doses than in Western countries (38), concern for safety events at higher exposures for a given dosage regimen certainly appears to have been an important consideration in regions where subjects weigh less. For example, the approved range of daily doses for imipenem/cilastatin is 500 to 1000 mg in Japan, whereas it is 1000 to 4000 mg in the US and EU. Since body weight has been shown to significantly affect drug clearance for certain agents, the evaluation of differences in body size may be an important consideration for successfully implementing a bridging strategy.

Genetic factors such as variations in drug targets, drug transporters, and drug-metabolizing enzymes also have the potential to influence PK variability (9,39–41). With respect to drug metabolism, genetic polymorphisms of cytochrome (CYP) P450 enzymes can represent a very important source of interindividual PK variability of drugs. The majority of drugs are metabolized by CYP3A4 and CYP2D6, with CYP2C9 and 2C19 also representing important isoenzymes. These enzymes serve to catalyze the Phase 1 biotransformation of many endogenous substrates and drugs. Phase 2 biotransformation reactions, the enzymes for which are also subject to genetic polymorphisms, include glucuronidation, sulfation, acetylation, methylation, conjugation with glutathione, and conjugation with amino acids. Figure 2 illustrates the relative percentages of Phase 1 and Phase 2 enzymes associated with metabolism of drugs. The contribution of each enzyme to drug metabolism is represented by the relative section size of the corresponding pie chart (41).

Differences in metabolic activity among ethnic groups have been observed for many different classes of drugs (5,6). Although many enzymes manifest ethnic variability, the most detailed information available relates to the differences between Asian and Caucasian populations in CYPs such as CYP2D6, CYP2C, CYP3A, CYP1A2, and CYP2A6 (6). For CYP2D6, the gene encoding for which is highly polymorphic, at least 70 alleles have been identified. Thus, the potential for interindividual variability in the rate of CYP2D6-mediated metabolism of many drugs is high (5). For example, the Cmax and AUC of nortriptyline, which is substrate of CYP2D6, were reported to be 23% and 58% higher in Japanese after half the dose administered to American Caucasians (42). With respect to CYP2C

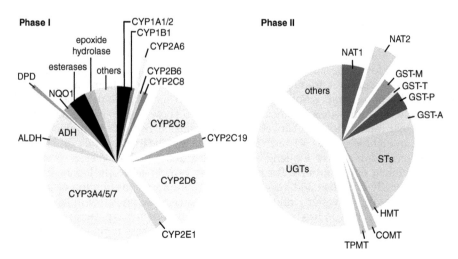

FIGURE 2 Relative quantities of Phase 1 and Phase 2 drug-metabolizing enzymes (41). *Abbreviations*: ADH, alcohol dehydrogenase; ALDH, aldehyde dehydrogenase; CYP, cytochrome P450; DPD, dihydropyrimidine dehydrogenase; NQO1, NADPH: quinone oxidoreductase or DT diaphorase; COMT, catechol *O*-methyltransferase; GST, glutathione *S*-transferase; HMT, histamine methyltransferase; NAT, *N*-acetyltransferase; STs, sulfotransferases; TPMT, thiopurine methyltransferase; UGTs, uridine 5′-triphosphate glucuronosyltransferases.

enzymes, all members of this subfamily exhibit genetic polymorphism. An example of this variability was seen when the PKs of omeprazole were compared between Chinese and American-Caucasian subjects. Mean values for AUC from zero to infinity were 70% higher in Chinese extensive metabolizers than in American Caucasian extensive metabolizers (43).

In the case of erythromycin, which is a substrate of CYP3A, CYP1A2, and CYP2A6, differences in AUC and Cmax have been found to be 32% and 48% higher, respectively, in Koreans than in Caucasians (44). Isoniazid is another example of an agent with demonstrated ethnic sensitivity. Isoniazid is primarily metabolized by arylamine *N*-acetyltransferase-2 (NAT2). Single nucleotide polymorphisms (SNPs) of the *NAT2* gene result in two human phenotypes (45), fast acetylators who have an isoniazid serum half-life ($t_{1/2}$) of 0.9 to 1.8 hours, and slow acetylators who have a $t_{1/2}$ of 2.2 to 4.4 hours (7). Due to varying distributions of NAT2 SNPs, populations will exhibit different proportions of fast and slow acetylators. For example, the proportion of slow acetylators in subjects in Shanghai, China has been reported to be 12% (8). In contrast, 67% of subjects from the US were found to be slow acetylators (7). Such variability in metabolic activity has the potential to impact the magnitude of exposure for a given dosage regimen, thereby impacting the probability of both the response and the development of bacterial resistance for antimicrobial agents such as erythromycin and isoniazid.

If a drug is known to be prone to such differences in PKs due to genetic polymorphisms of CYP P450, subjects should be screened and randomized to ensure homogeneity among populations for studies undertaken. Study designs that account for these differences will ultimately reduce the confounding due to differences in the distribution of response.

APPLICATION OF PKS-PDS AND BRIDGING STUDIES

Development of the ICH E5 guidelines, which began in 1992, proved to be among the most challenging, as evidenced by the extensive discussions that ensued. This guideline was finally adopted for implementation in March 1998 (2). In August 1998, the guideline received official notification in Japan (46). Since this time, bridging of data appears to have been successfully carried out in Japan. A review of 26 new drug applications (NDAs) approved in Japan from 1998 to 2003 demonstrated that drug approvals based on bridging strategies increased from 3.2% in 1999 to 25% in 2003 (47). Of note, 12 of these NDAs were approved with mandatory postmarketing requirements for collecting additional safety information. Given that the execution of a bridging strategy is associated with the examination of fewer patients in the new region compared to those examined in the original clinical development program, the requirement for additional safety data should be expected. It is of additional interest to note that for 7 of the 26 NDAs, major differences in labeling recommendations for dosage and administration were observed between Japan and the US and/or the EU at the time of approval in Japan. In the majority of these cases, these differences were due to higher exposures and/or occurrences of safety events at the same dose in Japanese versus US or EU populations.

In addition to the above-described experiences in Japan, a bridging strategy was recently described for linezolid, an oxazolidinone, in order to extrapolate data from the US and EU to Japan and Asia (48). Linezolid was approved in the US in 2000 for the treatment of infections associated with vancomycin-resistant *Enterococcus faecium* (VREF), including cases with bloodstream infection, and for the treatment of nosocomial pneumonia and uncomplicated and complicated skin and skin structure infections, including cases due to methicillin-resistant *Staphylococcus aureus* (MRSA). In order to gain regulatory approval in Japan and Asia, a bridging strategy based on four underlying hypotheses was executed (48). As shown in Figure 3, the latter three hypotheses represented a stepwise process upon which supporting data were assembled.

The four hypotheses were as follows: (*i*) linezolid susceptibility patterns of clinical isolates (including MRSA, methicillin-resistant *Streptococcus epidermidis* (MRSE), penicillin-resistant *Streptococcus pneumoniae* (PRSP), and VREF from Japan

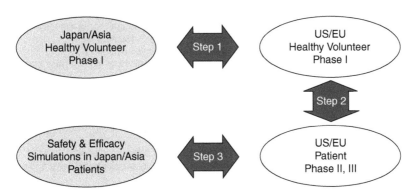

FIGURE 3   Schematic representation of linezolid bridging strategy. *Source*: From Ref. 48.

and Asia are similar to those from the US and EU; (*ii*) the PKs and occurrence of safety events of linezolid in Japanese healthy volunteers are similar to those seen in healthy volunteers from the US and EU (Step 1); (*iii*) the PKs and safety of linezolid in US and EU healthy volunteers are similar to those in US and EU patient populations with skin and soft tissue infections, pneumonia, and bacteremia (Step 2); and (*iv*) using simulation, that safety and efficacy of a given regimen in Japanese patients can be predicted using PK–PD relationships based on data from global Phase 2 and 3 studies (Step 3).

To address the hypothesis of similar susceptibility patterns by region, data from the Zyvox® Antimicrobial Potency Study Program, which consisted of clinical isolates collected before drug approval in the US, served to demonstrate that linezolid susceptibility patterns of staphylococci, *S. pneumoniae*, and enterococci (including MRSA, MRSE, PRSP, and VREF) from Japan and the Asia Pacific were similar to those from the US and EU (49,50). With respect to the second hypothesis, data from five Phase 1 studies (two conducted in Japan and three conducted in the US) in healthy Japanese (*n* = 47) and Caucasian (*n* = 57) males who received doses of linezolid ranging from 125 to 625 mg served to demonstrated similar PKs. Despite a higher mean weight-adjusted clearance in Caucasian subjects than Japanese subjects (1.38 mL/min/kg vs. 1.14 mL/min/kg, $P = 0.02$), the distributions for both groups overlapped (51). As shown in Table 1, this similarity would have been predicted based on properties of linezolid, which suggested minimal potential for ethnic sensitivity (48).

To assess the occurrence of safety events versus exposure (represented by cumulative dose) in Japanese and Caucasian subjects, laboratory data from healthy subjects enrolled in six multidose Phase 1 studies (two conducted in Japan, two conducted in the UK, and two conducted in the US) were evaluated (52). Absolute values and change from baseline in hematologic laboratory values including

TABLE 1 Linezolid Properties as Related to Properties Identified by ICH E5 Bridging Characteristics

Properties suggesting minimal ethnic sensitivity[a]	Linezolid properties[b]
Linear pharmacokinetics	At the recommended dosage, deviation from linearity is about 20%, which is not considered clinically relevant
Flat effect–concentration curve	Bridging analysis demonstrates lack of effect of exposure on safety laboratory parameters
Minimal metabolism or metabolism distributed among multiple pathways	Neither a substrate nor an inhibitor of the major human cytochrome P-450 isomers; nonenzymatic chemical oxidation
High bioavailability	Hundred percent bioavailability; no first-pass effect; no significant food effect
Low potential for protein binding	Protein binding, 31%
Little potential for drug–drug, drug–diet and drug–disease interactions	No cytochrome P-450–mediated interactions; mild, reversible inhibitor of MAO-A and MAO-B (This is well defined in terms of interacting food and drugs)
Nonsystemic mode of action	Yes-inhibits bacterial protein synthesis
Little potential for inappropriate use	Antibiotics are generally considered to have a low abuse potential

[a]As described in Ref. 2.
[b]Reproduced from Ref. 48.

hemoglobin, red blood cell, white blood cell, platelet and absolute neutrophil counts, as well as hepatic laboratory values including alanine and aspartate amino transferases concentrations were assessed in 65 Japanese and 56 Caucasian subjects. Although scatterplots of the absolute value of the safety end point and the end of treatment change from baseline value versus cumulative dose failed to reveal any clinically significant relationships, trends for mild decreases in platelet values with increasing exposure were apparent in both Japanese and Caucasian groups. These results served to validate the second hypothesis and complete Step 1 of the bridging strategy (Fig. 2).

To test the third hypothesis, PK and safety data from US and EU healthy volunteers from Phase 1 studies were compared to that of US and EU patients from Phase 2 and 3 studies with skin and soft tissue infections, pneumonia, and bacteremia (53). Using population PK analysis, sparse samples from Phase 2 and 3 data were best fit by a one-compartment model with first-order absorption and parallel first-order and Michaelis–Menten elimination. Bayesian estimates of PK parameters were compared to those from Phase 1 studies. The above-described difference of 20% in mean weight-adjusted clearance in Caucasian subjects and Japanese subjects (1.38 mL/min/kg vs. 1.14 mL/min/kg, $P = 0.02$) was within the range of clearance values observed for US/EU patients in Phase 2 and 3 studies. Additionally, the PK–PD analyses, which evaluated the relationship between 24-hour AUC values at steady state and the above-described hematologic or hepatic laboratory values failed to reveal any clinically significant differences between US and EU subjects from Phase 1 studies and US and EU patients from Phase 2 studies (Step 2) (48,53).

The final hypothesis was based on the assumption that safety and efficacy of a given regimen in Japanese patients could be predicted using PK–PD relationships based on data from global Phase 2 and 3 studies. Using a previously described PK–PD model for efficacy (54), simulations were conducted to evaluate the probability of achieving clinical success, given different dosing regimens and different point estimates for weight. As described by Cirincione et al., logistic regression was used to describe the relationship between the probability of clinical response and AUC: MIC ratios among 73 patients with gram-positive bacteremia who were evaluable for clinical response and from whom plasma concentrations were collected (54,55). Based on this model, the probability of clinical success increased with increasing AUC:MIC ratios ($P = 0.05$). An AUC:MIC ratio of 115 was associated with an 80% probability of clinical success (56), which is similar to the nature of the PK–PD relationships identified based on animal data (57) and clinical data from seriously ill bacteremic patients (58). Using this relationship, the population PK model, and Monte Carlo simulation (5000 patients), the probability of clinical success was assessed for patients weighing 65 or 82.2 kg who had received doses of linezolid of 500 or 600 mg twice daily (Table 2) (53).

For every 100 patients weighing 65 kg, the administration of the 600 mg rather than 500 mg twice daily regimen was associated with a successful clinical response in four to five additional patients. Based on the 95% confidence interval around the probability of response associated with each dose, this difference was statistically significant ($P < 0.05$). Thus, the simulations conducted in this final step, together with the demonstration of similar safety events at similar exposures based on data from Steps 1 and 2, supported the use of linezolid 600 mg twice daily in patients from Japan/Asia (Step 3), the same regimen that had been approved in the US.

TABLE 2 Percent Probability of Clinical Success for Patients Receiving Linezolid 500 and 600 mg Twice Daily, Stratified by Weight

Weight (kg)	Dose (mg)	Percent probability of clinical success	95% Confidence interval
82.2	500	70.0	69.8, 70.2
	600	74.1	73.8, 74.3
65	500	71.6	71.4, 71.8
	600	76.0	75.8, 76.3

Source: From Ref. 53.

Using the above-described strategy, data from an existing clinical data package together with Japanese PK bridging studies and PK–PD principles were successfully used to support the registration of linezolid in Japan. Approval was received from the Ministry of Health, Labour and Welfare in Japan in 2001 for the treatment of infections associated with vancomycin-resistant enterococcus. In 2006, based on these data and clinical data from Japanese patients, linezolid was approved in Japan for the treatment of infections associated with MRSA.

In the above-described bridging strategy for linezolid, simulations to assess adequacy of different dosage regimens were based on a prior PK–PD relationship for efficacy. Given that such relationships can be derived from in vitro or animal models of infection in the absence of clinical data and that these relationships are concordant with those from clinical data, the availability of a PK–PD relationship for efficacy can usually be anticipated for antimicrobial agents. Although not relevant in the above-described example, when apparent, PK–PD relationships for safety must also be considered. Using Monte Carlo simulation, different dosage regimens may be evaluated on the basis of the number of simulated patients who achieve optimal exposures which are defined as that which gives the largest difference between the probability of efficacy and a safety event (i.e., which maximizes the probability of efficacy and minimizes the probability of a safety event) or that which maximizes the probability of efficacy with an acceptable probability of a safety event.

FUTURE DIRECTIONS

The ICH E5 guidelines (2) were constructed to facilitate drug registration across the different ICH regions by recommending a framework for evaluating the impact of ethnic factors on safety and efficacy at a particular dosage and dose regimen, and, through appropriate evaluation of such factors, minimize unnecessary duplication of clinical studies and ultimately expedite the drug approval process in a new region. However, given the general nature of these guidelines, there is a potential for inconsistent interpretation in different regions, thereby defeating the original goals of the guidelines. Additionally, as evidenced by the results of survey conducted by the CMR International Institute for Regulatory Science among 13 pharmaceutical companies (59), Asian authorities outside of the three ICH regions (Japan, EU, and US) have been slow to officially implement the E5 guidelines. The companies surveyed identified the need for both a wider implementation of the guidelines to facilitate the acceptance of foreign clinical data and the need for a better understanding of the scientific basis for accepting foreign clinical data. Given that pharmaceutical companies are becoming more active in emerging Asian markets, harmonization of regulatory requirements and acceptance and implementation of the ICH E5 guidelines will be crucial to integrating Asian countries into global drug development plans and developing efficient bridging strategies. In addition, standardization of regulatory and good clinical practices in these countries will improve the feasibility of including significant numbers of patients from such countries in large international clinical trials, thereby allowing for valuable clinical experience to be gained in the countries where the new drug will be marketed.

CONCLUSION

One of the major factors in the current reluctance of large pharmaceutical companies to pursue development of antimicrobial agents has been the enormous resources required to meet regulatory standards versus the risk of failure and the current perception of regulatory uncertainty. As this reluctance to develop new antimicrobials comes at a time when rates of antimicrobial resistance are on the rise across the globe, the time may not be too far off when a crisis point will be reached in the treatment of bacterial infections.

In this review, great emphasis has been placed on the importance of intersubject variability in PKs, the potential for intrinsic and extrinsic factors to explain this variability, and the value of PK and PK–PD bridging studies. The use of PK–PD principles, both in the early development of an antimicrobial agent and in bridging strategies in accordance with ICH E5 guidelines (2), has the potential to help streamline the development process and improve the likelihood of regulatory success, making global drug development more economically viable and thereby ensuring that patients across multiple regions will have access to the best antimicrobial agents available with minimum delay.

REFERENCES

1. International Conference on Harmonisation of Technical Requirements for Registration of Pharmaceuticals for Human Use. (available at http://www.ich.org/cache/compo/276–259.1. html.) (Accessed on February 12, 2007)
2. ICH Harmonised Tripartite Guideline. Ethnic factors in the acceptability of foreign clinical data E5 (R1). Current Step 4 version. 5 February 1998. http://www.ich.org/LOB/media/MEDIA481.pdf. (Accessed February 12, 2007)
3. Lesko LJ, Atkinson AJ Jr. Use of biomarkers and surrogate endpoints in drug development and regulatory decision-making: criteria, validation, strategies. Annu Rev Pharmacol Toxicol 2001; 41:347–366.
4. Biomarkers Definitions Working Group. Biomarkers and surrogate endpoints: preferred definitions and conceptual framework. Clin Pharmacol Ther 2001; 69:89–95.
5. Bjornsson TD, Wagner JA, Donahue SR, et al. A review and assessment of potential sources of ethnic differences in drug responsiveness. J Clin Pharmacol 2003; 43:943–967.
6. Kim K, Johnson JA, Derendorf H. Differences in drug pharmacokinetics between East Asians and Caucasians and the role of genetic polymorphisms. J Clin Pharmacol 2004; 44:1083–1105.
7. Peloquin CA, Jaresko GS, Yong CL, Keung AC, Bulpitt AE, Jelliffe RW. Population pharmacokinetic modeling of isoniazid, rifampin, and pyrazinamide. Antimicrob Agents Chemother 1997; 41:2670–2679.
8. Ma QW, Lin GF, Chen JG, et al. Polymorphism of *N*-acetyltransferase 2 (*NAT2*) gene polymorphism in Shanghai population: occupational and non-occupational bladder cancer patient groups. Biomed Environ Sci 2004; 17:291–298.
9. Shah RR. Pharmacogenetics in drug regulation: promise, potential and pitfalls. Phil Trans R Soc B 2005; 360:1617–1638.

10. Nilchaikovit T, Hill JM, Holland JC. The effects of culture on illness behavior and medical care. Gen Hosp Psychiatry 1993; 15:41–50.
11. Craig WA. Pharmacokinetic/pharmacodynamic parameters: rationale for antibacterial dosing in mice and men. Clin Infect Dis 1998; 26:1–12.
12. Craig WA. Pharmacodynamics of antimicrobials: general concepts and applications. In: Nightingale CH, Murakawa T, Ambrose PG, eds. Antimicrobial Pharmacodynamics in Theory and Clinical Practice. New York: Marcel Dekker, Inc., 2002:1–22.
13. Ambrose PG, Bhavnani SM, Rubino CM, et al. Pharmacokinetics-pharmacodynamics of antimicrobial therapy: it's not just for mice anymore. Clin Infect Dis 2007; 44:79–86.
14. Drusano GL, Preston SL, Hardalo C, et al. Use of preclinical data for selection of a phase II/III dose for evernimicin and identification of a preclinical MIC breakpoint. Antimicrob Agents Chemother 2001; 45:13–22.
15. Bhavnani SM, Hammel JP, Cirincione BB, Wikler MA, Ambrose PG. Use of pharmacokinetic-pharmacodynamic target attainment analyses to support phase 2 and 3 dosing strategies for doripenem. Antimicrob Agents Chemother 2005; 49:3944–3947.
16. Lazor JA. Dose selection in antimicrobial drug development: incorporation of pharmacokinetics and pharmacodynamics. IDSA/ISAP/FDA Workshop, April 16, 2004. (Available at http://www.fda.gov/cder/drug/antimicrobial/FDAIDSAISAPPresentations/Lazor2.ppt. Accessed February 12, 2007)
17. Bonapace CR. In vitro/animal models to support dosage selection: FDA perspective. IDSA/ISAP/FDA Workshop, April 16, 2004. (Available at: http://www.fda.gov/cder/drug/antimicrobial/FDAIDSAISAPPresentations/Chuck%20Bonapace.ppt. Accessed February 12, 2007)
18. Lesko LJ. Proposal for end-of-phase 2A (EOP2A) meetings: Advisory Committee for Pharmaceutical Sciences; Clinical Pharmacology Subcommittee. November 17–18, 2003. (Available at: http://www.fda.gov/ohrms/dockets/ac/03/slides/3998S1_02_Lesko.ppt Accessed February 12, 2007)
19. Jones RN, Kirby JT, Beach ML, Biedenbach DJ, Pfaller MA. Geographic variations in activity of broad-spectrum β-lactams against *Pseudomonas aeruginosa*: summary of the worldwide SENTRY Antimicrobial Surveillance Program (1997–2000). Diag Microb Infect Dis 2002; 43:239–243.
20. Moet GJ, Jones RN, Biedenbach DJ, Stilwell MG, Fritsche TR. Contemporary causes of skin and soft tissue infections in North America, Latin America, and Europe: report from the SENTRY Antimicrobial Surveillance Program (1998-2004). Diag Microb Infect Dis 2007; 57: 7–13.
21. Johnson DM, Stilwell MG, Fritsche TR, Jones RN. Emergence of multidrug-resistant *Streptococcus pneumoniae*: report from the SENTRY Antimicrobial Surveillance Program (1999–2003). Diag Microb Infect Dis 2006; 56:69–74.
22. Boxenbaum H. Interspecies variation in liver weight, hepatic blood flow, and antipyrine intrinsic clearance: extrapolation of data to benzodiazepines and phenytoin. J Pharmacokinet Biopharm 1980; 8:165–176.
23. Wilkinson GR, Shand DG. A physiological approach to hepatic drug clearance. Clin Pharmacol Ther 1980; 18:377–390.
24. Abernethy DR, Greenblatt DJ, Divoll M, Shader RI. Enhanced glucuronide conjugation of drugs in obesity: studies of lorazepam, oxazepam, and acetaminophen. J Lab Clin Med 1983; 101:873–880.
25. Adolph EF. Quantitative relations in the physiological constitutions of mammals. Science 1949; 109:579–585.
26. Stokholm KH, Brochner-Mortensen J, Hoilund-Carlsen PF. Increased glomerular filtration rate and adrenocortical function in obese women. Int J Obes 1980; 4:57–63.
27. Sawyer M, Ratain MJ. Body surface area as a determinant of pharmacokinetics and drug dosing. Invest New Drugs 2001; 19:171–177.
28. Tod M, Lortholary O, Seytre D, et al. Population pharmacokinetic study of amikacin administered once or twice daily to febrile, severely neutropenic adults. Antimicrob Agents Chemother 1998; 42:849–856.
29. Green B, Duffull SB. What is the best size descriptor to use for pharmacokinetic studies in the obese? Brit J Clin Pharmacol 2004; 58:119–133.

30. Owen JS, Bhavnani SM, Fiedler-Kelly J, Loutit JS, Porter SB, Phillips L. Population pharmaco-kinetics of oritavancin. 44th Annual Interscience Conference on Antimicrobial Agents and Chemotherapy, Washington, DC, Oct 30–Nov 2, 2004. (Abstract number A-20).

31. Bhavnani SM, Passarell JA, Owen JS, Loutit JS, Porter SB, Ambrose PG. Pharmacokinetic-pharmacodynamic relationships describing the efficacy of oritavancin in patients with *Staphylococcus aureus* bacteremia. Antimicrob Agents Chemother 2006; 50:994–1000.

32. Yoshiike N, Seino F, Tajima S, et al. Twenty-year changes in the prevalence of overweight in Japanese adults: The National Nutrition Survey 1976-95. Obes Rev 2002; 3:183–190.

33. Flegal KM, Carroll MD, Kuczmarski RJ, Johnson CL. Overweight and obesity in the United States: prevalence and trends, 1960-1994. Int J Obes Relat Metab Disord 1998; 22:39–47.

34. Deurenberg P, Deurenberg-Yap M, Guricci S. Asians are different from Caucasians and from each other in their body mass index/body fat per cent relationship. Obes Rev 2002; 3:141–146.

35. Health Service Bureau, Ministry of Health and Welfare. National Nutrition Survey of Japan. Annual Report, 1996. Daiichi Shuppan: Tokyo, 1996.

36. Chern WS, Ishibashi K, Taniguchi K, Tokoyama Y. Analysis of the food consumption of Japanese households. FAO Economic and Social Development Paper No. 152. Food and Agriculture Organization of the United Nations. Rome, 2003. (ftp://ftp.fao.org/docrep/fao/005/y4475E/y4475E00.pdf. Accessed February 12, 2007)

37. Huang J, Bouis H. Structural changes in the demand for food in Asia. International Food Policy Research Institute. A 2020 Vision for Food, Agriculture, and the Environment. (Brief No. 41, December 1996.)

38. Naito C. Significance of dose setting studies in the bridging study strategy. Symposium on Global Drug Development Techniques, Kitasato University-Harvard School of Public Health, Tokyo, October 5–6, 2000. (http://www.pharm.kitasato-u.ac.jp/K-H_sympo2000/naito.html. Accessed February 12, 2007)

39. Guttendorf RJ, Wedlund PJ. Genetic aspects of drugs disposition and therapeutics. J Clin Pharmacol 1992; 32:107–117.

40. Daly AK. Pharmacogenetics of the major polymorphic metabolizing enzymes. Fundam Clin Pharmacol 2003; 17:27–41.

41. Evans WE, Relling MV. Pharmacogenomics: translating functional genomics into rational therapeutics. Science 1999; 286:487–491.

42. Kishimoto A, Hollister LE. Nortriptyline kinetics in Japanese and Americans. J Clin Psychopharm 1984; 4:171–172.

43. Caraco Y, Lagerstrom PO, Wood AJJ. Ethnic and genetic determinants of omeprazole disposition and effect. Clin Pharmacol Ther 1996; 60:157–167.

44. Yu KS, Cho JY, Shon JH, et al. Ethnic differences and relationships in the oral pharmacokinetics of nifedipine and erythromycin. Clin Pharmacol Ther 2001; 70:228–236.

45. Evans DA, Manley KA, McKusick VA. Genetic control of isoniazid metabolism in man. Brit Med J 1960; 2:485–491.

46. Pharmaceuticals Medical Safety Bureau, Ministry of Health, Labour and Welfare, 1998. Notification No. 739 (Kyokuchou-tsuuchi) and No. 672 (kachou-tsuuchi), August 11. Japan.

47. Uyama Y, Shibata T, Nagai N, Hanaoka H, Toyoshima S, Mori K. Successful bridging strategy based on ICH E5 guideline for drugs approved in Japan. Clin Pharmacol Ther 2005; 78:102–113.

48. Bergstrom T, Wong E, Antal E, Grasela T. Integration of pharmaceutical product development in Asia/Japan into the global program. Drug Information Association Meeting, Hong Kong Academy of Medicine, Hong Kong, China, Nov 16–19, 2000. (Poster number P-15)

49. Bell JM, Turnidge JD, Ballow CH, Jones RN; The ZAPS Regional Participants. Multicentre evaluation of the in vitro activity of linezolid in the Western Pacific. J Antimicrob Chemother 2003; 51:339–345.

50. Ballow CH, Jones RN, Biedenbach DJ; The North American ZAPS Research Group. A multicenter evaluation of linezolid antimicrobial activity in North America. Diag Microbial Infect Dis 2002; 43:75–83.

51. Cirincione B, Chiba K, Stalker D, et al. Comparison of linezolid pharmacokinetics from phase 1 studies of Japanese and Caucasians. 103rd Annual Meeting of the American Society for Clinical Pharmacology and Therapeutics, Atlanta, GA, Mar 24–27, 2002. (Poster number WPIII-84).

52. Cirincione B, Lobek F, Chiba K, et al. Comparison of laboratory safety data from Japanese and non-Japanese in linezolid phase 1 studies. 103rd Annual Meeting of the American Society for Clinical Pharmacology and Therapeutics, Atlanta, GA, Mar 24–27, 2002. (Poster number WPIII-33)

53. Antal E, Grasela T, Bergstrom T, Bruss J, Wong E. The role of population PK/PD analysis during the implementation of a bridging strategy for linezolid. Drug Information Association Meeting, Hong Kong Academy of Medicine, Hong Kong, China. Nov 16–19, 2000. (Poster number P-16).

54. Cirincione B, Grasela T, Sardella S, et al. Population pharmacodynamic assessment of linezolid efficacy in community-acquired pneumonia (CAP), skin and soft tissue (SST) infections and bacteremia (BAC). 40th Annual Meeting of Interscience Conference on Antimicrobial Agents and Chemotherapy, Toronto, Canada, Sep 17–20, 2000. (Abstract no. 1389).

55. Cirincione B, Grasela T, Caito K, Ludwig E, Antal E, Hafkin B, Bruss J. Influence of dose size on dose response relationships for linezolid efficacy using Monte Carlo simulations. American Association of Pharmaceutical Scientists 2000 Annual Meeting, Indianapolis, Indiana, Oct 29–Nov 2, 2000. (Presentation no. 2324)

56. Bullingham RES, Grasela T, Stalker D. Innovation in global development and registration. 40th Annual Drug Information Association Meeting, Innovations in Global Development and Registration, Washington, DC, June 13–17, 2004.

57. Andes D, van Ogtrop ML, Peng J, Craig WA. In vivo pharmacodynamics of a new oxazolidinone (linezolid). Antimicrob Agents Chemother 2002; 46:3484–3489.

58. Rayner CR, Forrest A, Meagher AK, Birmingham MC, Schentag JJ. Clinical pharmacodynamics of linezolid in seriously ill patients treated in a compassionate use programme. Clin Pharmacokinet 2003; 42:1411–1423.

59. Anderson C, Cone M, McAuslane N, Walker S. Bridging studies in Asia and the impact of the ICH E5 Guideline. Drug Information J 2003; 37:107S–116S.

22 Modeling of Toxicities Due to Antibiotics

Alan Forrest
Ordway Research Institute, Albany, New York and University at Buffalo School of Pharmacy, Buffalo, New York, U.S.A.

George L. Drusano
Ordway Research Institute, Albany, New York, U.S.A.

In order to provide optimal therapy for patients, a drug of choice and its accompanying dose and schedule must accomplish two ends. First and foremost, it must be efficacious. Secondly, it should optimally be nontoxic. While there are multiple types of toxicities, the kind that is subject to intervention is that where the likelihood of toxicity is related to the concentration of the drug. In this chapter, we will examine three different examples of toxicity linked to drug exposure and explore the implications for successful drug therapy.

THE AMINOGLYCOSIDES

These agents have fallen out of favor for use, particularly in the intensive care unit (ICU), primarily because of their toxicity profile. Investigators have shown (1) that ICU patients who develop renal failure have an in-hospital mortality that is significantly higher than patients with normal renal function. Consequently, there has developed an understandable reluctance to employ these agents in this setting. Unfortunately, high-level multiclass resistance among nosocomial pathogens is now common and for organisms such as *Pseudomonas aeruginosa* or Acinetobacter species, nephrotoxic agents like the aminoglycosides or polymyxins are often the only active agents. For this reason it is important to understand the determinants of nephrotoxicity and how to approach dose and schedule choices to minimize the probability of the occurrence of this toxicity.

Aminoglycoside nephrotoxicity is centered in the proximal renal tubular epithelial (PRTE) cells (2). These agents bind first to acidic phospholipids, but are rapidly passed to the aminoglycoside receptor, a transmembrane protein (megalin), on the tubular lumen of the PRTE cells and are then internalized. Internalization is by pinocytosis and is a saturable process. The saturability of the process is key to understanding the best way to dose to avoid nephrotoxicity. Once inside the PRTE cell, the drug is in an endosome, which fuses with a lysosome. Here, it binds to and inhibits phosphatidyl inositol phosolipases A1 and A2 and C. The inhibition of these enzymes results in the formation of myeloid bodies, which are associated with injury to the PRTE cell. If the cells die, they become unattached to the basilolateral membrane, causing signaling, resulting in the activation of the tubuloglomerular reflex. This is perceived as a decrease in glomerular filtration rate. Also, higher tubular pressures have been noted, decreasing filtering, so that the decrease in renal function is likely multifactorial. In this cascade, the key issue is the ability to saturate the uptake process.

Keeping drug from getting into the PRTE cell should, in theory, delay the onset of aminoglycoside nephrotoxicity. Since the uptake is a saturable process, the schedule of administration should have a definable impact. If the dose chosen markedly exceeds the Michaelis–Menten constant (K_m) of the uptake process (apparent K_m approximately 15 mg/L in rats) (3), then drug will be able to bypass the PRTE cell and the average daily uptake will decrease. Given the polycationic nature of aminoglycosides in the milieu of the phagolysosome, the charge status will keep the vast majority of the drug from effluxing. Therefore, continued uptake will only add to the concentration, until a threshold level is achieved, which will differ by drug and partly by individual and will ultimately result in nephrotoxicity.

We can markedly diminish the rate of uptake into the PRTE cell by administering large doses (usually 5–7 mg/kg of gentamicin or tobramycin and four times this for amikacin) intermittently (usually once-daily or less frequently). This will provide a broadened window for aminoglycoside administration without nephrotoxicity. If the drug administration continues unabated, it would be likely that a high fraction of the patients would demonstrate aminoglycoside nephrotoxicity, even if administered once-daily. Much of this hypothesis has been elegantly demonstrated by investigators employing animal models and, in one instance, in the clinic (4,5). There have been a large number of attempts at studying this phenomenon. Most have suffered from small size and employing nonoptimal mathematical methods. terBraak et al. did show a significant difference between once-daily and more frequent administration of the same daily aminoglycoside dose (6).

The only randomized, double-blind study of this mode of drug administration also identified a statistically significant impact on the probability of developing altered renal function (7). In this analysis, multivariate logistic regression was employed to examine a number of covariates that might influence the probability of nephrotoxicity. Ultimately, schedule of administration (once- vs. twice-daily), drug exposure (as indexed to daily drug AUC or area under the concentration-time curve), and concomitant administration of vancomycin could all be shown to have a significant impact of the probability of the patient developing nephrotoxicity. The final covariate model is displayed below in Table 1.

Vancomycin use and schedule are dichotomous variables; aminoglycoside AUC is a continuous variable.

Figure 1 demonstrates the impact of these three covariates on the likelihood of nephrotoxicity (7).

TABLE 1 Final Model from Multivariate Logistic Regression Analysis of Factors Affecting the Probability of Aminoglycoside Nephrotoxicity

Covariate	Constant	Estimate	Standard error	Pvalue
–	–	–	–	<0.0001
–	−37.239	–	2.482	–
Vancomycin				
Use−	–	0.0	–	–
Use+	–	3.531	1.411	–
Schedule				
QD	0.0	–	–	–
Q12hr	30.757	–	<0.001	–
AUC	0.049	–	0.026	–

Abbreviation: AUC, area under the concentration-time curve.

A

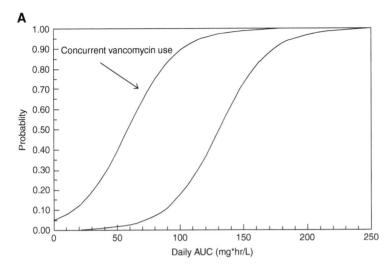

B

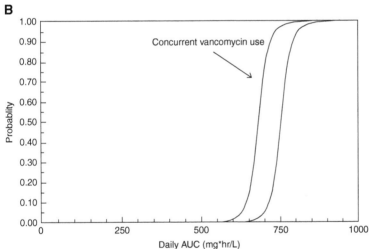

FIGURE 1 (**A**) Curve of probability of development of aminoglycoside nephrotoxicity for patients receiving drugs on a twice-daily basis as estimated by multivariate logistic regression analysis. The probability rises as a function of increasing daily exposure to aminoglycoside, as indexed to the AUC. Concurrent vancomycin use provides a marked increase in the probability of nephrotoxicity for equivalent aminoglycoside exposure. (**B**) Once-daily administration shifts the curves of probability of nephrotoxicity to the right. *Abbreviation*: AUC, area under the concentration-time curve.

It should be noted that once-daily administration markedly shifted the logistic function curves to the right. This is because there were no identified cases of nephrotoxicity in the once-daily group, irrespective of vancomycin use in a subset of these patients.

In this analysis, we also tested the hypothesis that schedule of administration would alter the time-to-nephrotoxicity. This was done by employing a stratified Kaplan-Meier analysis. Schedule was the stratification variable and was highly significant. Of the other covariates evaluated, only concurrent use of vancomycin affected the time-to-event in a Cox proportional hazards model. This is displayed in Figure 2.

Consequently, when one takes both analyses into account, the vast bulk of aminoglycoside nephrotoxicity can be avoided by simply dosing intermittently (once-daily, or less frequently, depending on baseline renal function), stopping therapy in the "protected window" provided by once-daily dosing (stopping after 5–7 days) and avoiding concurrent vancomycin administration.

Avoiding toxicity is critical, but is only one half of optimizing drug therapy. In addition, we must provide adequate antimicrobial therapy. Kashuba et al. (8) examined the relationship between aminoglycoside AUC/minimum inhibitory concentration (MIC) ratio and the likelihood of fever resolution. This is displayed in Figure 3.

If one takes the day 7 relationship (still in the "protected widow"), an AUC/MIC of 156 will provide approximately a 90% probability of fever resolution. For an MIC of 0.5 mg/L, an AUC of 78 is required to hit the target and will generate a probability of nephrotoxicity of <10%, even with twice-daily therapy. On balance, high probability of effect and low probability of toxicity can be provided for such a daily aminoglycoside exposure. On the other hand, if the MIC was 1.0 mg/L, the exposure required would be a daily AUC of 156, which would drive an unacceptable probability of toxicity (Fig. 1A) with twice-daily therapy, but would be quite acceptable for daily (intermittent) therapy (Fig. 1B).

FLUOROQUINOLONES AND HYPERGLYCEMIA

Fluoroquinolones have been extremely valuable additions to the therapeutic armamentarium, particularly for community-acquired lower respiratory tract infections

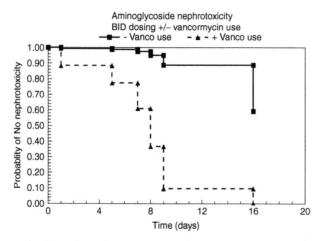

FIGURE 2 Effect of concurrent vancomycin use on the time-to-nephrotoxicity in the group receiving twice-daily aminoglycoside administration. Once-daily administration in 39 patients did not result in any nephrotoxicity and, therefore the graph is not displayed. It should be noted that, in the absence of vancomycin, only about 10% of patients had nephrotoxicity at day 9. Prolonging the duration of therapy to 16 days caused nephrotoxicity in a total of 41% of patients (with vancomycin, this occurred at day 8). Also, with concurrent vancomycin, 90% of patients were nephrotoxic by day 16.

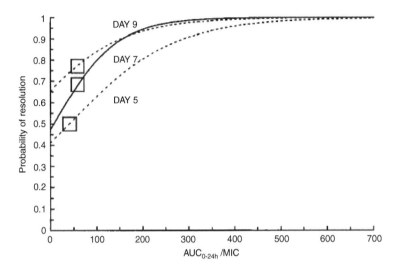

FIGURE 3 Probability of temperature resolution by days 5, 7, and 9 of aminoglycoside therapy as determined by logistic regression analysis. The squares are breakpoint values as determined by Classification and Regression Tree analysis. Drug was administered Q 12 hours and Q 8 hours. *Abbreviations*: AUC, area under the concentration-time curve; MIC, minimum inhibitory concentration.

(LRTIs). Unfortunately, many potent fluoroquinolones have failed due to unexpected toxicities, such as temafloxacin and trovafloxacin, among others. Most of the time, drug withdrawal was due to relatively infrequent (circa 1/6000 patients), but very serious adverse drug reactions. It is unlikely that these reactions were linked to the magnitude of drug exposure.

Gatifloxacin, on the other hand, had reports of both hypo- and hyperglycemia. Here, the concentration will be on the hyperglycemia patients, where this adverse reaction was concentrated mostly in the elderly (>60 years of age) population. Given that renal clearance accounts for a considerable portion of total gatifloxacin clearance, it is a reasonable hypothesis that the likelihood of this adverse event was influenced by the degree of drug exposure.

Ambrose et al. (9) employed a validated population pharmacokinetic model derived from patients being treated for respiratory tract infections to explore this problem. They operated under the hypothesis that total drug exposure, as indexed to AUC, influenced the likelihood of a hyperglycemic episode, particularly in the elderly population. They employed the pharmacokinetic model to estimate the range of gatifloxacin AUC values in 10 patients with a reported (Food and Drug Administration MedWatch) hyperglycemic event. This exposure range for these 10 patients was 57 to 100 mg*hr/L. They then employed Monte Carlo simulation using the parameter vector and covariance matrix from the population model to generate exposure (AUC) distributions for patients with ages ranging from 65 to 85 years of age. They then demonstrated the fraction of simulated subjects developing an AUC in excess of a value of either 60 or 70 mg*hr/L when treated with 400 mg (standard dose) or 200 mg (proposed age-reduction dose). The results are displayed in Table 2.

Clearly, dose reduction by half markedly reduces the likelihood of developing a gatifloxacin AUC in the range implicated to cause hyperglycemic events.

TABLE 2 Monte Carlo Simulation Results: Percentage of Patients Attaining a Given AUC_{0-24} Stratified by Age Cohort and Gatifloxacin Dose

	AUC_{0-24}			
	≥60 mg*hr/L		≥70 mg*hr/L	
Age (yr)	200 mg	400 mg	200 mg	400 mg
≥65	3.04	50.8	0.92	35.1
≥70	3.32	56.3	1.2	39.4
≥75	6.70	61.3	1.96	44.6
≥80	7.20	66.1	3.2	50.5
≥85	11.7	73.2	5.48	58.3

Abbreviation: AUC, area under the concentration-time curve.

However, all this is for naught if the reduced dose does not provide adequate coverage for the target pathogens for community-acquired LRTI.

In this analysis, a second Monte Carlo simulation was performed with the same two doses, with the same age stratifications and the ability to attain a therapeutic exposure (AUC_{0-24}/MIC ratio) of 30 determined. This therapeutic target had been previously shown to correlate with a good clinical outcome (10). In the analysis employing the 200 mg dose and determining the expected target attainment over a large (6700 isolate) contemporaneous distribution of *Streptococcus pneumoniae* with known gatifloxacin MIC values, the lowest target attainment seen was 98.9%. Here, again, it is possible to identify dose choices that simultaneously minimize the probability of obtaining a toxic drug exposure while still providing an exposure with a high likelihood of driving a good clinical outcome.

LINEZOLID AND THROMBOCYTOPENIA

With the amazing explosion of community-acquired methicillan-resistant *Staphylococcus aureus* (CA-MRSA) in the United States (11), linezolid has become one of the most valuable agents in our therapeutic armamentarium. One of the more serious toxicities attendant to linezolid administration has been the occurrence of thrombocytopenia, particularly with prolonged administration. Another risk factor is a low starting platelet count.

Such a toxicity is a serious issue and can lead to early drug discontinuation. It is important, therefore, to understand the balance between the effect provided by the drug and the toxicity. This is best accomplished by understanding the pharmacodynamics relationship between exposure and toxicity as well as between exposure and response. In the case of linezolid, it is critical to understand the pharmacokinetics of the agent as a prelude to understanding the effects, both therapeutic and toxicologic.

Meagher et al. (12) studied linezolid pharmacokinetics in a large population of patients ($n = 318$) receiving the drug in a compassionate use protocol. As part of the population pharmacokinetic analysis, an extensive model-building effort was undertaken and the authors were able to demonstrate that there were parallel first order and Michaelis–Menten kinetic clearance pathways present for the drug. The final model is shown in Figure 4.

The importance of the nonlinear clearance term is that it will ensure that, for any fixed dose, there will be a very broad range of linezolid exposures achieved, as

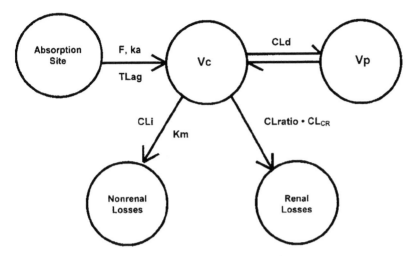

FIGURE 4 Two-compartment model used to fit linezolid oral and i.v. data. *Abbreviations*: Vc, volume of distribution of the central compartment; Ka, absorption rate constant; T_{lag}, lag time before onset of absorption; CLd, distributional clearance; K_m, Michaelis–Menten constant; CLi, intrinsic clearance; F, oral bioavailability; CL ratio*Clcr, renal clearance as a function of creatinine clearance.

indexed to AUC_{0-24}. To emphasize this point, we present Figure 5 the data and predicted concentration–time curves from two patients in this database.

Clearly, both efficacy and toxicity would be expected to be different for patients with such differences in concentration–time profiles.

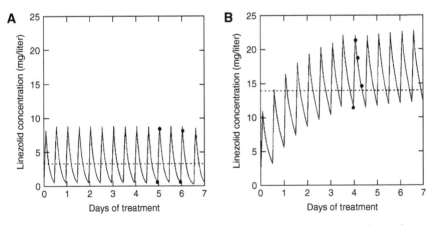

FIGURE 5 Fitted functions in two study patients. Solid circles represent the observed concentrations and the solid line represents the fitted function for two illustrative patients in the population. The dashed lines represent the average concentration during the first seven days of treatment. (**A**) Patient PK parameters typical of those found in this study (CL_i = 36.6 L/hr/65 kg; K_m = 1.8 mg/L; CL_{ratio} 0.38; Cl_{cr} = 82 mL/min; IBW = 75.3 kg). (**B**) Patient PK parameters similar to those found in healthy volunteers (CL_i = 21.3 L/hr/65 kg; K_m = 1.2 mg/L; CL_{ratio} = 0.33; Cl_{cr} = 62 mL/min; IBW = 70.0 kg). *Abbreviations*: CLi, intrinsic clearance; K_m, Michaelis–Menten constant; IBW, ideal body weight.

Rayner et al. (13) were able to demonstrate linkages between linezolid exposures in this patient population and the probability of a good outcome, as well as the time to event for some of these outcomes. Two of the indications bacteremia and LRTI are displayed below in Figure 6.

Given that the MIC distribution for *Staphylococcus aureus* is made up almost entirely of values of 1.0 and 2.0 mg/L, attaining AUC/MIC ratios of about 100

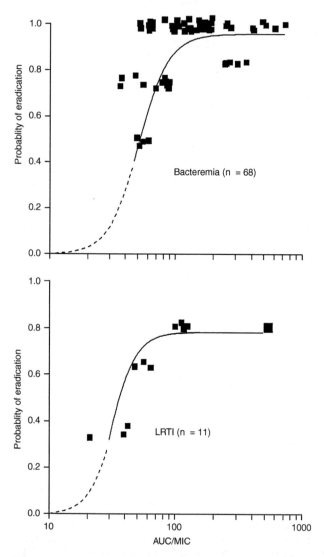

FIGURE 6 Probability of eradication versus AUC_{0-24}/MIC ratio fitted to modified Hill equations. The curve is the fitted relationship; each point represents one patient case at the calculated probability of eradication among three to five surrounding cases. Each patient course is represented once. *Abbreviation*: LRTI, lower respiratory tract infection; AUC, area under the concentration-time curve; MIC, minimum inhibitory concentration.

implies the need for an AUC of about 200 to be obtained. When one examines Figure 5, it is clear that any fixed dose will provide a very broad range of AUC values in the population. When seen against the background of the required AUC/MIC ratio target, it is clear that care must be given to assessing the target attainment rate for any specific dose.

The broad range of exposures also has major implications for occurrence of hematological toxicity. Here, we will concentrate on thrombocytopenia. The overall occurrence of hematological toxicity in the linezolid registrational trials by days of therapy was published by Gerson et al. (14) and is presented in Figure 7.

It is clear that the rate of thrombocytopenia with linezolid separates from the comparator rate after about 14 days of therapy and that this separation is quite different from that seen with anemia or neutrophil changes. The conclusion to be drawn is that the duration of therapy will have an impact on the occurrence of thrombocytopenia. It should also be noted that this analysis likely underestimates the rate of thrombocytopenia after day 14, as the denominator employed was the "n" for the entire group at time zero. If one were to do a risk analysis and employ the "n" of patients remaining on therapy after day 10 to 14, it is highly likely that the rate of thrombocytopenia would be substantially higher.

This still leaves open the question of whether the magnitude of the linezolid exposure, in addition to the duration of therapy, will have an impact on the likelihood of occurrence of thrombocytopenia. Forrest and coworkers (15) examined this issue and built a pharmacostatistical model for the occurrence of this toxicity. In Figure 8 using the population analysis for patients on a compassionate use protocol, these investigators demonstrated the range of AUC_{0-24hr} encountered as well as the range of duration of therapy.

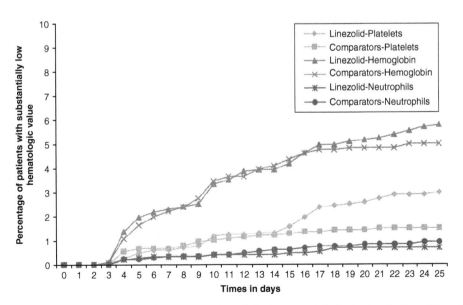

FIGURE 7 Patients with at least one substantially low hematological value in linezolid and comparator groups—cumulative percentage over time.

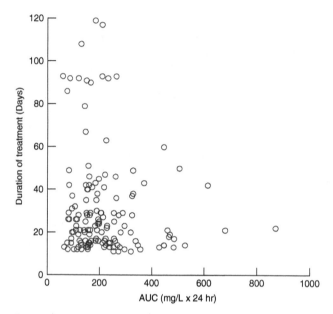

FIGURE 8 Range of linezolid exposure, indexed to AUC_{0-24hr} (x-axis) occurring in a large compassionate use protocol, versus duration of therapy (y-axis). *Abbreviation*: AUC, area under the concentration-time curve.

It can be straightforwardly seen that there is a huge range of cumulative exposure, ranging from circa 1120 mg*hr/L to slightly in excess of 27,000 mg*h/L for a complete course of therapy. This resulted in a two-independent variable model being fit to the data, with duration of therapy on the x-axis, AUC_{0-24hr} on the y-axis and the toxicity (% reduction in platelet count from baseline) on the z-axis. This is presented in Figure 9.

The model fit the data well, with the exception of three outliers who had very long durations of therapy and where percent reduction was greater than predicted by the model. Overall, the model fit is displayed below in Figure 10.

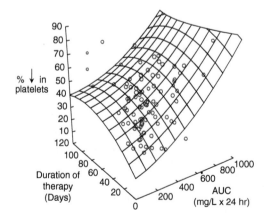

FIGURE 9 Percent reduction from base-line value of platelets as a function of both daily linezolid exposure and duration of linezolid therapy. *Abbreviation*: AUC, area under the concentration-time curve.

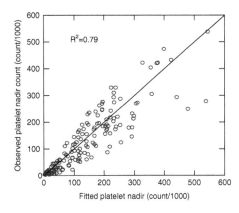

FIGURE 10 Fit of the model to the data for Figure 9 (predicted-observed plot).

Given the range of exposures, different therapy durations will provide different percent falls in platelet count. The percent fall for different durations for the expected AUC_{0-24hr} range is shown in Figure 11.

For a common short therapy duration (two weeks), the range of platelet reduction will range from about 10% to about 50%. Given the relatively low likelihood of AUC_{0-24hr} values exceeding 400 mg*hr/L (Fig. 8), two-week therapy durations are unlikely to cause clinically significant falls in platelet counts. As duration increases, percent decrease in platelets increase.

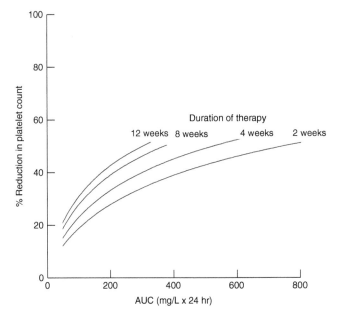

FIGURE 11 Predicted percent reduction in platelets from baseline as a function of the expected range of linezolid exposure, as indexed to AUC_{0-24hr} for therapy durations of 2, 4, 8, and 12 weeks. *Abbreviation*: AUC, area under the concentration-time curve.

Using Classification and Regression Tree analysis (CART), one can look for breakpoints in the data for both magnitude of exposure and therapy duration. The results of this analysis are presented in Figure 12.

The CART analysis, as one might expect, demonstrates that therapy duration has the greatest impact on the large subset of patients with lower drug exposures. The much smaller subset with very large linezolid exposures (11/158) had the appearance of thrombocytopenia made manifest earlier in the course and, therefore, the importance of duration would be made much harder to find.

For linezolid, then, as with prior examples, we can identify both exposure–response and exposure–toxicity relationships. Since this agent has well-defined relationships for both effect and toxicity and is used in many circumstances where the patient is critically ill, we can use both relationships, along with Monte Carlo simulation approaches, to identify the likelihood of achieving the true goal of anti-infective therapy, a high likelihood of clearing the infection with the lowest probability of the exposure engendering a concentration-related toxicity.

WHAT NEXT?

There has been less emphasis in antimicrobial chemotherapy in developing exposure–toxicity relationships relative to exposure–response relationships (16). It is likely that part of the explanation rests in the relatively wide toxic-therapeutic window we

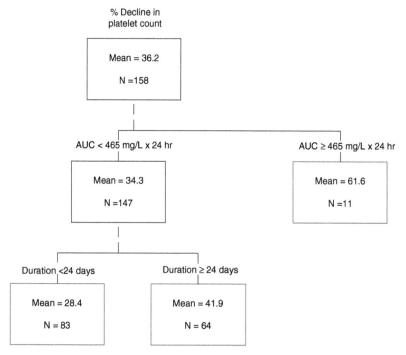

FIGURE 12 CART-derived breakpoints for linezolid exposure ($AUC_{0-24hr} \geq 465$) and linezolid therapy duration (≥ 24 days) and their impact on the percent platelet reduction from baseline. *Abbreviation*: AUC, area under the concentration-time curve.

shoot through with antibiotics. However, for seriously ill patients in the intensive care unit, toxicities such as nephrotoxicity, severe thrombocytopenia, or even severe hyperglycemia can be disastrous for the patient. Consequently, as we develop these exposure–response relationships for very ill patients, it will become increasingly important to be able to also develop the companion exposure–toxicity relationship.

Virtually no sponsor wishes to have an assay for their drug, because, in marketing terms, the need for therapeutic drug monitoring is seen as akin to suicide. However, as for aminoglycosides, there sometimes is an extant assay. Oftentimes, these assays are misused and the patient does not obtain the full benefit of the information developed. Recent work in the area of open-loop control with feedback by the group in the Laboratory of Applied Pharmacokinetics (17) has developed software to optimally choose doses to attain this goal when drug concentration information is available. This sort of optimal use of data is where we should be moving as clinicians.

Often, however, no assays will be available. Here, the use of Monte Carlo simulation for a specific dose and then generating probability of response by MIC relationships as well as probability of toxicity relationships will provide the clinician with guidance as to dose choice in the empirical therapy setting. In these ways, we can do the best job for our seriously ill patients.

REFERENCES

1. Uchino S, Kellum JA, Bellomo R, et al.; Beginning and Ending Supportive Therapy for the Kidney (BEST Kidney) Investigators. Acute renal failure in critically ill patients: a multinational, multicenter study. JAMA. 2005; 294:813–818.
2. Mingeot-LeClercq M-P, Tulkens PM. Aminoglycosides: nephrotoxicity. Antimicrob Agents Chemother 1999; 43:1003–1012.
3. Guiliano RA, Verpooten GA, Verbist L, Weeden R, DeBroe ME. In vivo uptake kinetics of aminoglycosides in the kidney cortex of rats. J Pharmacol Exp Ther 1986; 236:470–475.
4. Wood CA, Norton DR, Kohlhepp SJ, et al. The influence of tobramycin dosage regimens on nephrotoxicity, ototoxicity, and antibacterial efficacy in a rat model of subcutaneous abscess. J Infect Dis 1988; 158:113–122.
5. DeBroe ME, Verbist L, Verpooten GA. Influence of dosage schedule on renal accumulation of amikacin and tobramycin in man. J Antimicrob Chemother 1991; 27(suppl C):41–47.
6. terBraak EW, DeVries PJ, Bouter KP, et al. Once-daily dosing regimen for aminoglycoside plus β-lactam combination therapy of serious bacterial infections: comparative trial of netilmicin plus ceftriaxone. Am J Med 1990; 89:58–66.
7. Rybak MJ, Abate BJ, Kang SL, Ruffing MJ, Lerner SA, Drusano GL. Prospective evaluation of the effect of an aminoglycoside dosing regimen on rates of observed nephrotoxicity and ototoxicity. Antimicrob Agents Chemother 1999; 43:1549–1555.
8. Kashuba AD, Nafziger AN, Drusano GL, Bertino JS Jr. Optimizing aminoglycoside therapy for nosocomial pneumonia caused by Gram-negative bacteria. Antimicrob Agents Chemother 1999; 43:623–629.
9. Ambrose PG, Bhavnani SM, Cirincione BB, Piedmonte M, Grasela TH. Gatifloxacin and the elderly: pharmacokinetic and pharmacodynamics rationale for a potential age-related dose reduction. J Antimicrob Chemother 2003; 52:435–440.
10. Ambrose PG, Grasela DM, Grasela TH, Passarell J, Mayer HB, Pierce PF. Pharmacodynamics of fluoroquinolones against *Streptococcus pneumoniae* in patients with community-acquired respiratory tract infections. Antimicrob Agents Chemother 2001; 45:2793–2797.
11. Moran GJ, Krishnadasan A, Gorwitz RJ, et al.; Emergency ID Net Study Group. Methicillin-resistant *S. aureus* infections among patients in the emergency department. N Engl J Med 2006; 355:666–674.

12. Meagher AK, Forrest A, Rayner CR, Birmingham MC, Schentag JJ. Population pharmacoki-netics of linezolid in patients treated in a compassionate-use program. Antimicrob Agents Chemother 47:548–553.
13. Rayner CR, Forrest A, Meagher AK, Birmingham MC, Schentag JJ. Clinical pharmacody-namics of linezolid in seriously ill patients treated in a compassionate-use programme. Clin Pharmacokinet 2003; 42:1411–1423.
14. Gerson Sl, Kaplan Sl, Bruss JB, et al. Hematologic effects of linezolid: summary of clinical experience. Antimicrob Agents Chemother 2002; 46:2723–2726.
15. Meagher AK, Forrest A, Rayner CR, Birmingham MC, Schentag JJ. Population pharmacoki-netics of linezolid in seriously ill adult patients from a compassionate use protocol. 40th Interscience Conference on Antimicrobial Agents and Chemotherapy, Toronto, Ontario, Canada, September, 2000.
16. Ambrose PG, Bhavnani BM, Rubino CM, et al. Pharmacokinetics-pharmacodynamics of antimicrobial therapy: it's not just for mice anymore. Clin Infect Dis 2007; 44:79–86.
17. Laboratory of Applied Pharmacokinetics Research Activities. MMLQ. A Multiple Model Linear Quadratic control program which designs dosage regimens to minimize a therapeutic cost function, thus optimizing therapeutic precision about a specific selected therapeutic goal. It takes into account arbitrary process and measurement uncertainties inherent in the clinical environment of drug therapy. It uses the NPEM2 population model for the Bayesian prior. http://www.lapk.org/research.php

23 Pharmacodynamics and Antibacterial Resistance

Philip D. Lister and Nancy D. Hanson
Center for Research in Anti-infectives and Biotechnology, Department of Medical Microbiology and Immunology, Creighton University School of Medicine, Omaha, Nebraska, U.S.A.

Anton F. Ehrhardt
Department of Medical Affairs, Cubist Pharmaceuticals, Inc., Lexington, Massachusetts, U.S.A.

INTRODUCTION

Bacterial resistance to antibiotics is a serious consequence of the use and overuse of antibacterial agents in the environment, and the impact of antibacterial resistance on patients and society is staggering. Compared to infections caused by susceptible pathogens, infections caused by resistant pathogens are associated with higher rates of morbidity and mortality (1,2). Furthermore, microbial drug resistance has been projected to add between $100 million and $30 billion annually to health-care costs (3). Considering the distressing decrease in the number of novel agents entering the clinical arena, the judicious utilization of currently available antibiotics becomes essential for preservation of their clinical efficacy. Unfortunately, preventing antibacterial resistance problems is not as simple as reducing the use of antibiotics, but must also involve a more scientific approach to dosing strategies. This is especially true since inadequate exposure of bacteria to drugs during therapy is likely a key factor in selection of resistant mutants. The field of antimicrobial pharmacodynamics strives to establish relationships between the pharmacokinetics of a drug and the effective treatment of infections. However, just as important is the relationship between pharmacokinetics, pharmacodynamics, and the emergence of resistance during therapy. This chapter will review some of the most critical resistance problems facing clinicians today and the role of pharmacodynamic strategies in slowing the emergence of resistance during therapy, including the development of effective antibacterial combinations.

RESISTANCE PROBLEMS IN INFECTIOUS DISEASES
Gram-Positive Bacteria (Table 1)
Enterococcus spp.

Among the numerous species of enterococci, *Enterococcus faecalis* and *Enterococcus faecium* are the two most important clinically. In addition to causing serious hospital-acquired infections such as bacteremia and endocarditis (4), *E. faecalis* and *E. faecium* are two of the most intrinsically resistant bacteria challenging clinicians today (5). These two pathogens naturally exhibit low-level resistance to the β-lactams, aminoglycosides, clindamycin, trimethoprim-sulfamethoxazole, and

TABLE 1 Emerging Resistance Problems among Gram-Positive Pathogens

Bacterial species	Class of antibiotics	Possible molecular mechanisms
Enterococcus faecium	High-level β-lactam	Low-affinity PBPs
Enterococcus faecalis		β-lactamase inactivation
	Glycopeptides	Alteration of target D-ala-D-ala dipeptide to prevent binding
	Aminoglycosides	Inactivating enzymes
Staphylococcus aureus	Penicillins	Penicillinases
	All β-lactams	Acquired low-affinity PBP2'
	Glycopeptides	Molecular trapping of vancomycin in cell wall and/or acquired van genes from enterococci
Streptococcus pneumoniae	β-lactams	Acquired "resistant" PBP gene segments

other classes of antimicrobial agents, driving usage of drugs of last resort such as vancomycin. Data from intensive-care units of U.S. hospitals participating in the National Nosocomial Infections Surveillance (NNIS) program indicated that while vancomycin-resistant enterococci (VRE) were rare in 1989 (0.3%), their prevalence increased dramatically to over 25% in the next 10 years (6). More recent data suggest that VRE prevalence may have reached a plateau of near 28% by 2004 (7). When bacterial killing is required for clinical cure, the combination of a cell-wall active agent (penicillin or vancomycin) with an aminoglycoside provides one of the true examples of synergy between antimicrobial agents. Unfortunately, *E. faecalis* and *E. faecium* have not been satisfied with their intrinsic resistance and have been acquiring high-level resistance to these antibiotics as well (5). Once high-level resistance to either component of the combination is obtained, synergistic killing is lost, therapy of serious infectious is compromised, and the search for more effective antimicrobial agents or alternative combinations becomes essential. Alternative agents approved for use against enterococci include quinupristin/ dalfopristin (this drug is not active against *E. faecalis*), linezolid, daptomycin, and tigecycline. Quinupristin/dalfopristin is also approved for use against vancomycin-resistant strains of *E. faecium*.

Staphylococcus aureus
In contrast to the impressive intrinsic resistance displayed by *E. faecalis* and *E. faecium*, *S. aureus* represents a prototype bacterium for rapid acquisition and development of resistance in response to antibiotics in the environment. Soon after penicillin became available for clinical use, Spink and Ferris reported their isolation of a penicillin-resistant *S. aureus* that produced a penicillin-inactivating enzyme (8). Penicillinases in *S. aureus* are usually encoded on a plasmid and can spread easily among the genus. Currently, 70% to 90% of staphylococci are resistant to penicillin and the aminopenicillins (9). To circumvent this problem, penicillinase-resistant penicillins such as methicillin, oxacillin, and nafcillin were synthesized and introduced in the early 1960s. However, the first methicillin-resistant *S. aureus* (MRSA) was discovered as early as 1961 (10). High-level resistance to this class of penicillins is not mediated through enzymatic inactivation, but rather is mediated through the expression of a low-affinity penicillin-binding protein, PBP2' (11). By the late 1980s, 25% of all staphylococcal isolates in the United States were resistant to methicillin (12), with prevalence varying markedly between geographical areas and individual institutions.

Until the turn of the century, MRSA was predominantly considered to be a nosocomial issue. In the early years of the 21st century, however, community-associated strains (CA-MRSA) have raised significant concerns. These MRSAs isolate differ from classical nosocomial MRSAs in that they tend to bear a smaller *mec* determinant [Staphylococcal chromosomal cassette four (SCC*mec* IV)], that does not encode resistance to antibiotics other than β-lactams. Conversely, these CA-MRSA strains appear to be more virulent than their hospital-based relatives and have been associated with causing skin infections in children that progress to necrotizing pneumonia and even death within only a few days (13–15). The enhanced virulence may be due, at least in part, to the production of the panton-valentine leucocidin (PVL) toxin that is common in these isolates. The predominant strain of CA-MRSA found in the United States is called the USA300 clone, and it is now being found in many U.S. hospitals (16). This raises the disturbing possibility that over time, this new strain may crowd out older nosocomial strains (USA100, 200, and 500), and that genetic exchange or de novo selection under antibiotic pressure in the hospital setting may lead to predominance of USA300 strains with enhanced virulence, PVL production, and broader antibiotic resistance.

Hospital-associated MRSA are frequently resistant to multiple antibiotics, including the penicillins, cephalosporins, cephamycins, carbapenems, β-lactamase-inhibitor combinations, aminoglycosides, macrolides, tetracyclines, and sulfonamides (9). CA-MRSA bearing SCCmec IV has tended to remain susceptible to drugs such as tetracycline and trimethoprim-sulfamethoxazole, though susceptibility testing should be performed before relying on these compounds. The glycopeptides, vancomycin and teicoplanin, remain the preferred therapy for MRSA infections. However, strains of *S. aureus* exhibiting intermediate resistance to both vancomycin and teicoplanin have been reported (17,18). These strains range from those with slightly elevated minimum inhibitory concentrations (MICs) (1–4 μg/mL) that test as susceptible, yet contain glycopeptide-intermediate subpopulations (hVISA), to those with MICs above 8 μg/mL that test as intermediate (19,20). Although rare (fewer than 20 GISA strains known at the time of this writing), these isolates have been linked to therapeutic failure (21,22). Potentially more concerning though has been the isolation of frankly glycopeptide-resistant *S. aureus* from patients. Again exceedingly rare at this time, their spontaneous emergence illustrates that there is no genetic barrier to the enterococci transferring glycopeptide-resistance genes to *S. aureus*. Alternative drugs approved for use against MRSA infections include linezolid, daptomycin, and tigecycline.

Streptococcus pneumoniae

Like *S. aureus*, *S. pneumoniae* has shown a remarkable ability to evolve and develop resistance to antibiotics in the environment. In contrast to *S. aureus*, however, the rate of resistance development is much slower among pneumococci. Clinical isolates of *S. pneumoniae* remained susceptible to penicillin until the 1960s, when the first intermediate-resistant isolates were reported in Boston in 1965 (23). Thereafter, pneumococcal susceptibility to penicillin continued to decrease, with reports of intermediate resistance increasing globally, and the first reports of high-level resistance and clinical failures appearing in the literature (24–26). The prevalence of penicillin-nonsusceptible pneumococci in the United States has been steadily increasing through the 1980s and 1990s. Currently, more than 30% of pneumococcal clinical isolates in the United States lack susceptibility to penicillin, with more than 15% exhibiting full resistance with MICs ≥ 2 μg/mL (27).

The mechanism of penicillin resistance among S. pneumoniae involves the acquisition of "resistant" PBP gene segments from other streptococci in the environment (28,29). As susceptibility to penicillin decreases, a concurrent decrease in susceptibility to other penicillins, cephalosporins, cephamycins, and carbapenems, is also observed (30). In addition, penicillin-resistant pneumococci are increasing their resistance to the macrolides, tetracyclines, chloramphenicol, and trimethoprim-sulfamethoxazole (27,30), and the increased prevalence of multi-drug-resistant pneumococci presents a therapeutic dilemma for the treatment to serious pneumococcal infections.

Gram-Negative Bacteria (Table 2)

Antimicrobial resistance continues to evolve at a rapid pace among gram-negative bacteria, and new resistance mechanisms or enhancements of established mechanisms are being discovered almost daily. Since the scope of this chapter cannot include an in-depth discussion of all emerging resistance problems for all classes of antimicrobial agents, the following sections will focus specifically on the ever-evolving threat of β-lactamase–mediated resistance.

Haemophilus influenzae

Production of β-lactamase accounts for >90% of ampicillin resistance among H. influenzae, with TEM-1 and ROB-1 being the most common enzymes produced

TABLE 2   β-Lactamase–Mediated Resistance among Gram-Negative Pathogens

Bacterial species	Antibiotic class	Possible molecular mechanisms
Haemophilus influenzae	Penicillins	TEM-1 and ROB-1 β-lactamases
Klebsiella pneumoniae	Penicillins and narrow-spectrum cephalosporins	TEM-1 and SHV-1 β-lactamases
Escherichia coli	β-lactamase-inhibitor combinations	Inhibitor-resistant β-lactamases or hyperproduction of TEM-1/SHV-1
	Penicillins, cephalosporins, inhibitor-penicillin combinations, and monobactams	Increased production of bla_{shv-1} or ampC Plasmid-encoded AmpC β-lactamases Extended-spectrum β-lactamases
	Carbapenems	KPC carbapenemases
Enterobacter cloacae E. aerogenese Serratia marcescens Citrobacter freundii	Penicillins, cephalosporins, inhibitor-penicillin combinations, cephamycins, and monobactams	Chromosomal AmpC β-lactamases
	Carbapenems	KPC carbapenemases Metallo-β-lactamases OXA carbapenemases
Pseudomonas aeruginosa	Penicillins, cephalosporins, inhibitor-penicillin combinations, and monobactams	Chromosomal AmpC β-lactamase Extended-spectrum β-lactamases
	Carbapenems	Metallo-β-lactamases

(31, 32). Worldwide, up to 38% of *H. influenzae* isolates produce β-lactamase (32–34). Although a recent survey conducted in the United States indicates that our national average has dropped to 26.2%, the range was 19% to 36% depending on the geographical location (35). The majority of β-lactamase–producing strains remain susceptible to the β-lactamase-inhibitor combinations and extended-spectrum cephalosporins, but permeability and/or PBP changes can provide resistance to these agents as well (32).

Escherichia coli and Klebsiella pneumoniae

E. coli and *K. pneumoniae* have shown remarkable ability to evolve and adapt to the threat of antibiotics in the environment, especially to the β-lactam class. Similar to *H. influenzae*, β-lactam resistance is mediated primarily through the production of β-lactamases. However, the diversity of β-lactamases produced by *E. coli* and *K. pneumoniae* is far greater.

TEM-1 and SHV-1 are considered broad-spectrum β-lactamases and account for the majority of *E. coli* and *K. pneumoniae* resistance to the penicillins and early narrow-spectrum cephalosporins (36). Historically, the strategies used to circumvent these broad-spectrum β-lactamases have been to develop "enzyme-resistant" cephalosporins or to combine inhibitors of the enzymes with penicillins. *E. coli* and *K. pneumoniae* have evolved impressively in response to these two approaches and some strains have now become resistant to virtually all β-lactams and inhibitor-penicillin combinations.

In response to inhibitor-penicillin combinations, *E. coli* and *K. pneumoniae* have either mutated to increase production of their β-lactamases (37,38) or decreased the sensitivity of their plasmid-encoded β-lactamases to the inhibitory effects of clavulanate, tazobactam, and sulbactam (39,40). This latter strategy led to the evolution of inhibitor-resistant TEM and SHV enzymes. In response to the extended spectrum cephalosporins (ceftazidime, cefotaxime, and ceftriaxone), *E. coli* and *K. pneumoniae* have responded in three ways. (*i*) The first response has been to mutate the promoters of their chromosomal β-lactamase genes (*ampC* for *E. coli* and bla$_{shv-1}$ for *K. pneumoniae*) as a mechanism to increase production and provide resistance to the extended-spectrum cephalosporins and aztreonam. In addition, overexpression of the plasmid-encoded broad-spectrum β-lactamases (TEM-1 or SHV-1) can provide resistance to these drugs. (*ii*) The second strategy has been to acquire plasmid-encoded AmpC cephalosporinases, which are capable of conferring resistance to nearly all β-lactams. Plasmid-encoded AmpC cephalosporinases originated from the movement of chromosomal genes from many species of gram-negative organisms encoding an inducible *ampC* gene (41). To date, there have been at least 32 unique plasmid-encoded AmpCs identified, with the *ampC* coming from *Enterobacter cloacae*, *Citrobacter freundii*, Aeromonas spp., *Enterobacter asburiae*, *Morganella morganii*, and *Hafnia alvei* (42). (*iii*) Finally, some strains have responded by mutating the active sites of their older broad-spectrum β-lactamases (TEM-1 and SHV-1) to extend their hydrolytic capabilities to include the extended-spectrum cephalosporins and aztreonam. These enzymes are classified as extended-spectrum β-lactamases (ESBL), and their production increases the resistance profile of *E. coli* and *K. pneumoniae* to include all penicillins, cephalosporins, and aztreonam. The TEM- and SHV-associated ESBLs have evolved dramatically from the original broad-spectrum TEM-1, TEM-2, and SHV-1 enzymes to now include at least 150 unique TEMs and 99 unique SHVs (43).

Although ESBLs were first reported as members of the TEM and SHV families, the most predominate ESBLs worldwide today are the CTX-M β-lactamases (44). However, the genes encoding for CTX-M enzymes did not originate through alterations in existing enzymes like other ESBLs, but rather originated from the chromosome of *Kluyvera* spp. Nevertheless, CTX-M β-lactamases are referred to as ESBLs because they provide *K. pneumoniae* and *E. coli* with the same β-lactam resistance profile that TEM- and SHV-associated ESBLs do. There have been at least 53 CTX-M enzymes reported in the literature (43).

Among the β-lactam class of antibiotics, the carbapenems are the most reliable drugs against *E. coli* and *K. pneumoniae* that harbor ESBLs or plasmid-encoded AmpC cephalosporinases. However, the production of Class A carbapenemases (KPC) by *E. coli* and *K. pneumoniae* is becoming problematic, particularly in the United States (45–48). In fact, KPC-2–producing *K. pneumoniae* have become endemic in New York City (45). These enzymes are capable of hydrolyzing a variety of β-lactams including penicillins, cephalosporins, aztreonam, and the carbapenems, rendering the β-lactam class useless. A further complication in the selection of treatment options for ESBL, AmpC, and/or KPC-producing *E. coli* and *K. pneumoniae* is that these enzymes are encoded on plasmids, which carry genes encoding for resistance mechanisms to multiple classes of antimicrobial agents (49, 50). Therefore, it is not uncommon to find ESBL-, AmpC-, and/or KPC-producing strains that are resistant to the aminoglycosides, trimethoprim-sulfamethoxazole, chloramphenicol, and tetracyclines. In fact, there are reports of *K. pneumoniae*, which exhibit resistance to all available antimicrobial agents, and these strains pose a serious threat for the treatment of a pathogen, which was once relatively easy to treat (45,51).

Other Enterobacteriaceae

Although the predominant ESBL-producing bacteria are *E. coli* and *K. pneumoniae*, ESBLs have also been found among species of *Enterobacter*, *Citrobacter*, *Proteus*, *Salmonella*, *Serratia*, *Acinetobacter*, and other *Klebsiella* (36,52). However, it is the inducible chromosomal AmpC cephalosporinase that is most commonly involved in the emergence of high-level β-lactam resistance during the course of treating some species (53).

Rapid development of resistance to nearly all β-lactams can be associated with mutant subpopulations that constitutively produce high levels of their AmpC cephalosporinase (54). In most cases, emergence of resistance is associated with mutations within the structural gene, *ampD*. This gene encodes for AmpD, a cytosolic amidase involved in the pathway, which regulates *ampC* expression (55). In addition, it has recently been shown that a decrease in *ampD* expression can also lead to significant increases in *ampC* expression (56). Isolates that overproduce the AmpC β-lactamase as a result of *ampD* mutations are referred to phenotypically as derepressed. Such mutations occur in 1 out of every 10^6 to 10^7 viable bacteria in a culture or infection. Therefore, resistant mutants are likely to be present at the start of therapy for infections, which are associated with a large bacterial load. If these mutants are not eliminated by the host or killed by the antibiotic, they can eventually become the predominant population and lead to therapeutic failure. Clinical failure due to the emergence of AmpC-mediated resistance has been observed with 19% to 80% of patients infected with bacteria possessing an inducible AmpC cephalosporinase (36). Although most β-lactams can select for these mutants, there appears to be a particular association with the extended-spectrum cephalosporin ceftazidime (1).

In contrast to other β-lactams, carbapenem resistance is rare among members of the Enterobacteriaceae. However, when it does occur, it can be due to several different families of enzymes, including KPC-type β-lactamases (48), chromosomal class A KPC of *E. cloacae* and *Serratia marcescens* (57,58), metallo-β-lactamases, and OXA-KPC. The latter two groups of enzymes have been generally associated with *Acinetobacter baumannii* and *Pseudomonas aeruginosa* (59–61). However, these enzymes are also found in *K. pneumoniae* (61–63).

Pseudomonas aeruginosa

Among the gram-negative pathogens, *P. aeruginosa* is one of the most threatening in terms of rapidly developing resistance to β-lactam antibiotics. Similar to the Enterobacteriaceae, production of β-lactamase is the primary mechanism employed, and emergence of high-level β-lactam resistance during therapy is due to the selection of mutants that overproduce their chromosomal AmpC cephalosporinase (64). Emergence of AmpC-mediated resistance has been reported to cause clinical failure for 14% to 56% of patients with *P. aeruginosa* infections, with even higher rates of clinical failure when the infections are outside the urinary tract or in patients with cystic fibrosis or neutropenia (64–66).

P. aeruginosa can also develop β-lactam resistance through the acquisition of mobile genetic elements (plasmids or integrons) carrying β-lactamase genes. The major groups of enzymes, which fall under this category, are the OXA-type ESBLs, PSE, and the metallo-β-lactamases. There have been at least 88 OXA-like enzymes reported in the literature (43). In contrast to OXA-type enzymes, which possess a serine in the active site of the enzyme, metallo-β-lactamases require a divalent cation, usually zinc, in the active site (60). Among the metallo-β-lactamases identified thus far, the VIM and IMP families are the most prevalent worldwide. Within the United States, however, the production of metallo-β-lactamases remains rare, has only been identified among *P. aeruginosa* isolates, and represents only three different enzymes (VIM-7, VIM-2, and IMP-18) (59).

PHARMACODYNAMICS AND ANTIMICROBIAL RESISTANCE

The field of antimicrobial pharmacodynamics strives to establish relationships between the pharmacokinetics of a drug, its interaction with target bacteria, and the effective treatment of infections. Since emergence of "mutational" resistance during therapy can result in therapeutic failures, the relationship between antimicrobial pharmacodynamics and resistant subpopulations is critical for optimizing therapy of many infections.

Resistance to antimicrobial agents can develop in a bacterial pathogen through two distinct pathways: (*i*) acquisition of new genes encoding resistance-mediating proteins and (*ii*) evolutionary mutations within inherent genes leading to changes in level of expression of or activity of resistance-mediating proteins. The impact of optimizing antimicrobial pharmacodynamics on each of these pathways is very different. While optimization of antimicrobial pharmacodynamics can prevent or slow the evolution of most mutation-mediated resistance mechanisms, there are many acquired resistance mechanisms, which cannot be treated with safe doses of antibiotic. Therefore, the goal of pharmacodynamics should be to optimize therapy such that mutational resistance is prevented or reduced. This is especially true since it is the emergence of mutational resistance during therapy that causes the greatest concern for therapeutic failures.

Acquired Vs. Mutational Resistance

Acquired Resistance

Acquired resistance, as the name suggest, represents the development of resistance in a bacterium through the acquisition of a new gene or set of genes from other bacteria in the environment. Examples of acquired resistance mechanisms include plasmid-encoded β-lactamases, plasmid-encoded aminoglycoside-inactivating enzymes, vanA-mediated vancomycin resistance among enterococci and staphylococci, and mecA-mediated methicillin resistance among staphylococci. The common theme with each of these resistance problems is that they are passed from bacterium to bacterium on transferable genetic elements, i.e., plasmids or transposons. Once a bacterium acquires one of these resistance mechanisms, the level of resistance that is expressed can oftentimes exceed the levels of antibiotic that can be safely achieved in serum or other sites of infection. For example, acquisition of the TEM-1 β-lactamase by an E. coli can increase the MIC of ampicillin from 2 to 256 µg/mL (67). Similarly, acquisition of the genes responsible for vancomycin resistance can increase the MIC of vancomycin against enterococci from 2 to 512 µg/mL (68).

Mutational Resistance

In contrast to acquired resistance, mutational resistance is mediated through genetic changes within a bacterium's own genetic material. Such mutations may involve changing a single nucleotide (point mutation), insertions of extrabases, or deletions, and may occur within the coding sequences of a gene, its regulatory regions, or in noncoding sequences. If the mutation occurs within coding sequences, it may variously result in a change of a single amino acid, a frameshift that alters the encoded protein massively, or no change in the protein at all due to the redundancy of the genetic code. If the mutation locus is within regulatory sequences, it may cause higher or lower production of proteins being regulated (or no change at all). These evolutionary mutations can alter the binding of a drug to its target, decrease the accumulation of a drug at the site of action, or alter the production of, enzymatic efficiency of, or substrate spectrum of resistance-mediating enzymes.

Similar to acquired resistance mechanisms, mutational resistance can result in instantaneous high-level resistance that is untreatable with maximum antibiotic doses or more moderate decreases in susceptibility. Whether high-level resistance is achieved with a single point mutation or requires multiple mutational events depends upon the specific drug and target bacterium. In fact, difference between drugs within the same family can be observed. This point is best illustrated below using the example of β-lactam resistance mediated through derepression of ampC among gram-negative bacteria.

In contrast to the single ampC-derepression mutational event required for ceftazidime resistance among Enterobacter spp., it takes two independent mutational events in to provide cefepime resistance (69). First, the level of AmpC production must be increased through derepression of the ampC operon and then the penetration of cefepime through the outer membrane must be slowed or prevented through a second mutational event. Since each of these mutational steps occurs in 1 out of every 10^6 to 10^7 viable bacteria, it is unlikely that resistance will develop in a ceftazidime-susceptible Enterobacter spp. during the course of therapy with cefepime. This hypothesis is supported by data from a murine infection model (70). However, when Enterobacter mutate to a state of ampC derepression and numbers of these derepressed mutants exceed 10^6, the chances of selecting the second permeability mutation during therapy with cefepime increases and emergence of cefepime resistance becomes a threat.

Differences between antibiotics in the number of mutational steps required to achieve clinically relevant resistance is not unique to the β-lactam class, as similar observations have been reported with other drugs classes as well, particularly the fluoroquinolones (71). Although appropriate dosing of antibiotics does not often impact problems associated with acquired resistance, pharmacodynamics can play an influential role in preventing or slowing the development of mutational resistance, especially when first step mutations do not result in high-level resistance. Therefore, the relationship between antimicrobial pharmacodynamics and the emergence of mutational resistance is essential for developing optimum therapeutic strategies.

Antimicrobial Pharmacodynamics and Emergence of Resistance

The three most well-characterized pharmacodynamic parameters influencing the clinical efficacy of antibiotics are the time antibiotic concentrations remain above the MIC ($T > MIC$), ratio of peak concentrations to the MIC (peak/MIC), and ratio of the area under the concentration curve to the MIC (AUC/MIC). Which pharmacodynamic parameter is most predictive of clinical efficacy varies depending upon the class of drugs used for therapy. For example, with β-lactam antibiotics, the $T < MIC$ is the pharmacodynamic parameter that most influences clinical outcome, whereas with fluoroquinolones, the most important parameters can be either the peak/MIC ratio or the AUC/MIC ratio. Further complicating therapeutic strategies is the understanding that the pharmacodynamic parameter, which most influences overall efficacy, may not be the most important parameter impacting emergence of resistance during therapy. Whereas there are published studies evaluating the impact of AUC/MIC ratio on emergence of resistance (72, 73), the following discussion will focus on the relationship between peak/MIC ratios and emergence of resistance during therapy and the role pharmacodynamic research can play in identifying mechanism-based combinations for preventing the emergence of resistance.

Peak/MIC Ratio and Emergence of Resistance During Therapy

It is unlikely that two or more mutations required to move MICs from the highly susceptible wild-type range to high-level resistance will occur simultaneously and already be present in a population of bacteria causing an infection. Therefore, to prevent the emergence of resistance during therapy, the pharmacodynamic focus must be on effectively treating both the original isolate and any "first-step" or "next-step" mutants that could become the predominant population and cause therapeutic failure. This requires that the concentration of antibiotic at the site of infection exceed the MIC for the mutant subpopulations. Since AUC/MIC does not directly address this requirement, the peak/MIC ratio becomes the pharmacodynamic parameter that can most impact selection of resistance. Indirect evidence to support this hypothesis has come from studies in an animal model of infection (74). Using a neutropenic rat model of *Pseudomonas aeruginosa* sepsis, Drusano and coworker evaluated the effects of dose fractionation of lomefloxacin on survival. Although the emergence of resistance was not directly evaluated in this study, peak/MIC ratio most closely linked to survival when the ratios exceeded 10/1. Their hypothesis to explain these results was that sufficient peak/MIC ratios resulted in suppression of mutant subpopulations, thus preventing death due to the emergence of resistance. Similar conclusions have been obtained from human clinical data levofloxacin (75).

Additional evidence supporting the importance of peak/MIC ratios in preventing the emergence of resistance comes from studies using in vitro pharmacokinetic models. The flexibility provided by these models allows for the design of experiments to specifically address this issue. Using a two-compartment pharmacokinetic model, Blaser et al. varied the pharmacokinetics of enoxacin and netilmicin such that bacteria were exposed to a range peak/MIC ratios while maintaining a constant total drug exposure over 24 hours (76). These investigators observed the outgrowth of resistant subpopulations when simulated peak/MIC ratios failed to exceed 8/1. Similarly, Marchibanks et al. used a two-compartment pharmacokinetic model to simulate ciprofloxacin doses of 400 mg t.i.d. (peak/MIC ratio = 4/1) and 600 mg b.i.d. (peak/MIC ratio = 6/1) and observed outgrowth of resistant subpopulations of *P. aeruginosa* (77). In this study, peak/MIC ratios of only 6/1 were sufficient to prevent the emergence of resistant subpopulations of *S. aureus*. These data suggest that although the peak/MIC ratio exerts an important influence on the emergence of resistance during therapy, the minimum peak/MIC needed to prevent this problem will likely vary depending upon the drug studied and the target bacterium. This conclusion is supported with data from our own pharmacodynamic studies with levofloxacin and β-lactam antibiotics against *P. aeruginosa,* as detailed below.

Pharmacodynamics of Levofloxacin Against P. aeruginosa

A two-compartment pharmacokinetic model was used to simulate the serum pharmacokinetics of a once-daily 500 mg dose of levofloxacin and evaluate pharmacodynamic interactions against a panel of 6 *P. aeruginosa*. Three of the strains evaluated were susceptible to levofloxacin with MIC = 0.5 μg/mL, and the other three strains were borderline susceptible with MICs = 2 μg/mL. The peak concentration simulated for the 500 mg dose of levofloxacin was 6 μg/mL, which provided peak/MIC ratios of 12/1 for the three most susceptible strains and 3/1 for the borderline-susceptible strains. At the start of each experiment, logarithmic-phase cultures (5×10^7 cfu/mL) were inoculated into the extracapillary space of hollow-fiber cartridges and peak concentrations of levofloxacin were dosed into the central compartment of the model at 0 and 24 hours. The pharmacodynamics of levofloxacin against two representative strains of *P. aeruginosa* are shown in Figure 1. In studies with the most susceptible strains, levofloxacin exhibited rapid and significant bacterial killing, with viable counts falling >6 logs to undetectable levels (<10 CFU/mL) within 4 to 12 hours after the first dose. In studies with the borderline-susceptible strains, levofloxacin also exhibited significant bacterial killing. However, against these three strains, killing was only observed over the initial six to eight hours, with viable counts increasing rapidly thereafter due to the selection and outgrowth of resistant subpopulations. By 24 hours, viable counts in the levofloxacin-treated cultures were approaching those in the drug-free control cultures. The MICs of levofloxacin against the resistant subpopulations ranged from 8 to 16 μg/mL. Therefore, emergence of resistance could have been predicted since peak concentrations with each dose of levofloxacin never exceeded the MICs for these mutant subpopulations.

Pharmacodynamics of Ceftazidime and Ticarcillin Against P. aeruginosa

A two-compartment pharmacokinetic model was used to investigate the relationship between mutational frequencies, pharmacodynamic interactions, and the

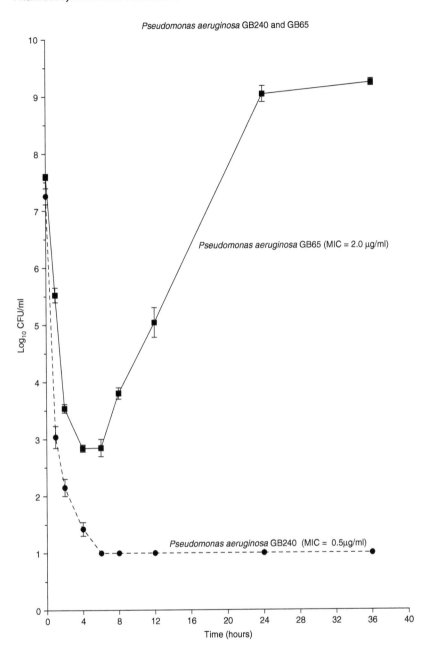

FIGURE 1 Pharmacodynamics of levofloxacin against *Pseudomonas aeruginosa* GB240 and *P. aeruginosa* GB65 in an IVPM. Levofloxacin MICs were 0.5 µg/mL for *P. aeruginosa* GB240 and 2.0 µg/mL for *P. aeruginosa* GB65. The human serum pharmacokinetics of the 500 mg dose of levofloxacin was simulated with the IVPM. Peak concentrations of levofloxacin were dosed into the IVPM at 0 and 24 hours, and elimination pharmacokinetics was simulated with the IVPM. Each datum point represents the mean number of viable bacteria per milliliter of culture for duplicate experiments. Error bars show standard deviations. *Abbreviations*: IVPM, in vitro pharmacodynamic model; MIC, minimum inhibitory concentration.

emergence of resistance during the treatment of *P. aeruginosa* with ticarcillin and ceftazidime. The initial MICs for ticarcillin and ceftazidime against *P. aeruginosa* 164 were 16 and 2 µg/mL, respectively. Mutational frequencies were measured in agar by exposing logarithmic-phase cultures of *P. aeruginosa* 164 to ticarcillin and ceftazidime at concentrations two-, four-, and eightfold above their respective MICs. The mutational frequency for ticarcillin was 10^{-7}, compared to 10^{-6} for ceftazidime. MICs of ticarcillin and ceftazidime against these mutants increased to >512 and 32 µg/mL, respectively. Using a two-compartment pharmacokinetic model, the kinetics of a 3.0 g q.i.d. dose of ticarcillin and 2.0 g t.i.d. dose of ceftazidime were simulated and their pharmacodynamics against *P. aeruginosa* 164 was evaluated over 24 hours. With peak concentrations of 260 µg/mL for ticarcillin and 140 µg/mL for ceftazidime, peak/MIC ratios against the original clinical isolate were 16/1 for ticarcillin and 70/1 for ceftazidime. Pharmacodynamic interactions are shown in Figure 2. In studies with ceftazidime, viable counts decreased over 3 logs throughout the 24-hour experimental period and no resistant mutants were detected on drug selection plates. The lack of resistance emergence was not unexpected since peak concentrations achieved with the 2.0 g of ceftazidime (140 µg/mL) were fourfold above the 32 µg/mL MIC of first-step mutants selected in the mutational frequency studies. In contrast, resistance emerged rapidly in studies with ticarcillin (Fig. 2). Although the simulated peak/MIC ratio of ticarcillin was 16/1 for *P. aeruginosa* 164, peak concentrations remained at least twofold below the MIC for first-step mutants.

These pharmacodynamic data and data from other investigators highlight the impact of the peak/MIC ratios on the potential for resistance to emerge during the course of therapy. Unfortunately, these studies also demonstrate that optimum peak/MIC ratio needed to prevent the emergence of resistance varies between drug classes and between specific bacteria-drug combinations. Therefore, a greater understanding of the quantitative influence different resistance mechanisms exert on susceptibility is needed to optimally apply the pharmacodynamic knowledge that has been gained thus far.

Preventing of Emergence of Resistance with Antibacterial Combinations

There are some pathogens such as *P. aeruginosa* for which a combination of two drugs is recommended to effectively treat infections. *P. aeruginosa* has the ability to rapidly mutate to a resistant phenotype, even during the course of therapy. Thus, the primary reason for using two antibacterial drugs against *P. aeruginosa* is to prevent resistance from emerging. A good example is the risk of AmpC-mediated resistance emerging due to derepression of *amp*C expression and the use of aminoglycosides/β-lactam combination to prevent this event from happening. This resistance problem has been shown to emerge during the course of therapy with antipseudomonal penicillins, antipseudomonal cephalosporins, inhibitor-penicillin combinations, and aztreonam, and to result in clinical failure for 14% to 56% of patients with *P. aeruginosa* infections (64,66,78–80). Mechanistically, a combination of an aminoglycoside and an antipseudomonal β-lactam makes sense since their cellular targets are different and since their pharmacodynamic interactions with target bacteria are different. One would predict that each drug in the combination would be able to eliminate mutant subpopulations affecting the susceptibility of the bacterium to the companion drug. However, clinical data have suggested that this combination is not always effective in preventing the emergence of resistance, since AmpC-mediated resistance to β-lactams has been shown

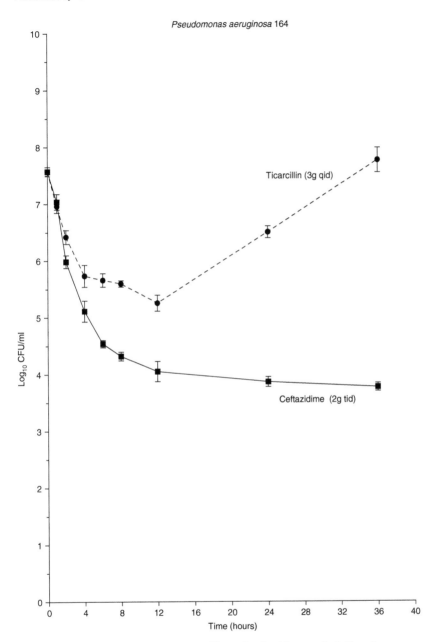

FIGURE 2 Pharmacodynamics of ticarcillin and ceftazidime against *Pseudomonas aeruginosa* 164 in an IVPM. Ticarcillin and ceftazidime MICs were 16 and 2 µg/mL, respectively. The human serum pharmacokinetics of the three gram dose of ticarcillin and two gram dose of ceftazidime was simulated with the IVPM. Peak concentrations of ticarcillin were dosed at 0, 6, 12, and 18 hours, whereas peak concentrations of ceftazidime were dosed at 0, 8, and 16 hours. Elimination pharmacokinetics was simulated with the IVPM. Each datum point represents the mean number of viable bacteria per milliliter of culture for duplicate experiments. Error bars show standard deviations. *Abbreviations*: IVPM, in vitro pharmacodynamic model; MIC, minimum inhibitory concentration.

to emerge during the course of therapy with the combination (36). Therefore, the search for more effective antipseudomonal combinations must continue. Pharmacodynamic research models can play an important role in the identification and development of novel combinations, especially when prevention of resistance is the goal. Whereas a great deal of attention is focused on finding combinations that exhibit synergy, it is important to recognize that the prevention of resistance during therapy may not be dependent upon a synergistic interaction between the two drugs, i.e., enhanced antibacterial activity. Therefore, standard susceptibility and synergy assays (checkerboard titration assay) are not really designed to assess the potential of an antibacterial combination to prevent the emergence of resistance during therapy. These studies require a more in-depth pharmacodynamic evaluation of the interactions of the drugs alone and the combination against target bacteria. Below are two examples of novel antipseudomonal combinations that have been proposed through a mechanism-based approach and proven to be effective in preventing the emergence of resistance among *P. aeruginosa*.

Cefepime-Aztreonam Combination

One problem associated with treatment of *P. aeruginosa* is the lack of effective inhibitors of the AmpC cephalosporinase. None of the commonly used β-lactamase inhibitors, i.e., tazobactam or clavulanate, exhibit sufficient inhibition of AmpC to be clinically useful (81) and attempts to develop clinically useful inhibitors of AmpC have been disappointing. However, the β-lactamase inhibitory activity of aztreonam is oftentimes overlooked, as this compound is one of the most potent competitive inhibitors of the chromosomal AmpC of Enterobacteriaceae and *P. aeruginosa* (82–84). Not only has aztreonam been shown to be a strong competitive inhibitor of AmpC in vitro, but data from studies with cystic fibrosis patients have demonstrated the ability of this compound to inhibit the *P. aeruginosa* AmpC in patient sputum (65).

The potential of aztreonam to serve as an AmpC-inhibitor and enhance the pharmacodynamics of cefepime against an isogenic panel of *P. aeruginosa* was evaluated in a two-compartment in vitro pharmacodynamic model (IVPM) with hollow-fiber cartridges (85). The isogenic panel of *P. aeruginosa* included a "wild-type" clinical isolate, an isogenic mutant partially depressed for AmpC production, and an isogenic mutant fully derepressed for AmpC production. Characteristics of the isogenic panel are shown in Table 3. Routine susceptibility tests failed to show any positive or negative MIC interactions between cefepime and aztreonam against these *P. aeruginosa* (Table 3). For pharmacodynamic experiments, logarithmic-phase cultures of the *P. aeruginosa* ($\sim 1 \times 10^7$ cfu/mL) were treated

TABLE 3 Characteristics of Isogenic *Pseudomonas aeruginosa* 164 Panel

Pseudomonas aeruginosa strain	AmpC phenotype	Uninduced Amp expression[a]	Cefepime MIC (μg/mL)	Aztreonam MIC (μg/mL)	Cefepime + Aztreonam MIC (μg/mL)[b]
164	Wild-type	5	8	8	8/4
164PD	Partially derepressed	38	16	16	32/16
164FD	Fully derepressed	1640	128	>128	128/64

[a]Cephalosporinase activity (nanomoles of cephalothin hydrolyzed per minute per milligram of protein) in sonic extracts of logarithmic-phase cultures.
[b]Cefepime-aztreonam combination tested at a ratio of two parts cefepime to part part aztreonam.
Abbreviation: MIC, minimum inhibitory concentration.

with simulated human doses of 1 g cefepime, 1 g aztreonam, and a 1 g combination of two β-lactams. The potential of this combination to prevent the emergence of resistance was best illustrated during pharmacodynamic experiments with the partially derepressed, borderline-resistant strain 164PD (Figure 3). Treatment of 164PD with both cefepime and aztreonam alone in the IVPM resulted in bacterial overgrowth and the selection of a highly resistant, fully derepressed mutant with MICs exceeding 128 µg/mL. In contrast, treatment with the combination of cefepime-aztreonam prevented this resistant subpopulation from emerging and provided almost 4 logs of killing over 24 hours. The ability of cefepime-aztreonam to prevent the emergence of resistance during therapy was shown to be associated with aztreonam's significant inhibition of extracellular AmpC in the IVPM, similar to clinical observations in the sputum of cystic fibrosis patients (65).

Levofloxacin-Imipenem Combination

In contrast to the direct inhibition of a resistance mechanism associated with the cefepime-aztreonam combination described above, the combination of levofloxacin-imipenem attacks the emergence of resistance problem with a different approach. Levofloxacin-imipenem takes advantage of the knowledge that mutational events selected for by imipenem (OprD-mediated resistance) do not affect susceptibility to fluoroquinolones (86,87). Similarly, mutational events selected for by fluoroquinolone resistance (target mutations and efflux pump overexpression) do not affect imipenem susceptibility, especially since imipenem is not a substrate for any of the efflux pumps characterized to date (88,89).

The potential of levofloxacin-imipenem to prevent the emergence of resistance during therapy of *P. aeruginosa* was evaluated pharmacodynamically against three clinical isolates in a two-compartment IVPM (90). The three clinical isolates chosen for this study were selected based on preliminary studies, which demonstrated the reliable emergence of resistance during course of therapy with each drug alone. Logarithmic-phase cultures (1×10^8 CFU/mL) of each strain were introduced into the peripheral compartment of the IVPM and were treated with simulated human doses of 750 mg of levofloxacin alone, 250 mg of imipenem alone, and a combination of levofloxacin-imipenem. The lowest recommended dose of 250 mg of imipenem was chosen for these studies to promote the emergence of resistance and to provide a greater therapeutic challenge for the combination. The peak concentrations of each drug targeted in the peripheral compartment of the IVPM were adjusted for protein binding (25% for levofloxacin and 20% for imipenem) and were 6.5 µg/mL for levofloxacin and 15 µg/mL for imipenem (based on the median total peak concentration of 19 µg/mL). Levofloxacin was dosed into the model at zero hours, whereas imipenem was dosed into the model at 0, 8 and 16 hours.

When mutant subpopulations were selected for and resistance emerged by 24 hours, 10 colonies were selected from drug-selection plates and these potential mutants were evaluated for changes in susceptibility by agar dilution methodology. Five confirmed mutants were then evaluated further for potential involvement of multidrug efflux pumps in resistance to levofloxacin. Using a reverse-transcriptase polymerase chain reaction methodology, five mutants selected with levofloxacin were evaluated for changes in the transcriptional expression of *mexAB-oprM, mexCD-oprJ, mexEF-oprN, and mexXY* by measuring relative amounts of transcription for the first gene in each operon (*mexA, mexC, mexE, and mexX*).

The general pharmacodynamics of levofloxacin, imipenem, and the combination levofloxacin-imipenem were similar against all three *P. aeruginosa* in the study,

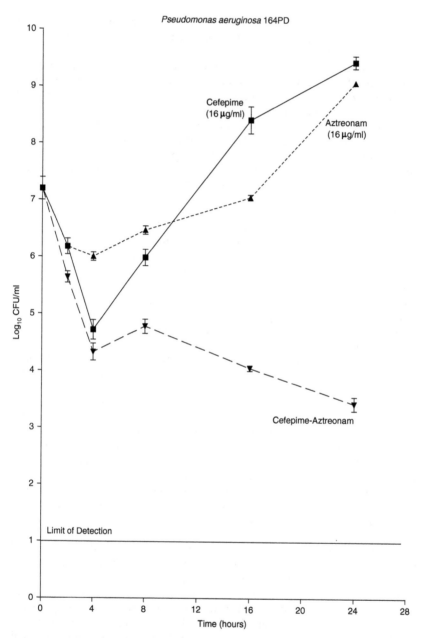

FIGURE 3 Pharmacodynamics of cefepime, aztreonam, and cefepime-aztreonam against partially derepressed *Pseudomonas aeruginosa* 164PD in a two-compartment IVPM. The human serum pharmacokinetics of the 1 g doses of cefepime, aztreonam, and the combination of cefepime-aztreonam (1 g each) was simulated with the IVPM. Individual drugs or the combination were dosed into the model at 0, 12, and 24 hours, and elimination pharmacokinetics was simulated with the IVPM. Each datum point represents the mean number of viable bacteria per milliliter of culture for duplicate experiments. Error bars show standard deviations. *Abbreviation*: IVPM, in vitro pharmacodynamic model.

and pharmacodynamic interactions observed with one representative strain (*P. aeruginosa* GB2) are shown in Figure 4. Data from this study demonstrated that resistance emerged rapidly during treatment of all strains with either levofloxacin or imipenem alone. As predicted, imipenem-selected resistance was associated with the decreased expression of *opr*D, whereas levofloxacin was found to select for a variety of mutant phenotypes, some of which were associated with the overexpression of multidrug efflux pumps. In contrast to the emergence of resistance and failure of each drug alone, the levofloxacin-imipenem combination rapidly eradicated all three *P. aeruginosa* from the IVPM and prevented the emergence of resistance. Furthermore, the combination was effective in preventing the emergence of resistance associated with overexpression of the MexEF-OprN efflux pump, a phenotype that is characterized by loss of susceptibility to both fluoroquinolones and imipenem. Although imipenem is not a substrate of the MexEF-OprN efflux pump, overexpression of MexEF-OprN is associated with a decrease in expression of the outer membrane porin OprD and loss of susceptibility to imipenem (91, 92). Subsequent studies have extended these initial observations and demonstrated that the combination of levofloxacin-imipenem can effectively eradicate *P. aeruginosa* that has already lost susceptibility to one or both drugs in the combination, and prevent the emergence of higher levels of resistance (93).

Our studies with the combinations of cefepime-aztreonam and levofloxacin-imipenem demonstrate the importance of pharmacodynamic research in the evaluation of potential antibacterial combinations for preventing the emergence of resistance. Neither of these combinations exhibited true synergistic interactions with regard to their bacterial killing dynamics and therefore would not have been identified as effective combinations in routine susceptibility or synergy assays. However, the use of pharmacodynamic methods provided insight into the potential of these combinations to prevent the emergence of resistance during therapy, a problem that is a primary concern for the treatment of *P. aeruginosa*.

CONCLUDING REMARKS

Antibacterial resistance is an inevitable consequence of the use and overuse of antibiotics in the environment and its inevitability is heightened by suboptimal pharmacodynamics. The emergence of multidrug resistance among prominent gram-positive and gram-negative pathogens presents serious therapeutic problems. As the development of stronger or novel antimicrobial agents decline, we may be faced with the feared "postantibiotic era." Already clinicians are facing the dilemma of treating bacteria, which are resistant to virtually all antibacterial agents. Now is the time for scientists to begin searching for ways to prevent or slow the emergence of resistance in the environment. One approach is to control the use of antibiotics in the environment and rotate antibiotics before resistance becomes a problem. This theory is referred to as antibiotic cycling (94). A further step, which can be taken is to scientifically optimize antimicrobial therapy such that the evolution of mutational resistance is slowed. This requires dosing antibiotics such that peak concentrations at the site of infection exceed the MICs of the parent and any resistant subpopulations that might be present. A final approach is to identify combinations of antibacterial agents that can prevent, or slow, the emergence of resistance during therapy. Through the combination of judicious antimicrobial use and pharmacodynamically based optimization of dosing strategies, the evolution of antibiotic resistance can be curved and the "postantibiotic era" need not become a reality.

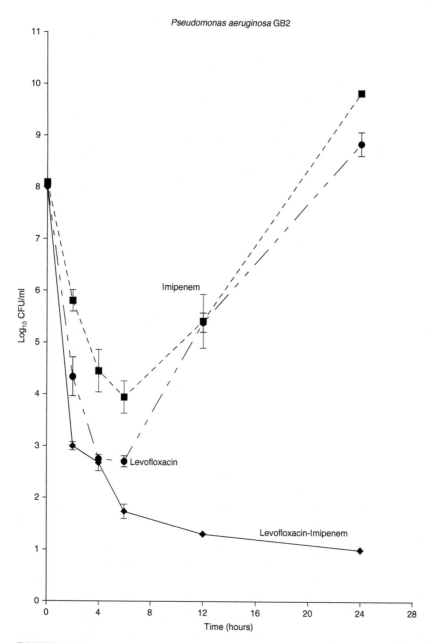

FIGURE 4 Pharmacodynamics of levofloxacin, imipenem, and levofloxacin-imipenem against *Pseudomonas aeruginosa* GB2 in an IVPM. The pharmacokinetics of the 750 mg dose of levofloxacin and 250 mg dose of imipenem was simulated with the IVPM. Peak concentrations of levofloxacin were dosed into the model at 0 and 24 hours, whereas peak concentrations of imipenem were dosed into the model at 0.8, 16, and 24 hours. Elimination pharmacokinetics was simulated with the IVPM. Each datum point represents the mean number of viable bacteria per milliliter of culture for duplicate experiments. Error bars show standard deviations. *Abbreviation*: IVPM; in vitro pharmacodynamic model.

REFERENCES

1. Chow JW, Fine MJ, Schlaes EADM. *Enterobacter* bacteremia: Clinical features and emergence of antibiotic resistance during therapy. Ann Int Med 1991; 115:585–590.
2. Holmberg SD, Solomon SL, Blake PA. Health and economic impacts of antimicrobial resistance. Rev Infect Dis 1987; 9:1065–1078.
3. Phelps CE. Bug/drug resistance: sometimes less is more. Med Care 1989; 27:194–203.
4. Robert J, Moellering C. *Enterococcus* species, *Streptococcus bovis*, and *Leuconostoc* species. In: Mandell GL, Bennett R, Dolin R, eds. Principles and Practices of Infectious Diseases. 4th ed. New York: Churchill Livingston, 1995:1826–1835.
5. Eliopoulous GM. Antibiotic resistance in *Enterococcus* species: an update. In: Remington JS, ed. Current Clinical Topics in Infectious Diseases. Cambridge, MA: Blackwell Science, 1996:21–51.
6. CDC. National nosocomial infections surveillance (NNIS) system report, data summary from January 1992–April 2000. Am J Infect Control 2000; 28:429–448.
7. Tenover FC, McDonald LC. Vancomycin-resistance staphylococci and enterococci: epidemiology and control. Curr Opin Infect Dis 2005; 18:300–305.
8. Spink WW. Quantitative action of penicillin inhibitor from penicillin-resistant strains of staphylococci. Science 1945; 102:221.
9. Jones RN, Kehrberg EN, Erwin ME. Prevalence of important pathogens and antimicrobial activity of parenteral drugs at numerous medical centers in the United States. I. Study on the threat of emerging resistances. Diagn Microbiol Infect Dis 1994; 19:203–215.
10. Barber M. Methicillin-resistant staphylococci. J Clin Pathol 1961; 14:385.
11. Hartman BM, Tomasz A. Low-affinity penicillin binding protein associated with beta-lactam resistance in *Staphylococcus aureus*. J Bacteriol 1984; 158:513–516.
12. Jones ME, Barry AL, Gardiner RV. The prevalence of staphylococcal resistance to penicillinase-resistant penicillins. A retrospective and prospective trail of isolates from 40 medical centers. Diagn Microbiol Infect Dis 1989; 12:383–394.
13. Al-Tawfiq JA, Aldaabil RA. Community-acquired MRSA bacteremic necrotizing pneumonia in a patient with scrotal ulceration. J Infect 2005; 51:241–243.
14. McAdams RM, Mazuchowski E, Ellis MW, Rajnik M. Necrotizing staphylococcal pneumonia in a neonate. J Perinatol 2005; 25:677–679.
15. Obed A, Schnitzbauer AA, Bein T, Lehn N, Linde HJ, Schlitt HJ. Fatal pneumonia caused by Panton-Valentine Leucocidine-positive methicillin-resistant Staphylococcus aureus (PVL-MRSA) transmitted from a healthy donor in living-donor liver transplantation. Transplantation 2006; 81:121–124.
16. Tenover FC, McDougal LK, Goering RV, Killgore G, Projan SJ, Dunman PM. Characterization of a strain of community-associated methicillin-resistant Staphylococcus aureus widely disseminated in the United States. J Clin Microbiol 2006; 44:108–118.
17. CDC. *Staphylococcus aureus* with reduced susceptibility to vancomycin – United States. MMWR 1997; 46:765–766.
18. Kiramatsu K, Hanaki H, Ino T, Yabuta K, Oguri T, Tenover FC. Methicillin-resistant *Staphylococcus aureus* clinical strain with reduced vancomycin susceptibility. J Antimicrob Chemother 1997; 40:135–136.
19. Howden BP. Recognition and management of infections caused by vancomycin-intermediate Staphylococcus aureus (VISA) and heterogenous VISA (hVISA). Int Med J 2005; 35(suppl 2): S136–S140.
20. Turner J, Howe RA, Wootton M, et al. The activity of vancomycin against heterogeneous vancomycin-intermediate methicillin-resistant Staphylococcus aureus explored using an in vitro pharmacokinetic model. J Antimicrob Chemother 2001; 48:727–730.
21. Koehl JL, Muthaiyan A, Jayaswal RK, Ehlert K, Labischinski H, Wilkinson BJ. Cell wall composition and decreased autolytic activity and lysostaphin susceptibility of glycopeptide-intermediate Staphylococcus aureus. Antimicrob Agents Chemother 2004; 48:3749–3757.
22. Sakoulas G, Eliopoulous GM, Fowler VG, et al. Reduced susceptibility of Staphylococcus aureus to vancomycin and platelet microbicidal protein correlates with defective autolysis and loss of accessory gene regulator (agr) function. Antimicrob Agents Chemother 2005; 49:2687–2692.

23. Kislak JW, Razavi LMB, Daly AK, Finland M. Susceptibility of pneumococci to nine antibiotics. Am J Med Sci 1965; 250:262–268.
24. Applebaum PC, Bhamjee A, Scragg JN, Hallett AJ, Bowen AF, Cooper RC. *Streptococcus pneumoniae* resistant to penicillin and chloramphenicol. Lancet 1997; ii:995–997.
25. Howes VJ, Mitchell RG. Meningitis due to relatively penicillin-resistant pneumococcus. Br Med J 1976; 1:996.
26. Naraqi S, Kirkpatrick GP, Kabins S. Relapsing pneumococcal meningitis: Isolation of an organism with decreased susceptibility to penicillin G. J Pediatr 1974; 85:671–673.
27. Johnson DM, Stilwell MG, Fritsche TR, Jones RN. Emergence of multidrug-resistant Streptococcus pneumoniae: report from the SENTRY Antimicrobial Surveillance Program (1999–2003). Diagn Microbiol Infect Dis 2006.
28. Dowson CG, Coffey TJ, Kell C, Whiley RA. Evolution of penicillin resistance in *Streptococcus pneumoniae*: The role of *Streptococcus mitis* in the formation of a low affinity PBP2B in *Streptococcus pneumoniae*. Molec Microbiol 1993; 9:635–643.
29. Dowson CG, Hutchison A, Brannigan JA, et al. Horizontal transfer of penicillin-binding protein genes in penicillin-resistant clinical isolates of *Streptococcus pneumoniae*. Proc Nat Acad Sci 1989; 86:8842–8846.
30. Lister P. Multiply-resistant pneumococcus: therapeutic problems in the management of serious infections. Eur J Clin Microbiol Infect Dis 1995; 14:18–25.
31. Farrell DJ, Morrissey I, Bakker S, Buckridge S, Felmingham D. Global distribution of TEM-1 and ROB-1 beta-lactamases in *Haemophilus influenzae*. J Antibiot 2005; 56:773–776.
32. Fluit AC, Florijn A, Verhoef J, Milatovic D. Susceptibility of European beta-lactamase-positive and -negative Haemophilus influenzae isolates from the periods 1997/1998 and 2002/2003. J Antimicrob Chemother 2005; 56:133–138.
33. Johnson DM, Sader HS, Fritsche TR, Biedenbach DJ, Jones RN. Susceptibility trends of *Haemophilus influenzae* and *Moraxella catarrhalis* against orally administered antimicrobial agents: five-year report from the SENTRY Antimicrobial Surveillance Program. Diagn Microbiol Infect Dis 2003; 47:373–376.
34. Mortensen JE, Jacobs MR, Koeth LM. Presented at the Interscience Conference on Antimicrobial Agents and Chemotherapy, Washington, DC, 1997.
35. Heilmann KP, Rice CL, Miller AL, et al. Decreasing prevalence of beta-lactamase production among respiratory tract isolates of Haemophilus influenzae in the United States. Antimicrob Agents Chemother 2005; 49:2561–2564.
36. Sanders CC, Sanders EC. Beta-lactam resistance in gram-negative bacteria: global trends and clinical impact. Clin Infect Dis 1992; 15:824–839.
37. Sanders CC, Iaconis JP, Bodey GP, Samonis G. Resistance to ticarcillin-potassium clavulanate among clinical isolates of the family *Enterobacteriaceae*: Role of PSE-1 beta-lactamase and high levels of TEM-1 and SHV-1 and problems with false susceptibility in disk diffusion tests. Antimicrob Agents Chemother 1988; 32:1365–1369.
38. Thomson KS, Weber DA, Sanders CC, Sanders WE. Beta-lactamase production in members of the family Enterobacteriaceae and resistance to beta-lactamase-enzyme inhibitor combinations. Antimicrob Agents Chemother 1990; 34:622–627.
39. Blazquez J, Baquero M, Canton R. Characterization of a new TEM-type beta-lactamase resistant to clavulanate, sulbactam, and tazobactam in a clinical isolate *Escherichia coli*. Antimicrob Agents Chemother 1993; 37:2059–2063.
40. Thomson CJ, Amyes SGB. Emergence of clavulanic acid-resistant TEM beta-lactamase in a clinical strain. FEMS Microbiol Lett 1992; 91:113–118.
41. Hanson ND. AmpC beta-lactamases: what do we need to know for the future? J Antimicrob Chemother 2003; 52:2–4.
42. Perez F, Hanson ND. Detection of plasmid-mediated AmpC beta-lactamase genes in clinical isolates by using multiplex PCR. J Clin Microbiol 2002; 40:2153–2162.
43. Jacoby GA. Beta-lactamase nomenclature. Antimicrob Agents Chemother 2006; 50:1123–1129.
44. Bonnet R. Growing group of extended-spectrum beta-lactamases: the CTX-M enzymes. Antimicrob Agents Chemother 2004; 48:1–14.
45. Bradford PA, Bratu S, Urban C, et al. Emergence of carbapenem-resistant Klebsiella species possessing the class A carbapenem-hydrolyzing KPC-2 and inhibitor-resistant TEM-30 beta-lactamases in New York City. Clin Infect Dis 2004; 39:55–60.

46. Bratu S, Mooty M, Nichani S, et al. Emergence of KPC-possessing Klebsiella pneumoniae in Brooklyn, New York: epidemiology and recommendations for detection. Antimicrob Agents Chemother 2005; 49:3018–3020.

47. Bratu S, Tolaney P, Karumudi U, et al. Carbapenemase-producing Klebsiella pneumoniae in Brooklyn, NY: molecular epidemiology and in vitro activity of polymyxin B and other agents. J Antimicrob Chemother 2005; 56:128–132.

48. Woodford N, Tierno PM, Young K, et al. Outbreak of Klebsiella pneumoniae producing a new carbapenem-hydrolyzing class A beta-lactamase, KPC-3, in a New York Medical Center. Antimicrob Agents Chemother 2004; 48(12):4793–4799.

49. Jacoby GA. Genetics of extended-spectrum beta-lactamases. Eur J Clin Microbiol Infect Dis 1994; 13(suppl 1):2–11.

50. Philippon A, Labia R, Jacoby GA. Extended-spectrum beta-lactamases. Antimicrob Agents Chemother 1989; 33:1131–1136.

51. Bradford PA, Urban C, Mariano N, Projan SJ, Rahal JJ, Bush K. Imipenem resistance in *Klebsiella pneumoniae* is associated with the combination of ACT-1, a plasmid-mediated AmpC beta-lactamase, and the loss of an outer membrane protein. Antimicrob Agents Chemother 1997; 41:563–569.

52. Abdalhamid B, Pitout JD, Moland ES, Hanson ND. Community-onset disease caused by Citrobacter freundii producing a novel CTX-M beta-lactamase, CTX-M-30, in Canada. Antimicrob Agents Chemother 2004; 48:4435–4437.

53. Sanders CC. Beta-lactamases of gram-negative bacteria: new challenges for new drugs. Clin Infect Dis 1992; 14:1089–1099.

54. Sanders CC. Chromosomal cephalosporinases responsible for multiple resistance to new beta-lactam antibiotics. Ann Rev Microbiol 1987; 41:573–593.

55. Hanson ND, Sanders CC. Regulation of inducible AmpC beta-lactamase expression among Enterobacteriaceae. Curr Pharm Des 1999; 5:881–894.

56. Schmidtke AJ, Hanson ND. A model system to evaluate the effect of *ampD* mutations on AmpC-mediated beta-lactam resistance. Antimicrob Agents Chemother 2006; In Press.

57. Naas T, Noordmann P. Analysis of a carbapenem-hydrolyzing class A beta-lactamase from Enterobacter cloacae and of its LysR-type regulatory protein. Proc Nat Acad Sci 1994; 91:7693–7697.

58. Rasmussen BA, Bush K, Keeney D, et al. Characterization of IMI-1 beta-lactamase, a class A carbapenem-hydrolyzing enzyme from Enterobacter cloacae. Antimicrob Agents Chemother 1996; 40:2080–2086.

59. Walsh TR. The emergence and implications of metallo-beta-lactamases in Gram-negative bacteria. Clin Microbiol Infect 2005; 11(suppl 6):2–9.

60. Walsh TR, Toleman MA, Poirel L, Nordmann P. Metallo-beta-lactamases: the quiet before the storm? Clin Microbiol Rev 2005; 18:306–325.

61. Walther-Rasmussen J, Hoiby N. OXA-type carbapenemases. J Antimicrob Chemother 2006; 57:373–383.

62. Bradford PA. Extended-spectrum beta-lactamases in the 21st century: characterization, epidemiology, and detection of this important resistance threat. Clin Microbiol Rev 2001; 14: 933–951.

63. Yan JJ, Ko WC, Tsai SH, Wu HM, Wu JJ. Outbreak of infection with multidrug-resistant Klebsiella pneumoniae carrying bla(IMP-8) in a university medical center in Taiwan. J Clin Microbiol 2001; 39:4433–4439.

64. Sanders EC, Sanders CC. Inducible beta-lactamases: clinical and epidemiological implications for use of newer cephalosporins. Rev Infect Dis 1988; 10:830–838.

65. Giwercmann B, Lambert PA, Rosdahl VT, Shand GH, Hoiby N. Rapid emergence of resistance in *Pseudomonas aeruginosa* in cystic fibrosis patients due to in-vivo selection of stable partially-derepressed beta-lactamase-producing strains. J Antimicrob Chemother 1990; 26:247–259.

66. Sanders CC, Sanders EC. Emergence of resistance during therapy with newer beta-lactam antibiotics: role of inducible beta-lactamases and implications for the future. Rev Infect Dis 1983; 5:639–648.

67. Bradford PA, Sanders CC. Development of test panel of beta-lactamases expressed in a common *Escherichia coli* host background for evaluation of new beta-lactam antibiotics. Antimicrob Agents Chemother 1995; 39:308–313.

68. Shlaes DM, Bouvet A, Devine C, Schlaes JH, Al-Obeid S, Williamson R. Inducible, transferable resistance to vancomycin in *Enterococcus faecalis* A256. Antimicrob Agents Chemother 1989; 33:198–203.

69. Sanders CC. Cefepime: the next generation? Clin Infect Dis 1993; 17:369–379.

70. Marchou B, Michea-Hamzehpour M, Lucain C, Pechere JC. Development of beta-lactam-resistant *Enterobacter cloacae* in mice. J Infect Dis 1987; 156:369–373.

71. Thomson KS, Sanders CC. Dissociated resistance among fluoroquinolones. Antimicrob Agents Chemother 1994; 38:2095–2100.

72. Croisier D, Ettiene M, Piroth L, et al. In vivo pharmacodynamic efficacy of gatifloxacin against *Streptococcus pneumoniae* in an experimental model of pneumonia: impact of the low levels of fluoroquinolone resistance on the enrichment of resistant mutants. J Antimicrob Chemother 2004; 54:640–647.

73. Firsov AA, Vostrov SN, Lubenko IY, Drilica K, Portnoy YA, Zinner SH. In vitro pharmacodynamic evaluation of the mutant selection window hypothesis using four fluoroquinolones against *Staphylococcus aureus*. Antimicrob Agents Chemother 2003; 47: 1604–1613.

74. Drusano G, Johnson D, Rosen M, Standiford M. Pharmacodynamics of a fluoroquinolone antimicrobial agent in a neutropenic rat model of *Pseudomonas* sepsis. Antimicrob Agents Chemother 1993; 37:483–490.

75. Preston SL, Drusano G, Berman AL, et al. Pharmacodynamics of levofloxacin: a new paradigm for early clinical trials. JAMA 1998; 279:125–129.

76. Blaser J, Stone BB, Groner MC, Zinner SH. Comparative study with enoxacin and netilmicin in a pharmacodynamic model to determine importance of antibiotic peak concentration to MIC for bactericidal activity and emergence of resistance. Antimicrob Agents Chemother 1987; 31:1054–1060.

77. Marchibanks CR, McKiel JR, Gilbert DH, et al. Dose ranging and fractionation of intravenous ciprofloxacin against *Pseudomonas aeruginosa* and *Staphylococcus aureus* in an in vitro pharmacokinetic model. Antimicrob Agents Chemother 1993; 37:1756–1763.

78. Letendre ED, Mantha R, Turgeon PL. Selection of resistance by piperacillin during *Pseudomonas aeruginosa* endocarditis. J Antimicrob Chemother 1988; 22:557–562.

79. Masterton RG, Garner PJ, Harrison NA, et al. Timentin resistance. Lancet 1987; 2:975–976.

80. Scully BE, Ores CN, Prince AS, Neu HC. Treatment of lower respiratory tract infections due to *Pseudomonas aeruginosa* in patients with cystic fibrosis. Rev Infect Dis 1985; 7(suppl): S669–S674.

81. Akova M, Yang Y, Livermore DM. Interactions of tazobactam and clavulanate with inducibly- and constitutively-expressed class I beta-lactamases. J Antimicrob Chemother 1990; 25:199–208.

82. Bush K. Beta-lactamase inhibitors from laboratory to clinic. Clin Microbiol Rev 1988; 1: 109–123.

83. Bush K, Freudenberger JS, Sykes RB. Interaction of azthreonam and related monobactams with beta-lactamase from gram-negative bacteria. Antimicrob Agents Chemother 1982; 22:414–420.

84. Sakuri Y, Yoshida Y, Saitoh K, Nemoto M, Yamaguchi A, Sawaii T. Characteristics of aztreonam as a substrate inhibitor and inducer for beta-lactamases. J Antibiot 1990; XLIII:403–410.

85. Lister PD, Sanders CC, Sanders EC. Cefepime-aztreonam: a unique double beta-lactam combination for *Pseudomonas aeruginosa*. Antimicrob Agents Chemother 1998; 42:1610–1619.

86. Masuda N, Sakagawa E, Ohya S. Outer membrane proteins responsible for multiple drug resistance in *Pseudomonas aeruginosa*. Antimicrob Agents Chemother 1995; 39:645–649.

87. Yoneyama H, Nakae T. Mechanism of efficient elimination of protein D2 in outer membrane of imipenem-resistant *Pseudomonas aeruginosa*. Antimicrob Agents Chemother 1993; 37:2385–2390.

88. Kohler T, Michea-Hamzehpour M, Epp SF, Pechere JC. Carbapenem activities against *Pseudomonas aeruginosa*: Respective contributions of OprD and efflux systems. Antimicrob Agents Chemother 1999; 43:424–427.

89. Masuda N, Sakagawa E, Ohya S, Gotoh N, Tsujimoto H, Nishino T. Substrate specificities of MexAB-OprM, MexCD-OprJ and MexXY-OprM efflux pumps in *Pseudomonas aeruginosa*. Antimicrob Agents Chemother 2000; 44:3322–3327.

90. Lister PD, Wolter DJ. Levofloxacin-imipenem combination prevents the emergence of resistance among clinical isolates of *Pseudomonas aeruginosa*. Clin Infect Dis 2005; 40:S105–S114.
91. Fukuda H, Hosaka M, Iyobe S, Gotoh N, Nishino T, Hirai K. *nfx*C-type quinolone resistance in a clinical isolate of *Pseudomonas aeruginosa*. Antimicrob Agents Chemother 1995; 39: +790–792.
92. Kohler T, Michea-Hamzehpour M, Henz U. Characterization of MexE-MexF-OprN, a positively regulated multidrug efflux system of *Pseudomonas aeruginosa*. Mol Microbiol 1997; 23:345–354.
93. Lister P, Black J, Wolter DJ. Presented at the European Congress of Clinical Microbiology and Infectious Diseases, Prague, Czech Republic; 2004.
94. Sanders CC, Sanders WE. Cycling of antibiotics: an approach to circumvent resistance in specialized units of a hospital. Clin Microbiol Infect 1998; 1:223–225.

24   The Principles of Pharmacoeconomics

Craig I. Coleman
University of Connecticut School of Pharmacy, Storrs, Connecticut, and
Pharmacoeconomics and Outcomes Studies Group, Hartford Hospital, Hartford,
Connecticut, U.S.A.

Effie L. Kuti
University of Connecticut School of Pharmacy, Storrs, Connecticut, U.S.A.

Joseph L. Kuti
Clinical and Economic Studies, Center for Anti-infective Research and Development,
Hartford Hospital, Hartford, Connecticut, U.S.A.

INTRODUCTION

In 2001, the United States spent a total of 1.4 trillion dollars on health care (1). This comprised approximately 14.1% of the nation's gross domestic product (GDP). Much of this spending has resulted from the increased development and utilization of sophisticated health-care therapies, technology, and services. While these advances in care empower health-care providers with new and effective means with which to prevent and treat disease, health-care resources are not unlimited (2). Fifty-two million Americans are without health-care coverage due primarily to rising costs and subsequent employer curtailment of health-care benefits, and many other Americans are sharing a high burden of the cost of healthcare in the forms of higher deductibles, copayments, and fees (1,3). As our nation struggles to provide more comprehensive care for all its citizens, the health-care system will have to find ways to meet the increased demands with the limited resources available. One of the first such hurdles that will have to be addressed will result from the approval of the Medicare Modernization Act (MMA), which calls for the provision of a basic prescription benefit to Medicare beneficiaries starting in 2006 (4). With the creation of the MMA, the impetus to provide cost-conscious healthcare, particularly relating to pharmaceutical products and services, has never been greater.

Until recently, new potential pharmaceutical products or services only needed to demonstrate efficacy and safety (5). These objectives were achieved through a series of randomized trials in which a new pharmaceutical product or service was compared to a placebo or an alternative agent commonly used to treat the disease in question. However, if we are to provide cost-conscious healthcare, efficacy and safety alone can no longer be the sole criterion used. Many countries including New Zealand, Australia, and Canada have mandated that considerations of cost be addressed in the development and use of new pharmaceutical products and services (6–11). As a result, health-care providers are being asked to incorporate considerations of cost into their decisions regarding the care they provide to their patients, creating a challenging work environment.

In response to this challenge, many decision-makers have developed strategies focusing on determining the least-expensive alternative. However, the least-expensive alternative is not always the alternative that provides the most "bang for the buck." For example, imagine a new oral antibiotic is being marketed for the treatment of skin and soft tissue infections and a hospital's pharmacy and therapeutics committee has to make the decision as to whether the drug should be available to its prescribers. The acquisition cost of this new antibiotic is $250 more per treatment course than the commonly used alternative, which happens to be an agent requiring i.v. administration. A formulary decision-maker would never allow such a drug onto formulary if the decision were solely based upon determining the least-expensive alternative. However, let us now consider additional costs and outcomes of the new antibiotic. While the newer agent has a higher acquisition cost, the absence of the need for i.v. administration might result in lesser costs associated with i.v. administration and admixture and nursing time. Furthermore, a large portion of the cost associated with room and board might be averted by decreasing a patient's total length of stay (LOS) since the patients would not be required to remain in the hospital to receive their i.v. antibiotic, but rather could be treated for a greater proportion of the treatment course as an outpatient. It is fairly easy to imagine that when one considers additional costs and outcomes, the total cost associated with utilizing the new antibiotic could be less than the alternatives. But then again, what if the new antibiotic was more or less efficacious? What about the cost of treating adverse drug events that may vary between the two agents?

In recognizing the need for a better way to make difficult decisions about the use of pharmaceutical products and services, the health-care community has increasingly discovered the utility of various pharmacoeconomic methodologies as aids in making informed allocation decisions.

The purpose of this chapter is to provide the reader with a basic understanding of pharmacoeconomic principles and to identify the inherent strengths and weaknesses of the various pharmacoeconomic methodologies. This chapter is not intended to make a decision for the reader regarding any single antimicrobial. Rather, it should enable the reader to apply such methodologies in their own evaluation of old and new drug entities, as well as in their critique of the available literature.

DEFINING MAJOR PHARMACOECONOMIC TERMS

The fundamental goal of pharmacoeconomic evaluation is to identify and measure relevant costs and outcomes, determine the value of these parameters, and then employ them to ascertain which alternative produces the best outcome for the resources invested (12). The inclusion of all relevant costs and outcomes when evaluating competing pharmaceutical products or services differentiates pharmacoeconomics from simple cost-containment strategies.

Costs are the value of resources consumed as a result of a pharmaceutical product or service. Costs included in pharmacoeconomic evaluations can generally be stratified into one of three categories: direct, indirect, or intangible costs. Direct costs are incurred as a result of preventing, detecting, or treating a disease. Costs related directly to the consumption of medical products or services (e.g., drugs, laboratory tests, and office visits) can be further described as direct medical costs. Alternatively, costs that result from preventing, detecting, or treating a disease, but do not involve the purchasing of medical services or products (e.g., transportation to site of treatment) can be classified as direct nonmedical costs.

Indirect costs are the end result of decreases in a person's productive contribution to society stemming from premature morbidity or mortality, whereas intangible costs are those associated with pain and suffering. Indirect and intangible costs are considerably more difficult to value than direct costs but are by no means less important. Neither indirect nor intangible costs are easily expressed in terms of dollars, although methods for doing so currently exist and are briefly discussed later in this chapter.

Costs can also be classified as opportunity, fixed, or incremental costs. Opportunity costs include the cost of benefits not realized because an alternative intervention was chosen. For example, your institution only has the staff and resources to initiate either an influenza or a pneumococcal vaccination clinic. When the resources available are devoted to the influenza clinic, the potential benefits of the pneumococcal clinic are lost. Fixed costs are those that generally do not change based upon changes in utilization. Good examples of these are labor costs. The labor costs of a nurse to administer an i.v. antibiotic will remain the same regardless of whether he/she administer a drug four times a day or once daily as a continuous infusion as the nurse is paid an hourly rate regardless (13). That is of course unless administering the drug once daily leads to less overtime or less need to hire temporary nurses because the staff nurse's time can now be spent providing other services. This situation is specific to each medical center and if applicable, could be added to the pharmacoeconomic analysis as a type of direct medical cost. Finally, incremental costs represent the additional cost to gain additional outcomes when choosing one treatment alternative over another. Determining this incremental cost is particularly useful since newer products and services entering the market generally provide added benefit for an additional cost.

In addition to the types of costs, the other important variable in a pharmacoeconomic analysis is the outcome, defined as the final consequence of the selection of a particular drug or service. Outcomes are often categorized as being clinical, humanistic, or economic (12). Clinical outcomes are those commonly seen evaluated in most randomized controlled trials of new pharmaceutical products and services. For a new antiretroviral agent to treat patients with human immunodeficiency virus, clinical outcomes might include changes in viral load, increases in CD4 cell counts, or the absence of an opportunistic infection. Humanistic variables include those related to patient functional status, quality-of-life, and satisfaction. Economic outcomes are defined as the direct, indirect, and intangible costs resulting from an intervention (e.g., a decrease in LOS, health-care utilization, or drug requirements).

When assessing costs and outcomes, it is important to note that the choice of what costs and outcomes will be measured, how they will be valued, and over what timeframe will depend on the perspective (i.e., viewpoint) to which the evaluation is conducted. The most common perspectives that are utilized include, but are not limited to, society's, the provider's [e.g., hospital, heath maintenance organization (HMO)], the payer's (e.g., government and employer), or the patient's. When in doubt as to the most appropriate perspective, the societal perspective should be used, as it is the broadest in scope including direct, indirect, and intangible costs and outcomes (6–11). While an evaluation is most commonly conducted from a single perspective, multiple perspectives can be utilized when necessary (6–11). This allows decision-makers to determine whether the results of their analyses hold true in multiple different perspectives, therefore, strengthening their confidence in the results.

Evaluations are most helpful when they are conducted from the perspective of the decision-maker. As such, a hospital would benefit most directly from utilizing a provider's perspective, an HMO would be more interested in the payer's perspective, and the Infectious Disease Society of America might be most interested in multiple perspectives, including that of the societal, provider, and patient. For example, if a pharmacoeconomic analysis was undertaken to determine the costs and outcomes of patients taking various highly active antiretroviral therapy regimens, certain costs and outcomes (e.g., reductions in viral load, increases in CD4 cell counts, and drug costs) would without question be included regardless of the perspective. However, an evaluation from an HMO's perspective would probably not include indirect costs such as decreased productivity and intangible costs and outcomes such as pain and suffering in the analysis since these factors do not impact the company's bottom line. However, if the same evaluations were conducted from the perspective of an employer or a patient, these indirect and intangible costs and outcomes would be of much greater concern and would need to be included. Additionally, while the evaluation will likely include costs relating to the drugs, the value assigned will also vary from perspective to perspective.

VALUING
Assigning Value to Costs and Outcomes
As previously mentioned, there are a number of different types of costs and outcomes that need to be assigned a value for the purposes of conducting a pharmacoeconomic evaluation. Costs and outcomes related to drug and resource utilization, labor, morbidity, and mortality are the most commonly measured and valued and therefore will be focused upon in this chapter.

Drug Costs
Valuation of drug costs and other direct costs can be done through a variety of methods. Most commonly, drug cost is assigned a value based either upon the average wholesale price (AWP) or actual acquisition cost (AAC). The AWP is the price paid for a specific drug by retailers to a wholesaler (14). The AWP will remain relatively constant for a given wholesaler regardless of whom they provide drug to and therefore provides a more externally valid estimation of drug cost. However, institutions and HMOs often purchase drugs at a discounted rate through a drug wholesaler. In this case, the AAC is the actual purchase cost of the drug by the institution in question and can vary markedly from the AWP (Table 1) (14,15).

The choice of using AWP or AAC in the pharmacoeconomic analyses is dependent upon the perspective or the extent of external validity sought by the analyst. When an evaluation is conducted from a patient's perspective, the AAC of a drug could be estimated based upon the out-of-pocket cost paid by the patient. When an analysis is conducted from the perspective of providers and they used

TABLE 1   Comparison of AWP and AAC of Selected Antibiotics

Drug	AWP	Hospital AAC	Percent difference
Amoxicillin 500 mg tablet	$0.37	$0.09	−311%
Levofloxacin 500 mg tablet	$10.14	$8.46	−19%
Linezolid 600 mg tablet	$64.41	$51.53	−25%

Abbreviations: AWP, average wholesale price; AAC, actual acquisition cost.
Source: From Ref. 15.

AWP, it is important to check your institution's AAC. By using your AAC, you reduce the ability to extrapolate your results to hospitals across the country but you increase the internal validity of the evaluation. To address this issue, pharmacoeconomic analysts should conduct sensitivity analysis (6–11) where they use either AWP or AAC but then reevaluate their results to examine the effect of varying the drug cost from AWP to AAC, or vice versa, or by a certain percentage (usually 25–50%) to see if it still results in the same conclusion (sensitivity analysis is discussed in greater detail later). In other cases, drug cost can be varied to its highest and lowest extremes or a breakeven point can be determined to demonstrate the cost a drug would have to exceed to no longer be advantageous over its competitor.

Labor and Personnel Costs

The time spent by health-care providers to complete such tasks as admixing or administering drugs or performing other procedures and services also needs to be considered in pharmacoeconomic evaluations. Traditionally, time-motion studies have been conducted as part of pharmacoeconomic evaluations to estimate the costs associated with labor costs (13,16). Time-motion studies involve timing multiple observations of a health-care provider performing a task in order to gain a representative value of the time it takes to perform. This time is then multiplied by the hourly wage rate of the provider in order to estimate the cost of performing the given task. For example, in a time-motion study conducted by Florea et al. (13), it was demonstrated that continuous infusions of piperacillin/tazobactam required less nurse administration time than intermittent bolus dosing, ultimately resulting in the continuous infusion regimen being a most cost-efficient option. Again, careful interpretation of such analyses is required, as just because the time spent is less, thus leading to lower costs, it is highly unlikely that a decision-maker would choose to reduce the number of nurses on that hospital unit. Consideration, however, may be given to the ability of that nurse to spend their time efficiently completing other patient responsibilities, which could improve patient care and thus indirectly reduce costs.

Health-Care Utilization Costs

The cost of other resource utilization such as hospitalizations, emergency room visits, and office visits must be valued as well. Unfortunately, providers frequently do not know the exact costs of providing these resources (17). What is commonly known is the charge generated per unit of a resource consumed. While these charges have been used as proxies for costs in the past, it has been found that charges may bear little resemblance to costs and therefore use of charges as proxies for costs may lead decision-makers to draw unwarranted conclusions from evaluations using such a methodology. To approximate the cost of hospitalizations and resources utilized during a hospitalization, charges from individual patient bills can be converted to costs using a cost-to-charge ratio (CCR) (18). At each institution, a CCR for each department is calculated by taking the institution's total accounting costs and dividing it by its total billing costs for a finite period of time. The resultant ratio approximates the percentage "mark up" of products or service. By taking the charges for each department (or product within a given department) and multiplying it by the CCR, one can calculate an estimate of the cost of that product or service. The use of this methodology will provide the most accurate information for a specific hospital's decision-maker; however, as the CCRs are hospital-specific, the external validity of the results should be considered prior to extrapolating the evaluation's conclusions to other institutions. When patient bills or CCRs are not available, the use of national average diagnosis-related group payments and Medicare Fee Schedules can be used to approximate hospitalization, emergency room, and office visit costs (18). Regardless of how resource consumption is valued, the methodology should be used consistently in valuing all costs throughout an evaluation (19).

Indirect and Intangible Costs

The most common methods for valuing indirect and intangible costs include the human capital and willingness-to-pay (WTP) approaches. These approaches will be briefly addressed in this chapter; more thorough discussion of their application can be found elsewhere (18).

The human capital approach assumes an individual's productive contribution can be measured based upon their earning capacity. While this may seem like a fair and reasonable approach, the use of the human capital approach results in an underestimation of the value of the young or elderly's contributions to society. Additional issues with the human capital approach include underestimating indirect costs due to job discrimination, failure to capture nonmarket work such as housekeeping and volunteer work, and its failure to consider intangible costs.

WTP requires that patients be asked the hypothetical question, "How much are you willing to pay to reduce the likelihood of morbidity or mortality related to a disease state?" The advantage of WTP is that it considers both indirect and intangible costs. However, it assumes that patients are capable of comprehending and valuing even small changes in risks. Additionally, one's WTP is likely related to one's earnings. Therefore, patients who earn a great deal of money may be willing to pay more to reduce their risk of an adverse health event than a patient who earns significantly less or is unemployed.

Discounting and Inflation

Once the costs and outcomes to be measured are identified and values assigned, these values often must be adjusted to represent their present-day value. To this end, discounting or inflating of the costs and outcomes is often undertaken (18). Discounting is used in pharmacoeconomic evaluations to express costs and outcomes that occur in future years in present-day value (14). The reasoning for discounting is based upon the premise that individuals prefer to receive benefits today rather than in the future and that resources invested today could earn a return overtime. Discounting should be conducted when an evaluation spans a time period greater than one year using the following equation:

$$FV = PV/(1 + r)^t;$$

where FV = future value; PV = present value; r = the discount rate; and t = time. Consider a hypothetical treatment that costs $3000 in the first year, $2000 in the second, and $1000 each year thereafter. Assuming a discount rate of 6%, the cost in year 2 will be $1886, the cost in year 3 will be $892, and the total cost over three years will be $5778. Choosing a discount rate can dramatically affect the conclusions of the evaluation and the choice of discount rate is controversial (18). Discount rates used in pharmacoeconomic analyses have varied anywhere between 2% and 10%, with 5% being most common (6–11).

Alternatively, evaluations often have cost or outcome data that were assigned a value at a point in the past. For example, when a retrospective evaluation is undertaken, the value of both costs and outcomes was likely assigned at the time the resource was consumed or outcome realized. When conducting a study in 2005, the cost of a hospitalization occurring in 2002 must be adjusted to reflect 2005 dollars. In these situations, the cost must be "inflated" to represent present-day value. Many investigators have chosen to use the consumer price index (CPI) for medical care to represent this inflation rate (20). The CPI, which is calculated by the Department of Labor Statistics, represents the percentage change in the prices paid by urban consumers for a medical care–related products and services (20).

TYPES OF PHARMACOECONOMIC EVALUATIONS

Different methodologies for pharmacoeconomic evaluation including cost minimization, cost-benefit, cost-effectiveness, and cost-utility analyses exist to aid decisionmakers in making valid and detailed comparisons (Table 2). Each compares both costs and outcomes of alternatives, measuring costs in dollars. However, each differs in that they measure outcomes in different ways and thus, no single methodology can be universally applied to answer all pharmacoeconomic queries that may arise. Asking a few simple questions up front can usually make the decision as to which methodology to utilize more clear (Fig. 1).

Cost-Minimization Analysis

Cost-minimization analysis (CMA) is a pharmacoeconomic tool for comparing all relevant costs and outcomes of two or more therapeutic interventions. A requirement of CMA is that the outcomes of the intervention are equivalent and therefore it is unique among pharmacoeconomic methodologies. Since outcomes must be

TABLE 2 Comparison of Commonly Used Pharmacoeconomic Methodologies

Methodology	Cost	Outcomes	Calculation	Application
CMA	Dollars	Equivalent	$Costs_1 - Costs_2$	To compare alternative products or services that have equal efficacy and safety (e.g., brand versus generic substitutions)
CBA	Dollars	NB defined in dollars	$NB = \Sigma$ Benefits $- \Sigma$ Costs	To compare programs with different outcomes (e.g., a vaccination program and a colon cancer screening clinic)
CEA	Dollars	Natural units (e.g., infection cure rate)	$\dfrac{Costs_1 - Costs_2}{Cure_1 - Cure_2}$	To compare alternatives with similar indications or goals measured in the same unit of consequence (e.g., number of patients cured of an infection)
CUA	Dollars	QALY	$\dfrac{Costs_1 - Costs_2}{QALY_1 - QALY_2}$	To compare alternatives, which results in changes in both quality-of-life (QoL) and quantity-of-life (e.g., chemotherapy regimen)

Abbreviations: CMA, cost-minimization analysis; CBA, cost-benefit analysis; CEA, cost-effectiveness analysis; CUA, cost-utility analysis; QALY, quality-adjusted life years; NB, net benefit.

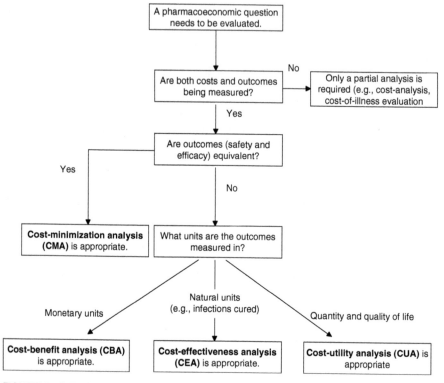

FIGURE 1 Selection of pharmacoeconomic methodology.

equivalent in CMA, the effectiveness term in these pharmacoeconomic evaluations drops out of the equation, and the results of CMA analyses are reported in terms of cost only. When measuring costs, all costs (direct, indirect, and intangible) should be included. Decision-makers will then choose the alternative with the lowest total cost.

CMA is most commonly used for comparisons of brand versus generic drugs, different routes of administration of the same drug, and different settings for administration of the same drug. For brand versus generic comparisons, often published equivalency ratings, such as those seen in the "Orange Book" (Approved Drug Products with Therapeutic Equivalence Evaluations), can be utilized to establish equivalency. When evaluating different products in a similar therapeutic class, evidence to support equivalent effectiveness of the treatment alternatives can be gathered from sources such as published studies or meta-analyses. Clinical intuition should not be the sole basis to establish equivalency. Incorrect assumptions concerning equivalent outcomes may lead to biased and misleading conclusions because no cost adjustment was made for differences in effectiveness. Many cost-minimization studies conducted and published with claims of equivalency are based on clinical assumptions rather than actual findings. Others measure outcomes as part of the analysis, yet fail to find a significant difference in outcomes across alternatives. When practices such as these are undertaken, considerations as to whether the statistical power of the test was sufficient to detect a difference in effectiveness and safety are warranted.

As CMA is one of the more simple forms of pharmacoeconomic analysis, numerous examples exist in the literature. Kotapati et al. evaluated the pharmacoeconomics of altering the dose of the carbapenem antibiotic, meropenem, from 1000 mg every eight hours to 500 mg every six hours (21). The 500 mg dosage regimen was derived through pharmacodynamic modeling that demonstrated the 2000 mg broken up four times a day would result in similar time above the MIC exposure as 3000 mg broken up three times daily, while reducing costs by approximately one-third, even with the additional cost of admixture supplies (22). After a protocol was put into place at the hospital, a retrospective review was conducted to evaluate first the clinical outcomes of each dosage regimen (21). Clinical success rates were 78% and 82% for the 500 mg every six hour and 1000 mg every eight hour regimens, respectively, ($p = 0.862$); therefore, a cost-minimization design was chosen for the economic analyses. These investigators evaluated the costs of the regimens on three separate levels, the first assessing only the AAC of both products, the second adding in costs of failing therapy or using concomitant antibiotics, and the third, evaluating the entire cost of hospital stay (i.e., bed costs and laboratory costs). The 500 mg every six hour regimen used less antibiotic over the course of treatment (13 g vs. 18 g, $p = 0.012$), thus leading to a significant difference in AAC between regimens. Additionally, when including costs of failures and the use of combination antibiotic therapy, the 500 mg every six hour regimen reduced costs by a median of $762 per patient ($p = 0.008$). The investigators did not consider the labor costs of administering one additional dose per day in their analysis, but provided reasoning that these fixed costs would not change the overall economics of each regimen and the time for nurses to administer one extra ADD-Vantage bag was insignificant based on previous time-motion studies (13). While the study was small, it does demonstrate the practicality and simplicity of applying pharmacoeconomic study techniques in a clinical setting.

Cost-Effectiveness Analysis

Cost-effectiveness analysis (CEA) identifies and compares costs and outcomes of competing interventions when outcomes are measured in the same units. In CEA, costs are measured in dollars while outcomes are measured in natural units such as lives saved or cases cured. Conducting CEA is most useful for decision-makers when a new product or intervention is more expensive and more effective than the alternatives or when it is less expensive and less effective. In these situations, CEA can determine the additional cost that must be invested to obtain one additional unit of outcome. When product or intervention is either less costly and more effective (dominant) or more costly and less effective, CEA is clearly not required, as a decision-maker should have no difficulty in choosing to accept or reject these scenarios.

Results of CEA are expressed as a ratio of costs to outcomes and can be reported as either the average cost-effectiveness ratio (ACER) or the incremental cost-effectiveness ratio (ICER). The ACER describes for each alternative the cost required for one unit of outcome. For example, antibiotic A costs $160 and yields a cure rate of 0.96 and antibiotic B costs $100 and yields a cure rate of 0.90. In this case, antibiotic A's ACER would be $160/0.96 or $167/infection cured and antibiotic B's ACER would be $100/0.90 or $111/infection cured. While having these results do yield some valuable information including the fact that it costs less to cure an infection with antibiotic B, it does not provide us with information concerning how much more we must invest to gain the additional efficacy achieved with antibiotic A. Therefore, it is often more useful to decision-makers to present the results of CEA as an ICER instead of separate ACERs for each alternative. The ICER can be calculated by dividing the difference in cost between the two alternatives by the difference in efficacy ($160–$100/0.96–0.90 = $1000/additional infection cured by using antibiotic A over B). Once provided with the ICER, decision-makers must determine whether the additional cost required is worth the additional outcome achieved. Although somewhat arbitrary, some have concluded that an intervention that results in ≤$20,000 per additional outcome gained would be considered cost-effective, whereas an intervention with an ICER of ≥$100,000 per additional outcome gained would not be (2).

Angus et al. conducted a CEA of a high-cost drug, drotrecogin alfa (activated), indicated for the reduction of mortality in adult patients with severe sepsis (23). Severe sepsis affects 750,000 patients each year, a third of whom die. Clinicians are motivated to prescribe drotrecogin alfa (activated) in severe sepsis given the statistically significant reduction in mortality rate found during the PROWESS trial (24), although the economic consequences of its use must be considered. Observed mortality rates were 30.8% for placebo compared with 24.7% for drotrecogin alfa (activated) ($p = 0.005$). The cost-effectiveness of drotrecogin alfa (activated) therapy was evaluated in incremental health-care costs for survival at 28 days. Incremental costs were measured as the difference in health-care costs (i.e., hospital, physician, study drug, and postdischarge costs) between drotrecogin alfa (activated) and placebo during the first 28 days. Study drug cost was estimated by multiplying the price of a vial of drotrecogin alfa (activated) by the minimum number of vials required based on actual patient dosage. Drotrecogin alfa (activated) increased costs by $9800 ± $2900 and survival by 0.061 ± 0.022 lives saved per treated patient, thus drotrecogin alfa (activated) cost $160,000 per life saved. The investigators further examined the cost-effectiveness of this agent over a survivor's lifetime as measured by the Quality-Adjusted Life Year (QALY); this is a form of CEA called cost-utility analysis (CUA) and will be described later. Drotrecogin alfa (activated) resulted in an incremental cost of $48,800 per QALY gained. Despite the high acquisition cost (i.e., approximately $7000), the investigators concluded that drotrecogin alfa (activated) had a favorable cost-effectiveness profile from a societal perspective and compared with many accepted health-care strategies. However, differences in mortality reduction, additional treatment modalities, drug acquisition costs, and number of years gained by sepsis survivors could change the cost-effectiveness ratio (25). Failure to address these factors as well as lack of follow-up beyond day 28 (i.e., long-term costs, quality-of-life, and duration of treatment) may worsen the CEA. Currently, drotrecogin alfa (activated) is the only drug currently approved for the treatment of severe sepsis, and hospitals struggling to determine in whom it should be used are encouraged to conduct their own CEA, specifically to assess if cost-effectiveness is maintained among patients with a lower risk of mortality as measured by severity scores such as APACHE II or SOFA (26).

A CEA performed by Dresser et al. determined the pharmacoeconomics of sequential i.v. to oral gatifloxacin therapy versus i.v. ceftriaxone with or without i.v. erythromycin to oral clarithromycin therapy to treat community-acquired pneumonia (CAP) (27). CAP, the most common cause of death from an infectious disease, is responsible for 64 million days of decreased activity, 39 million days of bed rest, and 10 million lost work days per year. Annual costs attributed to CAP approximate $23 billion; furthermore, antimicrobial costs are a small percentage of the overall cost for treatment. The predominant cost driver is hospital LOS. Clinical success rates and hospital stay were extrapolated from a randomized, controlled Phase III clinical trial. Patients randomized to the gatifloxacin arm received 400 mg i.v. with a switch to 400 mg oral gatifloxacin once a day. The ceftriaxone treatment group received 1 g i.v. once a day or 2 g i.v. once a day. Thirty-nine percent of patients in the ceftriaxone treatment group also received concomitant erythromycin 500 mg or 1 g i.v. four times per day for coverage of atypical organisms. The calculated cost-effectiveness ratio (CER) for gatifloxacin was $5236:1 compared with $7047:1 for ceftriaxone, a difference of $1811 per successful outcome achieved in favor of gatifloxacin. Based on these results, the investigators concluded that gatifloxacin was cost-effective compared with ceftriaxone plus or minus erythromycin. While this conclusion may in fact be true, several methodological problems should be pointed out with this study. In the Phase III trial, the clinical success rates appeared to differ between gatifloxacin and ceftriaxone treatment groups (97% vs. 91%, respectively), although the difference was not statistically significant (95% confidence interval ranging from –2.5% to 17.6%). Additionally, neither hospital length of stay (LOS) nor antibiotic-related LOS (LOSAR) differed significantly between the gatifloxacin (4.2 days vs. 4.1 days, respectively) and ceftriaxone (4.9 days vs. 4.9 days, respectively). Because of the lack of differences among all measured outcomes, a CMA model should have been applied. These insignificant differences were due to lack of power in the clinical trial, but contributed toward gatifloxacin's resulting cost-effectiveness secondary a trend toward a longer LOS and LOSAR in both groups among treatment failures. Additionally, if a significant difference in outcomes was originally present, CERs for each intervention should never be subtracted to determine the difference in cost-effectiveness. A more appropriate method would have been to calculate the ICER. Finally, depending on the cost of these antibiotics at one's medical center, gatifloxacin may be "simply dominant" in that it is both cheaper and more effective. In this case, a CEA would clearly be unnecessary.

Cost-Benefit Analysis

Cost-benefit analysis (CBA) is an important tool for comparing and assessing health-care programs and technologies. This methodology compares the value of the resources consumed (costs) in implementing a program or intervention against the value of the outcome (benefits) realized, both in dollars. Therefore, CBA is most useful in the comparison of products, services, or intervention that have different outcomes (e.g., a vaccination clinic and a colon cancer screening program) and when financial resources are limited, requiring a decision to be made as to which program to fund.

The results of CBAs have been presented in the literature in a number of ways. Net benefit (also referred to as net present value when the project spans greater than one year and discounting of costs and benefits is performed) is calculated by simply subtracting the value of the costs of administering the program, service, or intervention from the benefits realized. When reviewing the results of a CBA in this fashion, a decision-maker would choose the project where the net benefits are positive. When more than one alternative is being compared and resources are limited, the project with the greatest net benefit should be chosen.

Alternatively, the results of a CBA can be calculated as dollar-to-dollar comparisons of costs and benefits as a ratio. Therefore, if an intervention costs

TABLE 3 Reporting the Results of CBA as C/B Ratios or Net Benefit

Alternative	Costs	Benefits	C/B ratio	Net benefit
A	$100,000	$200,000	1:2	$100,000
B	$1000	$3000	1:3	$2000

Abbreviations: CBA, cost-benefit analysis; C/B, cost-to-benefit ratio.

$10,000 and yields $20,000 in benefits, the cost-to-benefit (C/B) ratio would be $10,000/$20,000 = 0.5:1 and would be interpreted as for every 50 cents invested, one gets $1 in benefits. Alternatively the benefit-to-cost (B/C) ratio could also be calculated as $20,000 ÷ $10,000 = 2:1 and interpreted as for every $2 in benefits realized, one must invest $1. When reviewing the results of a CBA in this fashion, a decision-maker would choose projects with a C/B < 1 or B/C > 1. The reporting of CBA results as ratios is less frequently used in the literature as ratios do not provide any indication of the magnitude of the benefit to be gained or the cost incurred.

For example, two competing interventions A and B were found to have C/B ratios of 1:2 and 1:3, respectively (Table 3). If data were only viewed in ratio form, a decision-maker's initial reaction would be to choose the latter that provided $3 in benefit for every dollar invested. However, if the decision-makers then reviewed the results for each alternative as the net benefit for each program, assuming the decision-makers had $100,000 to invest, they would likely choose program A as it would yield an additional $98,000 in benefits to their institution compared to program B.

Haberland et al. examined the potential health benefits, costs, and savings from the societal perspective associated with three strategies for identifying mothers at risk of passing Group B β-hemolytic Streptococcus (GBS) to their infants (28). The costs and benefits of a group B streptococci screening strategy using a new rapid polymerase chain reaction (PCR) assay at the time of labor were compared with two standard screening tests, the risk-factor method, which targets high risk mothers but fails to identify colonized mothers or at-risk infants, and the traditional maternal screening test (i.e., rectovaginal culture between weeks 35 and 37), which does not accurately identify genital tract colonization at the time of labor. Investigators focused on key perinatal outcomes (i.e., averted infant infection, infant death, infant disability, cost, and societal benefit of healthy infants). Rapid PCR resulted in a net societal benefit of $7 and $6 per infant compared with the risk-factor and traditional maternal screening methods, respectively. The PCR screening strategy provided greater net benefits because it is better able to identify mothers at-risk and prevent infection, death, and disability in infants. Unlike previous examples, these investigators did not acquire their assumptions (i.e., input data) from a randomized control trial comparing the interventions, as no such study currently exists. Therefore, they made assumptions on success rates, risks, and costs based on numerous studies and applied decision analysis to test a hypothetical population of patients. This approach is useful when comparative studies are lacking, but careful interpretation of what data were included is necessary to determine application of the results to one's own institution. For example, in the current study, the investigators assumed all mothers who tested positive were treated during labor with an initial dose of 2 g of i.v. ampicillin, followed by 1 g every four hours at a cost of $63 per course of therapy, as this was the regimen advocated at their hospital. Certainly, a different antibiotic regimen could be used that might result in an improvement or worsening of response rate, as well as an increase or decrease in cost. Furthermore, the investigators did not consider the indirect costs of overuse of antibiotics (i.e., resistance of GBS) or parental disutilities for losing an infant or raising a disabled child. Finally, PCR is less likely to be attractive in low-volume hospitals because these investigators assumed an on-site laboratory capable of conducting the test would be available, a characteristic of their hospital. Such assumptions may negate the positive net benefit of PCR at a different hospital, and these significant assumptions should always be discussed in a pharmacoeconomic paper so that the reader can judge external validity.

Cost-Utility Analysis

CUA is an evaluation method similar to that of CEA. CEA costs are measured in dollars and outcomes are measured in natural units; however, in CUA the natural units are by design a combined estimate of both quantity-of-life and quality-of-life (QoL) or what is often referred to as "utility." (29) CUA is generally the most difficult and costly pharmacoeconomic methodology to utilize and therefore is often reserved for evaluations of products or interventions that have significant effects on both QoL and quantity-of-life. In addition, as it measures outcomes in quantity-of-life and QoL, it can be used in a similar fashion to CBA to compare products or interventions that result in different outcomes, provided that they both can be measured in quantity-of-life and QoL.

Detailed discussion regarding the measurement of utility is outside the scope of this chapter and therefore only a brief explanation will be provided. Utilities are defined as a patient's preference for experiencing an outcome. Utility values range from 0 to 1, where 0 represents death and 1 represent perfect health. Utilities are elicited from patients through a number of methodologies including questionnaires, visual analog scales, and standard gamble and time trade-off theories. All assume the simple premise that patients when given a choice will chose the health state with the highest utility value assigned to it. In addition, values previously used in the literature or those based upon expert opinion may be used; however, they should be used with caution as these values may not be representative of every patient population.

Once determined, these utility values are then multiplied by the extent of time a patient is expected to be in that health state. Any unit of time can be utilized, but most commonly years of life are used and the resultant outcomes are reported as QALYs. For example, you are evaluating two possible treatments for a life-threatening foot infection. Treatment A involves amputation, which costs $20,000 but increases life expectancy by 4.5 years at a utility value of 0.60. Treatment B is an aggressive medical treatment with antibiotics, which costs $10,000 and increases life expectancy by only 1.5 years with a utility value of 0.90. Treatment A would therefore cost $20,000/ (4.5*0.4) or 1.8 QALY = $11,111/QALY, where treatment B would cost $10,000/1.5*0.9) or 1.35 QALY = $7,407/QALY. Similar to CEA, the incremental increase in cost per unit of outcome could also be calculated ($20,000–$10,000/1.8–1.35 = $22,222 additional dollars will be spent to obtain one additional QALY with treatment A compared with treatment B). Again, as with CEA, when provided with the incremental costs required for one additional outcome, decision-makers must determine whether the additional cost required is worth the additional outcome achieved.

While CUA is a powerful tool among pharmacoeconomic analyses, it is often difficult to conduct because of the lack of data required to calculate a QALY in a specific population. Often, significant assumptions must be made in this case. Shorr et al. conducted an incremental CEA (specifically a CUA) of linezolid compared with vancomycin for the treatment of ventilator-associated pneumonia (VAP) (30). Recent data have shown that empiric use of linezolid in patients with VAP significantly increases the rate of survival compared to vancomycin therapy, although linezolid is considerably more expensive than vancomycin (31). The investigators evaluated a hypothetical cohort of 1000 patients. Incremental cost-effectiveness was calculated in terms of added QALY gained divided by the sum of the incremental costs of drug use. The incremental cost-effectiveness of linezolid versus vancomycin was $29,945 per QALY (95% confidence interval ranging from $23,637 to $42,785). Multivariate sensitivity analysis and Monte Carlo simulation demonstrated that linezolid was a financially attractive alternative to vancomycin because the incremental cost-effectiveness was less than the accepted standard in health care of $100,000 per QALY for cost-effectiveness. Although linezolid is significantly more expensive than vancomycin, the increased efficacy of linezolid found in the study compensated for the cost differential. Thus, what may appear to be a more expensive therapy in the intensive care unit could actually be cost-effective. Importantly, this analysis relied on several critical assumptions. Health utilities (QALYs) were set at 0.83 because of the uncertainty in long-term VAP outcomes. Because survival data for VAP currently do not exist, the investigators extrapolated survival from those with severe sepsis and assumed that long-term VAP survivors lived an additional nine years, which may be significantly lower than that of the young trauma patient who develops VAP. Additionally, the positive clinical outcomes associated with linezolid therapy are based upon post hoc analyses, of which conclusions are not hypothesis proving. To address these concerns, the investigators did lower the predicted effectiveness of linezolid by 10% and biased the model against it when possible. Linezolid remained the cost-effective alternative in 99.8% of simulations. Thus, as previously mentioned, when provided with the incremental costs of a CUA, decision-makers must determine whether the analysis is applicable to their patient population and whether the additional cost is worth the additional outcome achieved. In an intensive care unit with a high prevalence of VAP caused by MRSA and an elderly population, a switch to linezolid may produce similar cost outcomes.

Decision Tree Analysis

Simulation models are commonly used to conduct pharmacoeconomic evaluations. The most common simulation study design found in the pharmacoeconomic literature is decision tree analyses (DTAs). They combine data from the medical literature (either from clinical trials or from meta-analyses) and expert opinion and data from one's own institution in order to estimate costs and outcomes when a more rigorous pharmacoeconomic methodology cannot be practically completed (32). Simulation models are relatively inexpensive compared to the cost of conducting a full pharmacoeconomic evaluation and require less time to complete; however, the model will only be as good as the data entered into it and relying on incomplete or questionable data may result in misleading conclusions.

DTAs are most useful for examining products or interventions for acute or short-term illnesses. For each alternative product or intervention, downstream costs and outcomes and their probabilities of occurring (each arm must total 100%) are determined and displayed graphically (Fig. 2). Each product or intervention is then

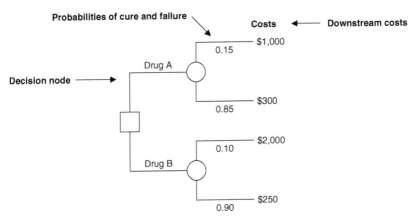

FIGURE 2 Decision tree analysis of two hypothetical antibiotics.

"folded back" (starting from the right and working back toward the decision node) to a single value by mathematically combining the costs, outcomes, and probabilities to determine the most favorable alternative.

Sensitivity Analysis

Sensitivity analysis is the process in which the soundness or robustness of the results and conclusions of pharmacoeconomic evaluations are tested by varying the underlying assumptions and variables over a range of plausible values. Sensitivity analysis is utilized when there is any uncertainty about the data or how it were valued, to determine the effect of data uncertainty (assumptions) on study conclusions, and to identify the most important study assumptions. It helps to answer questions such as, "How applicable are these study results to my institution which may have a different patient population?" and "Can data from this clinical trial be applied to my real life setting?"

There are multiple varieties of sensitivity analyses that can be utilized as part of pharmacoeconomic evaluations including simple sensitivity analysis, threshold sensitivity analysis, analysis of extremes, and probabilistic sensitivity analysis. Currently, there are no established rules as to which type of sensitivity analysis to use or how to interpret the results; however, the ranges for variables chosen should be well justified and the methodology transparent to decision-makers (6–11).

Simple sensitivity analysis is the most commonly used and involves varying study assumptions (either cost or outcome) within the range of plausible values to determine if the original conclusions remain sound. When only one variable is altered at a time, this is called one-way simple sensitivity analysis. For example, an antibiotic, which was valued at an institution's AAC in the original evaluation, might be valued as high as its AWP and as low as 50% of its AAC to see how the conclusions of the evaluation would be altered. If the conclusions remained similar, they would be robust to changes in antibiotic cost. If more than one variable is altered at a time, this is called multiple-way simple sensitivity analysis and conducted and interpreted in the same fashion as above.

Analysis of extremes is conducted by reassigning values of both cost and effectiveness variables concurrently to their highest and lowest extremes in order to determine the robustness of the conclusions in the best- and worst-case situations. The best-case scenario is depicted by the highest effectiveness and all lowest cost estimates and worst-case is the lowest effectiveness and all highest costs estimates. In order for this analysis to be useful to decision-makers, sufficient detail of the extremes must be provided to determine whether the extremes being evaluated are near their own health plan's resource cost. The usefulness of analysis of extremes is questionable because it is doubtful that all worst-case or best-case findings will occur at the same time.

Threshold analysis requires that a single assumption in the analysis be varied until the alternative treatment option has the same outcome and there is no advantage between the treatment options. The value determined by conducting such an analysis is often referred to as the breakeven point. This breakeven point can then be compared by decision-makers to their value to make decisions concerning the validity of the evaluation's results to their institution.

Probabilistic sensitivity analysis allows researchers to assign plausible ranges for variables and an estimate of the distribution of the data points for each variable. The most common type of probabilistic sensitivity analysis is Monte Carlo simulation, which randomly assigns values from inputted variable ranges and estimating outcomes from models with large numbers of hypothetical patients.

SUMMARY

Pharmacoeconomic methods endeavor to assess the value of products, services, and interventions in terms of economic, clinical, and humanistic terms (12). The continuing development of newer pharmaceutical products and services increased patient access to care and lack of unlimited resources will make it imperative that decision-makers have all the tools necessary to make important health-care allocation decisions. This chapter was designed to enable the reader to apply pharmacoeconomic methodologies in their own evaluation of new drug entities, as well as in their critique of the available literature (33,34). It is important to remember that while pharmacoeconomic evaluations can help facilitate the decision-making process, the responsibility placed upon the health-care provider to make patient-specific decisions should not be forgotten.

REFERENCES

1. National Center for Health Statistics. Health expenditures. Accessed at: www.cdc.gov/nchs/fastats/hexpense.htm (March 2004)
2. Weinstein MC, Siegel JE, Gold MR, Kalmet MS, Russell LB, for the Panel on Cost-Effectiveness in Health and Medicine. Recommendations of the panel of cost-effectiveness in health and medicine. JAMA 1996; 276:1253–1253.
3. Braden BR, Cowan CA, Lazenby HC, et al. National health expenditures, 1997. Health Care Fin Rev 1998; 20:83–83.
4. Stefanacci RG. The Medicare Modernization Act (MMA): how will it affect P&T committee members? P&T 2004; 29:95–95.
5. Boyer JG, Pathak DS. Establishing value through Pharmacoeconomics: the emerging third objective in clinical trials. Top Hosp Pharm Manage 1994; 13:1–1.
6. Australia Commonwealth Department of Health, Housing and Community Services. Guidelines for the pharmaceutical industry on preparation of submission to the Pharmaceutical Benefits Advisory Committee. Canberra: Commonwealth Department, 1992.
7. Langley PC. The role of pharmacoeconomic guidelines for formulary approval: the Australian experience. Clin Ther 1993; 1154–1175.

8. Canadian Coordinating Office for health Technology Assessment (CCOHTA). Guidelines for economic evaluation of pharmaceuticals: Canada. 1st ed. Ottawa: CCOHTA, 1994.

9. Jacobs P, Bachynsky J, Baladi JF. A comparative review of pharmacoeconomic guidelines. Pharmacoeconomics 1995; 8:182–182.

10. Task Force on Principles for Economic Analysis of Health Care Technology. Economic analysis of health care technology. Ann Intern Med 1995; 123:61–61.

11. Clemens K, Townsend R, Luscombe F, Mauskopf J, Osterhaus J, Bobula J. Methodological and conduct principles for pharmacoeconomic research. Pharmacoeconomics 1995; 8:169–169.

12. Kozma CM, Reeder CE, Schulz RM. Economic, clinical and humanistic outcomes: a planning model for pharmacoeconomic research. Clin Ther 1993; 15:1121–1121.

13. Florea NR, Kotapati S, Kuti JL, Geissler EC, Nightingale CH, Nicolau DP. Cost analysis of continuous versus intermittent infusion of piperacillin-tazobactam: a time-motion study. Am J Health-Sys Pharm 2003; 60:2321–2321.

14. ISPOR Lexicon; Pashos CL, Klein EG, Wanke LE, eds. First edition, 1998.

15. Cardinal Health website. Accessed at: www.cardinal.com (March 2004)

16. Hitt CM, Nightingale CH, Quintiliani R, Nicolau DP. Cost comparison of single daily i.v. doses of ceftriaxone versus continuous infusion of cefotaxime. Am J Heath-Sys Pharm 1997; 54:1614–1614.

17. Finkler SA. The distinction between costs and charges. Ann Intern Med 1982; 96:102–102.

18. Gold MR, Siegel JE, Russel LB, Weinstein MC, eds. Cost-Effectiveness in Health and Medicine. New York: Oxford University Press, 1996:176–176.

19. Copley-Merriman C, Lair TJ. Valuation of medical resource units collected in health economic studies. Clin Ther 1994; 16:553–553.

20. US Bureau of Labor Statistics. CPI Detailed Report. Washington, DC: Department of Labor; March 2004. Accessed at http://data.bls.gov/servlet/survey_output_servlet (March 2004)

21. Kotapati S, Nicolau DP, Nightingale CH, Kuti JL. Clinical and economic benefits of a meropenem dosage strategy based on pharmacodynamic concepts. Am J Health-Syst Pharm 2004; 61:1264–1270.

22. Kuti JL, Maglio D, Nightingale CH, Nicolau DP. Economic benefit of a meropenem dosage strategy based on pharmacodynamic concepts. Am J Health-Syst Pharm 2003; 60:565–565.

23. Angus DC, Linde-Zwirble WT, Clermont G, et al. Cost-effectiveness of drotrecogin alfa (activated) in the treatment of severe sepsis. Crit Care Med 2003; 31:1–1.

24. Bernard GR, Vincent JL, Laterre PF, et al. Efficacy and safety of recombinant human activated protein C for severe sepsis. N Engl J Med 2001; 344:699–699.

25. Chalfin DB, Teres D, Rapoprt J. A price for cost-effectiveness: implications for recombinant human activated protein C (rhAPC). Crit Care Med 2003; 31:306–306.

26. Betancourt M, McKinnon PS, Massanari RM, Kanji S, Bach D, Delvin J. An evaluation of the cost-effectiveness of drotrecogin alfa (activated) relative to the number of organ system failures. Pharmacoeconomics 2003; 21:1331–1331.

27. Dresser LD, Niederman MS, Paladino JA. Cost-effectiveness of gatifloxacin vs ceftriaxone with a macrolide for the treatment of community-acquired pneumonia. Chest 2001; 119:1439–1439.

28. Haberland CA, Benitz WE, Sanders GD, Pietzsch JB, Yamada S, Nguyen L, Garber AM. Perinatal screening for group B streptococci: cost-benefit analysis of rapid polymerase chain reaction. Pediatrics 2002; 110:471–471.

29. Pathak DS. QALYs in health outcomes research: representation of real preferences or another numerical abstraction? J Res Pharm Econ 1995; 6:3–3.

30. Shorr AF, Susla GM, Kollef MH. Linezolid for treatment of ventilator-associated pneumonia: a cost-effective alternative to vancomycin. Crit Care Med 2004; 32:137–137.

31. Kollef MH, Rello J, Cammarata SK, Croos-Dabrera RV, Wunderink RG. Clinical cure and survival in Gram-positive pneumonia: retrospective analysis of two double-blind studies comparing linezolid with vancomycin. Intensive Care Med 2004; 30:343–343.

32. Sanchez LA. Pharmacoeconomics and formulary decision making. Pharmacoeconomics 1996; 9(suppl 1):16–25.

33. Sanchez LA, Lee JT. Applied pharmacoeconomics: modeling data from internal and external sources. Am J Heath-Sys Pharm 2000; 57:146–146.

34. Gouveia WA, Bungay KM. Incorporating pharmacoeconomic principles into hospital pharmacy practice. Top Hosp Pharm Manage 1994; 13:31–31.

Index